Vinko V. Dolenc

Microsurgical Anatomy and Surgery of the Central Skull Base

in collaboration with Larry Rogers

SpringerWienNewYork

Vinko V. Dolenc
Department of Neurosurgery, University Medical Centre Ljubljana, Slovenia

Typesetting: SNP Best-set Typesetter Ltd., Hong Kong

SPIN: 10637891

With 189 partly coloured Figures

CIP data applied for

ISBN 978-3-211-83236-3 Springer-Verlag Wien New York

Preface

Parkinson's idea of crossing the dura from intradural to epidural space to repair arteriovenous fistulae and aneurysms of the parasellar space (PS) was a revolutionary step toward devising a surgical method for treating these difficult pathologies. Yet, owing to the difficulties and potential complications attendant to cardiac standstill and extracorporeal circulation, the idea was not universally accepted in the surgical community. The application of endovascular techniques, as introduced by Serbinenko, amounted to another major breakthrough in the treatment of these lesions. For a time, the endovascular method brought the development of surgical procedures for parasellar space (PS) lesions to a resounding halt. Combining, however, the results of anatomical studies by Taptas, Parkinson and Glasscock, the concept of opening the PS without manipulating heart function was born. Access could be gained simply by locally and temporarily interrupting internal carotid astery (ICA) flow. When this idea was presented as an alternative for treating PS tumors and vascular lesions at the 7th International Neurosurgical Congress in Munich in 1981, opposition was generalized but not universal. Neurosurgeons demanded proof that such methods afforded successful, safe treatment of PS lesions, that a sizable series of good postoperative results be established.

The presentation of these new anatomical concepts kindled the imaginations of neurosurgeons around the world, but not always to the benefit of their patients. In many cases actions were taken without appreciation of the technical difficulties involved. As neurosurgeons flocked to put theory into practice, poor results occurred. All too often the importance of gaining a full understanding of microsurgical anatomy through long hours of laboratory experience was overlooked. A new skepticism arose, in many instances voiced by nascent PS "experts"! It constituted a very disheartening chapter in the history of central skull base (CSB) surgery. Particularly sad is the fact that countless patients were forced to pay a stiff price for their surgeons' overextending themselves.

The first International Meeting of the Cavernous Sinus was convened in Ljubljana in 1986 to emphasize the importance of mastering a practical knowledge of CSB anatomy and the special surgical techniques required. Undoubtedly that meeting had a large impact on refining and disseminating safe methods of CSB surgery. Its reports were published in *The Cavernous Sinus. A Multidisciplinary Approach to Vascular and Tumorous Lesions*. After 1986, however, it became even more difficult to monitor and assess the numerous publications appearing. Most involved experiences with relatively few clinical cases, authors at times drawing conclusions based on inadequate statistical samples. *Anatomy and Surgery of the Cavernous Sinus*, published in 1989, was devoted to detailed CSB anatomy and its surgical implications. That book sold out within a year, leading to repeated requests for its re-printing, and finally, as experience grew, to update it. This volume is an attempt to describe concepts of CSB anatomy even more fully and to detail practical approaches to various skull base lesions now amenable to surgery. It emphasizes the evolution of surgical thought since the 1989 publication. Key among several changes is that whereas previously most lesions were exposed intradurally, currently we understand that they are ideally accessed by a totally extradural approach. At the present time lesions simultaneously inhabiting intradural and extradural compartments are approached first extradurally, then by opening dura only as a given lesion dictates. This affords maximum protection to intradural structures while allowing full extirpation of many tumors and vascular lesions. Gaining adequate dissection space at the expense of skull base bone (rather than retracting brain) amounts to a crucial advantage in dealing with such lesions. Where opening dura is absolutely required, it must be reconstructed in a watertight manner in order to avoid CSF leakage.

Here the most important vascular lesions and tumors are presented from the surgeon's point of view, and all results are analyzed in detail. The careful reader should be able to determine accurately what currently is possible with surgery and how to go about achieving good results. The current series of vascular lesions and tumors is the result of two decades' continuous experience. We believe it supports our belief that surgery for such lesions is useful, beneficial, and necessary. Continued advancing knowledge will lead to an even superior understanding of normal and pathological CSB anatomy, facilitating additional improvements and refinements.

Nowadays, surgery as therapy remains continuously and increasingly under fire. Surgeons are reminded daily of the unquestionable fact that to operate is not always to cure. Yet patients seek optimum treatment *today.* They require surgeons' care long before "less invasive" treatment modalities have proved their superiority in safety and efficacy. Surgeons will continue to face such patients. As a consequence, each is obliged to continue to improve his methods in order that much more is achieved, until every untoward effect and possible postoperative neurological deficit have been eliminated entirely. With respect to the challenging lesions of the CSB, such will be possible only if surgeons remain prepared to invest still more time and energy to advance anatomical knowledge and surgical skill with the demanding lesions occurring at the CSB. Surgeons must accept that no single treatment is best for long. The history of surgery details a never ending story of change and improvement. Our generation depends on the surgeons of tomorrow to continue to roll back the frontiers of what is possible. One may find satisfaction and pride only as an active participant in the process.

Acknowledgments

I like to extend a very special thanks to Mr. Alan McConnell Duff for his editorial expertise and the extensive time and effort he has devoted to the preparation of this manuscript.

I am deeply indebted to my dear friend and colleague, Larry A. Rogers, M.D., for reviewing the text discussing it with me in detail, and rephrasing elements when appropriate in order to make this work as readable as possible for the often weary young neurosurgeons and neurosurgical residents for whom it is primarily intended.

A special thanks goes to Ms. Martina Špeh, who transcribed the initial draft and patiently corrected subsequent versions as the manuscript gradually assumed its final form over the past two years.

I am particularly grateful to Ms. Jolanda Kofol, medical photographer, and to Mr. Ivan Mori, medical illustrator, for their preparation of operative photographs, drawings, and radiographic representations. They are talented, highly professional people indeed.

The indispensable assistance of my Ljubljana colleagues – Rado Pregelj, M.D., Borut Prestor M.D., Jasna Šušteršič, M.D., Marjan Koršič, M.D., Roman Bošnjak, M.D., Yousef Ardebili, M.D., Meta Fettich, M.D., Božidar Visočnik, M.D., Gvido Gumzej, M.D., Igor Tekavčič, M.D., Andrej Vranič, M.D., in surgical assistance, neuroanesthesia, and postoperative care of patients with central skull base lesions – is gratefully acknowledged.

And perhaps most of all, my heartfelt thanks go to the hundreds of patients who paid us the ultimate compliment by trusting our team with their lives and their futures.

Contents

Abbreviations

Cranial nerves

ON Optic nerve
III Oculomotor nerve
IV Trochlear nerve
V Trigeminal nerve
GG Gasserian ganglion
V_1 Ophthalmic division
V_2 Maxillary division
V_3 Mandibular division
Vm Motor branch of the trigeminal nerve
VI Abducens nerve
VII Facial nerve
VIII Vestibulocochlear nerve

Internal carotid artery

ICA Internal carotid artery
(AL) Anterior loop
(ML) Medial loop
(LL) Lateral loop
(PL) Posterior loop

Other intracranial arteries

ACA Anterior cerebral artery
ACom Anterior communicating artery
BA Basilar artery
CA Capsular artery
DAs Dawson's arteries
ILT Inferolateral trunk
MCA Middle cerebral artery
MHT Meningohypophyseal trunk
MMA Middle meningeal artery
OA Ophthalmic artery
PCom Posterior communicating artery
SCA Superior cerebellar artery
TA Tentorial artery (Bernasconi-Cassinari)

Other structures

ACP Anterior clinoid process
APB Apex of the petrous bone
BS Brain stem
CSB Central skull base
DR Distal (dural) ring
DS Dorello's space
FL Foramen lacerum
FO Foramen ovale
FR Foramen rotundum
FS Foramen spinosum
GPN Greater petrosal nerve
IAC Inner auditory canal
LM Liliequst membrane
LPN Lesser petrosal nerve
LR Lateral ring
MCF Middle cranial fossa
MM Mucous membrane
OC Optic canal
PB Pituitary body
PCL Petroclival ligament
PCP Posterior clinoid process
PR Proximal ring
PSt Pituitary stalk
PV Petrosal vein
SF Sylvian fissure
SN Sympathetic nerve(s)
SOF Superior orbital fissure
SPS Superior petrosal sinus
TM Temporal muscle
TTM Tensor tympani muscle

Introduction

However beautiful and unique, the complexity of three-dimensional anatomy of the "parasellar space" (PS), formerly known as the "cavernous sinus" (CS), has defied full understanding by generations of surgeons. Only recently, with the advent of visual magnification as provided by the operating microscope, have truly valid descriptions of this area begun to appear. The centuries old idea that the CS is primarily a venous channel similar to the corpora cavernosa of the penis, yet one complicated by the presence of the internal carotid artery (ICA) and nerves III, IV, VI, V1 and V2, is clearly in error. A more accurate concept is that this so-called "sinus" amounts to truly a "space" characterized by an intricate network of venous channels with identifiable walls surrounding the ICA and elements of four cranial nerves.

Examining injected fresh cadaver specimens microsurgically provides keys to precise PS anatomy. Yet formalin-fixed specimens are virtually useless for studying surgical anatomy since the relationships between individual structures are changed significantly – distorted, even sometimes obliterated. Injected fresh cadaver specimens, however, reveal three-dimensional relationships that approximate those of the living parasellar space. Though not a precise representation of the living anatomy, thus-treated specimens allow one to glean useful information about parasellar anatomy and that of adjacent regions, information required by any surgeon confronted with the anatomical variations and distortions resulting from disease at the central skull base (CSB).

Experiences gained in the treatment of 1721 CSB tumors and 533 vascular lesions over two decades will be presented in subsequent chapters. They will describe surgical strategies and specific techniques and confirm that such lesions can be approached directly with good results and minimum risk.

A descriptive system utilizing the simplest geometric figure, the triangle, facilitates orientation in the intricate maze of anatomical structures characterizing this region and is of particular value to the surgeon. It dictates that the parasellar region be divided into three subregions and ten surgical triangles which are defined by constant anatomical landmarks. The following describes a system of nomenclature, pertinent microsurgical anatomy, and specific surgical approaches to various parasellar lesions. Surgeons willing to master the anatomy of the CSB, in particular the PS, should be able to operate in this area with precision and relative impunity.

1

Central skull base anatomy

1.1

Anatomy of the parasellar space

The macroscopic and microscopic anatomy of the parasellar space (PS) and of the entire central skull base (CSB) region has been increasingly investigated over the past two decades. As a result, the parasellar anatomy of nerves III through VI, the ICA, its branches, the venous structures, and the related osseous structures and adjacent dural folds are well described [1–15].

The surgical anatomy of the parasellar and adjacent skull base regions can be divided into three subregions composed of ten triangles: (A) Parasellar subregion: (1) anteromedial triangle, (2) paramedial triangle, (3) Parkinson's triangle, and (4) oculomotor trigone; (B) Middle cranial fossa subregion: (1) anterolateral triangle, (2) lateral triangle, (3) posterolateral (Glasscock's) triangle, and (4) posteromedial (Kawase-Shiobara's) triangle; and (C) Petroclival subregion: (1) inferomedial triangle and (2) inferolateral (trigeminal) triangle: (i) osseous portion and (ii) tentorial portion.

The margins of each triangle are based on constant anatomical relationships of osseous, dural, vascular, and neural structures within the parasellar, middle cranial fossa and petroclival subregions. By describing surgical triangles defined by the osseous borders, dural walls and nerves III through VI of the PS one is able to indicate windows through which surgeons may attack various PS tumors and vascular lesions without damaging important vascular and neural structures.

Materials and methods

Specimens from fresh cadaveric skull base bone blocks (each consisting of the sphenoid bone, sella, both PSs, medial portions of both petrous bones, clivus, and related structures) were examined using opton OPMI1 and opton OPMI 6-CPC microscopes under 5–40 power. Only cadavers free of historical or autopsy evidence of intracranial disease were selected. The cranial nerves were transected as close as possible to their points of entry into and exit from the brainstem with great care to preserve adjacent dural folds, thus retaining accurate in situ anatomical relationships. After ligating both supraclinoid ICAs and both posterior communicating arteries (PCom), cannulation of the proximal petrous ICAs allowed injection of colored acrylic material into both carotid systems of each specimen. Similarly, venous compartments of the PS region were injected through the superior (SPS) and inferior petrosal sinuses (IPS). Photographs at appropriate stages of dissection were made with a Canon A-1 35 mm camera.

Anatomical relationships in the central skull base

Osseous and dural structures

The most important bony landmarks medial to the PS are the anterior clinoid process (ACP), the posterior clinoid process (PCP), and the lateral border of the clivus. Posteriorly the most prominent structure is the anterior surface of the petrous bone, and laterally, the foramen rotundum (FR), foramen ovale (FO), and foramen spinosum (FS) are important landmarks. Anteriorly, the essential identifying feature is the superior orbital fissure (SOF).

The tentorium cerebelli is attached to the petrous apex, the PCP, and ACP. Resulting dural folds form the limits of a triangular field (referred to as the oculomotor trigone), at the lateral border of which nerves III and IV enter the lateral wall of the PS. Margins of the oculomotor trigone include: (a) the anterior petroclinoid fold extending from the petrous apex to the ACP; (b) the posterior petroclinoid fold extending from the petrous apex to the PCP; and (c) the interclinoid fold extending between the ACP and the PCP. The lateral wall of the PS is consistently formed by two layers: the superficial, a dense layer formed by the dura mater, and an inner reticular layer, between which pass nerves III, IV, V1 and V2 (15). In 40% of the specimens the thin inner layer was incomplete, often absent between nerve III and IV, leaving a triangular opening into the cavity of the PS. Anterolateral to Meckel's cave, a layer of fibrous connective tissue lies over the ICA, fixing it to the skull base. The well-formed anterior border of this fibrous bridge delineates the fixed portion of the ICA, which runs over the foramen lacerum from its mobile parasellar segment. Peripheral to the Gasserian ganglion (GG) and along the lateral wall of the PS, the superficial dural layer can be separated from the inner reticular layer without entering the PS. At the anterior limit of the PS, the two layers of dura separate to envelop the ACP and a part of the anterior loop (AL) of the ICA. The inner layer passes inferior to the ACP and surrounds the AL of the ICA, forming its proximal ring (PR) which thus becomes the roof of the anterior PS, below and lateral to which lie the blood-filled venous trabeculated spaces medial to the ICA. The superficial dural layer continues over the superior surface of the ACP, tightly enveloping the ICA to form a separate ring referred to as the distal dural ring (DR). (According to this terminology, "proximal" and "distal" refer to paraclinoid rings with respect to their location along the AL of the ICA). It should be noted that the paraclinoid segment of the AL of the ICA between the PR and DR is extradural, not necessarily embedded in the venous compartments of the PS. The inner reticular layer continues medially, extending over the pituitary body (PB), and inferiorly along the medial wall of the PS. Openings within the inner reticular layer provide passage for venous channels between PSs on the left and right.

Oculomotor nerve

After exiting from the interpeduncular fossa, nerve III courses in an anterolateral and slightly inferior direction in the anterior incisural space between the posterior cerebral artery (PCA), and the superior cerebellar artery (SCA) to enter the roof of the PS lateral to the PCP. At this point, the nerve acquires its own sheath of dura in the lateral portion of the anterior petroclinoid fold (the occulomotor canal) and continues anteriorly within the lateral wall of the PS. In this dural envelope, nerve III is divided into the inferior and superior divisions anterior to the apex of the AL of the ICA. Further branching of the inferior and superior divisions of nerve III takes place more peripherally where additional divisions of nerve III occur within the apex of the orbit.

Trochlear nerve

Nerve IV exits from the dorsal surface of the midbrain just below the inferior colliculus at the level of the superior medullary velum. In the cerebello-mesencephalic fissure, nerve IV curves anteriorly around the lateral surface of the tectum and tegmentum in the quadrigeminal and ambiens cisterns, then continues in an anterolateral and inferior direction to enter the inferior surface of the tentorial edge lateral to the cerebral peduncle. It then proceeds anteriorly for several millimeters in a groove on the inferior surface of the tentorium before becoming completely encased in a dural canal. After entering the tentorium, nerve IV continues anteriorly, following the margin of the anterior petroclinoid fold to enter the lateral wall of the PS where, running inferior and lateral to nerve III, it is embedded between two dural layers initially, and between the outer dural layer and inner reticular layer of the lateral wall of the PS more anteriorly. In its anterior segment, while still coursing along the lateral wall of the PS, nerve IV angles toward nerve III, then courses from the inferolateral to the superomedial surface of nerve III. At this point nerve IV takes its place on the lateral surface of nerve III.

Trigeminal nerve

After arising from the lateral pons, the motor and sensory roots of nerve V pass in an anterior and lateral direction through the middle incisural space of the posterior cranial fossa, superior to the petrous apex to enter the arachnoidal and dural outpouching known as Meckel's cave. Several millimeters superior and posterior to nerve V's point of entry into Meckel's cave the petrosal vein enters the SPS. Ventro-lateral to Meckel's cave at the petrous apex lies the lateral loop (LL) of the ICA, which runs over the foramen lacerum (FL). Previous studies indicated that bone between the dura of Meckel's cave and the ICA was absent in 68% of specimens [2, 6]. In such cases, the GG was separated from the ICA only by a layer of fibrous connective tissue. The greater petrosal nerve (GPN), composed of pre-ganglionic parasympathetic fibers, after exiting from its hiatus on the anterior surface of the petrous bone, passes anteriorly and epidurally over the floor of the middle cranial fossa to join nerve V.

As the fascicles of nerve V lose their arachnoidal covering, they coalesce to form the GG. Although fascicles of the nerve V root are consistently parallel through much of their course, four or five millimeters medial to the GG this orientation changes abruptly, becoming more braided in appearance beneath the microscope, interwoven and diagonal. This short segment is known as the "plexiform segment" of the nerve V root, an element carrying great surgical significance. Dissection of lesions densely attached to this part of the nerve almost certainly will result in at least some neurological deficit since fibers crossing perpendicular to the longitudinal axis of the nerve are particularly vulnerable to injury.

Laterally the GG gives off V1, V2, and V3. V3 continues inferiorly, laterally, and slightly anteriorly to leave the middle cranial fossa via the FO several millimeters postero-lateral to FL. V2 runs anteriorly and slightly inferiorly between the two layers of the lateral wall of the PS to leave the middle cranial fossa through the FR. Before entering the SOF, V1 passes anteriorly between two layers of the lateral wall of the PS, where it is located inferolateral to nerve IV and superomedial to V2.

Abducens nerve

Nerve VI extends anterosuperiorly through the prepontine cistern after leaving the anterior pontomedullary sulcus. It pierces dura overlying the basal venous plexus of the clivus. The fascicles of nerve VI are sometimes divided, forming several individual nerve bundles as it enters clival dural opening. The length of nerve VI within Dorello's space (DS) is somewhat variable, depending on where it enters the dura. After entering DS, nerve VI continues superiorly and slightly medially in close apposition to the dorso-lateral clivus. It then courses anteriorly and slightly medially beneath the petroclinoid ligament, near the base of the PCP, to enter the PS lateral to the ascending segment of the ICA and medial to V1. Nerve VI proceeds parallel to and adjacent to the horizontal segment of the ICA and inferior to the meningo-hypophyseal (MHT) and the inferolateral (ILT) trunks, close against the ICA, tight with sympathetic fibers (originating from sympathetic cords on the ICA wall) joining it. Nerve VI is the only nerve fully within the PS, all the others (nerves III, IV, V1, and V2) coursing within its lateral wall.

Internal carotid artery

In order to understand the changing orientation of the ICA at the skull base one must imagine two adjacent cubes, one on the right, the other on the left, each housing half of the skull base. The ICA enters the skull base at the petrous carotid canal at the posterior, lateral, and inferior corner of the ipsilateral imaginary cube, then courses diagonally upward within it, ultimately entering the intradural space at the cube's anterior, superior, and medial corner. Within the imaginary cube this approximately 6 cm ICA segment makes four loops, each in a different plane.

The posterior loop (PL) lies entirely within the petrous carotid canal, arching upward to become the first horizontal segment of the intracranial ICA, then medially, as the LL, overlying the FL, then upward through the lateral ring (LR), a distinct band of connective tissue crossing its lateral surface, entering the PS beneath and medial to nerve V2 and on toward the PCP where it forms the medial loop (ML). From the ML, adjacent to the lateral surface of the PCP, the ICA courses again horizontally toward the ACP. Inferior and medial to the ACP, the vessel turns abruptly, making a U-turn over the lateral wall of the sphenoidal sinus and enters the intradural space. The ICA's U-turn inferomedial to the ACP, lateral to the lateral wall of the sphenoidal sinus, is marked proximally by bands of connective tissue, the proximal ring (PR) inferomedial to the ACP, and the distal (dural) ring (DR) as the extradural ICA finally becoms intradural.

The AL of the ICA is not necessarily embedded entirely within the venous plexus of the PS. The segment of PS-ICA ascending from the LL to the ML gives off a large arterial branch, the meningo-hypophyseal trunk (MHT), while the horizontal segment of the PS-ICA gives off the infero-lateral trunk (ILT). The ILT arises from the ICA inferolaterally and courses superior to nerve VI where it gives off branches to the lateral dural wall of the PS and adjacent cranial nerves coursing toward the SOF. The MHT gives off the inferior hypophyseal artery, the dorso-meningeal artery, and the tentorial artery known also as the artery of Bernasconi-Cassinari. The ICA is accompanied by sympathetic cords which give off branches to nerve VI, firmly fixing it to the ICA. These sympathetic fibers course along nerves VI and V1 into the orbit. Most of the sympathetic fibers – cords on the surface of the ICA – course with the ICA into the intradural space.

It is often useful for the surgeon's orientation to recognize that the three horizontal segments of the ICA lie in parallel planes. The first horizontal segment (in the petrous canal) is not only oriented parallel to the PS horizontal segment, both lie parallel to the intradural ICA, all three standing like three decks for a ship.

Neuro-vascular relationships in the parasellar space

Nerves III, IV, and V1 course anteriorly between the outer dural and inner reticular layers forming the lateral wall of the PS to enter the SOF. Throughout the posterior portion of the lateral wall of the PS these nerves maintain a constant superior-inferior relationship to each other, although the course of nerve IV, inferior to nerve III and superior to V1, has been shown in a previous study to be somewhat more variable anteriorly as these nerves enter the lateral wall of the PS [3]. However, 5–7 mm posterior to the bony margin of the SOF, nerve IV crosses over the lateral surface of nerve III, then continues anteriorly, superior and slightly lateral to nerve III, to enter the SOF.

Nerve VI runs infero-lateral to the petroclinoid ligament (PCL), then enters the posterior PS maintaining this direction throughout the PS, lateral to the ICA and immediately medial to V1. Sympathetic fibers, running with the ICA, coalesce into discrete fiber bundles, which leave ICA sympathetic cords to join nerve VI for several millimeters before crossing to join V1, with which they run peripherally to provide sympathetic innervation in the corresponding trigeminal territory. As nerve VI continues anteriorly, it assumes a slightly more inferior location within the PS relative to V1, so that, at the level of the SOF, nerve VI is both medial and inferior to V1.

In addition to nerve VI, sympathetic cords and fibers, venous trabecular channels, and the PS-ICA, variable amounts of fatty connective tissue are also found throughout the PS. The venous tributaries include the superior (SOV) and inferior ophthalmic veins (IOV), the SPS, the IPS, the sphenoparietal sinus, cortical veins, veins of the pterygoid plexus via the FO, and the basilar venous plexus.

1.2

The surgical triangles on and around the parasellar space

The constant anatomical relationships of the dural, osseous, neural, and vascular structures of the parasellar skull base region permit the organization of the relevant anatomy into a series of ten triangular windows – on each half of the CSB – each with particular surgical significance to various skull base lesions. One of these triangles has been subdivided further into osseous and tentorial portions. These ten triangles have been grouped into three subregions (parasellar, middle cranial fossa, and petroclival). Conceptually, the two-dimensional triangles may be expanded into three-dimensional figures to form a series of ten tetrahedrons, each containing superficial dural, osseous, and neuro-vascular structures. For the sake of simplicity and consistency, however, the term "triangle" will be retained throughout the book to include the three-dimensional extension of each triangle with its specific contents.

Parasellar subregion

Anteromedial triangle

The sides of this triangle include: the lateral surface of the optic nerve (ON) confined within the optic canal (medial border of the triangle); the medial surface of nerve III within the sheath of dura entering the SOF (lateral border); and the dura extending between the dural entry point of nerve III and the ON (posterior border), which forms a part of the DR around the paraclinoid ICA [3]. Exposure of the anteromedial triangle requires complete (extradural) removal of the ACP. Important structures within the anteromedial triangle include: (i) the horizontal segment of the PS-ICA, the AL of the ICA, (ii) the venous trabecular channels of the anteromedial portion of the PS, and, (iii) a thin layer of the fibrous connective tissue continuous with that overlying the AL of the ICA forming the proximal dural ring (PR) of the paraclinoid ICA. The anteromedial triangle therefore includes the extradural space occupied by the ACP as well as the anteromedial portion of the PS. These two adjacent regions are separated by the connective tissue of the PR extending toward the PCP along the lateral surface of the sella.

The anteromedial triangle of the left PS is shown in Figs. 1.1–1.5.

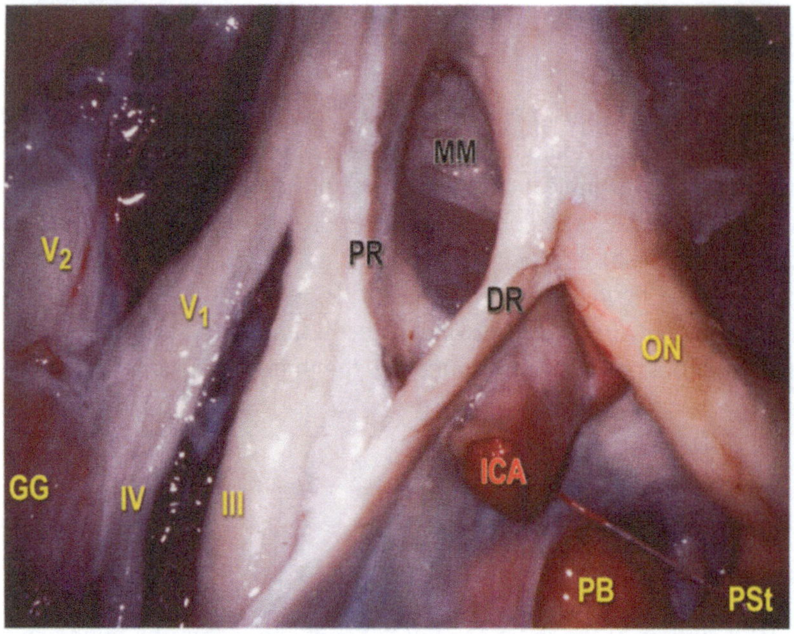

Fig. 1.1. A left-sided fresh cadaver specimen with vessels injected with acrylic, the arteries in red and venous compartments in "transparent blue." The anterior clinoid process (*ACP*) and the outer dural layer of the lateral parasellar space (*PS*) wall have been removed. The anteromedial triangle is bordered medially by the optic nerve (*ON*) enclosed in dura propria, laterally by nerve III, its medial surface covered by fibrous tissue continuous with the proximal ring (*PR*), and, posteromedially, by dura extending from nerve III to the ON. The anterior loop (*AL*) of the ICA is visible in the floor of the anteromedial triangle, covered by "venous blood" beneath a thin fibrous layer. Farther anteriorly is the mucous membrane (*MM*) of the sphenoidal sinus (its thin bony lateral wall inadvertently removed). Posterior to the AL, fibrous tissue courses between dura and nerve III

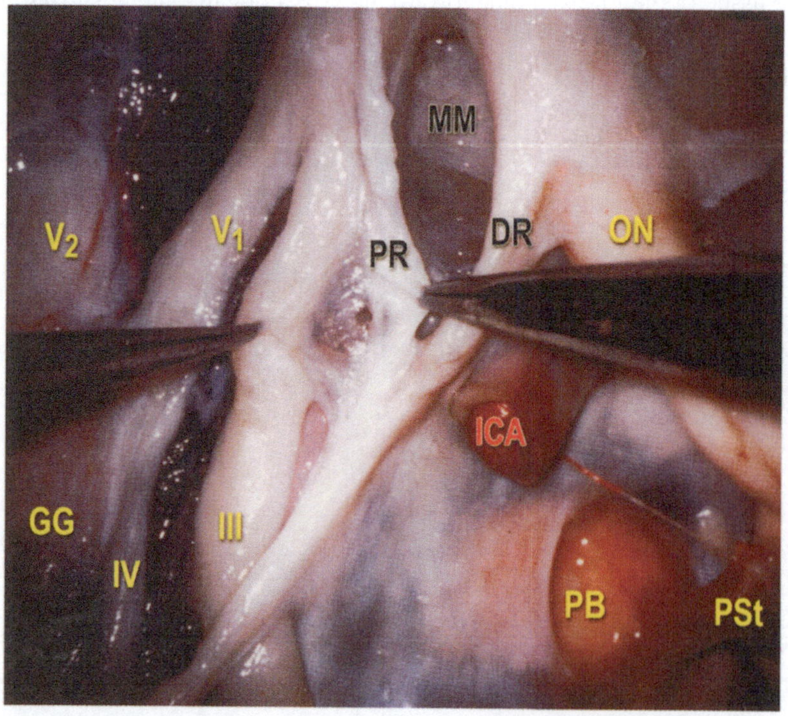

Fig. 1.2. In the specimen shown in Fig. 1.1, the fibrous tissue extending from dura along the medial surface of nerve III is being teased apart. This fairly strong fibrous layer divides extra-parasellar space from parasellar space (*PS*) compartments at the floor of the anteromedial triangle

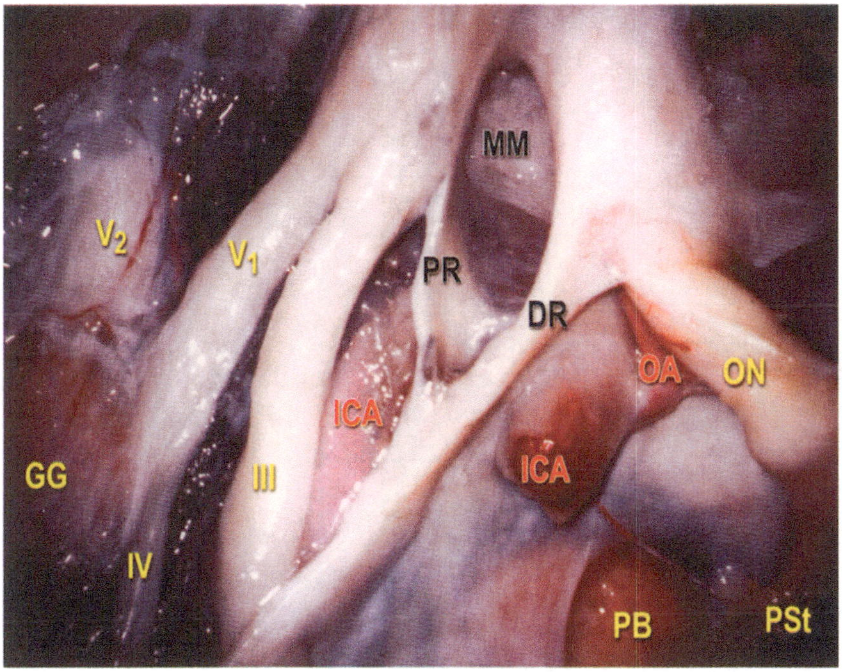

Fig. 1.3. Further dissection of the left parasellar space (*PS*). The proximal ring (*PR*) isolates parasellar space (*PS*) compartments from extra-parasellar space medial to it. The membrane extending from the PR to the distal (dural) ring (*DR*) on the floor of the anteromedial triangle covers the anterior, lateral, and posterior surfaces of the anterior loop (*AL*) of the ICA. Postero-lateral to the PR and medial to nerve III, much of the horizontal PS-ICA segment is apparent through the enlarged anteromedial triangle. It is embedded in "venous blood" beneath nerve III

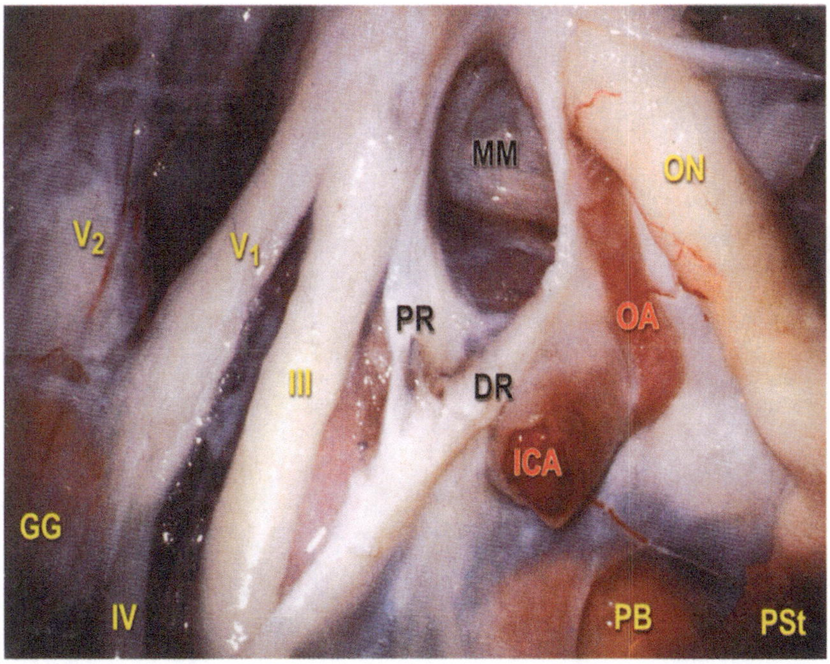

Fig. 1.4. Continued dissection of the left parasellar space (*PS*). The anteromedial triangle has been enlarged further by incising dura propria along the longitudinal axis of the optic nerve (*ON*). By dissecting the ON from its dura propria, it may be displaced medially, achieving additional space and nicely exposing the ophthalmic artery (*OA*) and the distal (dural) ring (*DR*) of the ICA, which may then be incised in a circumferential manner

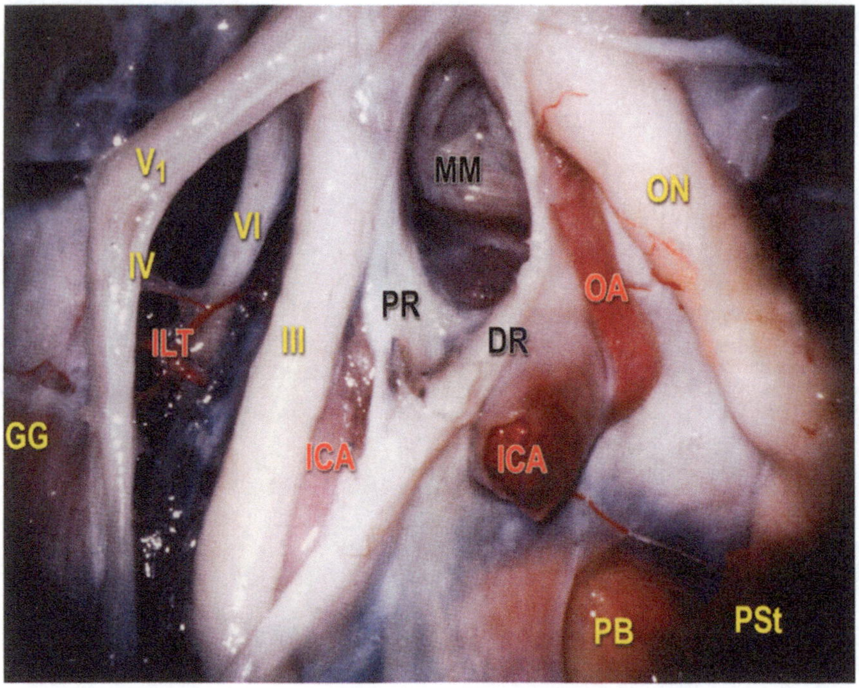

Fig. 1.5. The same specimen as in Figs. 1.1–1.4. The anteromedial triangle may also be enlarged laterally. Displacing V1 along with nerves IV and III offers nice access to the horizontal segment of the ICA. In this specimen nerve III remains in its normal position. The inferolateral trunk (*ILT*), crossing nerve VI, may be identified either from lateral to nerve III through the paramedial triangle (between nerves IV and III) as shown here, or from medial to nerve III through the anteromedial triangle (not shown in this photo)

Paramedial triangle

The lateral surface of nerve III (medial border of the triangle), the medial surface of nerve IV (lateral border), and dura extending between the points of entry of nerves III and IV (posterior border) define the paramedial triangle. Its anterior apex is formed by the point at which nerve IV crosses over nerve III. Structures of importance within this triangle include: the horizontal segment of the PS-ICA, branches of the ILT, PS venous trabecular channels, sympathetic fibers, and, in some cases, the ML of the ICA, the MHT and the proximal PS segment of nerve VI.

The paramedial triangle of the left PS is shown in Figs. 1.6–1.10.

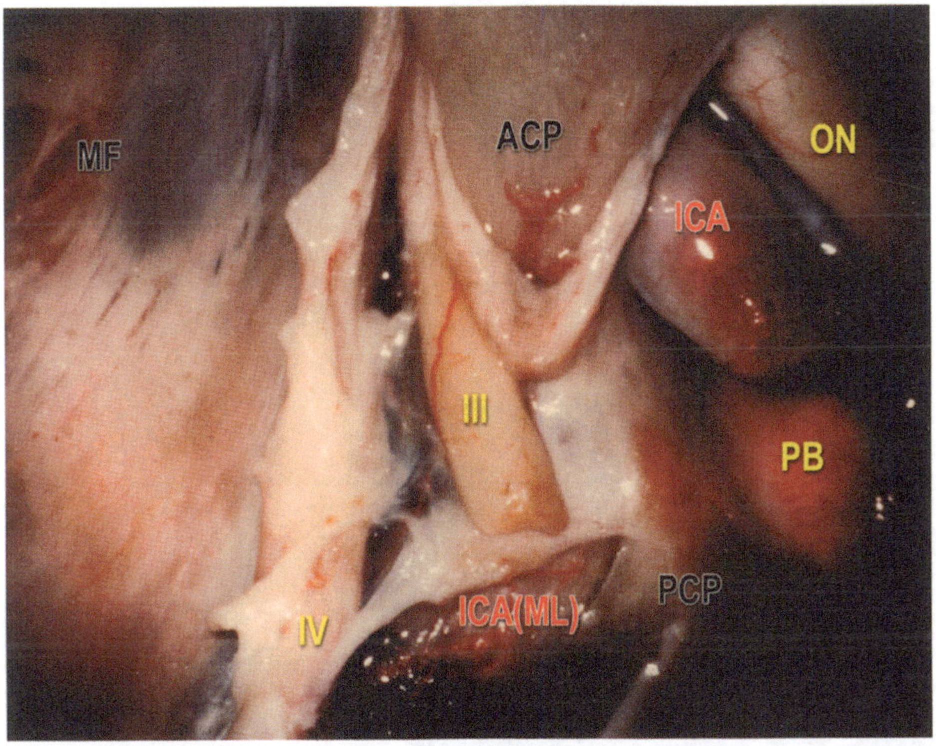

Fig. 1.6. The paramedial triangle. Nerve III in the left parasellar space (*PS*), coursing infero-lateral to the anterior clinoid process (*ACP*), constitutes the medial border of the paramedial triangle. As shown here, nerve III is covered by thick dura both laterally and proximally, and by a thin transparent reticular layer medially (and beneath it), which faces the PS. The lateral border of the paramedial triangle is formed by nerve IV. The dural edge between nerves IV and III makes up the base of this triangle, which is also part of the lateral border of the oculomotor trigone

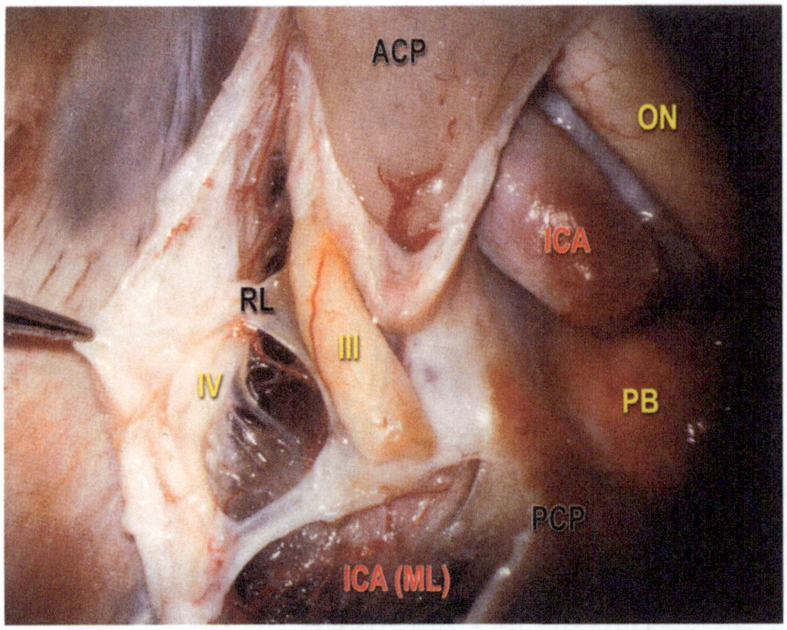

Fig. 1.7. The left paramedial triangle has been widened, exposing the delicate reticular layer (*RL*) beneath nerves III and IV. The arterial supply of nerves III and IV is provided primarily by the inferolateral trunk (*ILT*). A portion of the medial loop (*ML*) of the ICA is visible through virtually transparent venous compartments

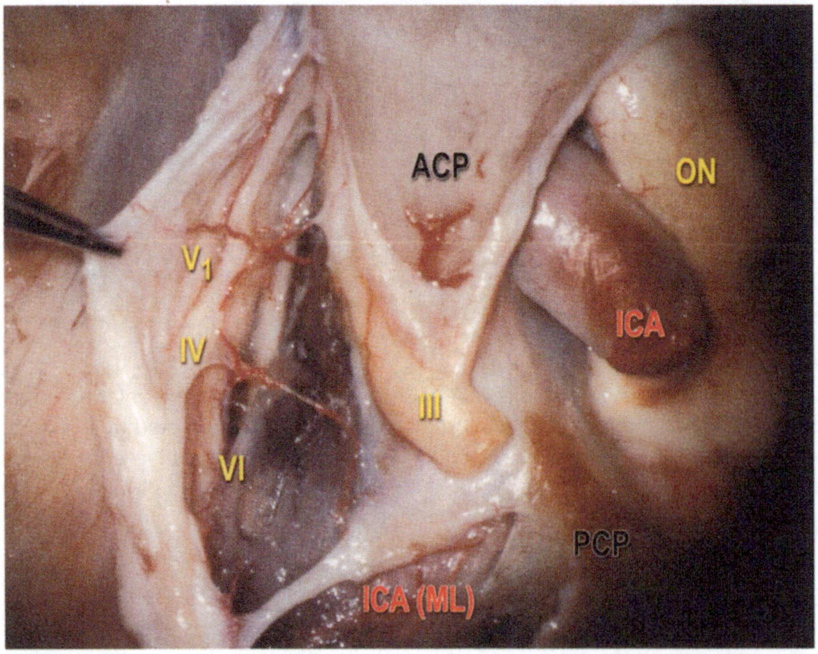

Fig. 1.8. The left paramedial triangle has been enlarged further by retracting the lateral wall of the parasellar space (*PS*), exposing V1, nerve VI, and the anterolateral surface of the ICA as it ascends from its lateral loop (*LL*) to its medial loop (*ML*). Note the rich arterial supply of these nerves and the large amount of "venous blood" in this region

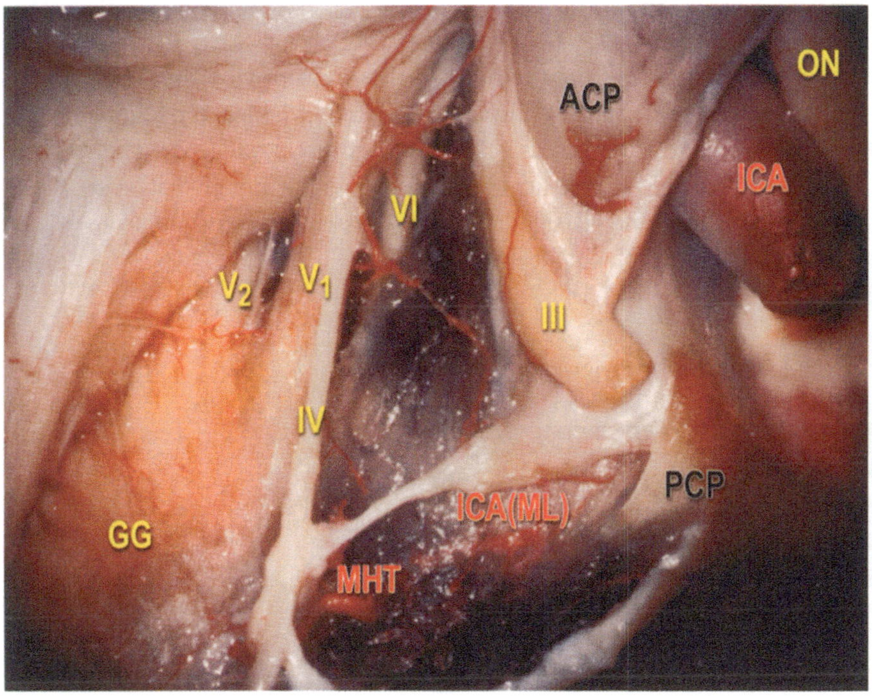

Fig. 1.9. By retracting the outer layer of the lateral wall of the left parasellar space (*PS*) farther laterally, the paramedial triangle is widened even more. This exposes nerve VI in the PS as it crosses the ICA and remains lateral to it and parallel to the horizontal segment of the ICA along its entire course. By incising dura at the base of paramedial triangle, the inferomedial triangle may be accessed easily. If the meningo-hypophyseal trunk (*MHT*) is not visible through the paramedial triangle, it may be exposed by further dissection through the paramedial triangle into the inferomedial triangle, posterior to the ICA as it ascends in the parsellar space (*PS*). By peeling away the outer layer of the lateral PS wall, lateral to nerve IV, the Gasserian ganglion (*GG*) and initial portions of V1 and V2 may be exposed

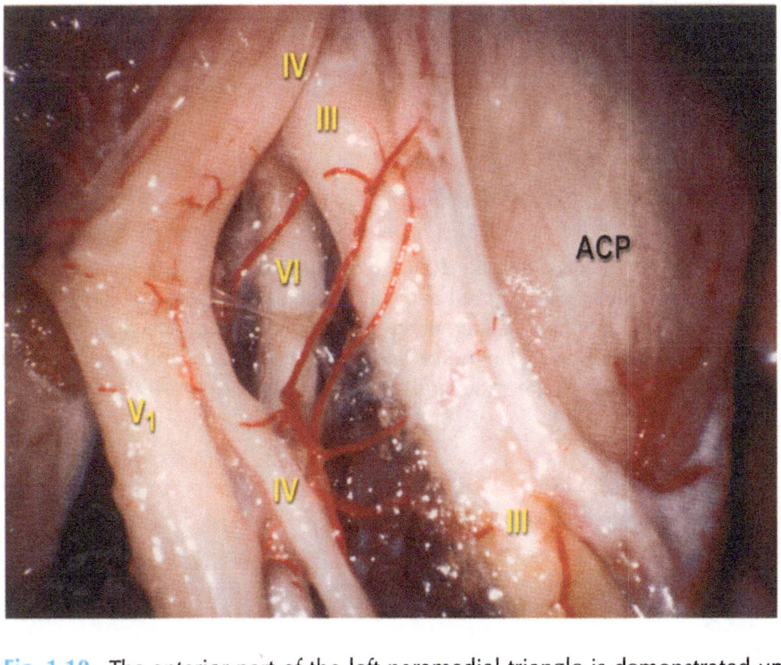

Fig. 1.10. The anterior part of the left paramedial triangle is demonstrated under higher magnification. Notice that nerves IV and V1 cross over nerve III before any of them enters the superior orbital fissure (*SOF*). Deep in the center of this corner of the parasellar space (*PS*), nerve VI remains embedded in "venous blood." Note also the rich arterial supply of nerves III, IV, and VI from branches from infero-lateral trunk (*ILT*)

Parkinson's triangle

Parkinson's triangle is delineated by nerve IV (medial border), the medial surface of V1 (lateral border), and dura between these two nerves posteriorly (posterior border). Through Parkinson's triangle, the following structures may be accessed: the ML of the ICA with, in most cases, the MHT, venous trabecular channels, the proximal and distal segments of nerve VI, and sympathetic fibers. It should be noted that even though Parkinson's triangle is fairly narrow, it can be enlarged significantly by dissecting nerve IV from V1 and displacing it medially, also by dissecting V1 and displacing it laterally.

The left Parkinson's triangle is shown in Figs. 1.11–1.14.

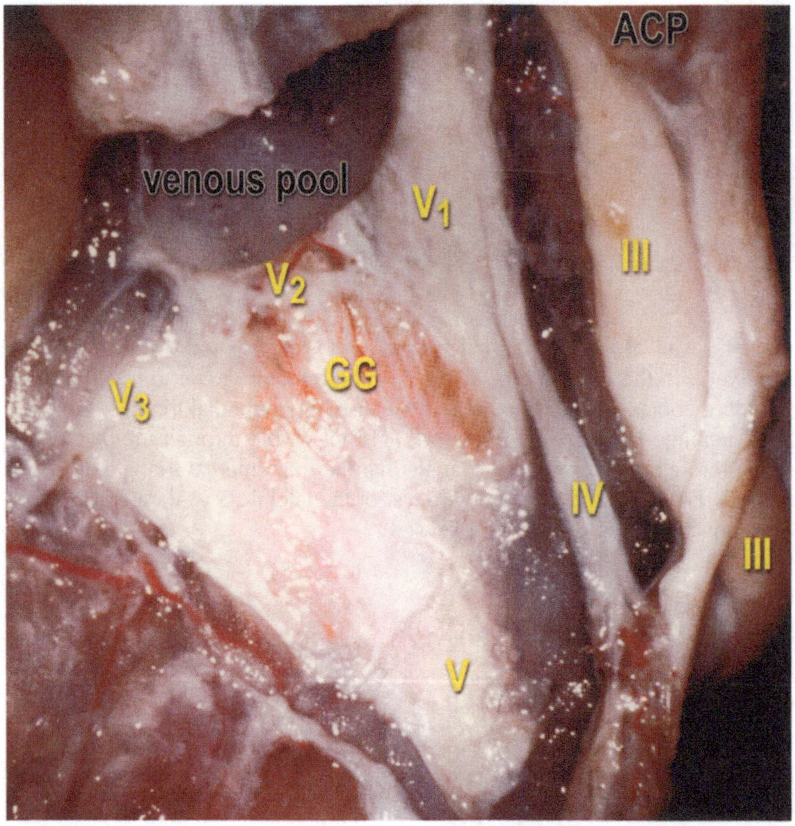

Fig. 1.11. With the outer dural layer of the lateral wall of the left parasellar space (*PS*) removed, one may appreciate the amount of "venous blood" in the PS. Note the large "venous pool" formed as it drains laterally across V1 and V2, also drainage posterior to the Gasserian ganglion (*GG*) and V3, both channels emptying into the superior petrosal sinus (*SPS*). Nerve IV is immediately medial to V1. In this specimen Parkinson's triangle lies somewhat posteriorly and is quite small. It is bordered by nerve IV medially, nerve V, the GG and nerve V1 laterally, and the dura at its base (posteromedially)

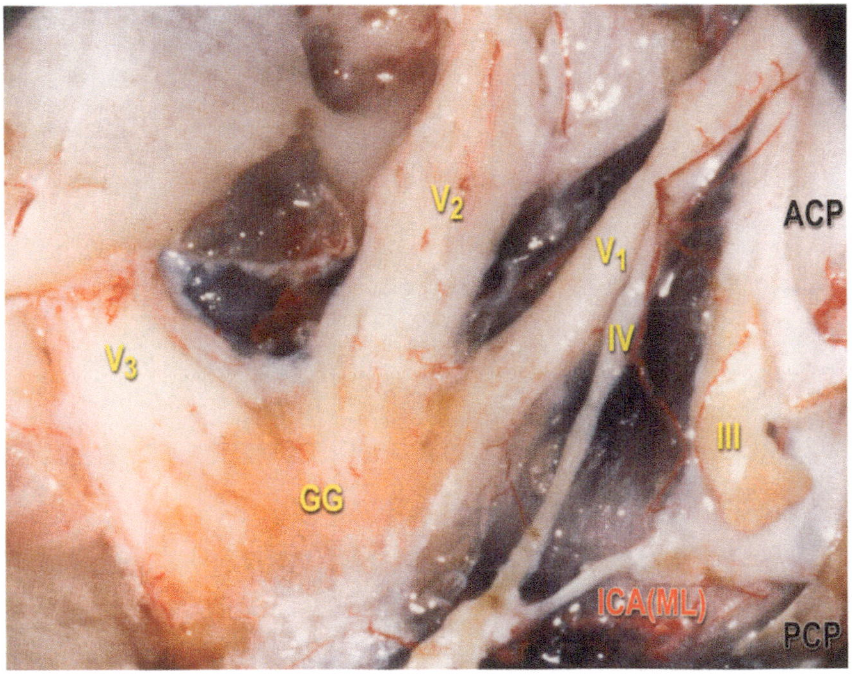

Fig. 1.12. Stripping away the outer dural layer of the left parasellar space (*PS*) exposes the Gasserian gangilion (*GG*), V3, V2, V1, and nerves IV and III. Note: Because of the close relationship between nerves IV and V1, Parkinson's triangle is much smaller than the paramedial triangle

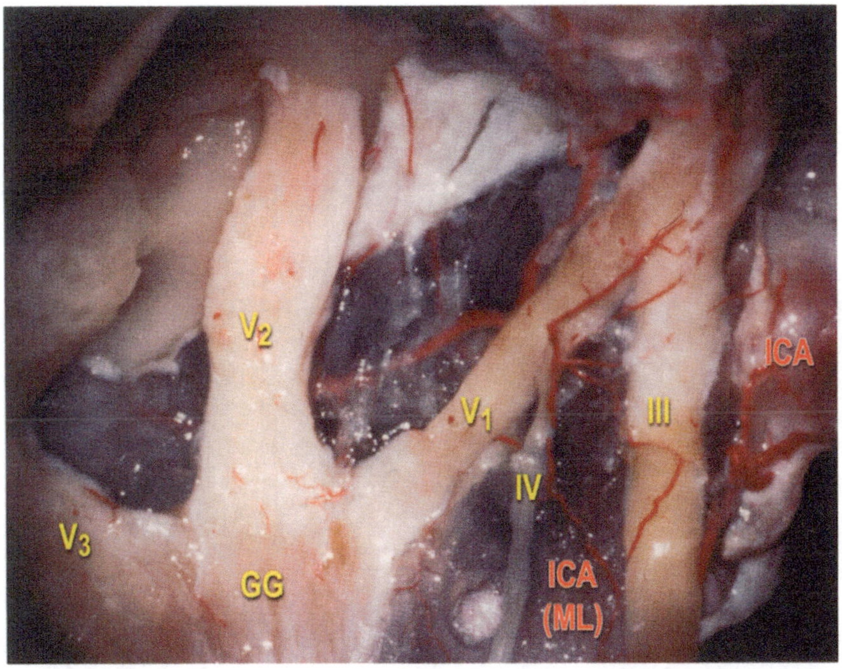

Fig. 1.13. By mobilizing nerve IV the left Parkinson's triangle is widened, giving good access to the meningo-hypophyseal trunk (*MHT*), the ICA, and nerve VI. The segment of ICA ascending from the lateral loop (*LL*) to the medial loop (*ML*) is well exposed through this triangle. Parkinson's "window" may be enlarged even further by separating nerves IV and V1 anterior ward. Note the parasellar space (*PS*) venous compartments and numerous small arteries supplying adjacent nerves

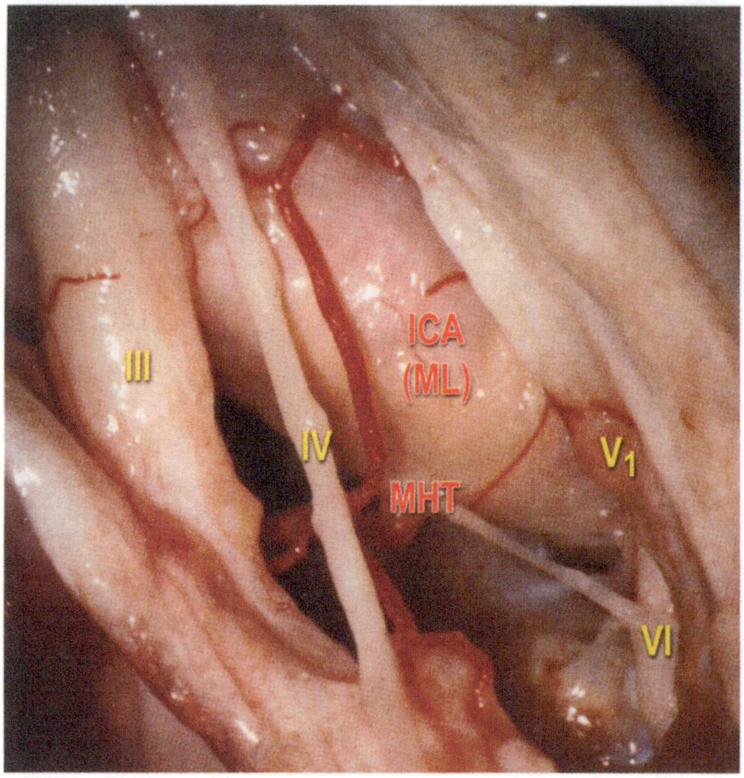

Fig. 1.14. The medial loop (*ML*) of the ICA, the meningo-hypophyseal trunk (*MHT*), and its three branches are exposed through the right Parkinson's triangle. Both the medial and lateral surfaces of the entire ML are demonstrated

Oculomotor trigone

The triangular area of the oculomotor trigone is defined by the anterior petroclinoid fold extending from the petrous apex to the ACP; the posterior petroclinoid fold extending from the petrous apex to the PCP; and the interclinoid fold extending between the ACP and the PCP. This triangle is located posterior to the base of the anteromedial and paramedial triangles, and anterolateral to the inferomedial triangle. On the lateral border of the oculomotor triangle lie the entry points of nerves III and IV into the lateral wall of the PS. In the depth of the oculomotor triangle, the apex of the ML of the ICA, the medial surfaces of the ML and horizontal segment of the ICA may be accessed. In surgery involving the anteromedial, paramedial or inferomedial triangle(s), the oculomotor trigone is also opened, allowing each of them (paramedial, anteromedial and/or Parkinson's, and inferomedial triangle) to be enlarged.

Middle cranial fossa subregion

Anterolateral triangle

The lateral edge of V1 (medial border), the medial edge of V2 (lateral border), and the anterolateral wall of the bony middle cranial fossa (anterior border) compose the anterolateral triangle. Through the anterolateral triangle, the following structures are seen: the inferolateral surface of the distal horizontal segment of the PS-ICA, the venous trabecular channels of the inferolateral PS, nerve VI, sympathetic branches, and the floor of the PS, which is covered by a fibrous dural membrane.

The anterolateral triangle of the left PS is shown in Figs. 1.15–1.20.

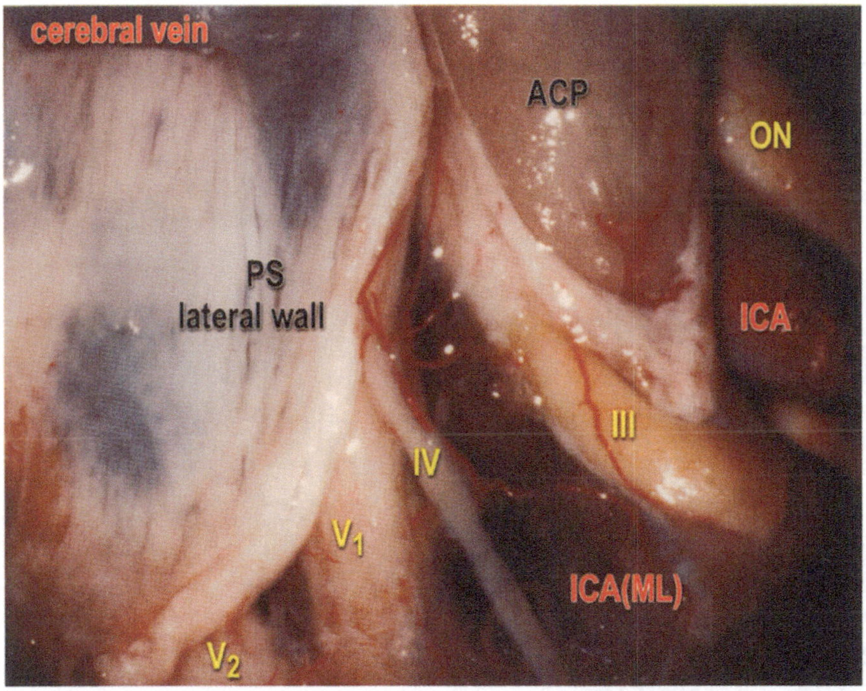

Fig. 1.15. A large cerebral vein enters the lateral wall of the left parasellar space (*PS*) in the anterior part of the anterolateral triangle (V1 and V2 still covered by dura). The intense blue color of the lateral PS wall confirms the presence of a large basin of venous blood between its layers

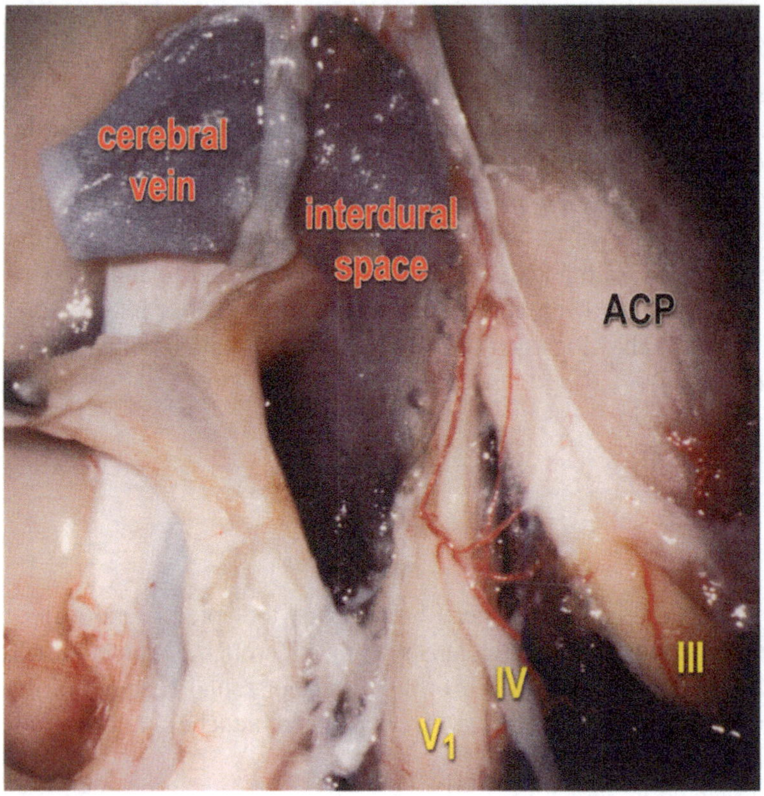

Fig. 1.16. The outer layer of the left parasellar space (*PS*) lateral wall has been peeled from the "venous pool" lying between its two layers of dura

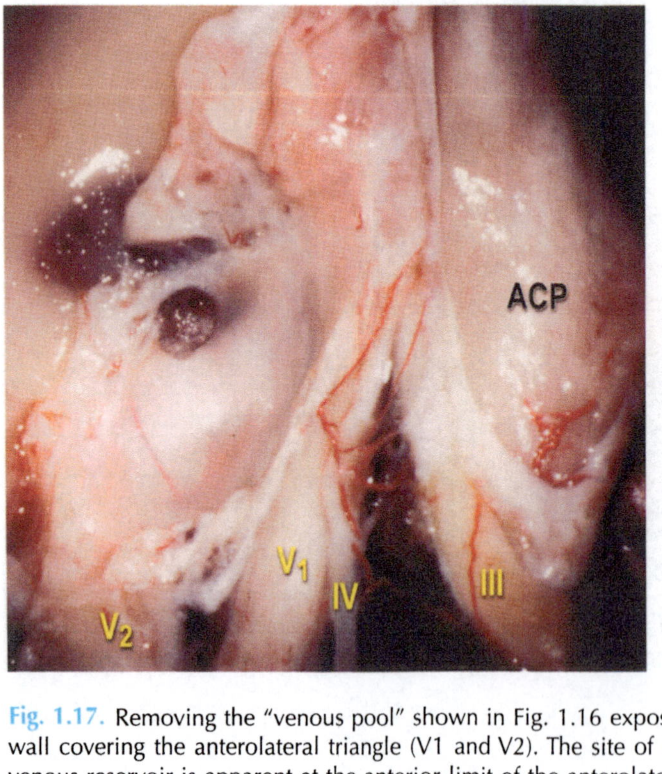

Fig. 1.17. Removing the "venous pool" shown in Fig. 1.16 exposes the inner layer of the left lateral parasellar space (*PS*) wall covering the anterolateral triangle (V1 and V2). The site of blood entering into the PS the now-removed inter-dural venous reservoir is apparent at the anterior limit of the anterolateral triangle

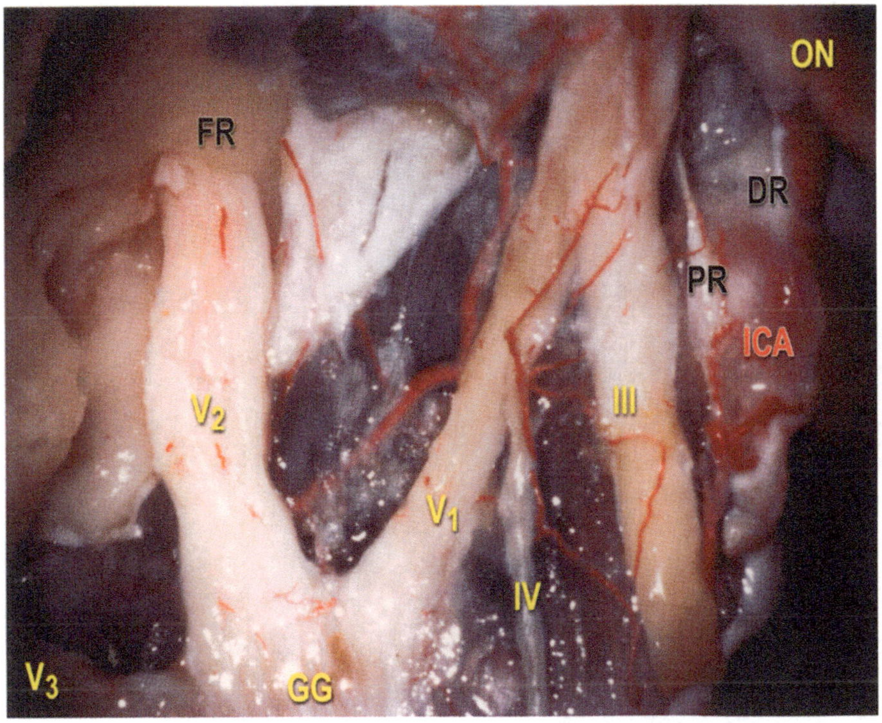

Fig. 1.18. The inner layer of the lateral wall of the left parasellar space (*PS*) has been removed, completely exposing the anterolateral triangle (V1 and V2). Note the PS "venous blood" and the preserved fibrous-dural tissue covering bone within the triangle anteriorly, also the arterial branches supplying the nerves and surrounding structures

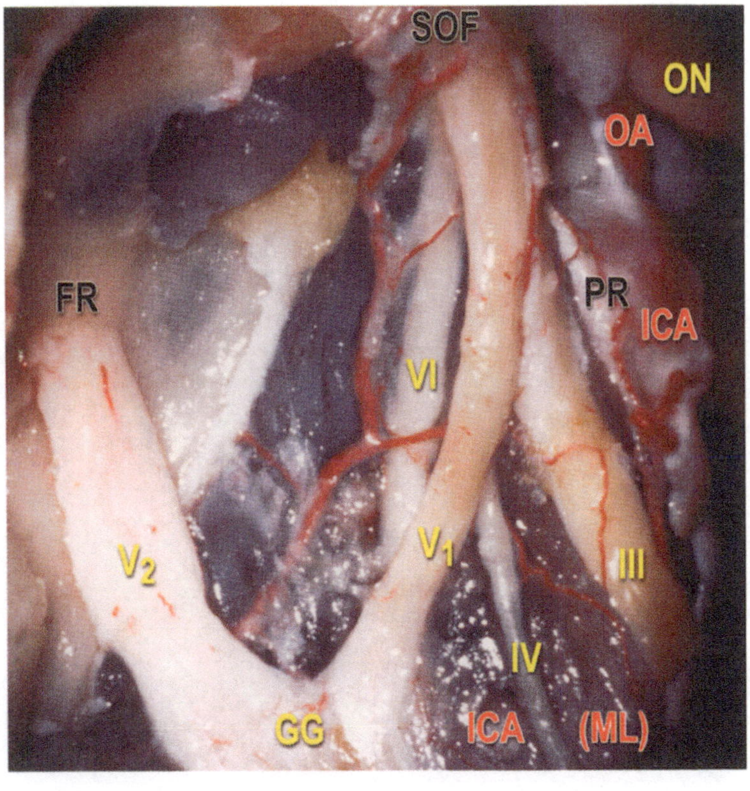

Fig. 1.19. The left anterolateral triangle, from its apex at the Gasserian ganglion (*GG*) to base – middle cranial fossa bone between the superior orbital fissure (*SOF*) and the foramen rotundum (*FR*) is shown. Nerve VI has been exposed by slightly displacing V1 medially. Note the "venous blood" deep in the parasellar space (*PS*) and arterial branches supplying adjacent nerves

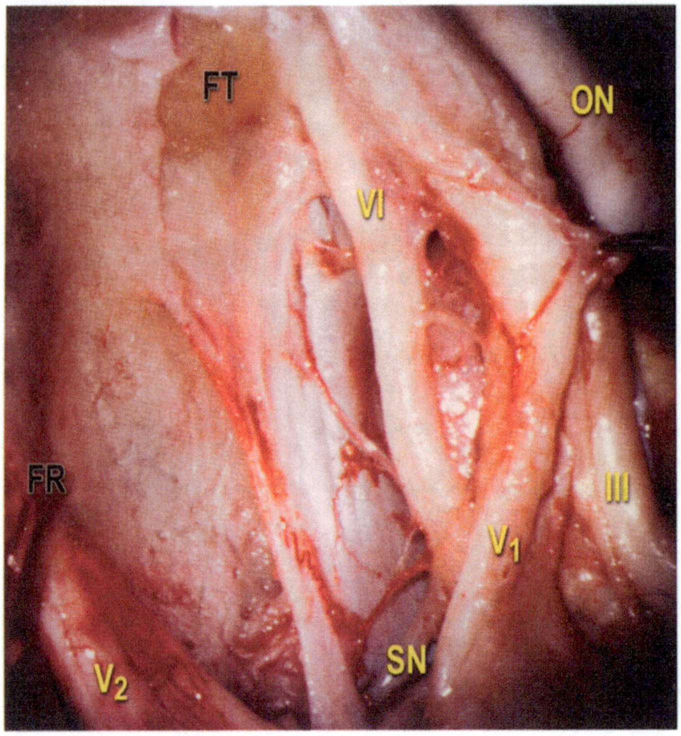

Fig. 1.20. A long segment of nerve VI is prominent within the left anterolateral triangle. By retracting V1 medially, nerve VI and the ICA (with sympathetic fibers running between them, also to V1) are better exposed. The fibrous membrane in the center of the photograph is an extension of dura over the petrous bone posteriorly. It forms the lateral ring (*LR*) around the ICA medial to the foramen lacerum (*FL*) – not seen here) – then proceeds further anteriorly to cover bone. The fatty tissue (*FT*) in the anteromedial corner of the anterolateral triangle is continuous with orbital fat. Note: Venous structures were not injected in this specimen

Lateral triangle

The lateral edge of V2 (anteromedial border), the anterior edge of V3 (posterior border), and the bone of the middle cranial fossa (lateral border) describe the lateral triangle. Through this triangle the anterior and lateral surfaces of the LL of the ICA, with the sympathetic cord(s) running over them, can be seen and these structures accessed. Owing to the short course of V3 and the proximity of the GG, only slight retraction of V3 and V2 is possible without endangering their function. By drilling middle fossa bone laterally, additional space is gained for better access to the LL of the ICA.

The lateral triangle of the left PS is shown in Figs. 1.21–1.25.

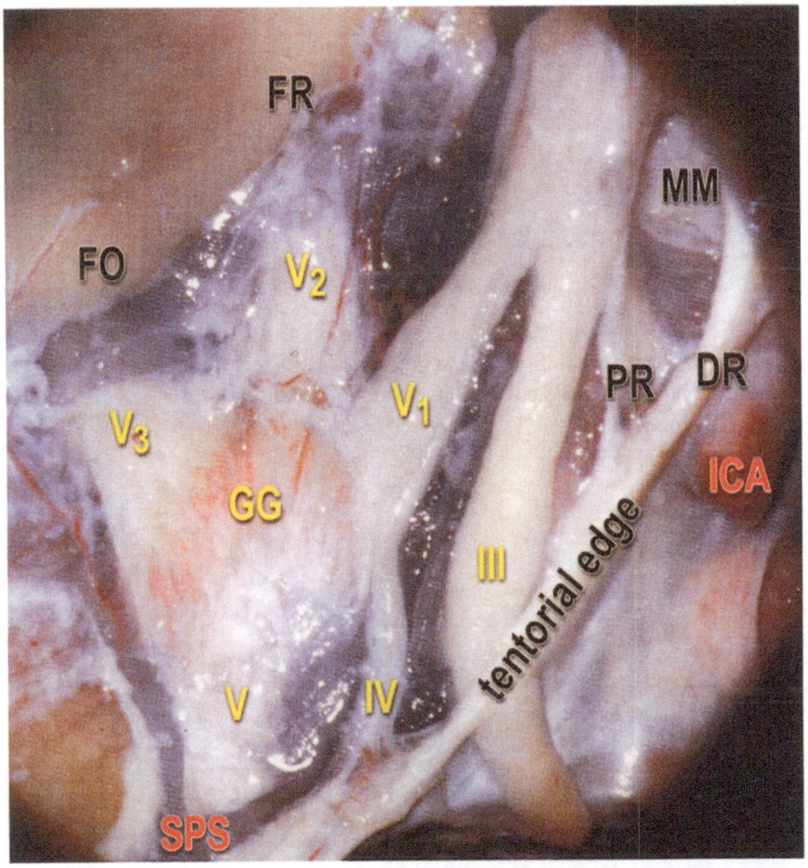

Fig. 1.21. The dural layer of the lateral wall of the left parasellar space (*PS*) and dura over the floor of the middle cranial fossa have been removed. Of all the anatomical triangles demonstrated here (anteromedial, paramedial, Parkinson's, anterolateral, lateral), the lateral triangle, circumscribed by V2, V3, and bone between the foramen rotundum (*FR*) and foramen ovale (*FO*), is the smallest. Like the other triangles, it is filled entirely with "venous blood." These venous structures drain, along with the one coursing posterior to V3 and the Gasserian ganglion (*GG*), into the superior petrosal sinus (*SPS*)

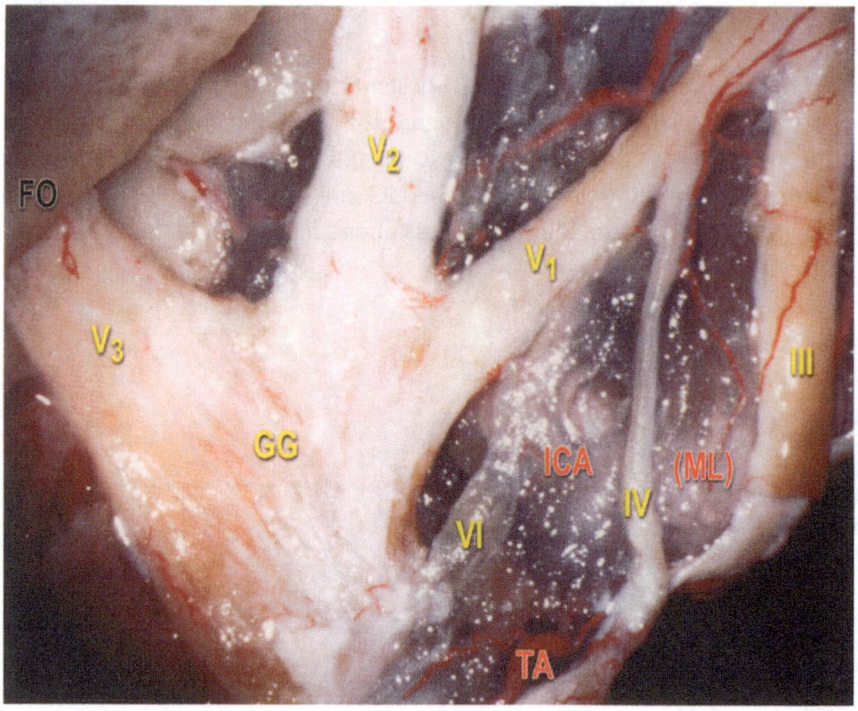

Fig. 1.22. The exposed left parasellar space (*PS*) demonstrates that the lateral triangle is much smaller than the anterolateral triangle. It is filled with "venous blood" through which tiny arteries are seen. Note that V3 is much shorter than V2, placing the Gasserian ganglion (*GG*) much closer to the foramen ovale (*FO*) than to the foramen rotundum (*FR*). The medial loop (*ML*) of the ICA embedded within the "venous blood" deep to V1 and nerves IV and III

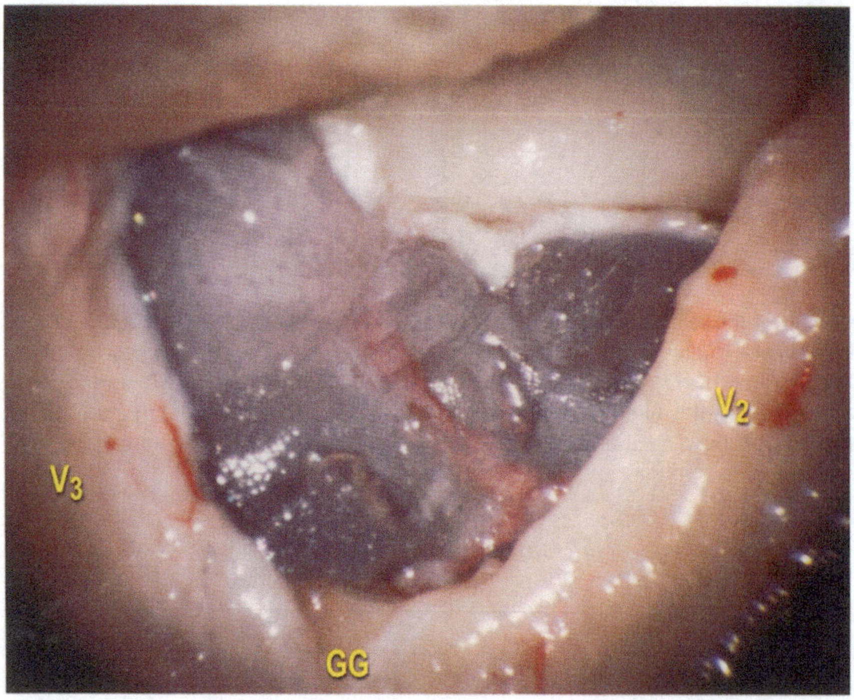

Fig. 1.23. The left lateral triangle under higher magnification. Only by drilling bone laterally may this triangle be enlarged, providing access to the space deep to the Gasserian ganglion (*GG*). Note the arterial branch visible through parasellar space (*PS*) "venous blood." "Surprisingly enough" it arises from the external carotid artery

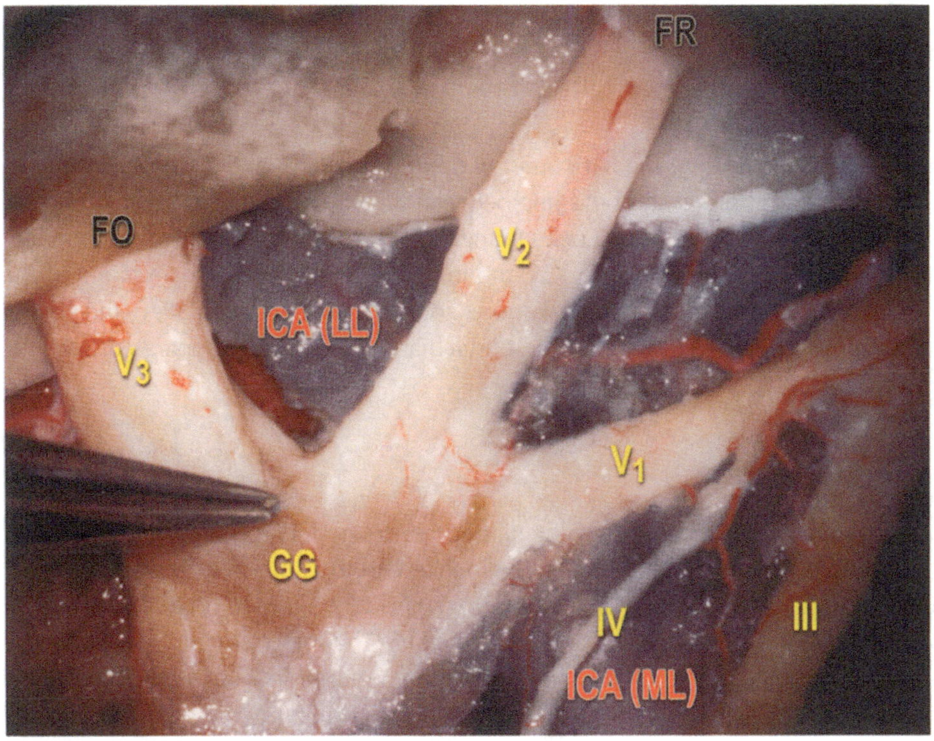

Fig. 1.24. Removing "venous blood" and slightly retracting the left Gasserian ganglion (*GG*) exposes the petrous horizontal ICA segment extending from the posterior loop (*PL*) to the lateral loop (*LL*). Since it provides safe access to the lateral loop (*LL*) of the ICA, the lateral triangle, though quite small, is very important

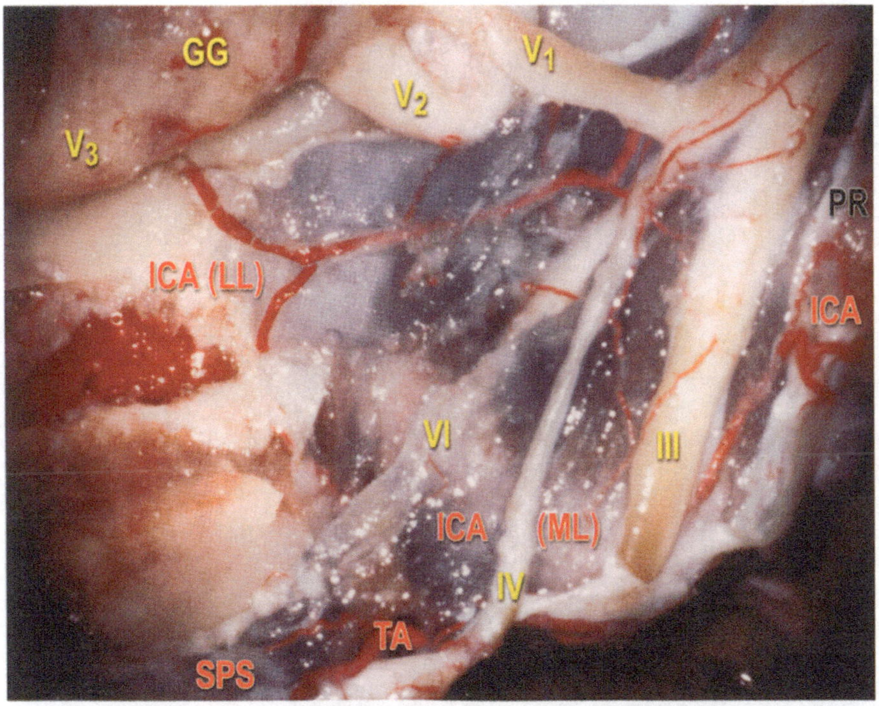

Fig. 1.25. For demonstration purposes nerve V and the Gasserian ganglion (*GG*) have been reflected lateralward and mobilized well laterally, exposing the lateral loop (*LL*) of the left ICA. The ICA is seen from the foramen lacerum (*FL*), its LL, to the posterior clinoid process (*PCP*), its medial loop (*ML*). (The surgeon, however, under no circumstances may displace nerve V in this manner, being forced to access the ICA only through appropriate triangles.) Note the following relationships: Nerve VI crosses the lateral surface of the ICA ascending from its LL to its ML, and nerve III lies parallel to its horizontal segment. Between them nerve IV crosses the ML. Anteriorly, nerves III, IV, V1 and VI coalesce tightly as they approach the superior orbital fissure (*SOF*)

Posterolateral (Glasscock's) triangle

The line from the arcuate eminence to the postero-medial corner of the GG (postero-medial border), the posterior surfaces of the GG and V3 (anterior border), and the line between the FS and the arcuate eminence of the petrous bone (postero-lateral border) constitute Glasscock's triangle [5]. Through it the following important structures may be accessed: the distal leg of the PL of the ICA, the initial segment of the proximal leg of the LL of the ICA, the labyrinthine branch of the middle meningeal artery (coursing parallel to the GPN), the greater and lesser petrosal nerves, the tensor tympani muscle, and the eustachian tube, which is separated from the ICA by a 1–2 mm thick bony wall.

Posteromedial (Kawase-Shiobara's) triangle

The line connecting the arcuate eminence and the postero-medial corner of the GG (postero-lateral border), nerve V (anterior border), and the line connecting the arcuate eminence and the posterior aspect of nerve V where it crosses the crest of the petrous apex (postero-medial border) delineate the posteromedial triangle [3]. It includes the dura and bone of the petrous apex, and a portion of the SPS. The importance of this triangle is twofold. By drilling bone posterior and medial to the GG and nerve V, the posterior fossa can be reached, the region behind nerve V explored, and the clivus visualized. By more extensive drilling of the petrous apex, the anterior limit of the internal auditory canal (IAC) may be exposed. The second important point is that by drilling the petrous apex, the medial surface of the PL of the ICA can be visualized (where the vessel may be prepared for temporary clipping, and/or grafting).

The posterolateral and posteromedial triangles of the left side are shown in Figs. 1.26–1.30.

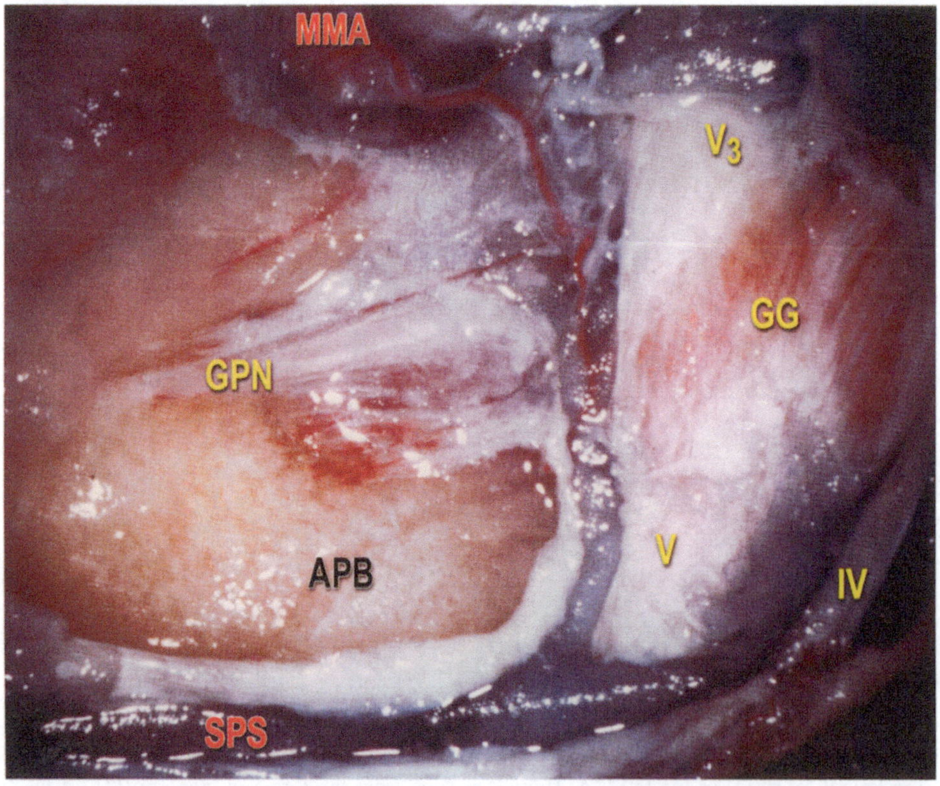

Fig. 1.26. Well posteriorly in the left middle cranial fossa, dura has been removed from the anterior surface of the petrous bone and from nerve V and the Gasserian ganglion (*GG*). Venous blood drains posteriorly from the parasellar space (*PS*) on either side of nerve V, into the superior petrosal sinus (*SPS*). The anterior surface of the apex of the petrous bone (*APB*) lies immediately posterior to the GG and the root of nerve V. The middle meningeal artery (*MMA*) exits the foramen spinosum (*FS*). The posterolateral and posteromedial triangles lie posterior to V3, the GG, and the root of nerve V respectively

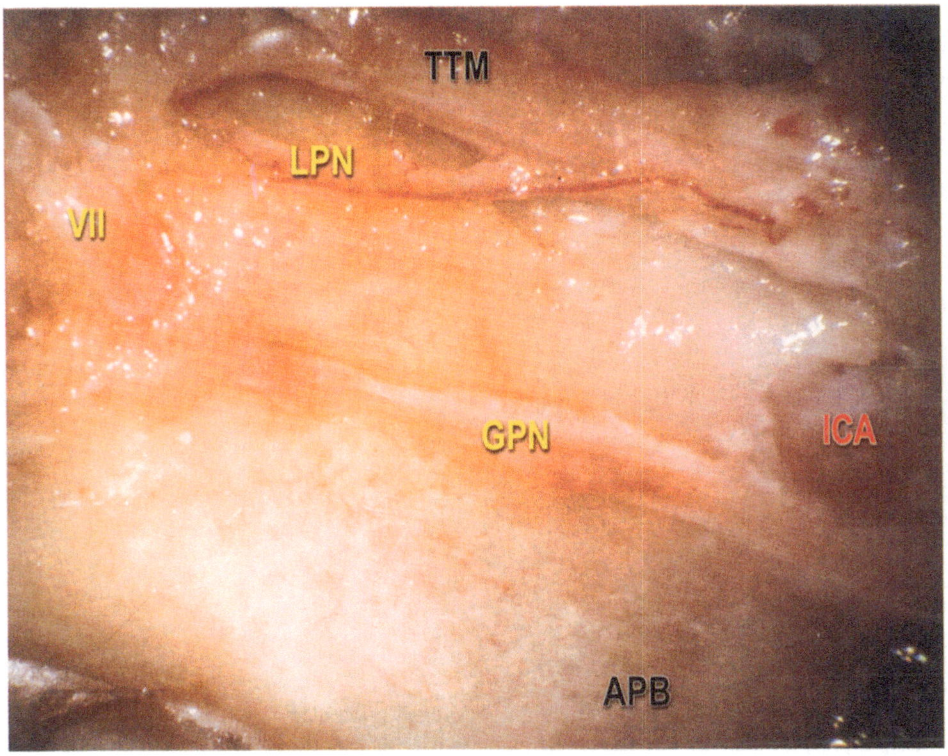

Fig. 1.27. Posterior to V3 and lateral to the sphenopetrous fissure and the greater petrosal nerve (*GPN*), the thin layer of bone covering the petrous ICA has been removed. The lesser petrosal nerve and the tensor tympani muscle (*TTM*) lie dorsal and lateral to the petrous carotid canal. The posteromedial triangle is medial to the GPN, well posterior to the trigeminal nerve. A word of caution: Here the geniculate ganglion of nerve VII is not covered by bone. At live surgery: traction on the GPN or coagulation in the vicinity of the geniculate ganglion will result in peripheral facial palsy

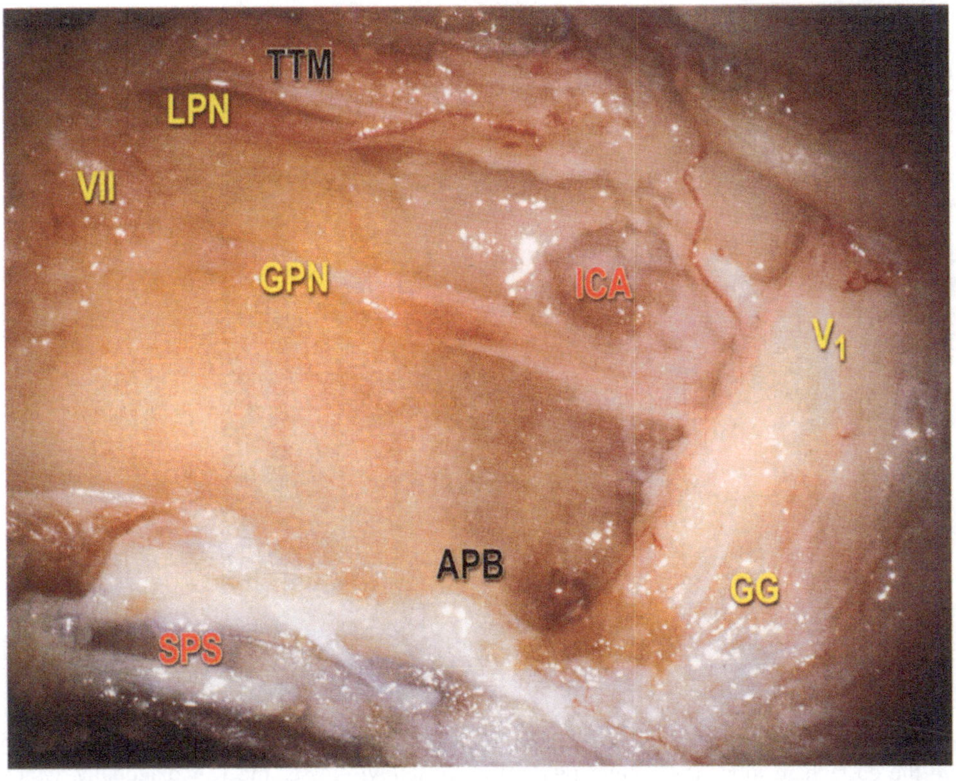

Fig. 1.28. The ICA in the petrous carotid canal has been partially unroofed and lies lateral to the greater petrosal nerve (*GPN*). The geniculate ganglion (*VII*) is not covered with bone. The posteromedial triangle lies medial to the GPN, enclosing the apex of the petrous bone (*APB*). The tensor tympani muscle (*TTM*) and the lesser petrosal nerve (*LPN*) are seen postero-laterally to the petrous ICA

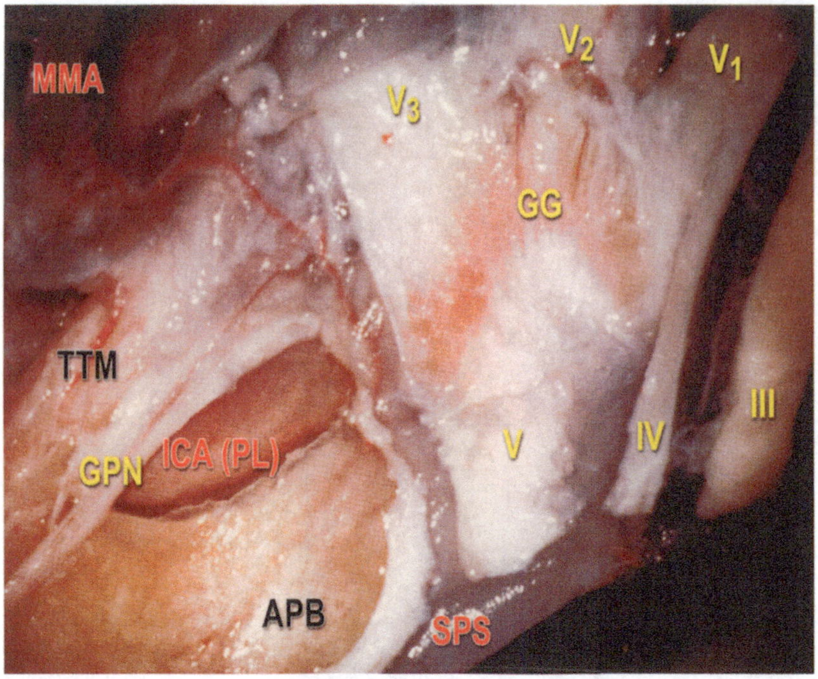

Fig. 1.29. The left ICA in the petrous bone has been unroofed. The greater petrosal nerve (*GPN*) is adjacent to the lateral surface of the posterior loop (*PL*), slightly medial to the tensor tympani muscle (*TTM*) and the eustachian tube. The intact apex of the petrous bone (*APB*) medial to posterior loop (*PL*) lies within the posteromedial triangle, posterior to the Gasserian ganglion (*GG*) and nerve V. Venous drainage is medialward and posteriorly, into the SPS

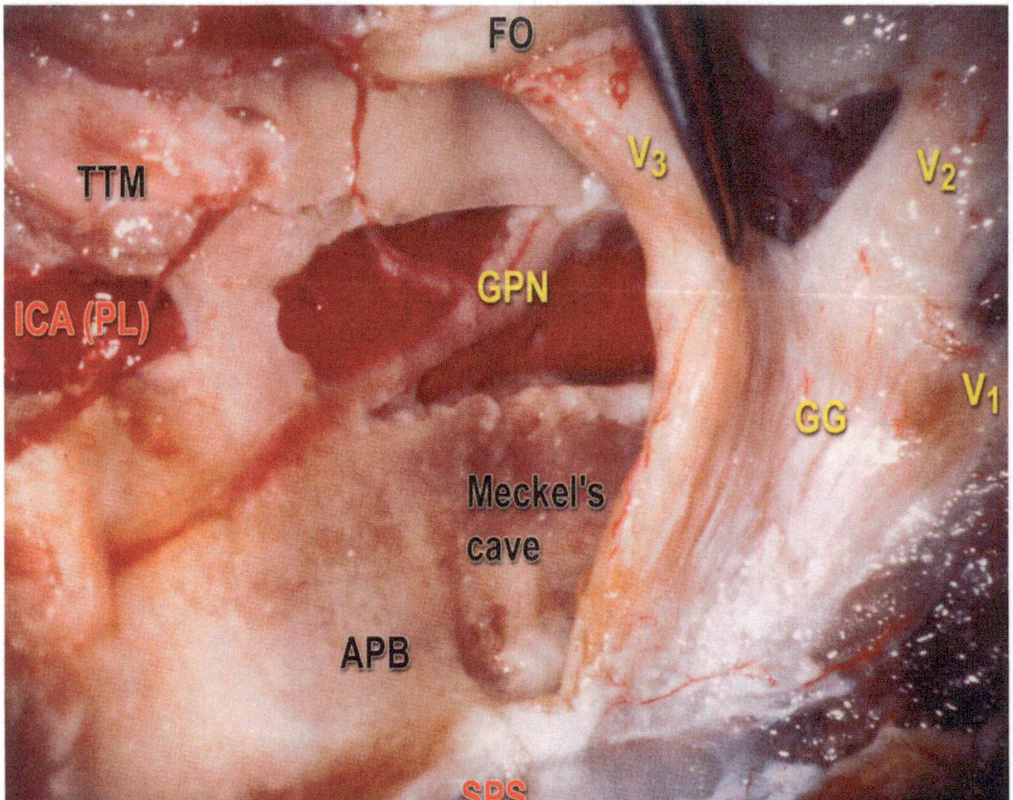

Fig. 1.30. The left petrous carotid canal has been unroofed, exposing the injected ICA. The tensor tympani muscle (*TTM*) lies lateral to the ICA, above the eustachian tube. The greater petrosal nerve (*GPN*) crosses the ICA diagonally, which courses beneath V3 and the Gasserian ganglion (*GG*), toward the foramen lacerum (*FL*) where it forms the lateral loop (*LL*). The superior petrosal sinus (*SPS*) is in continuity with venous compartments superficial to nerve V and in Parkinson's triangle. Note: By lifting the GG and V3 the Meckel's cave is exposed

Paraclival subregion

Inferomedial triangle

The triangle is formed by the apex of the PCP (medial point), the dural entry point of nerve IV into the tentorium (laterosuperior point), and the dural entry point of nerve VI into DS (lateroinferior point). The structures of importance within this triangle include: the basilar venous plexus and dura overlying it, dura forming the posterior wall of the PS, the posterior petroclinoid fold, the segment of nerve VI within DS beneath and medial to the PCL coursing toward the lateral surface of the ICA, the PCL, and the dorsal meningeal branch of the MHT.

The inferomedial triangle of the left side is shown in Figs. 1.31–1.35.

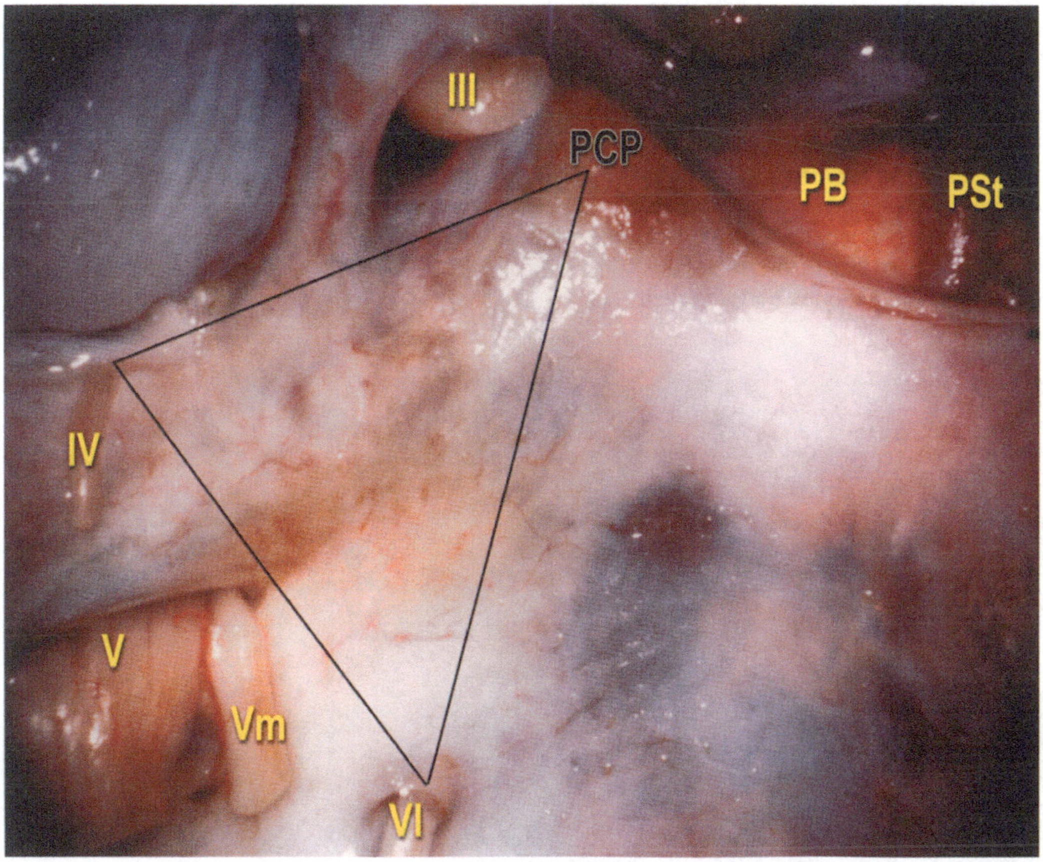

Fig. 1.31. The left paraclival area and inferomedial triangle are seen from the posterior cranial fossa. This triangle is defined by the posterior clinoid process (*PCP*), the site of nerve IV's entry into the tentorial edge, and nerve VI's entry into clival dura (into Dorello's space). The superior border of the triangle is the dural fold between where nerve IV enters the under surface of the tentorium and the tip of the PCP. The medial border is the line between where nerve VI pierces dura lateral to the clivus and the tip of the PCP, and the lateral border is the line connecting nerves VI and IV. In the middle cranial fossa, the medial half of the superior border of the inferomedial triangle constitutes the dorsal border of the oculomotor trigone. Note the large dural opening traversed by nerve III. The intense blue of the injected parasellar space (*PS*) venous compartments is visible through thin dura adjacent to the clivus. The pituitary body (*PB*) and the pituitary stalk (*PSt*) are seen anterior to the dorsum sellae

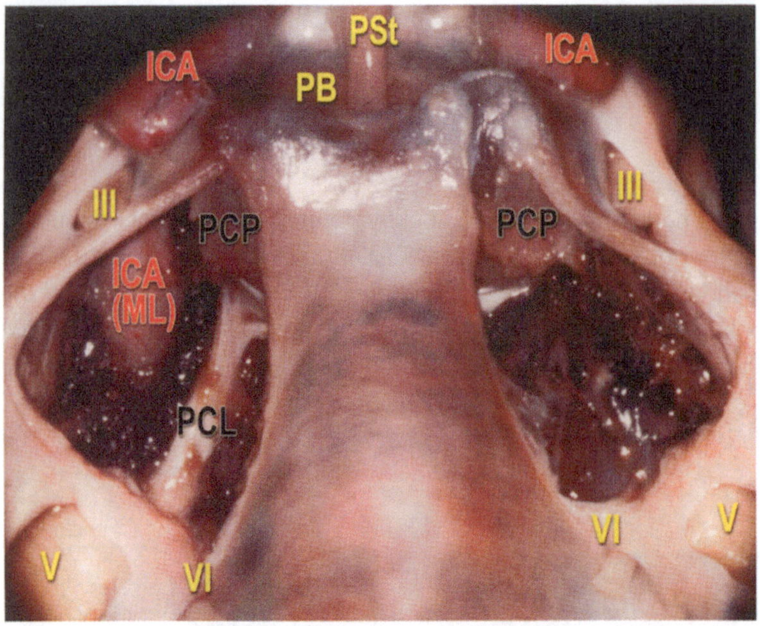

Fig. 1.32. Dura over the inferomedial triangles has been removed bilaterally. On the right side, the injected parasellar space (*PS*) venous compartments have not been disturbed. The dorsal surface of the posterior clinoid process (*PCP*) is adjacent to "venous blood" in the superomedial corner of the right inferomedial triangle. In the left inferomedial triangle, however, some "venous blood" has been removed, exposing the petroclinoid ligament (*PCL*) and the posterior surface of the medial loop (*ML*) of the ICA antero-lateral to it. Both nerves VI are seen intradurally but not yet extradurally within Dorello's space and the PS. Both nerves III are apparent in the oculomotor trigones. The pituitary stalk (*PSt*) and the pituitary body (*PB*) are seen anterior to the dorsum sellae

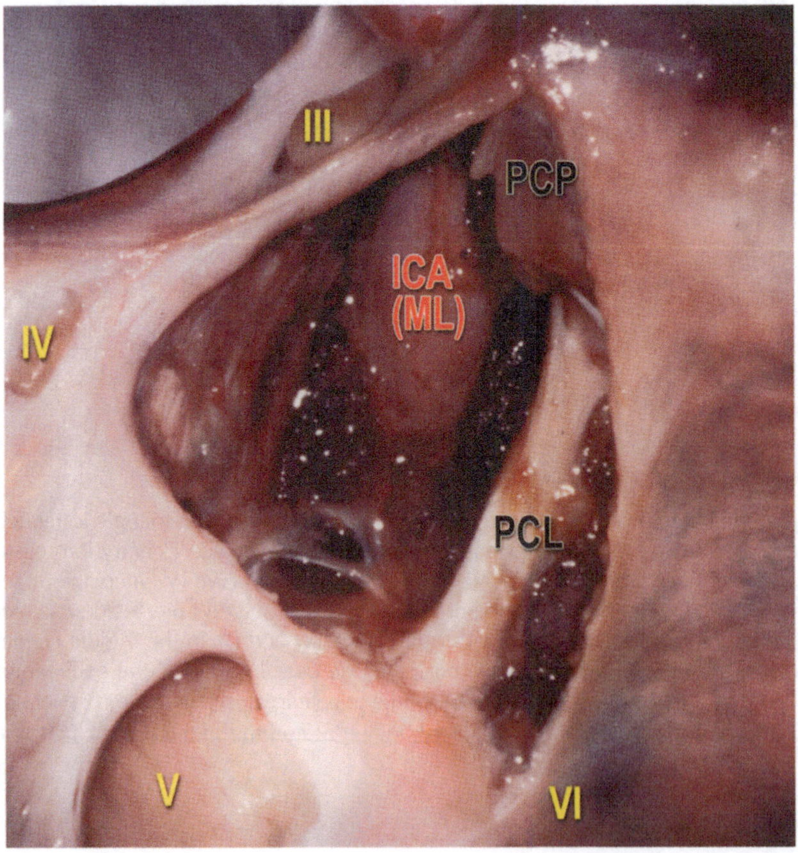

Fig. 1.33. After removing additional "venous blood" from the left inferomedial triangle shown in Fig. 1.32, the petroclinoid ligament (*PCL*) is demonstrated, extending from the petrous apex to the posterior clinoid process (*PCP*). "Venous blood" fills the space between the PCP and medial loop (*ML*) of the ICA. Delicate curvilinear venous trabeculae are visible lateral to the PCL and posterior to the ascending ICA

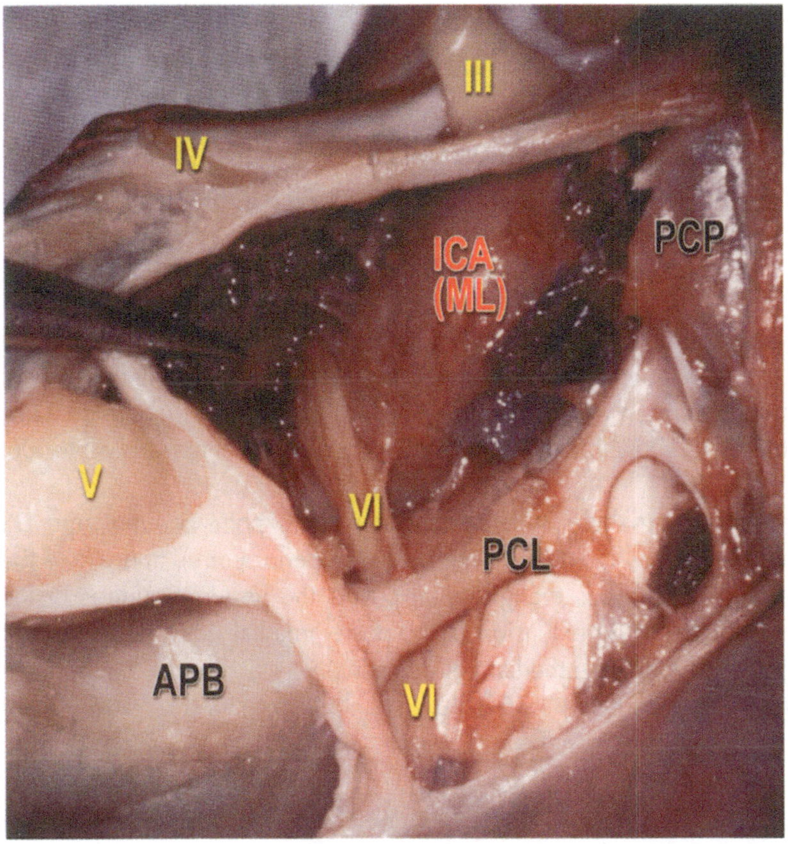

Fig. 1.34. The origin of the petroclinoid ligament (*PCL*) from the very apex of the petrous bone (*APB*) is seen on the left side. Adjacent to the petrous apex, medially to the PCL, nerve VI is exposed from where it climbs to the lateral surface of the ascending ICA. One is able make up to five indivudual bundles of nerve VI. Several curvilinear venous trabeculae are apparent medial to the anterior half of the PCL. The APB medial to the trigeminal nerve is posterior to the base of the inferomedial triangle

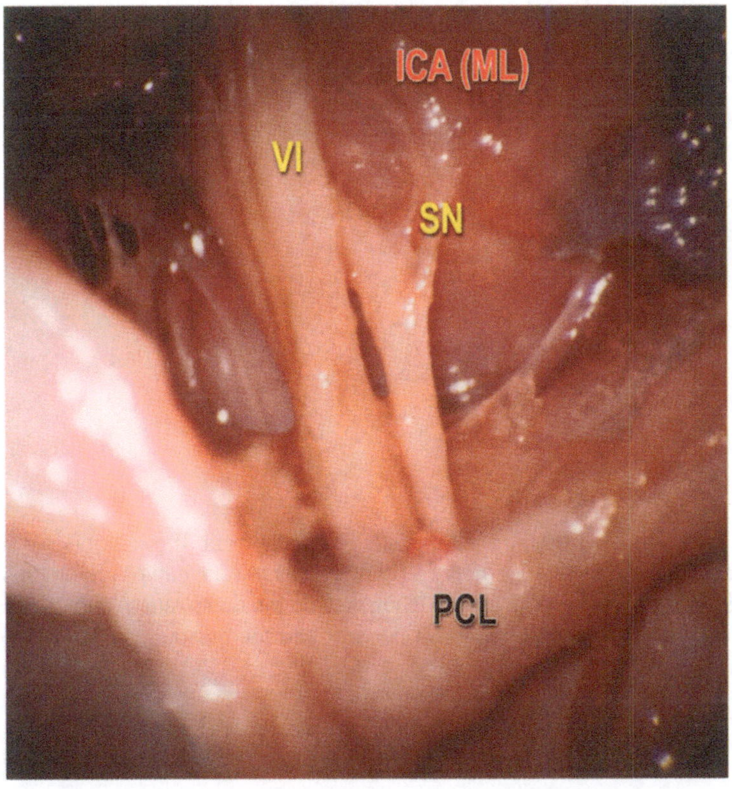

Fig. 1.35. Nerve VI is shown under higher magnification climbing from the left petroclinoid ligament (*PCL*) toward the lateral surface of the ascending ICA. All five fascicle are apparent

Inferolateral (trigeminal) triangle

The inferolateral triangle is defined by the dural entry point of nerve IV into the tentorium (superior point), the point of nerve VI entry into DS (medial point), and the entry point of the petrosal vein into the SPS (lateral point). By drawing a line from the entry point of the petrosal vein into the SPS to the anterior limit of the point of nerve V's exit from the posterior cranial fossa into Meckel's cave, the inferolateral triangle can be subdivided into two additional triangles.

The "tentorial triangle" within the inferolateral triangle is defined by the point of nerve IV's entry into the dural fold at the edge of the tentorium (superior point), the anterior border of nerve V as it enters Meckel's cave (inferomedial point), and where the petrous vein enters the SPS (posterior point). This triangle contains part of the SPS, the tentorial branch of the MHT known as the artery of Bernasconi Cassinari, and tentorium superior and posterior to Meckel's cave.

The "osseous triangle" within the inferolateral triangle is defined by nerve VI's point of entry into DS (inferomedial point), the anterior border of nerve V as it enters Meckel's cave (superior point), and where the petrosal vein enters the SPS (posterior point). Within this triangle, dura covers the posterior surface of the petrous apex. This triangle represents the posterior fossa extension of the posteromedial (Kawase-Shiobara's) triangle in the middle cranial fossa (projecting through the medial petrous apex).

The inferolateral triangle of the left side is shown in Figs. 1.36–1.40.

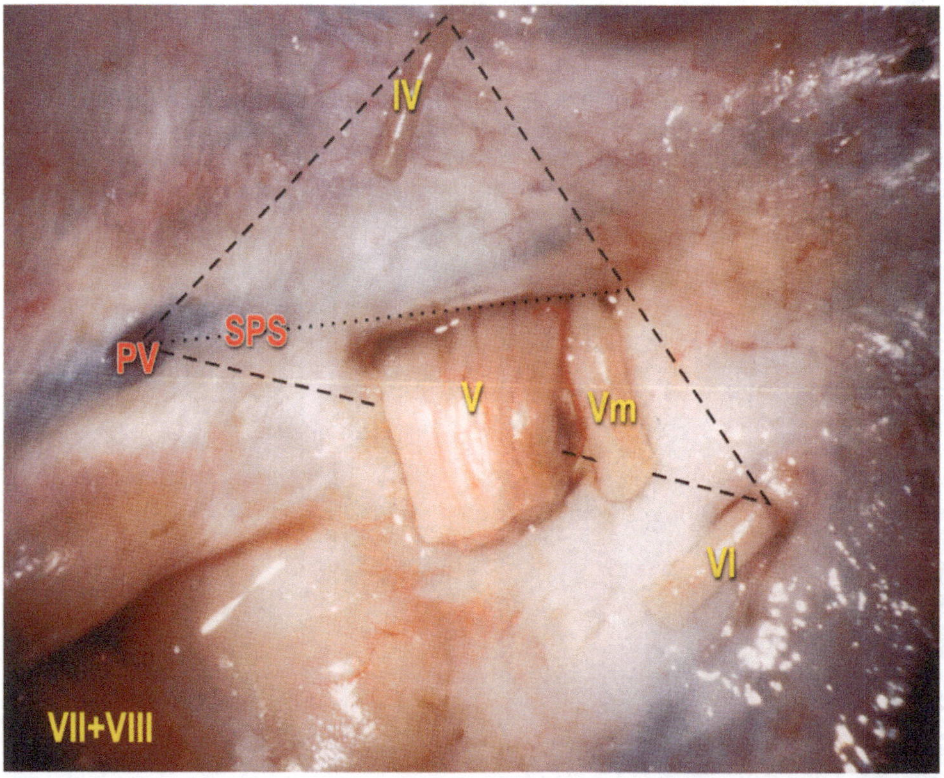

Fig. 1.36. The left inferolateral (trigeminal) triangle (interrupted lines), from the posterior cranial fossa. The site of nerve IV entering the tentorium composes the superomedial corner, and where nerve VI pierces clival dura is the inferomedial corner. The junction of the petrosal vein (*PV*) and the superior petrosal sinus (*SPS*) makes up the dorso-lateral corner. This triangle may be divided into two smaller triangles, the tentorial triangle above nerve V and the osseous triangle below nerve V (dotted line)

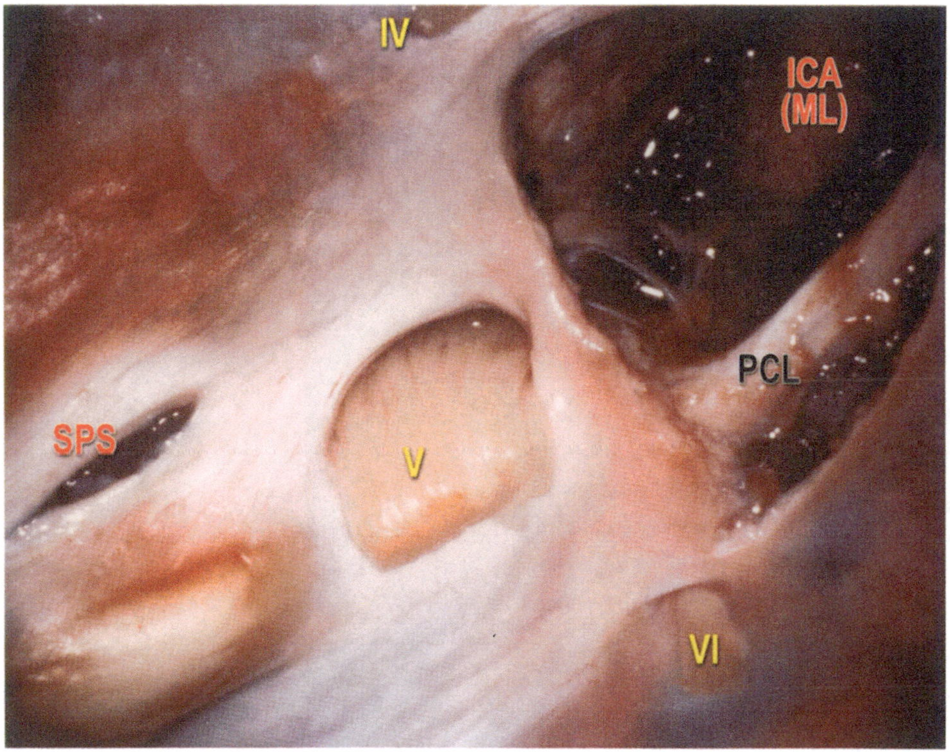

Fig. 1.37. The lateral corner of the left inferolateral triangle, shown here, is the petrosal vein's site of entry into the superior petrosal sinus (*SPS*). Nerve IV's site of entry into the tentorial edge and nerve VI's point of entry into dura lateral to the clivus constitute the upper and lower corners of this triangle, respectively, the line between them making up the medial border of the inferolateral triangle, also the lateral border of the exposed inferomedial triangle, shown here with dura removed. Nerve V is in the midportion of the inferolateral triangle, dividing it into two smaller triangles. The inferomedial and the inferolateral triangles possess a common base, the apex of inferomedial triangle projecting anteromedially and the apex of the inferolateral triangle projecting posterolaterally

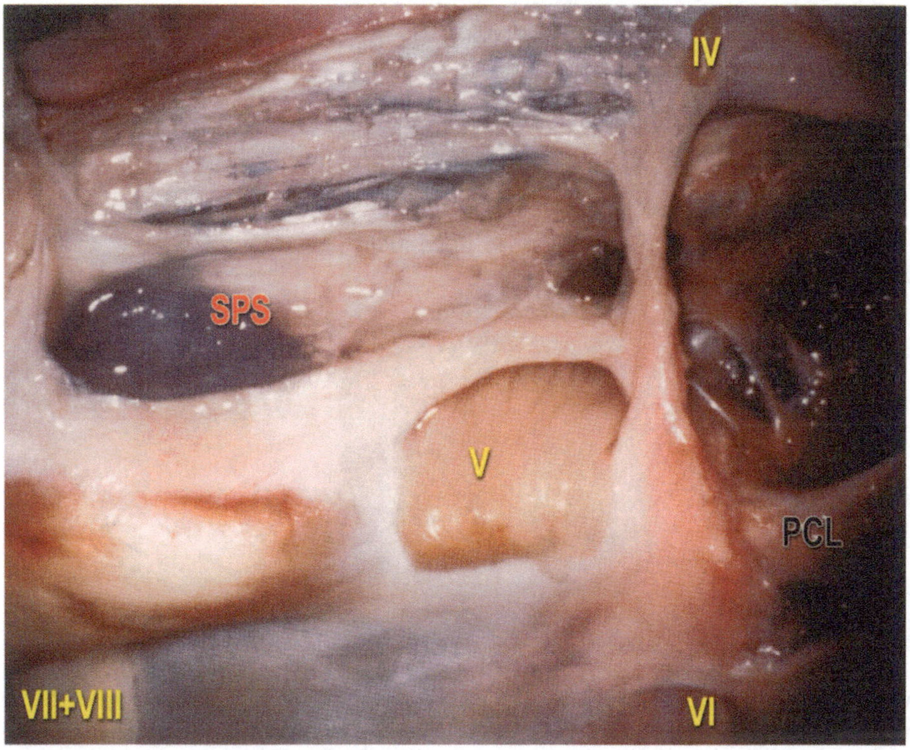

Fig. 1.38. The tentorial part of the left inferolateral triangle is partially open, but a narrow strip of tentorium, composing the base common to the inferolateral and inferomedial triangles, is preserved. Dura covering the osseous portion of the inferolateral triangle (postero-medial to nerve V) remains intact. The superior petrous sinus (*SPS*) has been opened and "venous blood" indicates the point of the petrosal vein's entry

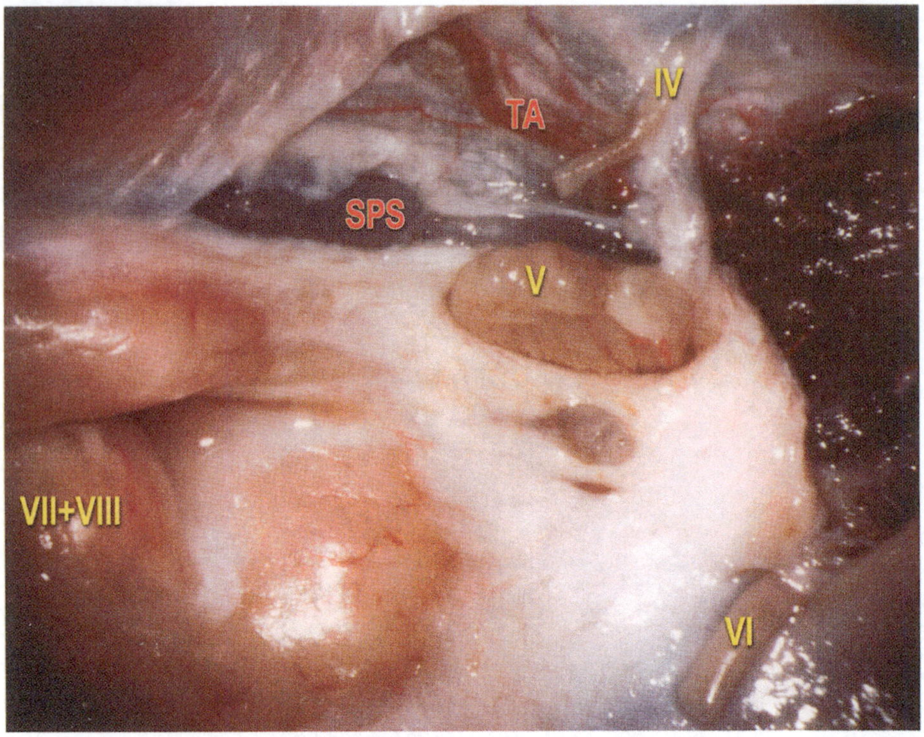

Fig. 1.39. The tentorial part of the left inferolateral triangle has been opened, and the "venous blood" of the superior petrosal sinus (*SPS*) is exposed, clearly contiguous with "venous blood" in the inferomedial triangle. The tentorial artery (*TA*), a branch of the meningo-hypophyseal trunk (*MHT*), called also Bernasconi-Cassinari artery, is seen. The narrow strip of tissue extending from nerve IV's site of entry into the tentorial edge to the medial surface of nerve V represents the medial border of the tentorial portion of the inferolateral triangle

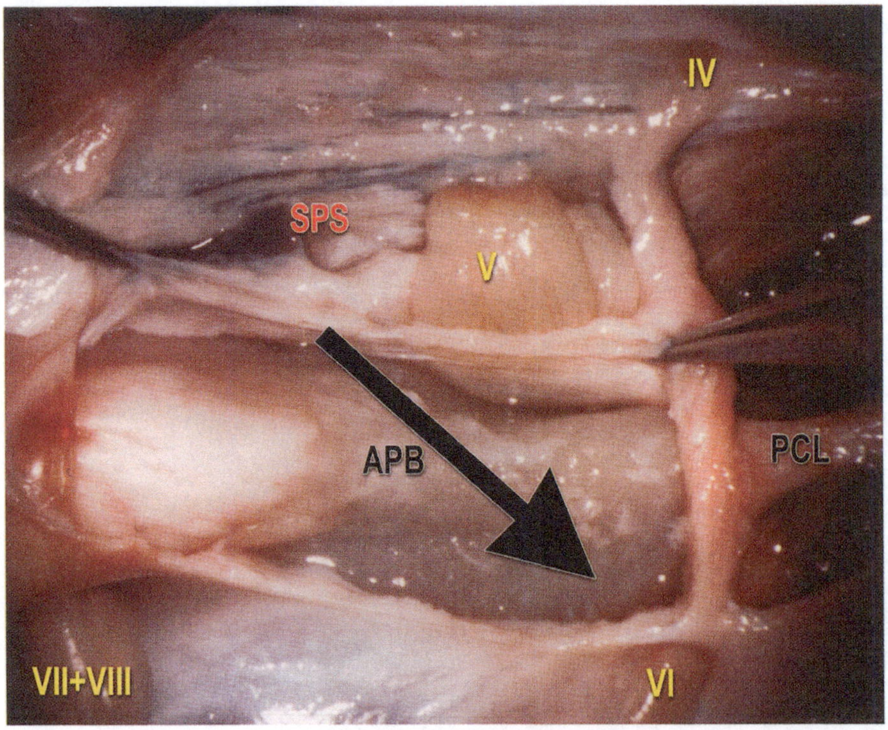

Fig. 1.40. Dura covering the apex of the petrous bone (*APB*) has been excised, exposing the osseous portion of the left inferòlateral triangle. Dura medially to nerve V, adjacent to the APB at the floor of Meckel's cave is shown. The narrow strip of preserved dura extending from tentorial edge adjacent to nerve IV to nerve VI's site of dural penetration represents the common boundary of the inferolateral and inferomedial triangles. The entrance into the inner auditory canal (*IAC*) indicated by nerves (VII + VIII) is near the dorso-lateral side of the infero-lateral triangle. Note: Surgically drilling the APB from the middle cranial fossa toward the posterior cranial fossa should begin posterior and lateral to nerve V, proceeding anteriorly and medially toward the clivus and nerve VI's point of entering the dura (as indicated by the arrow)

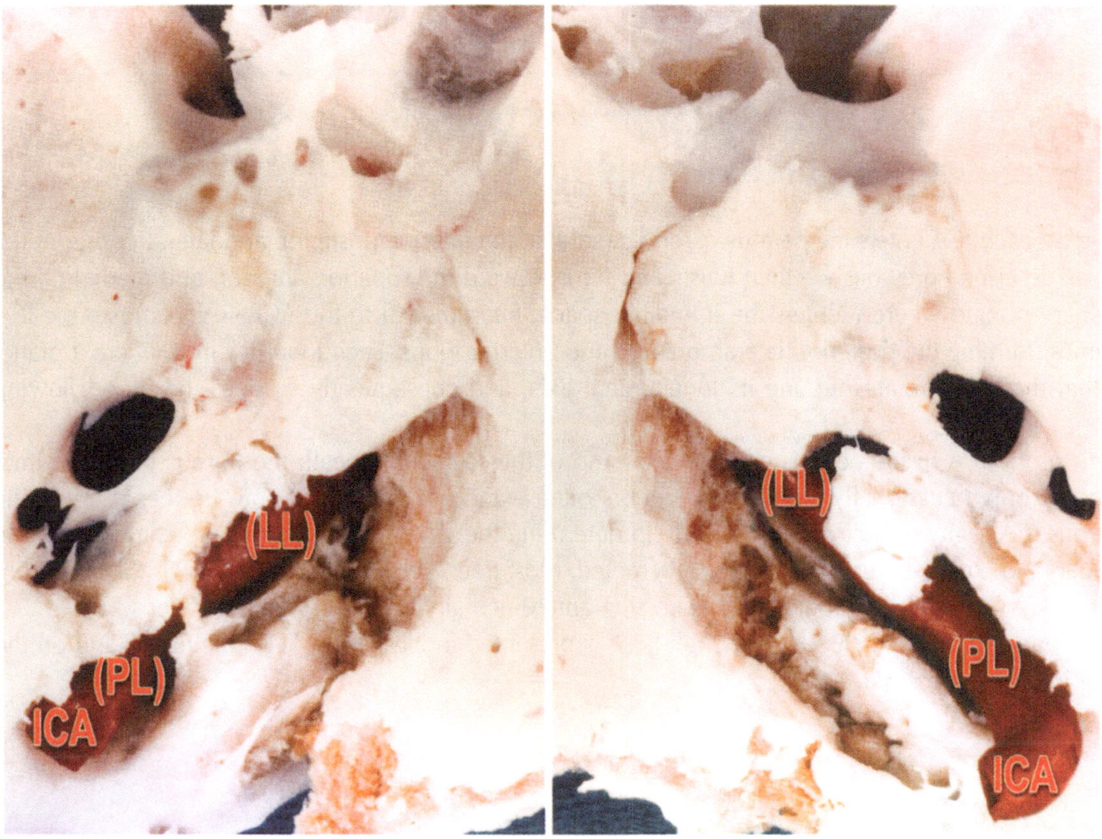

Fig. 1.41. This view from beneath bony structures of the central skull base (*CSB*) demonstrates both ICAs entering their respective petrous canals. The thin walls of the petrous carotid canals beneath the ICAs have been removed. Immediately after entering petrous bones, each ICA turns anteromedially, progressing in virtually the horizontal plane to the foramen lacerum (*FL*). The turn the ICA makes within the petrous carotid canal is called the posterior loop (*PL*). As its intrapetrous horizontal segment terminates, above the FL, the ICA turns upward and medialward, forming the lateral loop (*LL*)

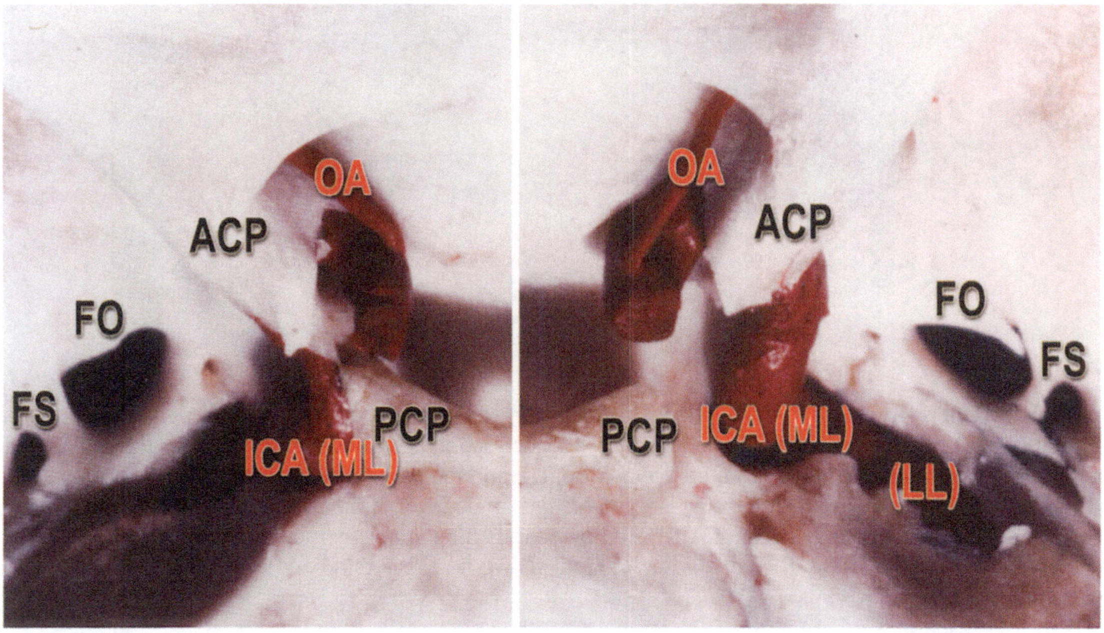

Fig. 1.42. At the foramen lacerum (*FL*), the ICA turns upward and medially toward the posterior clinoid process (*PCP*), forming the lateral loop (*LL*). Adjacent to the lateral surface of the PCP it turns forward, forming the medial loop (*ML*). Then it runs horizontally to the inferomedial surface of the anterior clinoid process (*ACP*) where it makes a U-turn, forming the anterior loop (*AL*)

1.3

Relation of the ICA to the anatomical triangles, and to the surgically relevant skull base structures

At the skull base each ICA enters the petrous carotid canal – at the inferior, posterior, and lateral corner of the imaginary box described previously – then twists and turns (toward the superior, anterior, and medial corner of the imaginary box) to where it enters the intradural space, inferomedial to the ACP. In its course, the ICA turns four times, forming the posterior, lateral, medial, and anterior loops. Each loop lies in a different plane, suggesting that the artery rotates around its longitudinal axis, and consequently so does the blood flowing within it.

Aneurysms of the ICA in the petrous carotid canal and of the segment beneath the fibrous layer proximal to the LR, occur only occasionally. Conversely, tumors often extend along the ICA into the canal, requiring that the petrous ICA segment be exposed in order to determine the extent of the lesion, especially when the tumor is benign and complete resection is to be attempted. Most traumatic aneurysms of the PS-ICA arise from the segment of ICA between the PR and DR, projecting anteromedially into the sphenoidal sinus.

The entire course of the ICAs through the skull base is presented in Figs. 1.41, 1.42 bilaterally, and on the left side only in Figs. 1.43–1.55.

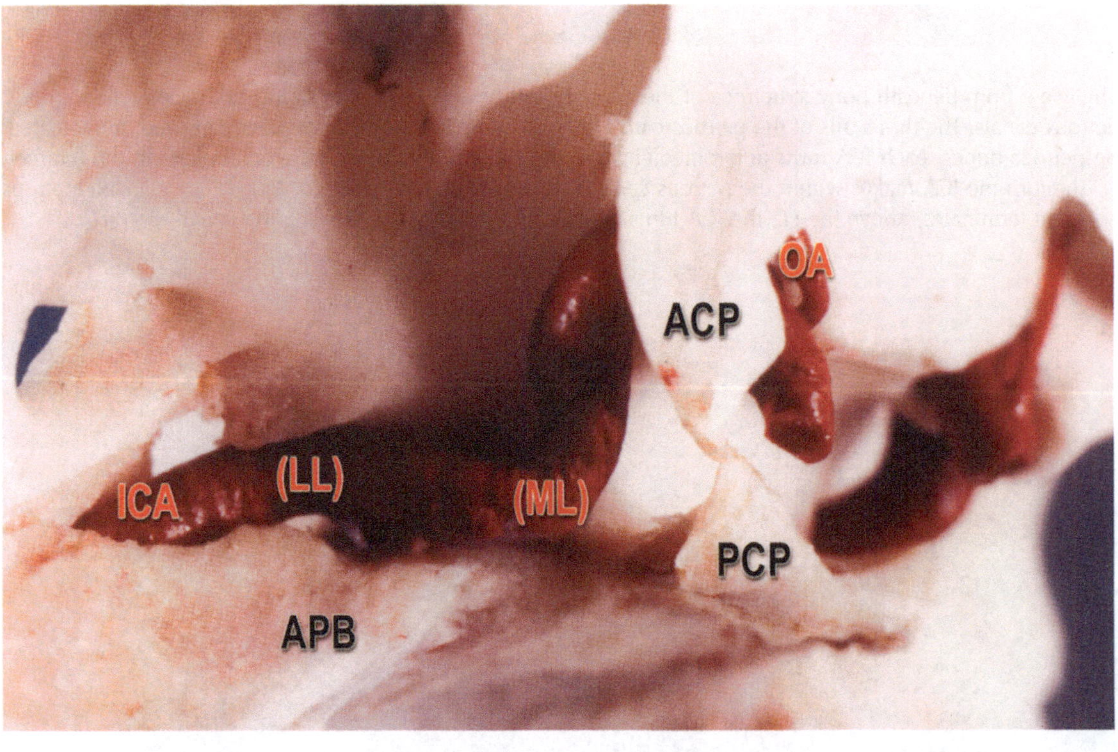

Fig. 1.43. A semi-lateral view of the left ICA – approximating the surgeon's perspective. The entire course of the ICA at the skull base may be seen as two consecutive S-shaped segments, the second a continuation of the first. The proximal S-segment contains the ICA's posterior loop (*PL*) and the lateral loop (*LL*), and the peripheral S-segment, its medial loop (*ML*) and anterior loop (*AL*). The junction between the two S-segments lies at the level of the apex of the petrous bone (*APB*), the site of lateral ring (*LR*), the junction between ICA covered by petrous bone (removed here) and ICA coursing through the parasellar space (*PS*) at the skull base. The lateral ring (*LR*) (not shown here) is a thin band of connective tissue covering the ICA on its dorsolateral surface

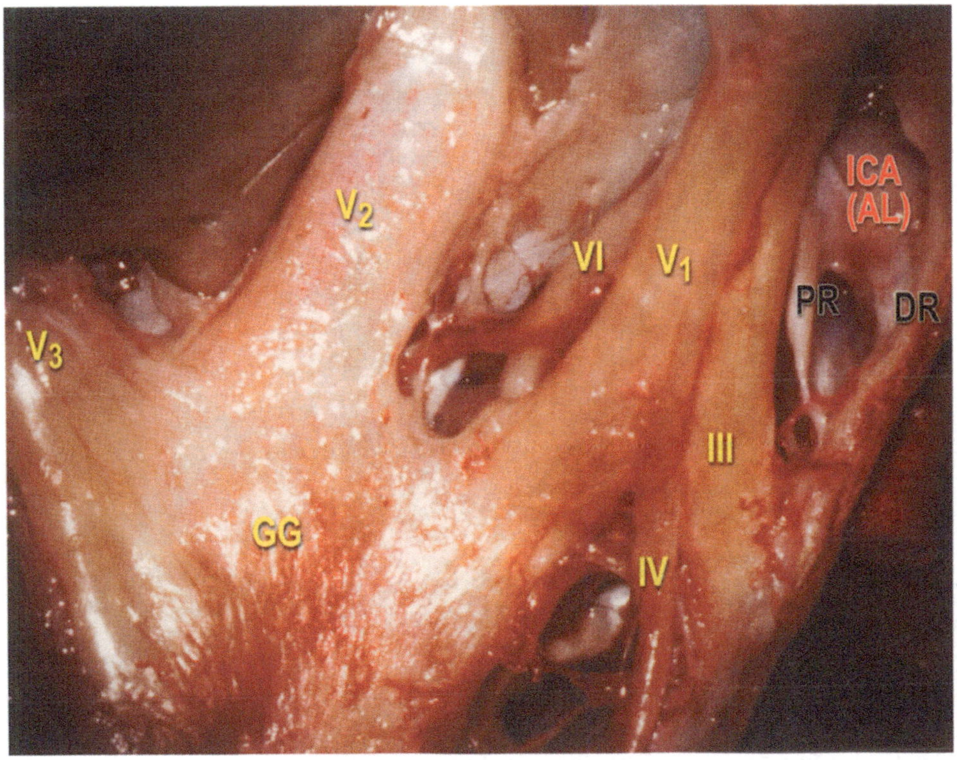

Fig. 1.44. The anterior loop (*AL*) of the left ICA is seen through the anteromedial triangle. Through the anterolateral triangle (V1 and V2), a small portion of the ICA is visible as it exits the petrous bone beneath the lateral ring (*LR*). A large branch of the infero-lateral trunk (*ILT*) crossing nerve VI is seen through the anterolateral triangle. The lateral ring (*LR*) of the ICA marks the junction of the two S-segments of the skull base ICA. The proximal S-segment consists of the posterior loop (*PL*) and lateral loop (*LL*) and lies within bone (the petrous carotid canal). The distal S-segment – consisting of ICA ascending to its medial loop (*ML*), its horizontal segment, and it's anterior loop (*AL*) – lies entirely within the parasellar space (*PS*) and is not covered by bone from the lateral ring (*LR*) to the proximal ring (*PR*)

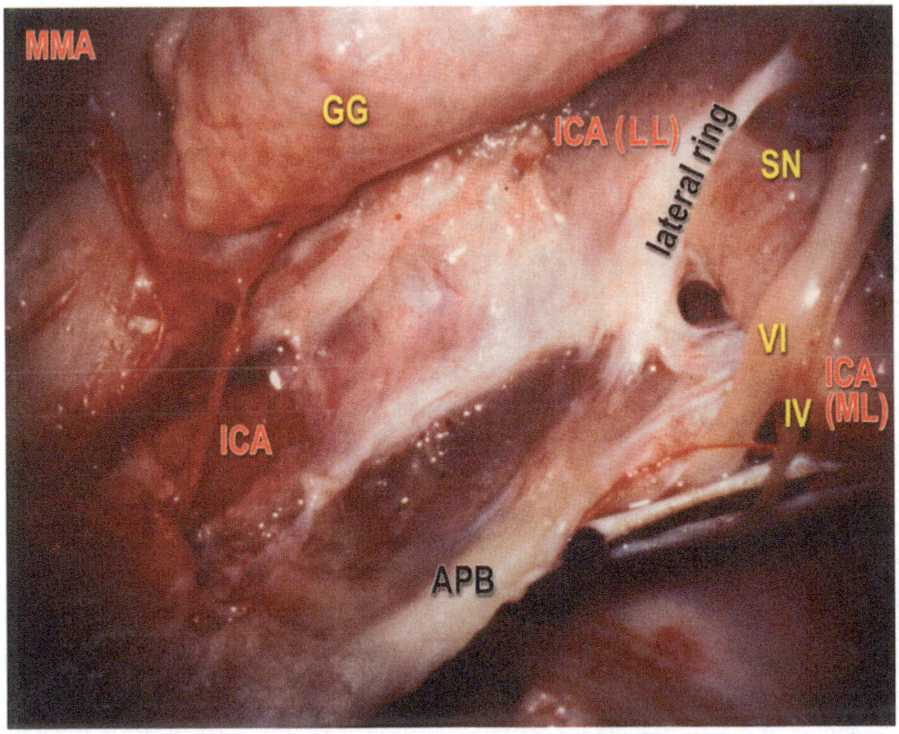

Fig. 1.45. Nerve V has been reflected out of view, exposing the lateral ring (*LR*) around the left ICA. The lateral loop (*LL*) lies adjacent to the Gasserian ganglion (*GC*) (beneath the path of nerve V), antero-lateral to Meckel's cave, and is covered by a fibrous membrane. The sympathetic nerve cords (*SN*) running along the ICA beneath the fibrous membrane, turn away from the vessel as it exits the (lateral ring) LR to join nerve VI

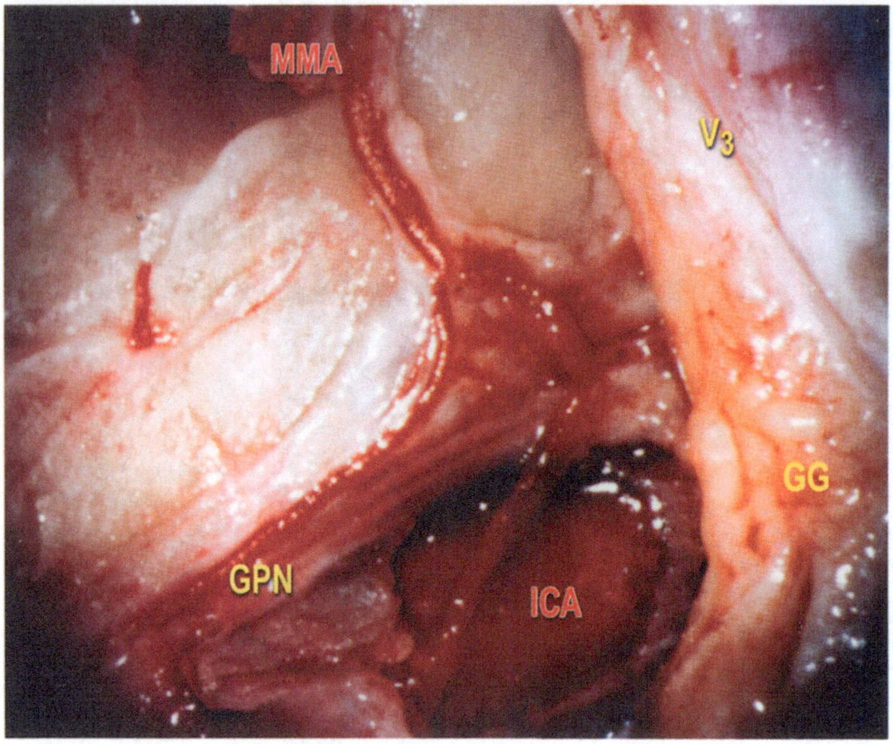

Fig. 1.46. The left posterolateral triangle. The lateral edge of the Gasserian ganglion (*GG*) is elevated, exposing the ICA in the petrous carotid canal. With the greater petrosal nerve (*GPN*) lateral to and posterior to it, the distal leg of the posterior loop (*PL*) courses beneath the GG toward the foramen lacerum (*FL*). A large branch of the middle meningeal artery (*MMA*) accompanies the GPN in the opposite direction

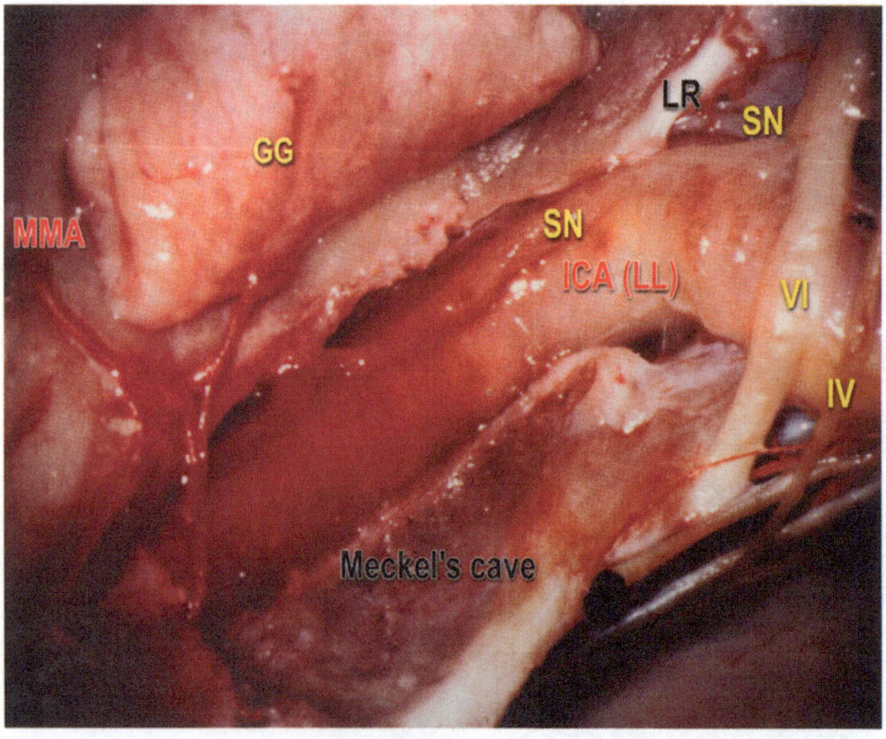

Fig. 1.47. The left nerve V has been reflected laterally. The fibrous membrane over the lateral loop (*LL*) of the ICA and the lateral ring (*LR*) has been cut and the horizontal petrous ICA and the LL nicely shown. Note the large sympathetic fibers (*SN*) from the ICA joining nerve VI. This Fig. and Fig. 1.45 offer an explanation why aneurysms rarely arise from the petrous segment of the ICA, occurring far more frequently between the LR and proximal ring (*PR*), where space is much less restricted

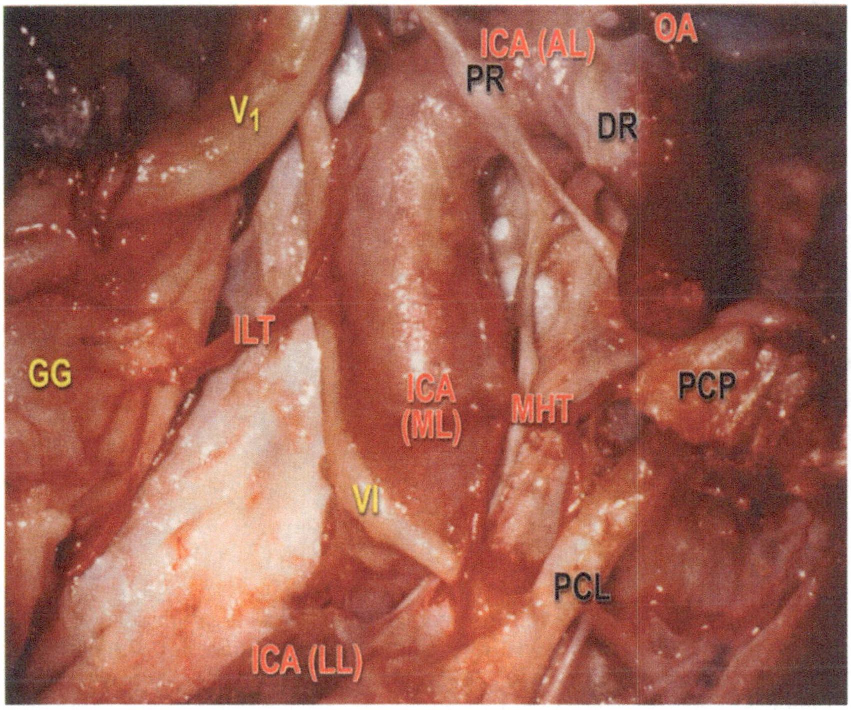

Fig. 1.48. The distal S-segment of the left ICA, fully within the parasellar space (*PS*), consisting of the ascending segment, the medial loop (*ML*), the horizontal segment, and the anterior loop (*AL*). The AL and the ML lie in the horizontal and sagittal planes, respectively. Cranial nerves within the lateral wall of the PS have been elevated with the exception of nerve VI, which remains in place. Nerve VI is consistently inferior and lateral to the origin of the meningo-hypophyseal trunk (*MHT*) and the infero-lateral trunk (*ILT*). The proximal ring (*PR*) and the distal (dural) ring (*DR*) surrounding the AL are exposed, the PR is fixed to the posterior clinoid process (*PCP*). Both intra-PS and extra-PS portions of the AL are demonstrated

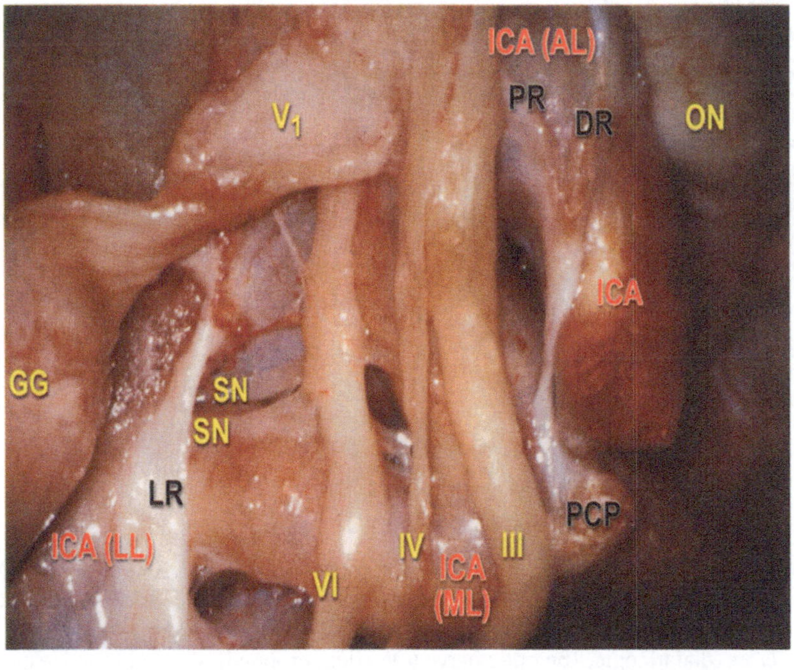

Fig. 1.49. The left ICA in the parasellar space (*PS*) from the lateral ring (*LR*) to the distal (dural) ring (*DR*) is seen. Sympathetic fibers tether nerve VI at several sites while adjacent nerves III and IV, coursing parallel to the horizontal segment of the PS-ICA, are not fixed to the artery but lie within the lateral PS wall

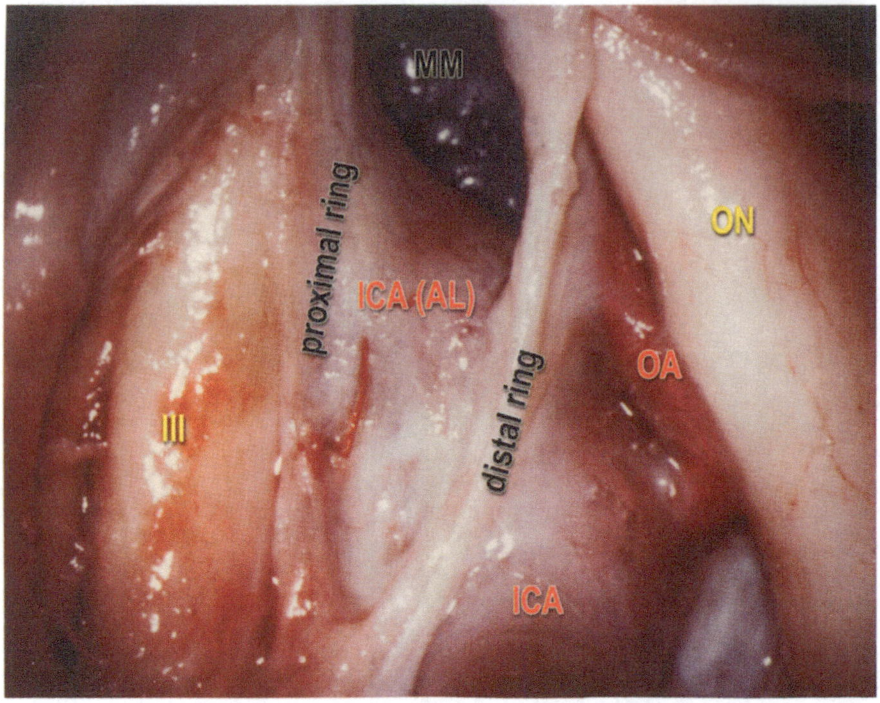

Fig. 1.50. The anteromedial triangle in the left parasellar space (*PS*). The proximal ring (*PR*) and distal (dural) ring (*DR*) fix the anterior loop (*AL*) of the ICA to the lateral wall of the sphenoidal sinus and surrounding structures. The fibrous PR, firmly attached to anteroinferior surface of the anterior clinoid process (*ACP*) on one side (completely removed here), the ICA on the other, prevents mobility of the ICA segment between the PR and the DR. Fractures of the lateral wall of the sphenoidal sinus (note its mucous membrane anterior to the AL) may injure the ICA, resulting in a traumatic aneurysm likely to bulge into the sphenoid sinus cavity. Note the origin of the ophthalmic artery (*OA*) several mm distal to the DR and the course of this vessel infero-lateral and parallel to the optic nerve (*ON*). The PR, fixed to nerve III, limits venous compartments to the PS

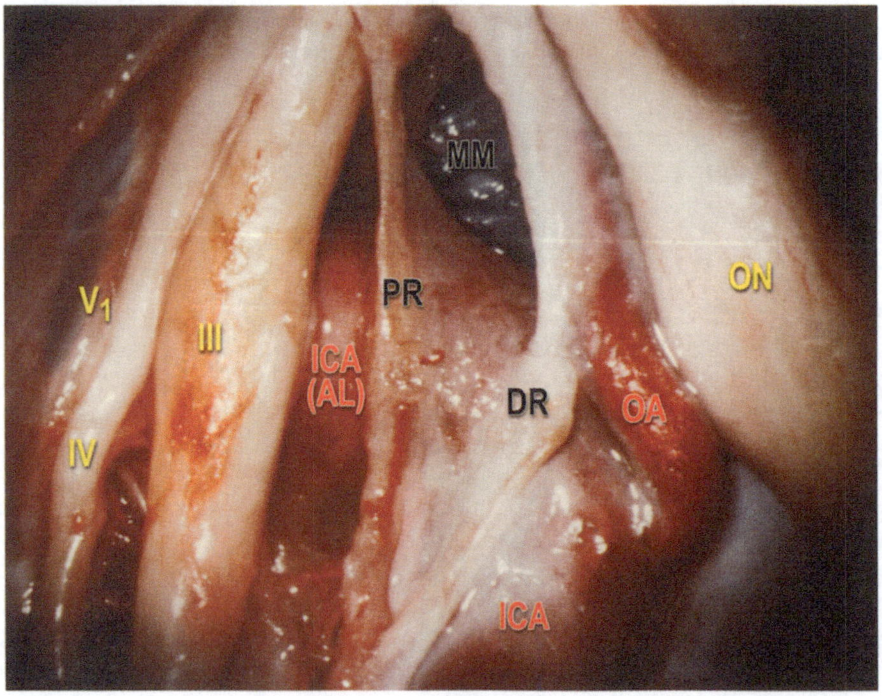

Fig. 1.51. Further dissection of the specimen shown in Fig. 1.50. The proximal ring (*PR*) has been separated from nerve III, opening the parasellar space (*PS*). The ICA segment between the PR and distal (dural) ring (*DR*) remains covered by the membrane between the ICA and the inferomedial surface of the anterior clinoid process (*ACP*). The PR and the DR join at the posterior half of the base-line of the anteromedial triangle. The optic nerve (*ON*) dura propria has been incised, exposing the ophthalmic artery (*OA*) inferior and lateral to the ON as it courses toward the orbit through the optic canal. The mucous membrane (*MM*) of the sphenoidal sinus lies anterior to the anterior loop (*AL*) at the apex of the anteromedial triangle. Note: To avoid a CSF leak the sphenoidal sinus MM should not be exposed surgically, certainly not lacerated. By incising dura propria along the longitudinal axis of the ON, and dissecting the PR from nerve III, the ON may be displaced farther medially, and nerves III and IV laterally, hence enlarging the anteromedial triangle and achieving complete exposure of anterior loop (*AL*)

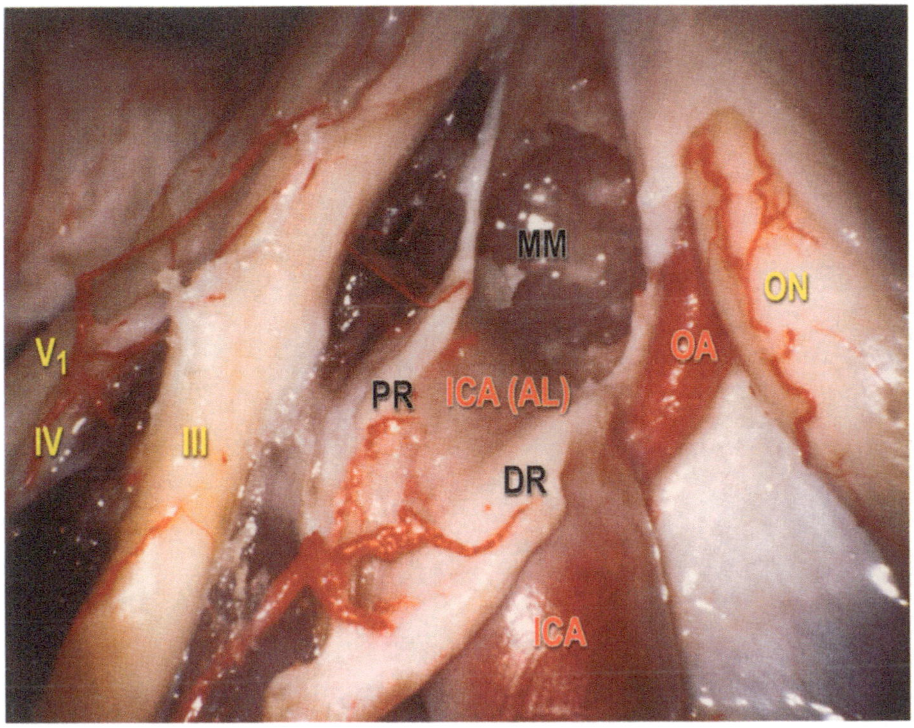

Fig. 1.52. The anterior loop (*AL*) of the left ICA is fixed by the proximal ring (*PR*) and distal (dural) ring (*DR*). Note: Parasellar space (*PS*) surgery – except in rare cases – requires the removal of the anterior clinoid process (*ACP*) and exposing the entire AL. It is important to realize that wall of the sphenoidal sinus medial to the AL is extremely thin, and therefore quite vulnerable. If it is violated and the mucous membrane (*MM*) perforated, a CSF leak will result if a watertight dural closure is not achieved at the time of closing the wound, a problem far better avoided than attempted to solve

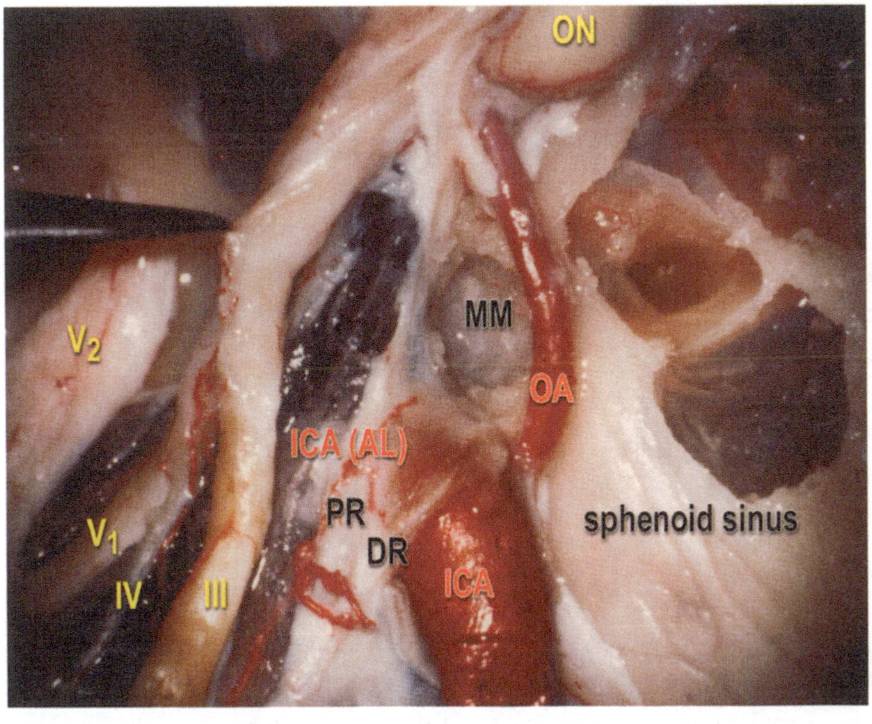

Fig. 1.53. The left parasellar space (*PS*) shown in Fig. 1.52 is dissected further. A segment of the optic nerve (*ON*) has now been elevated, exposing the ophthalmic artery (*OA*) better. The wall of the sphenoidal sinus has been opened medial to the ICA and the OA. "Venous blood" is apparent beneath nerves III, IV, VI, and V1. Note: The proximal (*PR*) and distal (dural) (*DR*) rings become united at the mid-base of the antero-meidal triangle

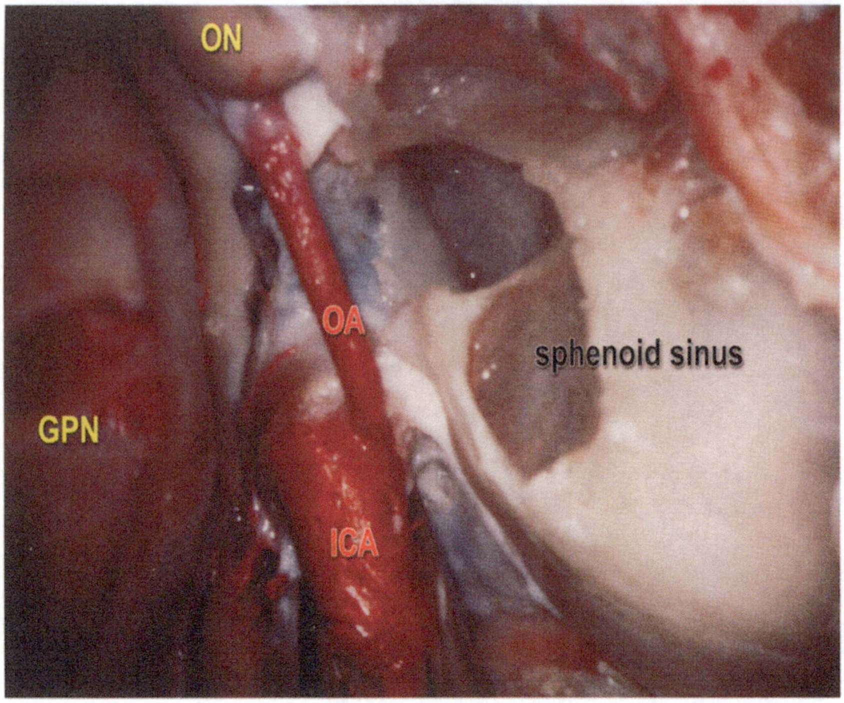

Fig. 1.54. Further dissection of the specimen presented in Figs. 1.52 and 1.53. The anterior loop (*AL*) of the ICA is firmly fixed to the very thin lateral wall of the sphenoidal sinus by the proximal (*PR*) and distal (dural) (*DR*) rings. Note: Its delicate wall makes the sphenoidal sinus quite vulnerable to cranial trauma and surgical dissection

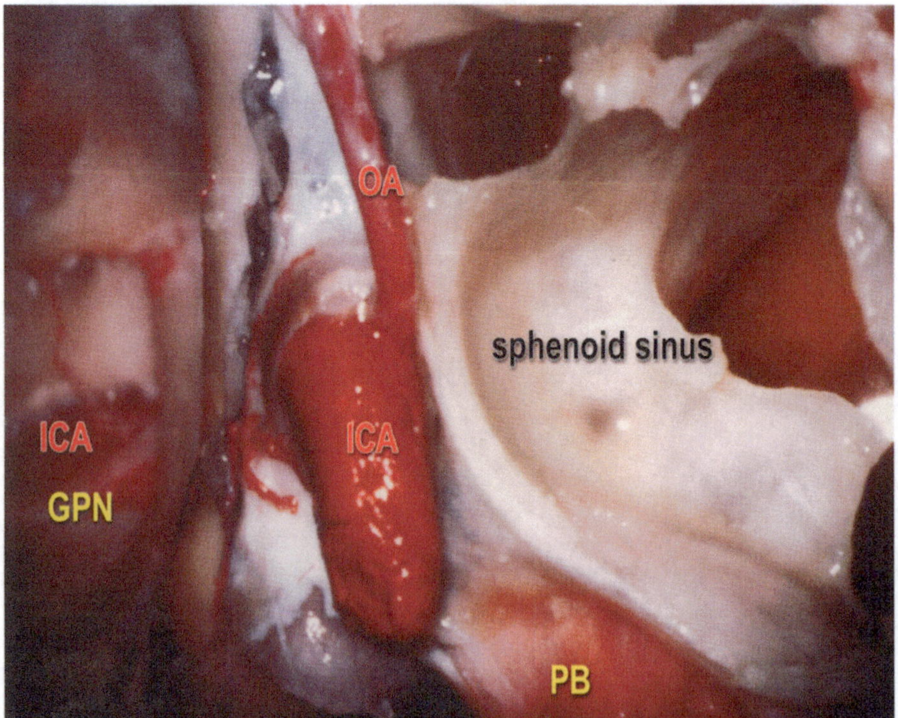

Fig. 1.55. The same specimen as in Figs. 1.52–1.54, emphasizing again the relationship of the ICA to the sphenoidal sinus. When approaching the anterior loop (*AL*) extradurally (transcranially or transsphenoidally), one must be acutely aware of the lateral wall of the sphenoid sinus. It must not be disrupted!

1.4

Relation of the parasellar space to the sella and through it to the contralateral parasellar space

The sella, containing the pituitary body (PB), is centrally located at the skull base, and the left and right PSs lie lateral to it. The PSs communicate with each other through venous sinuses inside the sella – on its floor and within its anterior and posterior walls. The arterial supply of the PB capsule is provided by PS-ICA branches while the pituitary stalk is supplied by branches of the intradural ICA (Dawson's arteries). PB hormones are released into the venous structures surrounding it. Tumors of the PS may extend into the sella and compress the PB, resulting in endocrine changes. Conversely, a PB tumor may compress or invade the parasellar space, creating symptoms and signs involving nerves III, IV, V1, V2, and VI.

The relation of the PS to the sella is presented in Figs. 1.56–1.61.

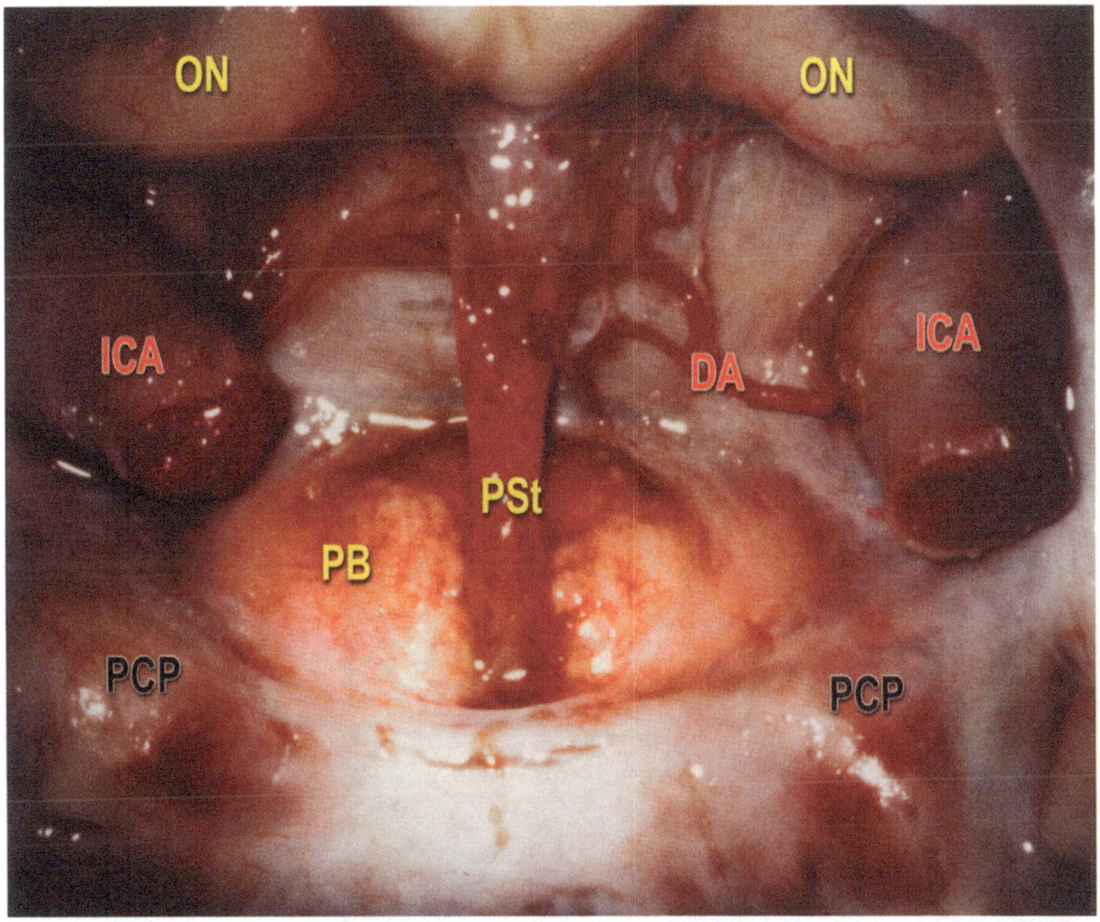

Fig. 1.56. Both ICAs with branches coursing medially toward the pituitary stalk (*PSt*), the optic chiasm, and the optic nerves (*ON*), both posterior clinoid processes (*PCP*) and pituitary body (*PB*) are seen. These intradural ICA segments are virtually parallel to their horizontal segments within the parasellar spaces (*PS*) coursing from their medial (*ML*) to anterior (*AL*) loops on either side

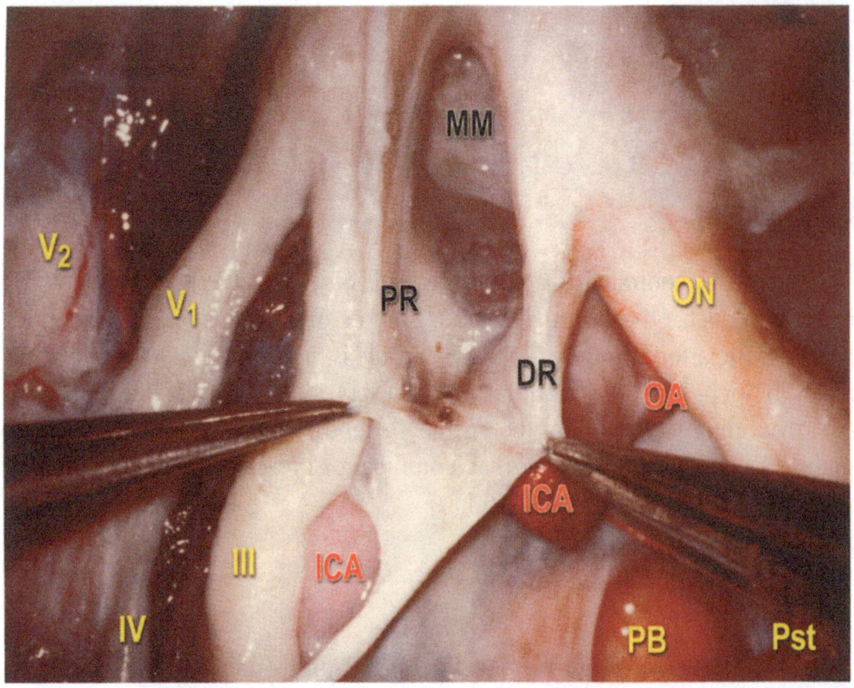

Fig. 1.57. After removing the apex of the left anterior clinoid process (*ACP*), "venous blood" is apparent in the anteromedial triangle, between the optic nerve (*ON*) and nerve III. This portion of the ACP may also lie medial to the horizontal segment of the parasellar space (*PS*) ICA, very close to the concave surface of the anterior loop (*AL*). When surgically removing the ACP, venous bleeding from the PS may occur only when the final sliver of bone is removed. Occasionally the ACP is very long, projecting far medially into the sella. In some cases it even joins the lateral wall of the sphenoidal sinus, forming an entirely bony canal housing the ICA. In this specimen "venous blood" is visible beneath membrane between the ICA and the pituitary body (*PB*), also in the paramedial (between nerves III and IV) and anterolateral (between V1 and V2) triangles

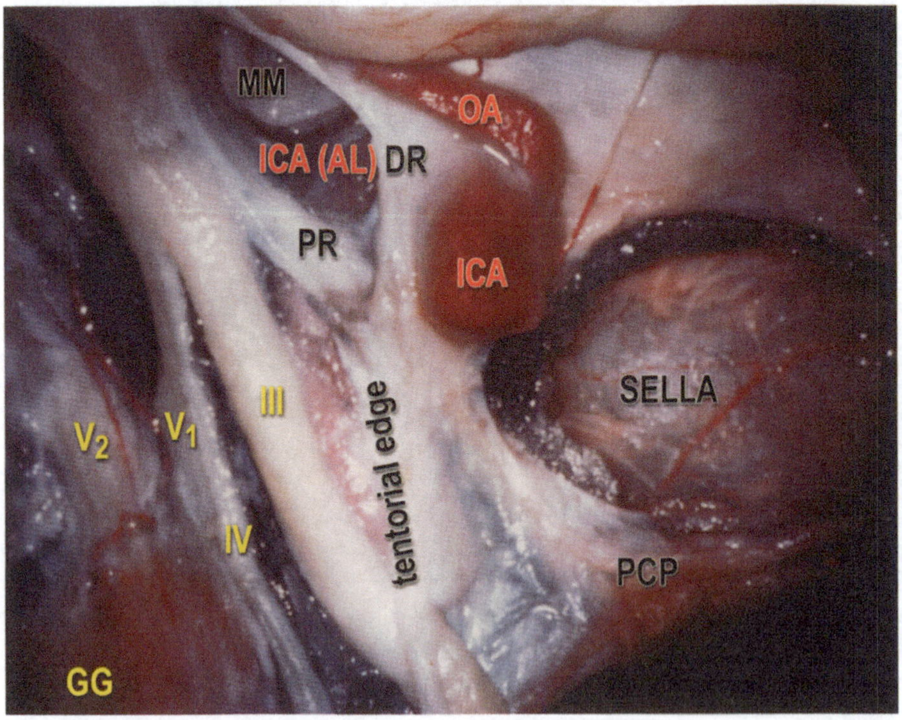

Fig. 1.58. The dural fold extending between the apex of the left posterior clinoid process (*PCP*) to the anterior clinoid process (*ACP*) – ACP removed in this specimen – constitutes the medial border of the oculomotor trigone. The pituitary body (*PB*) has been removed and "venous blood" is visible in the floor of the sella and along its anterior, lateral, and posterior walls. The "venous blood" lining the floor of the sella is covered by a thin membrane containing a network of tiny arterial branches which arise from the ICA inside the parasellar space (*PS*). Note how much the anteromedial triangle may be widened by dissecting the proximal ring (*PR*) medial to nerve III and dura propria along the optic nerve (*ON*), exposing the ophthalmic artery (*OA*). Thus enlarging the anteromedial triangle provides ample space for dealing with vascular lesions and tumors involving the horizontal and anterior loop (*AL*) segments of the ICA, also those in the vicinity of the intradural ICA, the tuberculum sellae, the dorsum sellae, the intrasellar, and at the upper clivus

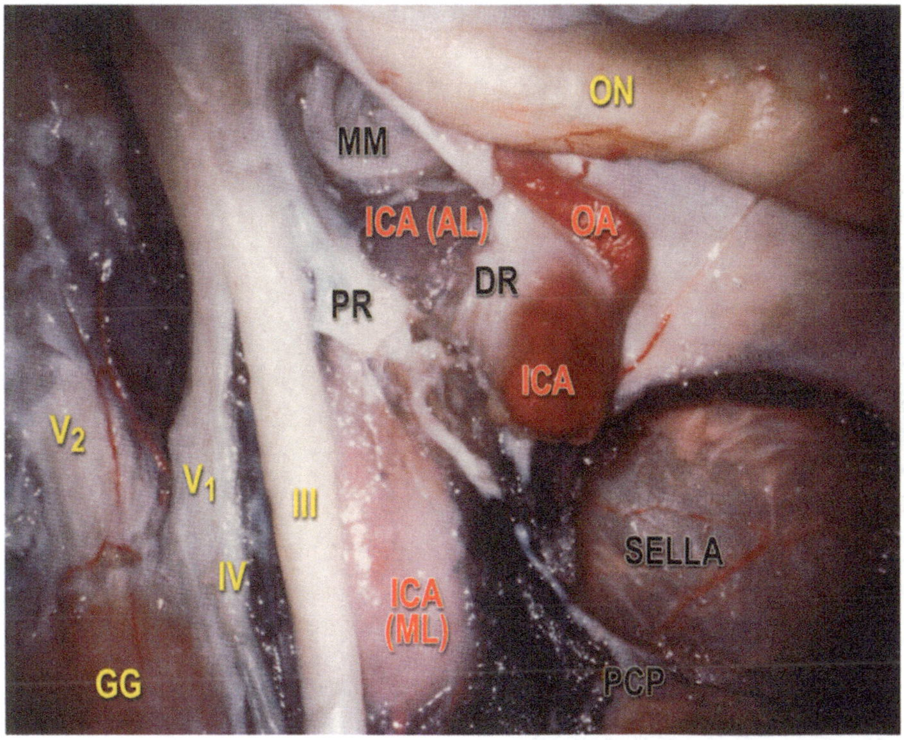

Fig. 1.59. Further resection of oculomotor trigone dura and dura between the left posterior clinoid process (*PCP*) and the proximal (*PR*) and the distal (dural) (*DR*) rings reveals a deep membrane forming a filmy curtain between the sella and the parasellar space (*PS*). "Venous blood" medial to the horizontal and the medial loop (*ML*) segments of the ICA is in continuity with venous compartments inside the anterior and posterior walls of the sella and along its floor. These venous channels connect the left and right PSs

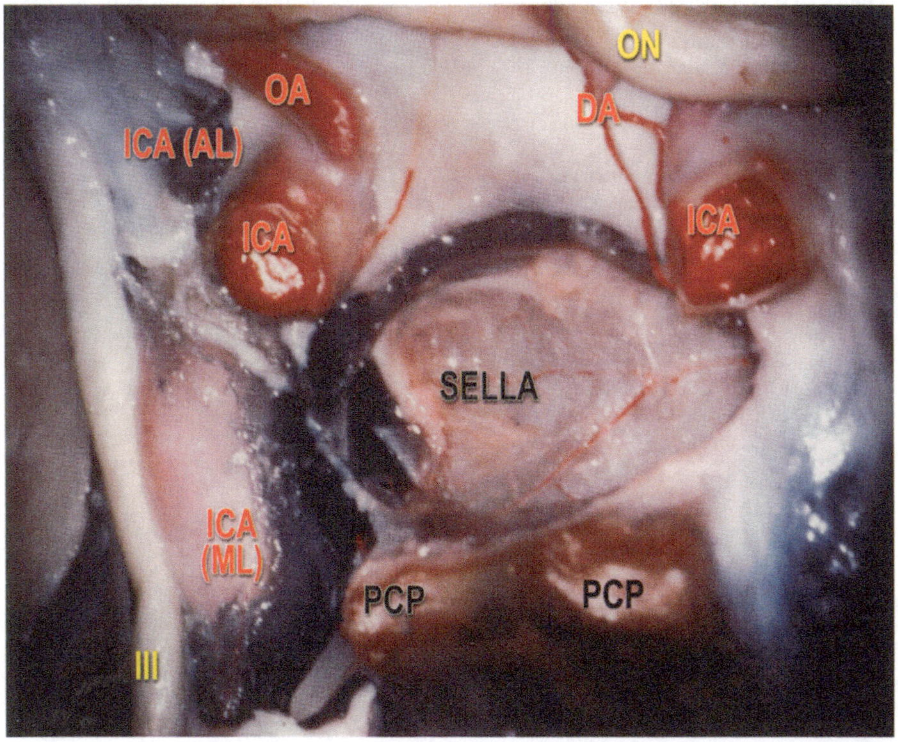

Fig. 1.60. Additional excision of the capsule surrounding the pituitary body (*PB*) – removed here – in the floor of the sella (as shown in Figs. 1.58 and 1.59) reveals abundant "venous blood" between the ICA and the left side of the sella. A small "window" is visible in the thin veil separating the sella from the parasellar space (*PS*). Openings like this one, afford connections between venous compartments of the left PS, to the right PS. Note: Venous flow in such channels is dictated by pressure gradients which are affected by tumors present on either side of the membrane

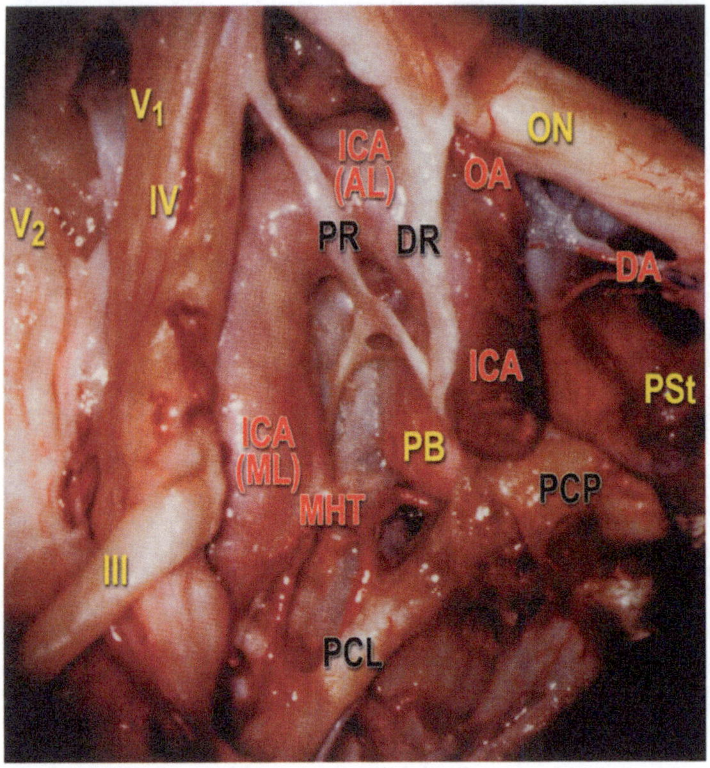

Fig. 1.61. The anterior (*AL*) and medial (*ML*) loops of the ICA in the left parasellar space (*PS*) are seen. The proximal (*PR*) and distal (dural) (*DR*) rings extend in the depth to the posterior clinoid proces (*PCP*) and the floor of the sella, and the meningo-hypophyseal trunk (*MHT*) arises from the ML. The optic nerve (*ON*) and the ophthalmic artery (*OA*) (arising from the intradural ICA, then passing between the ON and its dura propria) are seen. The relationship between the sella and pituitary body (*PB*) and the parasellar space is apparent. Note: The transcranial approach allows simultaneous access to sellar and PS compartments

1.5

Venous system of the parasellar space

The primary venous structures draining the PSs on either side are the SOV and IOV, the sphenoparietal sinus, and the somewhat variable veins draining the brain. Sometimes these veins are very large and drainage from the brain into the PS is considerable. In other cases this drainage is minimal or virtually non-existent. The primary route for venous flow from the PSs is through the paired SPSs and IPSs, but considerable venous drainage is via other skull base foramina. When one PS or the other is filled with a large or giant aneurysm or a tumor, typically it contains very little venous blood. When the PS is only partially blocked by a pathological lesion, the primary venous stream is displaced into one corner. Consequently, it is much easier to control venous bleeding in a PS harboring a large lesion than smaller lesions where most venous compartments continue to function normally. Experience has shown that gradual occlusion of the major venous channels of one PS by tumor, and subsequent surgical occlusion of those remaining, almost never results in ipsilateral exophthalmos. Acute and total occlusion of a PS (from acute thrombosis or infection) uniformly results in ophthalmoplegia and exophthalmos.

Drainage of the brain into the venous compartments of the PS is variable.Venous blood entering inter-dural compartments in the lateral wall of the PS may by-pass the PS altogether [4, 14], flowing directly into the SPS, or possibly in combination with PS venous compartments. It is important to understand which variation is present, since each results in a different clinical picture and has important implications with respect to the treatment of dural AVMs and fistulae.

The venous system of the left PS is shown in Figs. 1.62–1.65.

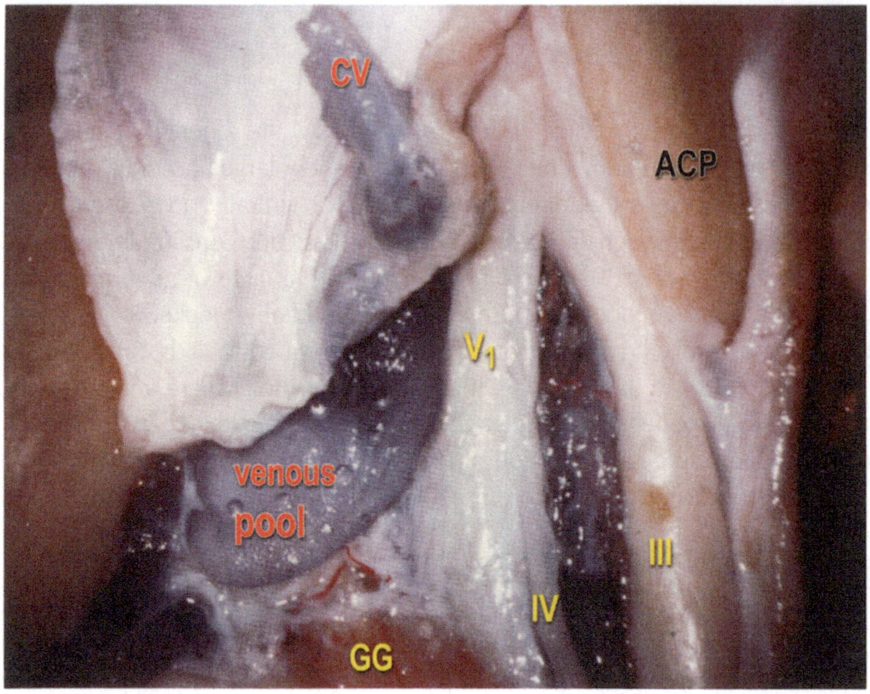

Fig. 1.62. Venous injection of the left parasellar space (*PS*) reveals a large vein (*CV*) draining the brain as it enters the lateral wall of the PS, forming a pool of "venous blood" between the two layers of the dura making up the PS wall

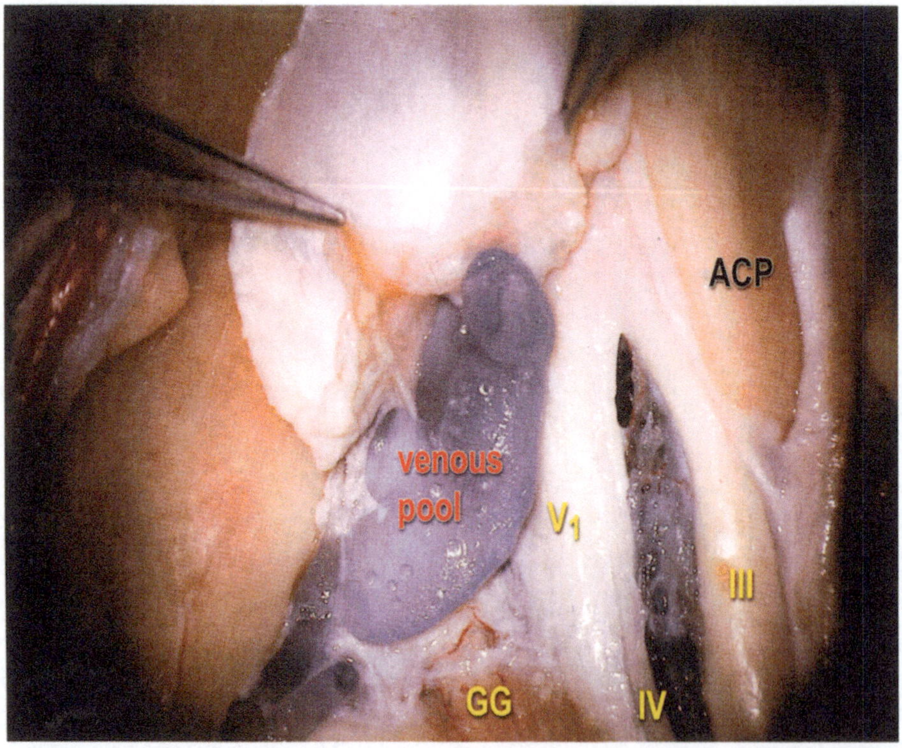

Fig. 1.63. The interdural venous pool within the lateral wall of the parasellar space (*PS*) is shown. Note: "Venous blood" is also visible in the paramedial triangle (nerves III and IV) and in the lateral triangle (V2 and V3)

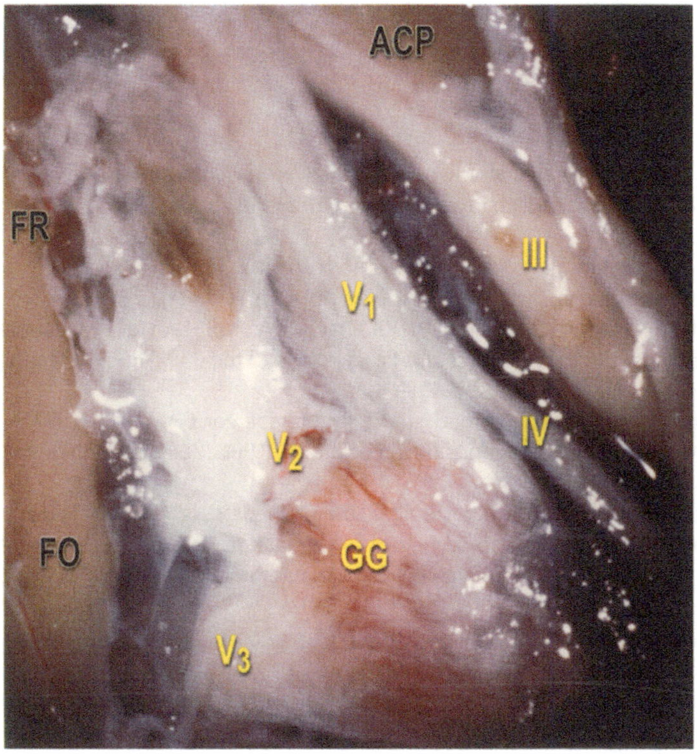

Fig. 1.64. "Venous blood" has been removed from the lateral parasellar space (*PS*) wall in the specimen shown in Figs. 1.62 and 1.63. "Venous blood" is seen in the paramedial triangle (nerves III and IV), Parkinson's triangle (nerve IV and V1), and over V3. The PS drains lateral and posterior to the Gasserian ganglion (*GG*) and nerve V, into the superior (*SPS*) and inferior (*IPS*) petrosal sinuses

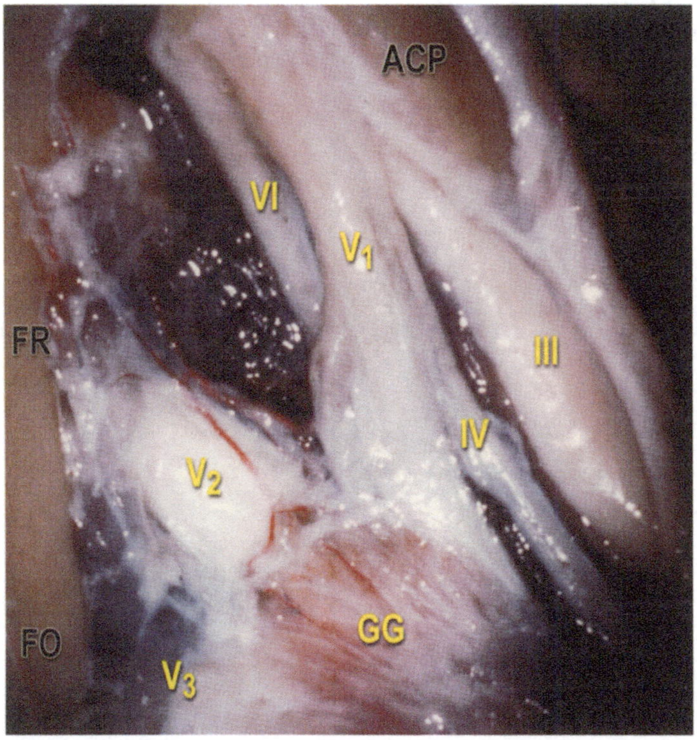

Fig. 1.65. The inner layer of the lateral wall of the parasellar space (*PS*) has been removed additionally from the specimen shown in Figs. 1.62–1.64, revealing dense "venous blood" in the anterolateral triangle (V1 and V2). Nerve VI lies inferolateral to V1. Note: The paramedial (nerves III and IV), Parkinson's (nerve IV and V1), and the lateral (V2 and V3) triangles are fairly small in their normal states, but "venous blood" is visible in all of them. The enlarged anterolateral triangle demonstrates how other triangles may be enlarged when necessary to access lesions within the depths of the PS

References

1. Brassier G, Lasjaunias P, Geugan Y, Pecker J (1987) Microsurgical anatomy of the intracavernous internal carotid artery. In: Dolenc VV (ed) The cavernous sinus. A multidisciplinary approach to vascular and tumorous lesions. Springer, Wien New York, pp 81–103
2. Dolenc VV (1983) Direct microsurgical repair of intracavernous vascular lesions. J Neurosurg 58: 824–831
3. Dolenc VV (1989) Anatomy and surgery of the cavernus sinus. Springer, Wien New York, pp 3–137
4. Gailloud P, San Millan Ruiz D, Nichols DA, Muster M, Fasel JHG, Dolenc VV, Rufenacht DA (1996) A venous channel in the lateral wall of the cavernous sinus: Anatomy and clinical relevance. In: Proceedings of the 34th Annual Meeting of the American Society of Neuroradiology, pp 251–251
5. Glasscock ME (1969) Exposure of the intra-petrous portion of the carotid artery. In: Hamberger CA, et al (eds) Disorders of the skull base region. Proceedings of the Tenth Nobel Symposium, Stockholm, 1968. Almqvist & Wiksell, Stockholm, pp 135–143
6. Harris FS, Rhoton AL Jr (1976) Anatomy of the cavernous sinus. A microsurgical study. J Neurosurg 45: 169–180
7. Knosp E, Meller G, Perneczky A (1987) The blood supply of the cranial nerves in the lateral wall of the cavernous sinus. In: Dolenc VV (ed) The cavernous sinus. A multidisciplinary approach to vascular and tumorous lesions. Springer, Wien New York, pp 67–80
8. Lang J (1987) Middle cranial base anatomy. In: Sekhar LN, et al (eds) Tumors of the cranial base. Diagnosis and treatment. Futura, Mont Kisco, NY, pp 313–334
9. Lie TA (1968) Congenital anomalies of the carotid arteries, including the carotid-basilar and carotid-vertebral anastomoses. An angiographic study and a review of the literature. Excerpta Medica, Amsterdam
10. Ono M, Ono M, Rhoton AL Jr (1984) Microsurgical anatomy of the region of the tentorial incisure. J Neurosurg 60: 365–399
11. Parkinson D (1964) Collateral circulation of cavernous carotid artery: anatomy. Can J Surg 7: 251–268
12. Renn WH, Rhoton AL Jr (1975) Microsurgical anatomy of the sellar region. J Neurosurg 43: 288–198
13. Rhoton AL Jr, Hardy DG, Chambers SM (1979) Microsurgical anatomy and dissection of the sphenoid bone, cavernous sinus and sellar region. Surg Neurol 12: 63–104
14. San Millan Ruiz D, Gailloud P, De Miquel Miquel MA, Muster M, Dolenc VV, Rufenacht DA, Fasel HDJ (1999) Laterocavernous Sinus: The anatomical record 254: 7–12
15. Umansky F, Nathan N (1987) The cavernous sinus. An anatomic study of its lateral wall. In: Dolenc VV (ed) The cavernous sinus. A multidisciplinary approach to vascular and tumorous lesions. Springer, Wien New York, pp 56–66

2

Elements of the epidural approach to the parasellar space and adjacent regions in the central skull base

Patient positioning

The patient is placed supine for surgery, with the head rotated 30–35 degrees toward the side opposite the lesion and stabilized in the three-point Mayfield fixation device. The neck is neither extended nor flexed, but the head positioned so that the superior orbital rim and malar eminence are uppermost. During the course of surgery, the long axis of the operating table is tilted from its initial position to the Trendelenburg or reverse Trendelenburg position of the patient (Fig. 2.1), and/or rotated right or left as necessary to expose the central skull base to best advantage (Fig. 2.2), and the position of the operating microscope is shifted accordingly in the opposite direction (Figs. 2.1 and 2.2).

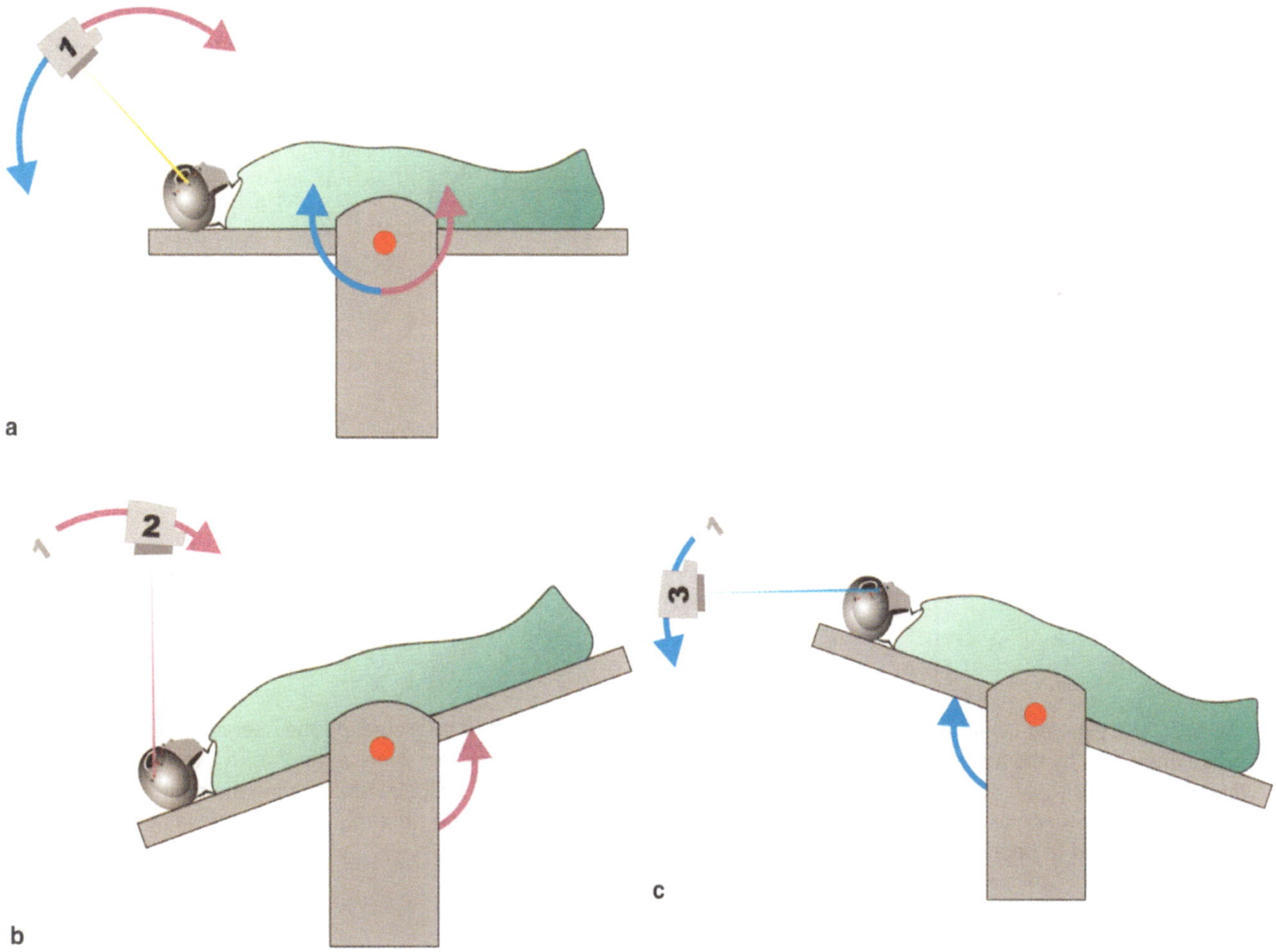

Fig. 2.1. The patient is positioned supine on the operating table with the head immobilized in three-point skeletal fixation (a). To view the posterior (b) or anterior (c) limits of the operating field to best advantage, the table is tilted into Trendelenburg or reverse Trendelenburg as indicated by the *arrows* on its stand, adjusting the operating microscope in the opposite direction as appropriate (indicated by the arrows on the microscope)

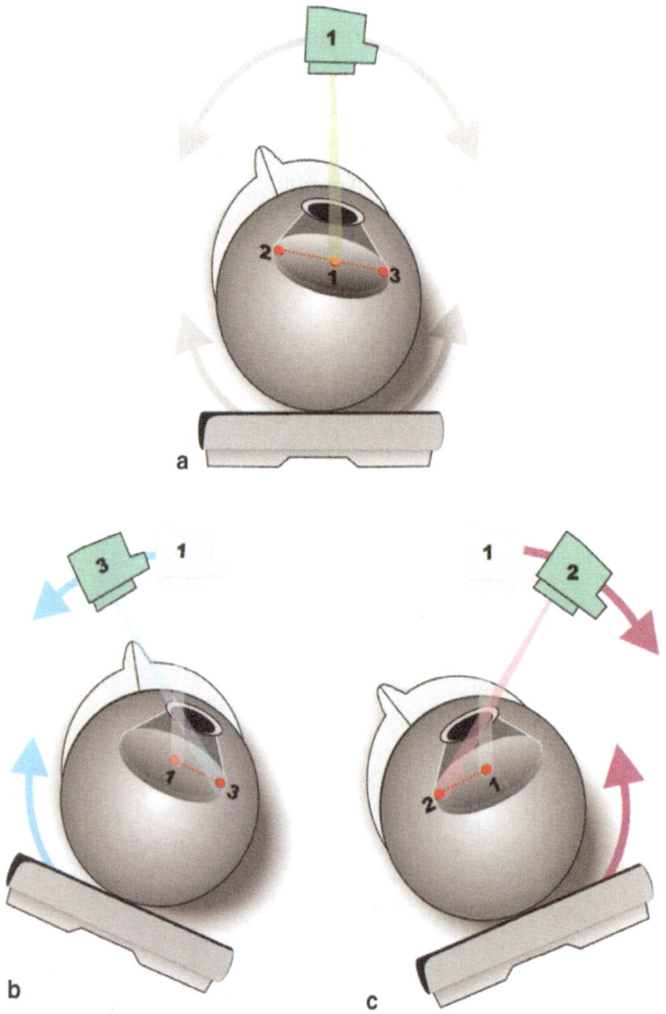

Fig. 2.2. The extreme lateral limits of the operative field may be visualized best by rotating the operating table as appropriate (indicated by the *arrow*), adjusting the microscope in the opposite direction as necessary (a). To expose the far right lateral limit of the operative field (b), the operating table is rotated to the right (clockwise), moving the microscope in the opposite direction (to the left, or counterclockwise). To expose the far left lateral limit (c), the operative table and microscope are rotated in opposite directions as shown

Craniotomy

A 3–4 cm strip of hair is shaved from the ear ipsilateral to the lesion in a curvilinear fashion to slightly beyond the midline behind the hairline. The scalp is then prepared with antiseptic solution, the line of the proposed incision marked (Fig. 2.3), and sterile drapes applied. The scalp incision extends from immediately anterior to the tragus of the ear, arching upward behind the hairline to slightly posterior to the upper part of the auricle, where it sweeps anteriorly, circumventing the belly and insertion of the temporalis muscle (TM), ending approximately 1–2 cm beyond the midline. Keeping the incision as close as possible to the tragus of the ear will avoid injury to the frontal branch of the facial nerve, the superficial temporal artery and vein, and the auriculotemporal nerve. The scalp flap is carefully separated from the periosteum and fascia of the TM, then reflected over the eye sufficiently far to expose the orbital rim where it joins the zygoma, then fixed anteriorly and laterally with fish hooks (Fig. 2.4a).

Periosteum is incised in a curvilinear fashion, exposing frontal bone, the coronal suture, and the parietal bone while preserving the insertion of the TM. The TM is then incised beneath the posterior limit of the scalp incision parallel to its fibers, scrupulously preserving as much of the muscle belly as possible, finally fixing its posterior edge to adjacent superficial fascia with stay suture(s) to allow easy access to underlying bone (Fig. 2.4b). At this point the first burr hole is made in the parietal bone 3–4 cm posteriorly, to the coronal suture, and a second, in the frontal bone up to 1 cm anterior to the coronal suture (Fig. 2.4c). At the pterion, bone is initially thinned with a cutting burr (toward the orbit, middle and anterior cranial fossae), then a diamond burr for the internal table in order to avoid injuring underlying periorbita and dura. The burr hole at the pterion is made only in elderly patiens in order to avoid laceration of the dura. In most of patients only two burr holes are needed (Fig. 2.4d).

Employing a craniotome, a linear cut is made from the first burr hole (behind the coronal suture) inferiorly, beneath the TM, into the squamosal portion of the temporal bone. A second, semicircular bone cut is made connecting the two burr-holes, then a third cut is made to the pterion from the second burr hole. In most younger patients only one burr (in the parietal bone) is made since their dura is normally well developed and not adherent to the internal table of the skull as in the elderly. Several smaller, matching holes are then drilled at appropriate angles through the outer table of adjacent bone for later suturing the bone plate to skull at the time of closure (Fig. 2.4d). As the bone flap is elevated, the squama of the temporal bone is broken (Fig. 2.5a). The TM remains fixed to bone, preserving a true osteoplastic (osteo-muscular) flap (Fig. 2.5b).

If the bone flap obscures the surgeon's view parallel to the floor of the middle cranial fossa, its central portion may be resected additionally in order to facilitate ideal positioning of fish-hooks into the TM Figs. 2.5(c and d). The TM is further separated from bone beneath the fracture line, and temporal squama still projecting from the base of the skull is resected first with craniotome (Fig. 2.6a) and in addition with a rongeur as close as possible to the floor of the middle cranial fossa (Fig. 2.6b). The osteoplastic flap is then reflected laterally and posteriorly (away from the operating field) and secured with fish hooks. If dura is lacerated during craniotomy, it must be repaired immediately in order to preserve the arachnoid and CSF protecting intradural vital structures.

At this point the exposed dura should be fixed with sutures to the small holes drilled previously along the cut edges of the frontal and parietal bones Figs. 2.6(b and c). This is important to prevent epidural bleeding after surgery, also troublesome oozing during the operation. Hemostasis of the exposed dura is maintained with the bipolar coagulator and of the cut edge of the temporal bone with a diamond burr.

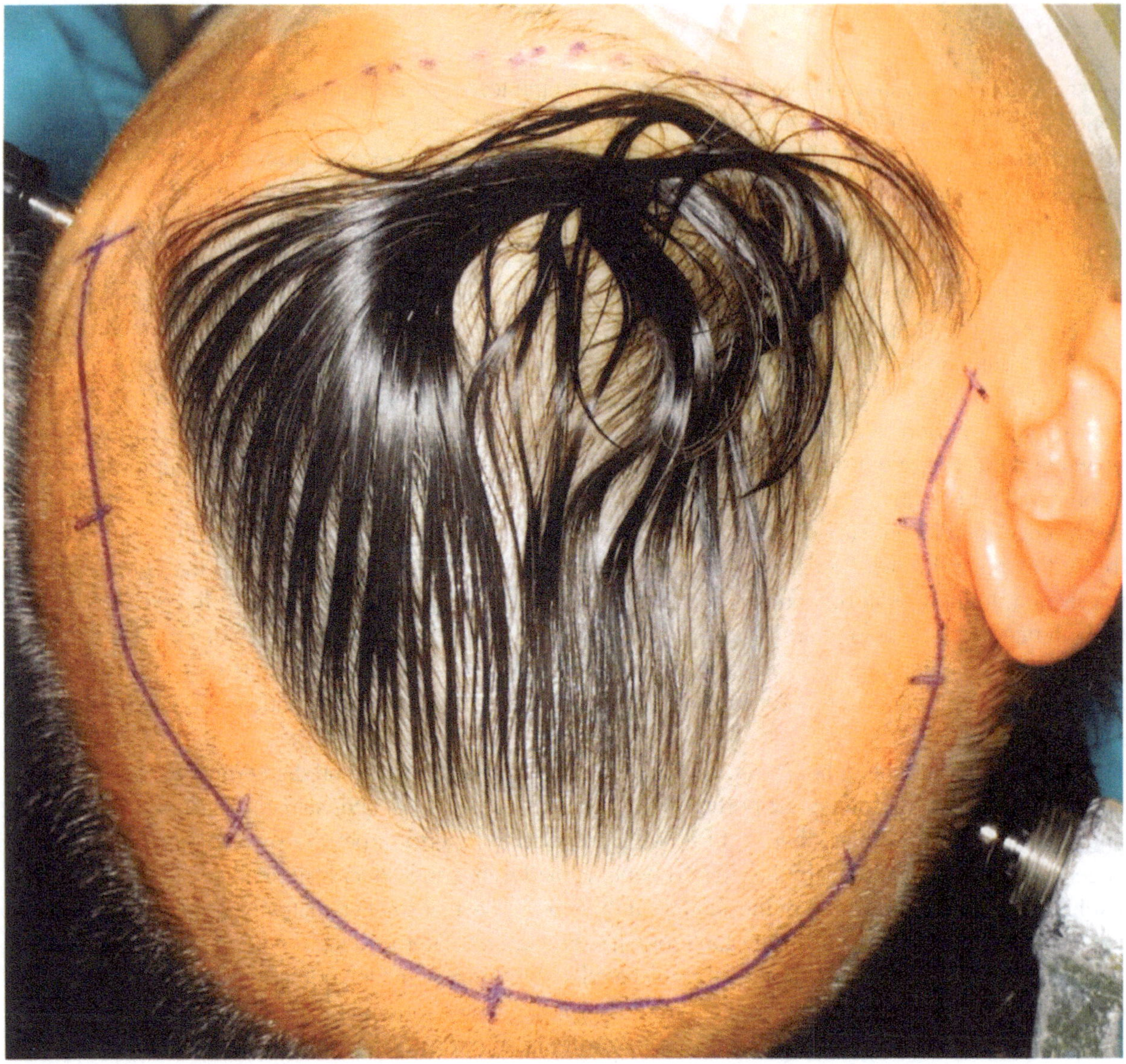

Fig. 2.3. Hair is removed as shown. The scalp incision extends behind the hairline from in front of the tragus of the ear to 1–2 cm beyond the midline. The dotted line indicates how far forward the scalp flap must be dissected in order to expose the rim of the orbit

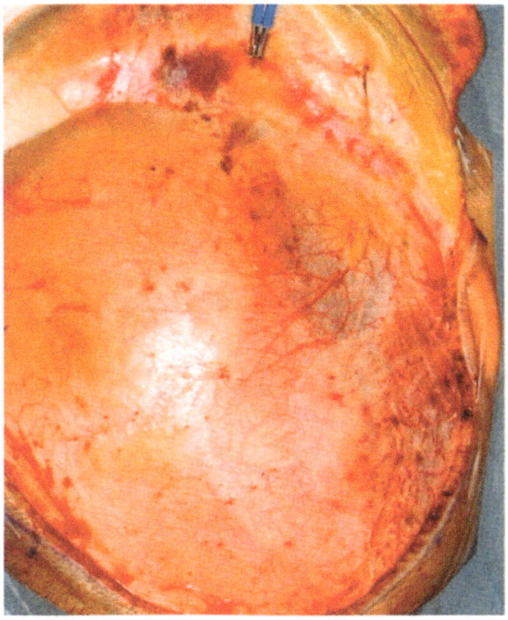

a

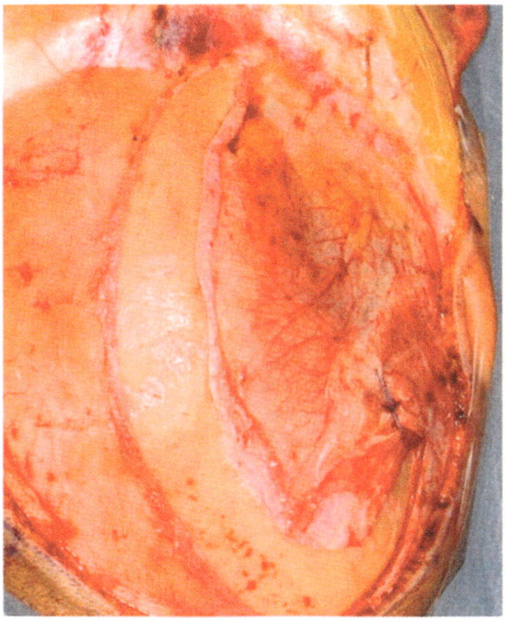

b

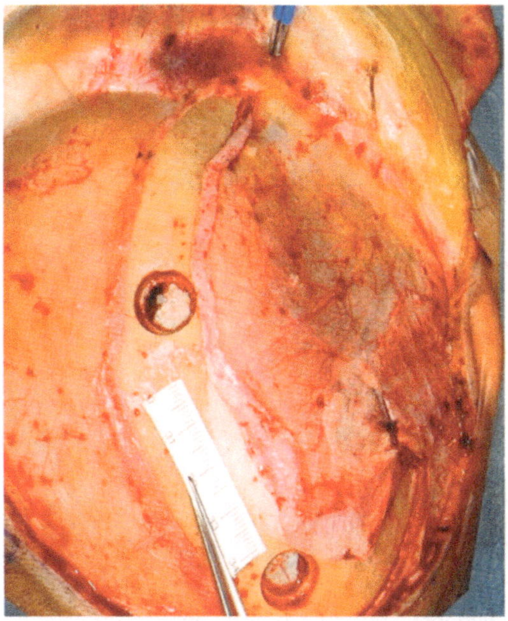

c

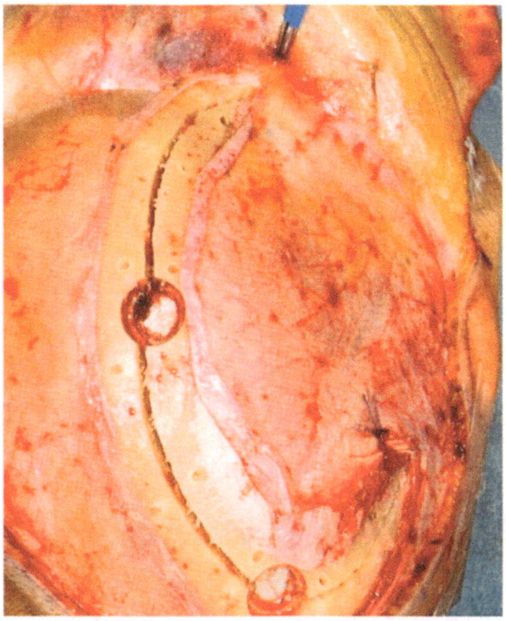

d

Fig. 2.4. In (a), a right-sided scalp flap has been reflected after dissection from periosteum and TM fascia. Note the proximity of the incision to the ear, anterior and proximal to it. At the top of the photograph the forceps points out the orbital rim. In (b), periosteum has been incised circumferentially from the pterion to the parietal bone well above and behind the insertion of the TM, then stripped from bone, exposing frontal and parietal bone anterior and posterior to the coronal suture. The posterior corner of the TM has been divided parallel to its fibers in order to minimize muscle injury. Note that the anterior edge of the divided TM has been fixed to its fascia with a stay suture in order to preserve muscle while cutting bone in its vicinity. (c) The first burr hole has been placed in parietal bone, proximal and posterior to the TM, 4 cm posterior to the coronal suture. A second burr hole is made approximately 1 cm anterior to the coronal suture. Note: The tips of the forceps indicate the orbital rim. If dura is easily separated from bone at both burr holes, a third burr hole, at the pterion, is not necessary. (d) Bone has been cut with a craniotome, initially from the parietal burr hole into the temporal squama, then connecting the burr holes, finally extending to anteriorly to the pterion from the frontal burr hole. Note the small matching holes drilled at intervals on either side of the bone trough. They enable tight, precise bone plate replacement at the time of closure while insuring virtually normal TM function very soon after surgery

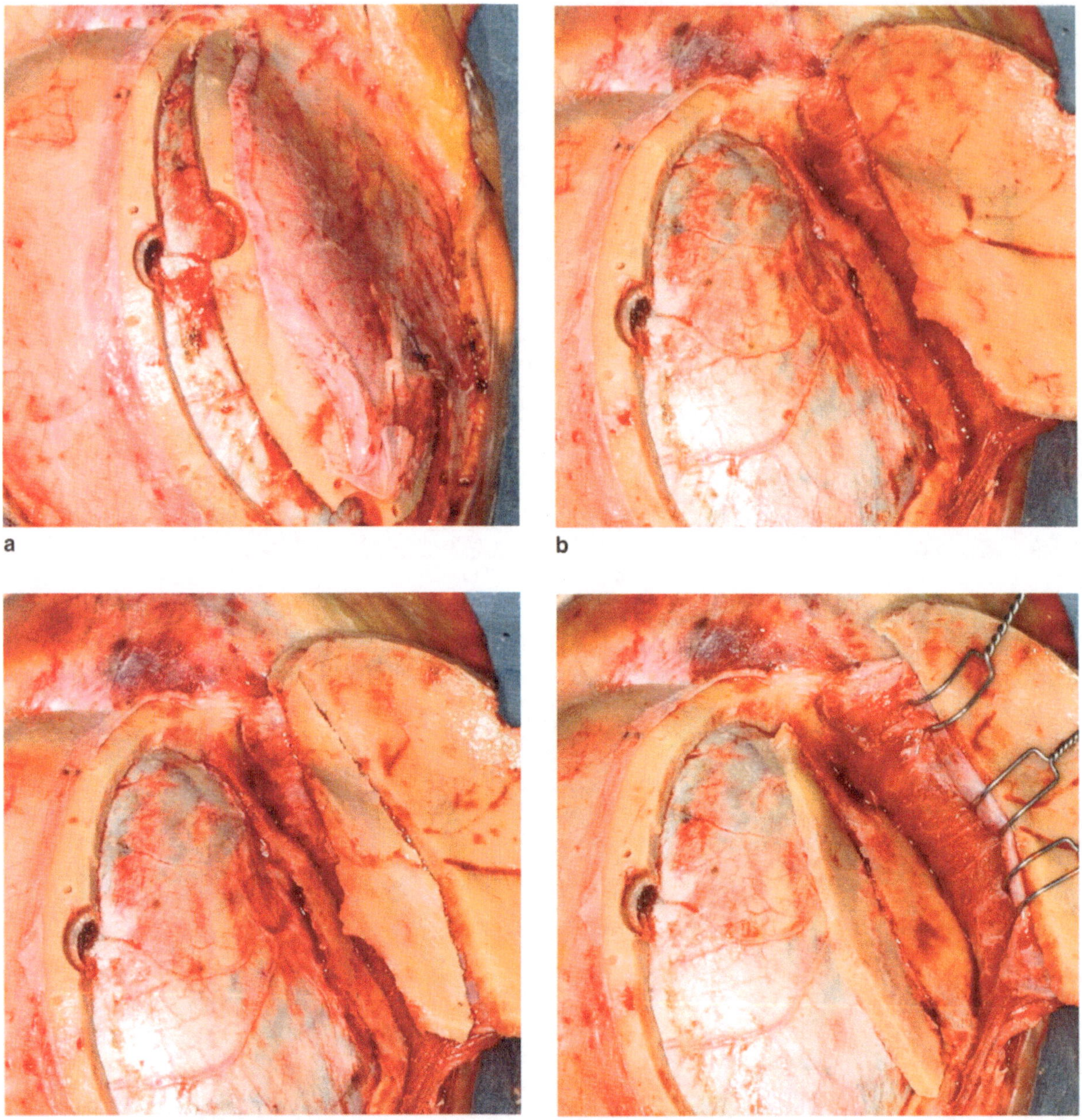

a b c d

Fig. 2.5. (a) By gently elevating the detached edge of the bone plate the temporal squama is broken, and dura is carefully separated from bone in order to avoid lacerations in cases where it is adherent to the under surface of the bone plate. Small matching holes not extending through the inner table of the skull are freshened at this point, with dura fully exposed. In (b), the bone plate is reflected laterally and TM is separated additionally from the squama of the temporal bone, allowing maximal reflection of the osteomuscular flap. (c) After TM has been separated from the central part of the bone flap, the flap is cut, allowing sufficient exposure of the deep surface of the TM for the ideal placement of fish hooks (d). Trimming the bone plate also enhances exposure of the under surface of the TM, allowing it to be elevated fully from the squama of the temporal bone. Note: In order to demonstrate the extent of its resection, the segment of bone plate cut away is placed against the squama of the temporal bone (d)

Unroofing the orbit

The orbital opening, which was made previously after thinning the pterion, is enlarged with a diamond burr. In cases where the pterional burr hole was not made prior to the bone cut Figs. 2.5(a–d) and 2.6(a,b), the orbit is open with a diamond burr after fixing the osteo-muscular flap (Fig. 2.6c). When the diameter of the opening is approximately 1 cm, periorbita is swept from the under surface of the orbital roof with a thin dissector. Similarly, dura is elevated from the intracranial surface of the orbital roof and bone is resected as far centralward that the lateral end of the superior orbital fissure (SOF) is met, and the dural duplicature is visualized Figs. 2.6d and 2.7(a and b).

If not already in use, the operating microscope should definitely be employed from this point forward. The orbital branch of the middle meningeal artery (MMA) is then identified, coagulated and transected, and dura covering the pole of the temporal lobe is dissected from the orbital roof on the postero-lateral surface of the SOF. Periorbita is further dissected from beneath the orbital roof, which is subsequently removed piecemeal with a small rongeur, exposing entirely the dura bulging dorsally from the SOF. More laterally and basally, bone medial to the foramen rotundum (FR) is somewhat thicker. No attempt should be made to remove it with a rongeur. Rather, it should be drilled away with a diamond burr, then bone anterior to the FR, fully opening the V2 canal Figs. 2.7(a and b). Care must be taken to avoid proceeding too far antero-basally in order to avoid entering the maxillary sinus.

Following unroofing the orbit dorsolateral to dura traversing the SOF, the orbital roof is removed antero-medial to the SOF dura. Great care must be taken at this point. If the frontal sinus is larger than normal, it may be entered inadvertently. Orbital roof bone must not be removed before inspecting it both inside and outside the orbit and, particularly, the peripheral end of the optic canal (OC), where there is considerable risk of injuring the optic nerve (ON). This point cannot be over-emphasized. All edges of cut bone, laterally as well as medially, should be reexamined, and precise hemostasis achieved with a diamond burr.

Resecting the sphenoid wing

Dura extending from the middle cranial fossa to the orbit should be separated very gently from the sphenoid wing (SW), exposing bone both medially and laterally. After resecting the lateral portion of the SW with a small rongeur, the sleeve of dura entering the SOF from the middle cranial fossa is fully exposed. It is composed of two distinct layers and extends laterally as the lateral wall of the parasellar space (PS). Doubled dura at the lateral end of the SOF dura is cut with curved microscissors directed posteriorly in order to avoid damaging nerves III, IV, V1 and VI (Fig. 2.7c). After identifying the cleavage plane between the layers of the lateral wall of the PS, the outer dural layer is separated from the inner layer, which is left intact over nerves III, IV, V1 and V2 (Fig. 2.7d). Great care must be taken to avoid injury to these nerves. After dura has been separated from the posterior surface of what SW remains, the bulk of this structure may be removed slowly and carefully with drilling or with a small rongeur, leaving the anterior clinoid process (ACP) intact as the surgeon remains continuously mindful of the ON medially and nerves III, IV, and V1 laterally, still covered by dura. Stripping the outer layer of the lateral wall of the PS in this manner, beginning at the lateral end of the SOF, enables the surgeon to approach the ACP from its infero-lateral side (Figs. 2.7d and 2.8a).

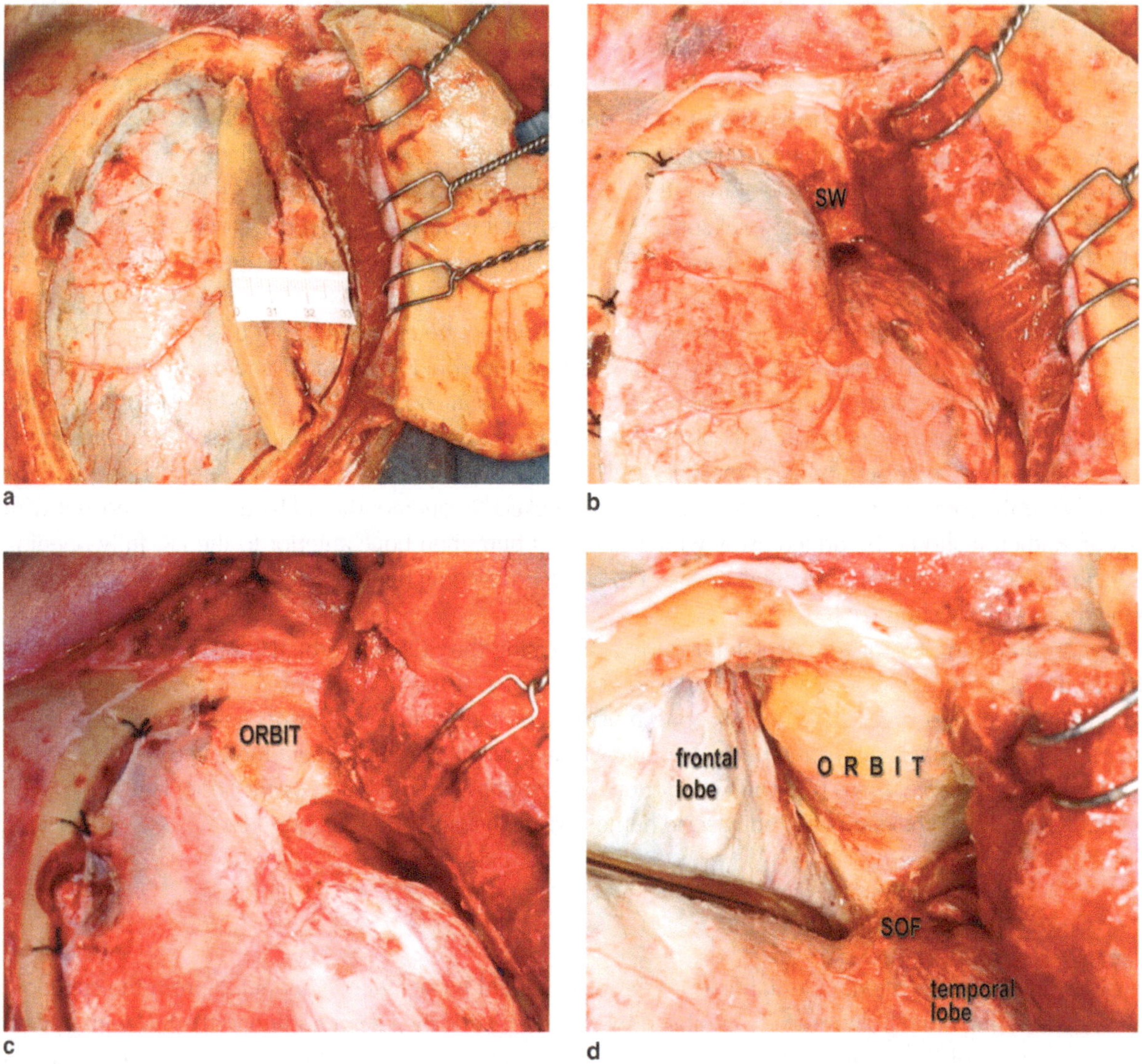

Fig. 2.6. (a) The temporal squama still attached to the skull base has been resected flush with the edge of the reflected TM. A total of approximately 3 cm (central bone plate plus remaining temporal squama) is normally removed with the craniotome. (b) Removing additional temporal squama will expose most of the temporal lobe dura, greatly simplifying the approach into the middle cranial fossa. Note: At this point the orbit has not been opened. Adjacent dura is fixed with sutures to cut surfaces of bone through the small holes drilled previously. (c) The orbit is now opened, leaving the orbital rim intact. This enables the bone plate to be replaced accurately after surgery, thus avoiding postoperative cosmetic deficits. Note: Dura has been peeled from the dorsolateral orbital roof (and the squamosal part of the temporal bone prior to its resection), allowing an appreciation of the depth of the floor of the anterior part of the middle cranial fossa. By rotating and tilting the operating table the floor of the middle cranial fossa may be exposed without retracting dura. (d) After resecting the roof of the orbit further posteriorly and medially, the lateral side of the sleeve of SOF dura between the intracranial cavity and the orbit comes into view

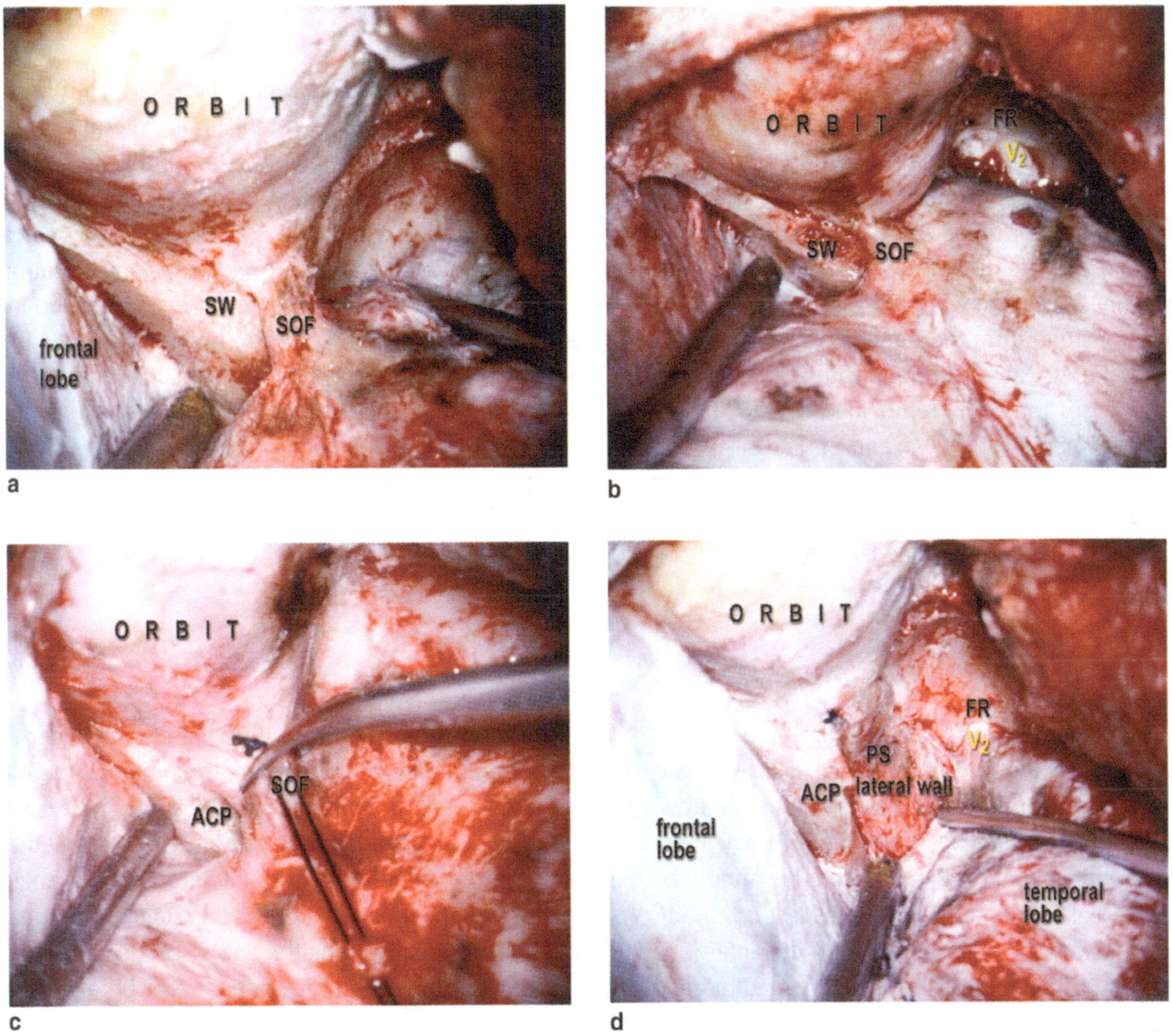

Fig. 2.7. (a) The orbit is now open, its roof removed well laterally, exposing the apex and lateral wall of the sleeve of dura coursing through the superior orbital fissure. For the sake of description this dural structure will be called the SOF. In this photograph the posterior limit of the orbital wall, overlying the SOF, remains intact. Dura over the frontal pole has been separated from bone. Immediately medial to the SOF, a remnant of the sphenoid wing and the medial limit of the orbital roof remain. (b) Here the orbital roof overlying the SOF has been resected and the FR has been slightly enlarged (by drilling), exposing V2. Medial to the SOF, the residual sphenoid wing and the medial limit of the orbital roof have not yet been completely resected. (c) The sphenoid wing remnant and remaining posterior orbital roof have been removed, exposing the medial surface of the SOF. A stitch has been placed in periorbita at the peripheral end of the SOF and a stay suture slightly proximally. The lateral surface of SOF may now be opened as indicated between the sutures with curved microscissors pointed posteriorly. Directing the scissors posteriorly insures the preservation of nerves III–VI, which course through the SOF from the PS into the orbit. (d) The inferolateral surface of the ACP is exposed after removing the sphenoid wing entirely and separating the two layers of the lateral PS wall anterior to the line between the ACP (medially) and V2 (laterally). Note: The knotted suture in the periorbita indicates the amount of exposure yet to be obtained by removing the ACP along the medial surface of the SOF. For maximum safety this must be accomplished from its inferolateral surface

Resecting the anterior clinoid process

Following the removal of the SW, dura on the inferolateral surface of the ACP is gently dissected from bone with a microdissector, and the ACP is carefully hollowed out with a diamond drill from its inferolateral surface medialward. This amounts to an important change from the technique of ACP removal described previously. By drilling the ACP from its lateral surface, disturbing any aneurysm medial to it is greatly minimized (Fig. 2.8b). The interior of the ACP is removed until only thin walls of bone remain (Fig. 2.8c), the medial of which makes up the lateral wall of the OC. Safety is enhanced by drilling in short bursts under constant irrigation with saline. After removing the thinned elements of the ACP with delicate instruments, the optic strut is encountered projecting from the inferior wall of the OC over the anterior loop (AL) (Fig. 2.8d). In rare instances the apex of the ACP is continuous with the lateral wall of the sphenoidal sinus, forming a canal around the ICA. In such cases forcefully removing the ACP may result in entering the sphenoidal sinus and/or damaging the ICA and/or nerves III, IV, V1 and VI. For this reason, the remainder of the ACP must always be dissected with extreme care, ultimately removing it with the least amount of dural traction possible (Fig. 2.9a). Cut surfaces of bone must be carefully inspected. Any sign of a fracture line must be taken as clear evidence that the sphenoidal sinus has been violated.

Opening the optic canal

Thinning the ACP at the lateral wall of the OC with a diamond burr must be done with great care in order to avoid injuring the ON or exposing it to excessive heat. The importance of drilling at short intervals while keeping the diamond burr beneath a steady flow of saline is extremely important. Once the ACP has been completely removed (Fig. 2.9b), the ICA, which courses beneath a membrane on the floor of the antero-medial triangle (between the ON and nerve III), should be exposed from the lateral limit of the anteromedial triangle to its medial border (adjacent to the ON). If the ICA cannot be visualized adequately at this stage, additional drilling toward the anterior corner of the antero-medial triangle – proximally along the course of the extradural ICA, toward the ON – will expose the membrane overlying the AL of the ICA, allowing the ICA to be dissected from the lateral wall of the sphenoidal sinus. In some cases the lateral wall of the sinus is paper-thin, requiring great care to avoid violating it.

Dura on the dorsal surface of the OC is elevated posteriorly, exposing dura posterior to the proximal limit of the OC to the falciform ligament, the tough fibrous band crossing the dorsal surface of the ON. All dura is preserved intact. Dura is also peeled from the tuberculum sellae medially to avoid injuring it while drilling the proximal and lateral walls of the OC.

Bone over the ON is removed slowly and carefully with a small diamond burr, essentially "rubbing" it away rather than boring through it. Bone is thus thinned to the point of transparency, allowing it to be removed with a microdissector, significantly reducing the risk of injuring the ON. Sufficient bone is removed to expose the full width and length of the ON. Care should be taken to avoid entering the sphenoidal sinus immediately medial to the OC. Removing the lateral wall of the OC is rarely difficult, but continuous saline irrigation is extremely important.

Resecting the optic strut

After completely opening the OC superiorly and laterally, a fine retraction suture is placed in the dura propria covering the lateral surface of the ON, allowing the ON to be gently elevated, affording maximal exposure of the optic strut beneath to it. To facilitate maximum exposure and safety, the operating table should be rotated away from the side of the PS lesion and the microscope adjusted more temporally, orienting it from lateral to medial (Fig. 2.2c). The optic strut is usually a thin sliver of bone within the narrow conical space between the ON and the AL of the ICA Fig. 2.9(b and c). To remove it, the smallest diamond burr (less than 2 mm) is selected and drilling proceeds from its lateral surface, from the ICA toward the ON. The thus thinned optic strut represent then the lateral wall of te sphenoid bony sinus.

Although extremely high speed drills are currently available, the art of drill design and construction remains in its infancy. For the purpose of these exposures drill speed offers no advantage. Ideal drill speeds in the vicinity of neural structures are slow, slower, and slower still! Far more important are burrs approaching 1 mm and less in diameter which function from drill assemblies that are sleek and streamlined, affording little obstruction of view within narrow working spaces. Frequently even the smallest burr available is larger than the bony element to be removed, further limiting continuous vision. Nevertheless, such bone elements may be thinned sufficiently for safe removal so long as the surgeon is extremely careful and patient. He must program himself to be satisfied with only a few turns of the burr before actually beginning to drill. It is essential that all drilling (here and elsewhere) be accomplished in very short bursts at relatively low speeds, with the tip of the drill kept under constant saline irrigation. So long as these techniques are employed scrupulously and dura remains intact, intradural structures are protected. Surgeons not familiar with these gentle drilling techniques are encouraged to practice thinning the shell of an egg with a diamond burr in the laboratory. By mastering the technique of thinning an eggshell without disrupting the egg white beneath it, one becomes accustomed to the sense of feel and delicacy necessary to drill bone overlying neural structures.

When drilling the optic strut special care is required to avoid damage to fascia overlying the ICA, the distal (dural) ring (DR), and the dura propria of the ON. The lateral wall of the sphenoidal sinus must be left intact to avoid a cerebrospinal fluid (CSF) leak following surgery. Completely removing the optic strut is necessary to deal with lesions both within the sella and in the area of the ACP and posterior clinoid process (PCP). Its removal allows full mobilization of the entire distal half of the AL of the ICA.

After complete removal of the ACP, the entire lateral and superior walls of the OC, and the optic strut, the ON, the anteromedial triangle, and anterior part of the inner layer of the lateral wall are fully exposed (Fig. 2.9d).

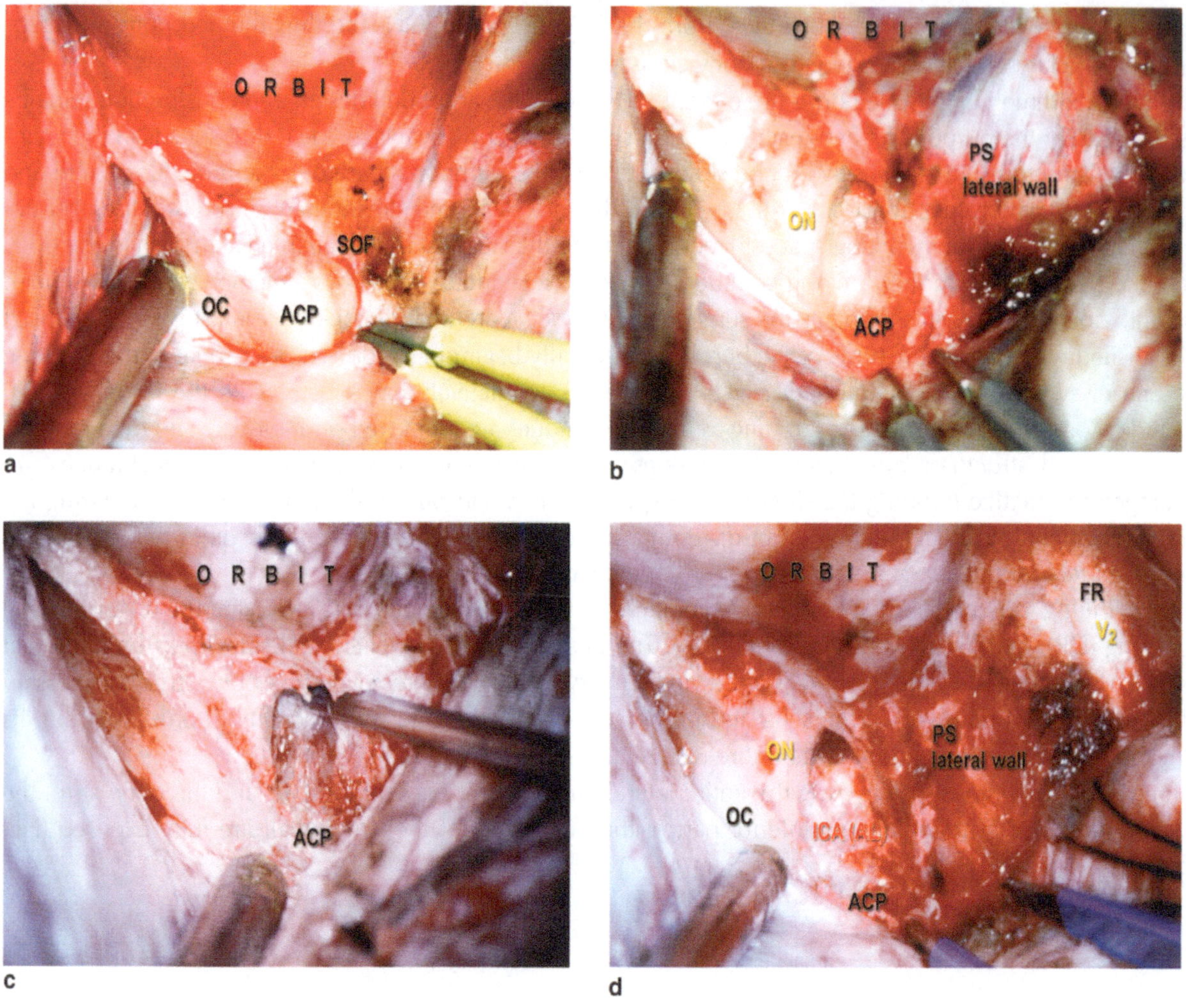

Fig. 2.8. (a) Following resection of the entire sphenoid wing (*SW*), the inferolateral, lateral, and posterior surfaces of the anterior clinoid process (*ACP*) are exposed. Slightly displacing dura from bone with forceps and a suction tube indicates the entire mass of ACP and superior wall of the optic canal to be removed. This requires only the slightest of dural tension, certainly not enough to disturb intradural structures, either neural or vascular. (b) The ACP has been hollowed from its inferolateral surface, through which the lateral and superior walls of the optic canal (*OC*) have been partially resected. Note: The apex of the ACP and the optic strut remain undisturbed. The ON in the OC remains completely covered although the lateral and superior walls of the peripheral OC have been thinned to some extent by careful drilling. (c) In this photograph the ACP has been drilled additionally posteriorly, toward the dura, and anteromedially toward the optic strut and the lateral wall of the sphenoidal sinus, beneath the OC, where the smaller suction tube points. The OC has not yet been opened completely, and the AL of the ICA (medially) remains covered by thin bone laterally and anteriorly. (d) Only the tip of the ACP remains, plus part of the optic strut (beneath the ON, above the ICA). The OC is still not completely open. The lateral wall of the parasellar space (*PS*) has been divided, its dural layer retracted by stay sutures and forceps. Medial to V2 Surgicel has been placed in the anterolateral triangle where a cerebral vein had entered the PS

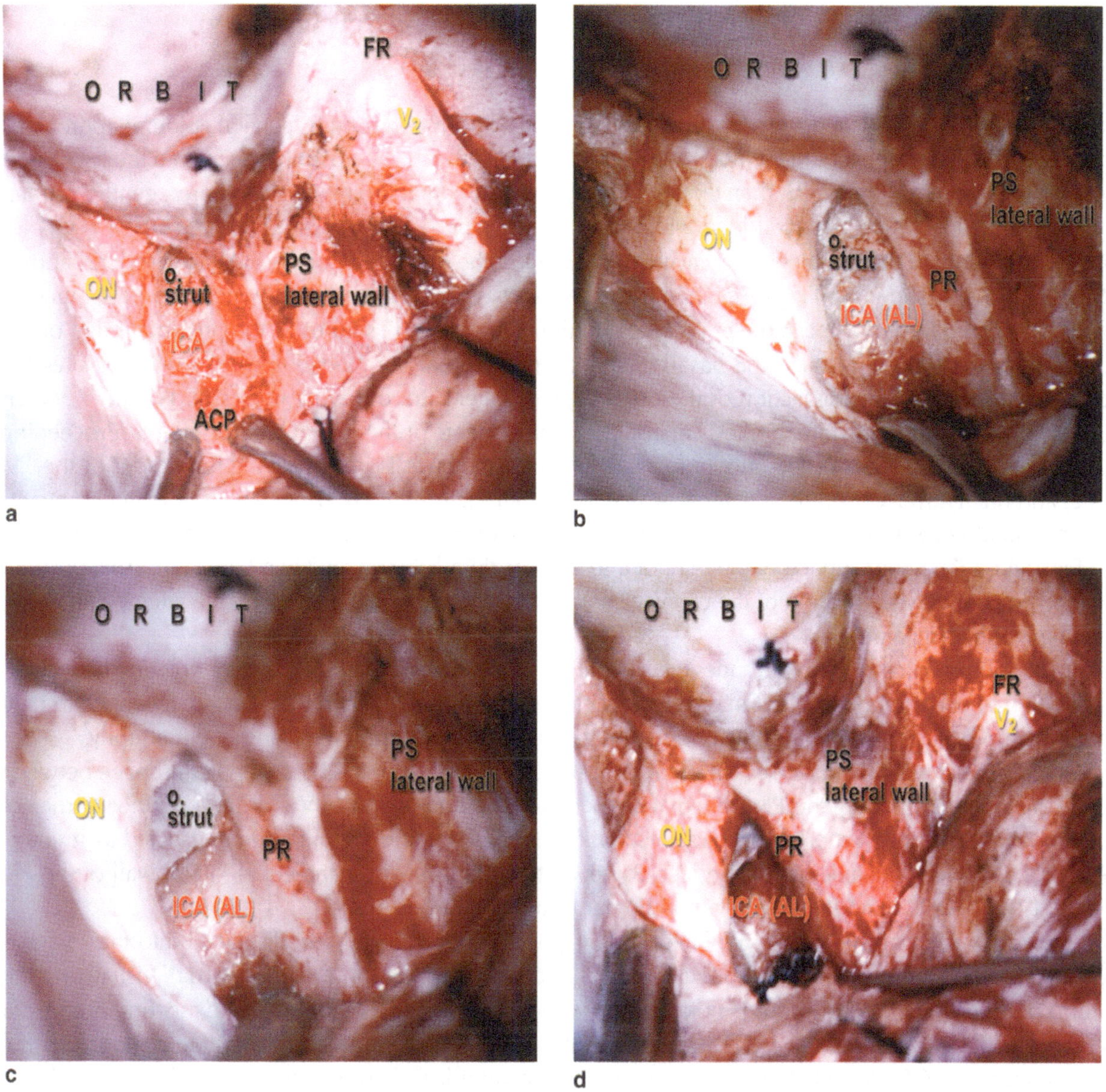

Fig. 2.9. (a) Complete exposure of the inner layer of the lateral wall of the right parasellar space (*PS*). Laterally, V2 is visible coursing from PS into the foramen rotundum (*FR*). Medially, the anterior loop (*AL*) is partially exposed, although the tip of the anterior clinoid proces (*ACP*) remains, and the optic strut has not yet been completely removed. The optic nerve (*ON*) is still partially covered by bone. The posterolateral side of the orbit is free of bone. (b) Further resection of the posterior and lateral walls of the optic canal (*OC*) exposes dura propria posterior and lateral to the ON as it courses from the intracranial cavity into the orbit. With the apex of the ACP now removed, the lateral surface of the AL is exposed, but the optic strut remains in place. (c) Partial resection of the optic strut, beneath the OC, facilitates dissection of the AL and exposure the proximal ring (*PR*), distal (dural) ring (*DR*), and medial wall of the ICA adjacent to the lateral wall of the sphenoidal sinus. The ON is sheathed in its dura propria, and the orbit anteriorly and inner layer of the lateral PS wall are free of bone. (d) The anteromedial triangle, beneath the ON, is exposed. Nerve III lies beneath the PR laterally and dura posteriorly. The ICA is seen at the bottom of the anteromedial triangle, looping from PR to the DR. The ON (in its dura propria) is exposed throughout its length, from dura to periorbita. The anterior part of the PS lies beneath the inner layer of its lateral wall, with V2 coursing into the the FR. Note: The stay suture posteriorly and the knotted suture at the edge of the orbit indicate the extent of exposure gained by separating the layers of the lateral PS wall

Dissecting the outer layer of the lateral wall of the parasellar space and exposure of nerves III, IV, V1, V2, V3, and GG

The following also represents a decided change from previous descriptions of the "general epidural approach". By extending the bony exposure to the floor of the middle cranial fossa, it is possible to separate the two layers composing the lateral wall of the PS, then tenting outward the outer dural layer with stay sutures, gently "retracting" the anterior pole of the temporal lobe within its dural envelope.

Prior to incision into the doubled outer layers of the dura at the lateral end of the SOF as described above, a suture is placed as shown in Fig. 2.7c. Then the outer layer of dura is sequentially incised and stripped to the floor of the middle fossa, the surgeon placing additional "retraction" sutures in the proximal edge of the dura as he proceeds more laterally and basally. Tenting the proximal dural edge upward gently elevates the temporal lobe and facilitates dissection of the distal edge of outer dura overlying nerves III, IV, and V, which are covered by an areolar membrane, the "inter-dural" components of the lateral wall of the PS. Proceeding laterally and toward the floor of the middle fossa, the outer dural layer is incised along the border of the PS, over V2, and finally over V3, preserving these structures within their inner covering while sequentially retracting the proximal edge. Tilting the operating table into a position of greater Trendelenburg at this point minimizes tension on the under surface of the temporal lobe as its dura is being retracted (Fig. 2.1b). This position also allows maximum visualization of the cleavage plane between the layers making up the lateral wall of the PS, facilitating their separation.

This technique is much preferred to opening dura first over the brain's convexity, then retracting brain with a spatula only to open dura beneath it in order to access the epidural space at the skull base. Evidence to date indicates that this method minimizes focal (often unrecognized) brain injury and subsequent atrophy.

A fairly large vein typically traverses the outer layer of the lateral wall of the PS, entering the space between the two layers, and ultimately the PS immediately medial to V2 and lateral to V1, at the posterior corner of the antero-lateral triangle. It should be coagulated within the temporal lobe dura, then transected as it enters the PS, which is finally occluded with Surgicel. By separating the layers over the entire lateral wall of the PS, nerves III, IV, V1, V2, V3 and the antero-lateral border of the GG, are exposed.

The proximal stump of the middle meningeal artery (MMA) is coagulated and displaced into the foramen spinosum (FS). The FS should be plugged with Surgicel and sealed with bone wax (Fig. 2.10a). Visualizing the plexiform portion of the nerve V root is not possible without transecting the MMA and dissecting dura from the anterior surface of the petrous bone. The dural outer layer is incised along the posterior surface of the Gasserian ganglion (GG) toward the apex of the petrous pyramid, then dissected posteriorly and medially, completing the exposure of the GG, ultimately leaving only the reticular membrane and arachnoidea covering its lateral surface and the nerve V root and its plexiform segment, both proximal to the ganglion.

As dura is being separated from the anterior surface of the petrous bone, care must be taken to avoid injuring the greater petrosal nerve (GPN), which exits the facial hiatus postero-medial to the FS and runs epidurally. The GPN must be preserved by all means in cases where parasellar tumors may compromise function of V1 or the GG or the root of nerve V. Loss of function of both V1 and the GPN leads to the loss of protection of the eye. Where V1 sensation is intact, and its function is not likely to be jeopardized by resecting the lesion being addressed, it is preferred to transect the GPN cleanly rather than to risk avulsing it from the geniculate ganglion and creating a nerve VII deficit.

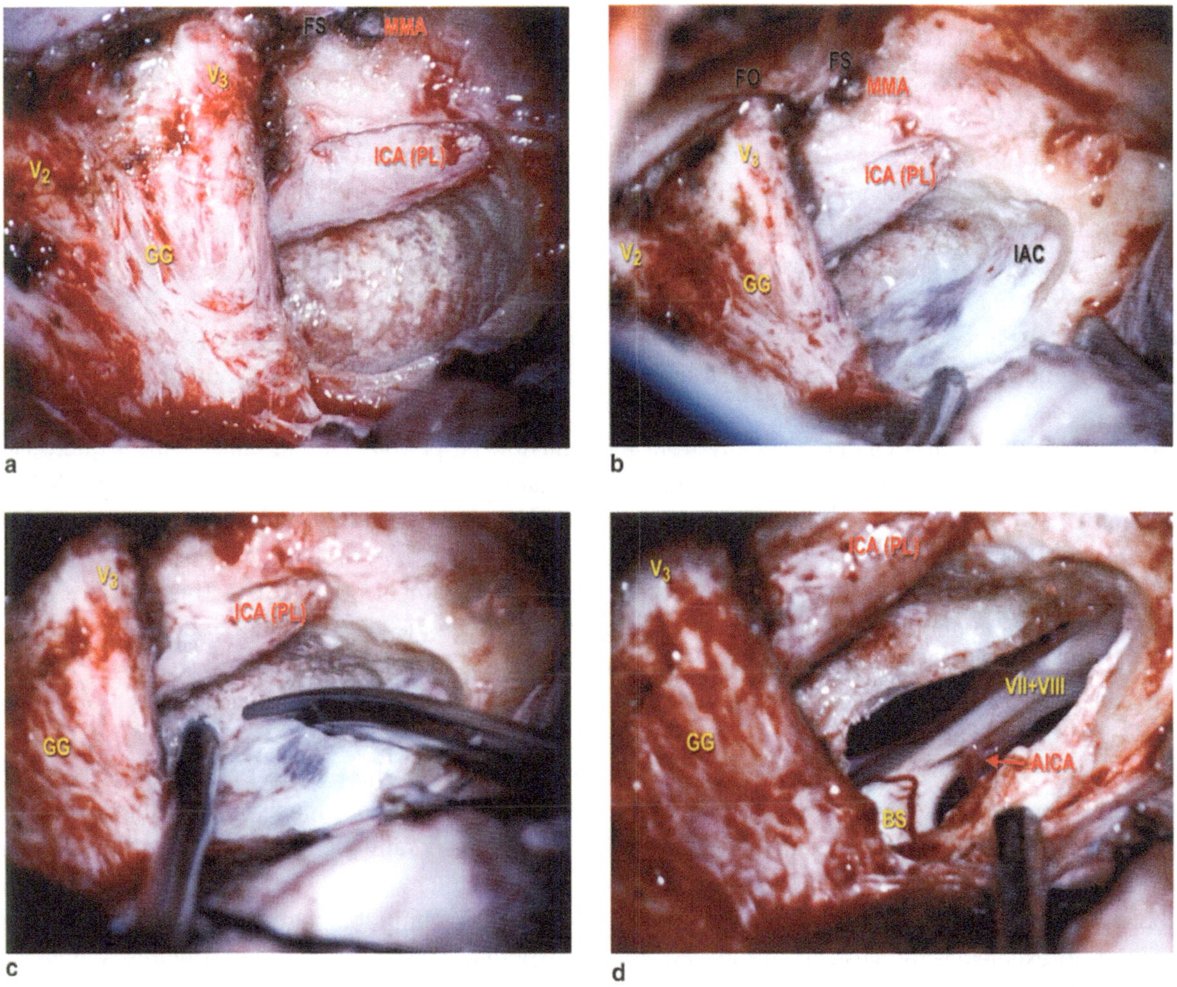

Fig. 2.10. (a) This photograph shows the gasserion ganglion (*GG*), V3, the foramen spinosum (*FS*) containing the stump of middle meningeal artery (*MMA*), and the posterior loop (*PL*) of the ICA within the right petrous carotid canal. Note the opening into the petrous bone medial to the ICA in the petrous canal and underneath Meckel's cave. (b) Further removal of petrous bone medially and toward its apex exposes periosteum (dura) within the internal auditory canal (*IAC*). More superficially, the distance from the FS to the lateral surface of the ICA in the petrous canal is approximately 5 mm, where the tensor tympani muscle and beneath it, the eustachian tube, are located. V3, GG, and the plexiform segment of nerve V remain covered by the inner layer of dura. In (c), after resecting bone from the petrous carotid canal to dura medial to the petrous apex, this dura may be incised and removed. (d) By removing posterior fossa dura, the surgeon's orientation is confirmed by exposing the brainstem and nerves VII and VIII, coursing toward the IAC. More superficially, the ICA is seen laterally, and V3 and GG, anteriorly. Note: In this patient the anterior inferior cerebellar artery (*AICA*) courses between nerves VII and VIII

Unroofing the petrous carotid canal

After transecting the MMA, dissecting dura from the anterior surface of the petrous pyramid, exposing the GPN, and mobilizing the dura dorsal to V3 and Meckel's cave, the surgeon can begin to remove the roof of the petrous carotid canal. This is done with a diamond burr, beginning 5–8 mm medial to the FS, drilling immediately adjacent and posterolateral to the GG or where V3 enters the GG. Encountering the tensor tympani muscle beneath bone indicates that the surgeon has begun drilling too far laterally and must work more medially in order to avoid damaging the Eustachian tube. After identifying the medial wall of the ICA, additional bone is removed along the vessel beneath V3, exposing the supero-medial surface of the horizontal segment of the ICA – between its posterior (PL) and lateral (LL) loops – essentially 1/2 of its total circumference (Fig. 2.10a). Drilling posterior to the ICA as exposed here runs the risk of entering the middle and/or inner ear and should be avoided.

Periosteum is then peeled circumferentially from the walls of the petrous carotid canal with a microdissector over a distance of approximately 1 cm in order to facilitate temporary ICA occlusion if necessary. At this point, whether or not ICA flow can be interrupted with a clip should be tested. Where interposing a graft between the petrous and supraclinoid segments of the ICA is anticipated, periosteum surrounding the vessel should be stripped from bone farther proximally into the unroofed segment of the canal. This provides sufficient space to insert a ballooned catheter along PL of the ICA between the vessel and the inner surface of the canal. Inflating such a balloon provides precise occlusion of the ICA, a method preferred over clip-occlusion when maximizing working space constitutes an advantage (avoiding the presence of a clip), as with the construction of an end-to-end anastomosis.

In approximately 20% of cases, the horizontal segment of ICA within the petrous carotid canal is either uncovered by bone, or only partially covered. In such circumstances the fibrous membrane (or thin, transparent bone) overlying the ICA may be gently and easily removed, but generally the canal requires at least some additional drilling in order to provide satisfactory access to the ICA.

Although initial descriptions of the technique for exposing the ICA within the petrous carotid canal include a suggestion to exenterate the Eustachian tube [5], further experience indicates that this structure should be preserved intact, even when the wall between the petrous carotid canal and the Eustachian tube canal has been damaged. Preserving the Eustachian tube and its bony canal reduces the risk of an ipsilateral hearing deficit even when injury may cause scarring likely to result in inhibiting the equilibration of pressure between the middle ear and the oral cavity.

Resecting the petrous apex and unroofing the inner auditory canal

Historically neurosurgeons have been reluctant to drill the petrous apex themselves, primarily out of fear of disrupting not only the contents of the inner auditory canal (IAC) but the delicate structures of the inner ear and hearing apparatus. For two generations many have depended on specially trained ENT surgeons to drill the petrous bone in the course of acoustic neurinoma resections [6]. For the transpetrous exposures required by some lesions of the PS and Meckel's cave, however, such fears are entirely unfounded. One need only to examine a skull (or a model of the skull base) to confirm that such exposures involve removing a section of the petrous pyramid well anterior to the IAC, the cochlea, and the semicircular canals, leaving them completely protected and undisturbed. By drilling the pyramid from a point posterior and medial to the GG, toward the anterior and medial petrous surface (toward the clivus), no important structure is placed at risk.

Although the orientation of anatomical structures will vary as the operating table is tilted Fig. 2.1 (a–c) and rotated Fig. 2.2(a–c), in general the following description applies to the head position at the time of performing the craniotomy. The surgeon begins drilling the posterior and lateral surface of the petrous pyramid at the exposed ICA segment as described above, medial to the FS, posterior and adjacent to the GG, proceeding anteriorly and medially beneath the GG and dura at the floor of Meckel's cave toward the posterior cranial fossa and the clivus. By thus creating a trough to posterior fossa dura the petrous apex is essentially resected, and the bone anterolateral to the IAC is removed, so that the dura (periosteum) of the IAC is exposed (Fig. 2.10b). After transecting the pyramid in this manner and opening the posterior fossa dura vertically (Fig. 2.10c), the brainstem and nerves VII and VIII will come into view posteriorly, confirming the surgeon's orientation (Fig. 2.10d). Then it is a simple matter of drilling away the posterior wall of the trough, beneath the superior petrosal sinus (SPS) (which is superior to the petrous apex), to the IAC. At this point the SPS is left intact, and drilling toward the IAC is done in very short bursts, keeping the diamond burr continuously bathed with saline to avoid thermic injury to nerves VII and VIII. When the dura of the IAC is exposed, bone between the IAC and the SPS is drilled away, unroofing a 5–8 mm segment of the superior and medial IAC. Similarly, bone inferior and medial to the IAC is drilled away, leaving bone only posterior and lateral to the IAC. At this point the anterior wall of the trough bisecting the petrous apex is removed as necessary and posterior fossa dura is opened, allowing visualization of nerve VI as it ascends in the prepontine cistern and finally enters Dorello's space (DS) at the clivus. Removing clival bone between Dorello's space and the horizontal segment of the ICA requires that its periosteum be meticulously preserved in order to prevent nerve VI injury.

This bone exposure, if carried out as described, is entirely safe. No important neural or vascular structures are present within the petrous pyramid antero-medial to the IAC.

In those cases where the petrous apex cannot be totally resected by this technique, the dural membrane anteromedial to Meckel's cave may be transected through Parkinson's triangle, and the root of nerve V retracted posteriorly. This affords ample space anterior to the nerve V root for resecting the remainder of the petrous apex.

2.1

Evolution from the initial intradural to the contemporary extradural approach to the parasellar space and adjacent regions in the central skull base

The epidural approach to the PS, published in its initial version two decades ago (1) and in improved version more than one decade ago [3], was completely different from the approaches published earlier elsewhere [7–12].

In the course of refining the extradural approach to parasellar space (PS) lesions over two decades certain elements have been altered. Although some of the changes may seem trivial, all are important to making these difficult operations simpler and safer, with an eye toward achieving the best results possible.

The initial patient positioning for surgery has not been altered. During the first 10 years of direct PS surgery, the entire anterior quadrant of the patient's head was shaved, but subsequently only a narrow strip of hair has been removed from the vicinity of the planned scalp incision (Fig. 2.3). Previously, the incision was made more anteriorly, from in front of the tragus of the ear, extending behind the hairline in a semicircular fashion to the midline. Currently the incision is begun closer to the tragus of the ear, posterior to the frontal branch of the facial nerve, the auriculo-temporal nerve, and the superficial temporal artery and vein, bending upward to just above and slightly around the auricle, then curving anteriorly to 1–2 cm beyond the midline (Fig. 2.3).

Different from earlier operations the current approach requires a larger scalp flap, allowing more complete exposure of the temporalis muscle (TM). The TM is no longer transected parallel to its insertion and detached from the bone flap. Incising the TM only at the posterior limit of the craniotomy exposure parallel to its fibers insures its more complete preservation. Anteriorly, the TM is not stripped from the pterion – where the orbit and the middle and anterior cranial fossae will not be opened with a diamond burr in contrast with previous techniques. A standard burr-hole is no longer made at this point. By not detaching the TM at its insertion, a true osteomuscular flap results, maximizing continued TM, and bone plate vascularization and the normal TM tension when the bone flap is replaced and sutured at the end of surgery. This affords superior TM function immediately after surgery and an excellent cosmetic outcome so long as the motor branch of the trigeminal nerve has been preserved. The craniotomy still extends well anteriorly, to the rim of the orbit.

The techniques of unroofing of the orbit anterior to the superior orbital fissuer (SOF) and resecting temporal squama close to the floor of the middle cranial fossa remain unchanged. Epidural exposure of the posterior side of the SOF, as well as the foramen rotundum (FR), the foramen ovale (FO), and the foramen spinosum (FS), are the same as described previously, also transection of the middle meningeal artery (MMA) and everting its proximal stump into the FS [1–4].

But resecting the proximal sphenoid wing, unroofing the optic canal (OC), and resecting the anterior clinoid process (ACP) and the optic strut have been modified. These changes came about when the technique of transecting the doubled dura mater covering neural structures coursing from the PS into the orbit (at the SOF) was adopted. Separating the two layers of the lateral wall of the PS greatly enhance the removal of the ACP opening of the OC and resecting the optic strut without damaging nearby cranial nerves and the ICA. The revised techniques of removing the lateral and superior walls of the OC and the optic strut assure that no tension is exerted on dura in the vicinity of the ICA or the ON intradurally, as may occur with the previous method. One of the primary shortcomings of the former approach was that there was insufficient epidural space to resect the ACP without exerting tension on dura and structures beneath it. For this reason, many neurosurgeons have had reservations about resecting the entire ACP extradurally, especially when carotid-ophthalmic aneurysms are present, for fear of causing premature rupture of the lesion.

By transecting the dural duplication at the lateral end of the SOF, and separating the layers of the lateral wall of the PS, extradural access to the ACP is achieved from infero-lateral side. This results in additional safety since dura in the vicinity of most aneurysms is not disturbed. This is true for both carotid-ophthalmic aneurysms and large and giant distal ICA lesions.

But the major advantage of the current approach occurs from in separating the two layers of the lateral wall of the PS, allowing an entirely epidural approach to all PS lesions. Though extradural in part, the original approach was predominantly intradural. In the course of critically evaluating the former method a number of questions arose. Why approach epidural lesions at all through the intradural space? Is opening the dura truly necessary? Opening the dura may only place anatomical structures at risk which are not directly involved with PS lesions! While the previous approach was primarily lesion-oriented, the current one is equally or even more oriented toward protecting the brain.

There little question that retracting the temporal lobe with a spatula, however padded with rubber or cottonoid patties, for long periods in the operating theater commonly results in contusions which sometimes are not fully appreciated until atrophy occurs. By splitting of the lateral wall of the PS this problem is circumvented. The current technique obviates direct temporal lobe retraction, allowing brain to be gently displaced upward while still enveloped within dura. To date this method has resulted in no cases of surgically related temporal lobe atrophy.

Even resecting the petrous apex is accomplished extradurally, also resulting in less trauma to the temporal lobe. The transdural method frequently required extensive temporal lobe retraction in order to expose the petrous apex, endangering important brain tissue and the vein of Labbé.

In the past, for the purpose of gaining proximal control, the ICA was exposed within the petrous carotid canal in all operations to treat PS tumors, PS-ICA aneurysms, and carotid-ophthalmic aneurysms. With additional experience, however, this technique has been abandoned for trigeminal neurinomas, pituitary tumors extending into the PS, carotid-ophthalmic aneurysms, small intracavernous PS-ICA aneurysms, and tumors within but not filling the PS (lesions less likely to invade the ICA wall). With tumors encasing the ICA, however, particularly when there is preoperative MRI or angiographic evidence of narrowing of the vessel lumen, the ICA must be exposed in the petrous carotid canal as described. As a practical matter, however, no surgeon not fully experienced treating skull base lesions should miss an opportunity to open the petrous carotid canal repeatedly, at least until he is fully adept and skilled with the technique. Furthermore, any surgeon operating on PS lesions involving the ICA will be made more comfortable by the added safety of having the petrous ICA exposed. And finally, despite Glasscock's early suggestion that the Eustachian tube be exenterated [5], we have always unroofed the petrous carotid canal with maximum effort to preserve the tensor tympani muscle, the Eustachian tube, and the GPN.

Conclusions

The current version of the epidural approach to the central skull base (CSB) lesions maximizes exposure by resecting the following bony structures: the roof of the orbit, the lesser sphenoid wing, the entire ACP, the superior and lateral walls of the OC, the optic strut, the roof of the petrous carotid canal, the petrous apex, and the roof of the IAC. This method allows safe exposure of nerves II, III, IV and V, as well as the AL and PL of the ICA without opening either the PS or the intradural space.

The epidural approach, as described here, either in its complete or partial form, should be selected when approaching CSB lesions. Recommendations for appropriately omitting specific elements of the method will be made in subsequent chapters. Only if anticipating limiting the procedure to the biopsy of a PS tumor or trapping a PS-ICA aneurysm or a carotid-parasellar space fistula should one chance a more limited exposure. Compared to our previous method, the current approach is safer, easier to perform, and provides superior surgical results for tumors and vascular lesions in this region. The described approach, in its complete or partial form, represent the first step toward the safe removal of tumors and/or vascular lesions of the PS and adjacent skull base regions.

References

1. Dolenc VV (1983) Direct microsurgical repair of intracavernous vascular lesions. J Neurosurg 58: 824–831
2. Dolenc VV (1985) A combined epi- and subdural direct approach to carotidophthalmic artery aneurysms. J Neurosurg 62: 667–672
3. Dolenc VV (1989) Anatomy and surgery of the cavernous sinus. Springer, Wien New York, pp 1–344
4. Dolenc VV, Škrap M, Šušteršič J, Škrbec M, Morina A (1987) A transcavernous-transsellar approach to the basilar tip aneurysms. Br J Neurosurg 1: 251–259
5. Glasscock ME (1969) Exposure of the intra-petrous portion of the carotid artery. In: Hamberger CA, et al (eds) Disorders of the skull base region. Proceedings of the Tenth Nobel Symposium. Almqvist & Wiksell, Stockholm, pp 135–143
6. Kawase T, Toya S, Shiobara R, Kimura C, Nakajima H (1987) Skull base approaches for meningiomas invading the cavernous sinus. In: Dolenc VV (ed) The cavernous sinus. A multidisciplinary approach to vascular and tumorous lesions. Springer, Wien New York, pp 346–354
7. Lesoin F, Pellerin P, Autricque A, Clarisse J, Jomin M (1987) The direct microsurgical approach to intracavernous tumours. In: Dolenc VV (ed) The cavernous sinus. A multridisciplinary approach to vascular and tumorous lesions. Springer, Wien New York, pp 323–331
8. Parkinson D (1965) A surgical approach to the cavernous portion of the carotid artery. Anatomical studies and case report. J Neurosurg 23: 474–483
9. Parkinson D (1973) Carotid cavernous fistula: direct repair with preservation of the carotid artery. J Neurosurg 38: 99–106
10. Parkinson D (1988) Surgical management of internal carotid artery aneurysms within the cavernous sinus. In: Schmidek HH et al (eds) Operative neurosurgical techniques. Indications, methods and results. Grune & Stratton, Orlando, FL, pp 837–844
11. Pool JL, Potts DG (1965) Aneurysms and arteriovenous anomalies of the brain: diagnosis and treatment. Harper & Row, Nex York, p 307
12. Taptas JN (1982) The so-called cavernous sinus: a review of the controversy and its implications for neurosurgeons. Neurosurgery 11: 712–717

3

Central skull base vascular lesions

3.1

A combined extradural-intradural approach to carotid-ophthalmic aneurysms

Introduction

The precise definition, classification, and ideal treatment of carotid-ophthalmic aneurysms (COAs) have remained controversial since Drake brought attention to them over three decades ago [13]. Generally, COAs are recognized as superomedial lesions of the most proximal intrathecal segment of the ICA, immediately distal to the origin of the ophthalmic artery (OA), well proximal to the origin of the posterior communicating artery (PCom) [5, 8, 10–14, 20, 25, 38–40, 42]. Not only have these lesions been described as "aneurysms of the ophthalmic segment of the ICA" [7], some authors have classified them as "paraclinoid aneurysms" (when they originate from the intradural inferomedial wall of the vessel) [16, 34, 42] and "carotid-cave aneurysms" [24]. Others have lumped them with all aneurysms arising between the OA and the ICA bifurcation [22, 25]. Such protean descriptions and definitions have resulted in a wide range of their reported incidence, varying between 1.5% and 8% of all intracranial aneurysms [13, 20, 25, 29]. Yet most agree that COAs' unique clinical presentation and difficulty of treatment, particularly when they are large or giant, justifies their being considered as a distinct group.

They are as likely to rupture as any of their intracranial brothers, but the proportion of unruptured lesions causing neurological deficits is greater than with any other group of aneurysms. Neurological symptoms and signs of unruptured COAs are quite variable, depending on their location, direction of projection, and, above all, their size.

The original report of direct surgery for COAs changed immediately neurosurgeons' attitudes toward treating them [13]. Because of the high incidence of large and giant aneurysms at this location, their penchant for causing symptoms though unruptured, and, especially, the technical difficulties they present surgeons trying to occlude them, some still favor treating them indirectly, with carotid artery ligation [25, 39, 40]. Others favor attacking them directly, completely excluding them from the cerebral circulation [1, 5, 7, 9, 10, 20, 22, 25, 37, 42, 43]. More recently endovascular techniques, specifically balloon and coil modalities, have challenged the direct surgical approach [21, 28]. Obviously, patients harboring COAs benefit from competition between interventionalists and microvascular surgeons. Since the primary goal of treating these lesions remains complete exclusion from the cerebral circulation while preserving ICA patency without creating new symptoms and signs, it is likely that the (Dallas technique) combined endovascular/surgical method represents the optimal treatment [3, 4].

Relevant anatomy and the definition of the COA

Prior to entering intradural space, the ICA courses through the skull base in loops and rotations along its longitudinal axis. It is interesting that the horizontal segment of the ICA within the parasellar space (PS), between the medial loop (ML), and the anterior loop (AL), is virtually parallel to the intradural ICA, although it is shorter than the intradural segment. The horizontal segment of the PS-ICA loops inferomedial to the anterior clinoid process (ACP) along the lateral wall of the sphenoidal sinus, making a complete U-turn to pierce the dura, the intradural ICA continuing to its terminal bifurcation, the origins of the middle cerebral artery (MCA) and the anterior cerebral artery (ACA). The intradural segment, varying between 1 and 2 cm in length, gives off, first, the OA, then the Dawson's arteries, finally the PCom and the anterior choroidal artery. The OA originates from the superomedial surface of the ICA 1–2 mm inside the dura, then courses upward and anteromedially beneath the optic nerve (ON) into the optic canal (OC). Dawson's arteries, arising inferomedially from the ICA distal to the OA, course toward the optic chiasm, the pituitary stalk (PSt), and the ipsilateral optic tract [6]. More peripherally, the PCom originates laterally from the ICA, and even more distally, the anterior choroidal

Intracranial vascular lesions (aneurysms and AVMs) are divided into two primary groups, those involving the carotid system and those of the vertebrobasilar system. In the following four sections, carotid and vertebrobasilar aneurysms will be discussed with respect to their relationships to the skull base, particularly how they may be treated surgically at the expense of bony structures, minimizing or entirely avoiding brain retraction. An additional section will describe a similar approach to carotid parasellar fistulae (CPF).

Most aneurysms of the internal carotid artery (ICA) in the parasellar space (PS) may be managed without opening the dura, a method affording maximum brain protection. Lesions immediately adjacent to dura, carotid-ophthalmic aneurysms (COA), necessitate dural opening in order to expose the ICA intradurally as well as extradurally. The safest, most efficient means of fully exposing the ICA as it pierces the dura is to incise the distal (dural) ring (DR), possible only after resecting elements of the central skull base (CSB), including the sphenoid wing (SW), anterior clinoid process (ACP), and the walls of the optic canal (OC). An extended approach – additionally resecting the posterior clinoid process (PCP) and dorsum sellae – facilitates exposure of distal BA aneurysms. Isolating the ICA in the petrous carotid canal provides the best means of obtaining proximal control of ICA flow, which is often necessary with PS-ICA and COA aneurysms and ICA parasellar fistulae (CPF). Resecting the petrous apex gives the surgeon excellent access to the posterior cranial fossa and basilar trunk aneurysms. Each of these lesions presented enormous difficulty prior to the advent of skull base techniques, brain injury virtually unavoidable in many instances. By approaching them through the skull base, brain retraction becomes required only occasionally if at all. The techniques to be described also facilitate exposures of large and giant aneurysms at more peripheral sites – the distal ICA, middle cerebral artery (MCA), anterior cerebral artery (ACA), and anterior communicating artery (ACom) – affording a more direct approach to their necks, circumventing cerebral tissue manipulations necessitated by approaching them from the brain's convexity.

The skull base method requires strict adherence to techniques detailed in the sections that follow, each a logical extension of the general extradural approach method described in Chapter 2. Focus must be directed on protecting the brain as much as excluding the lesion. Among the treatment modalities currently available for CSB vascular lesions, direct surgery continues to play a vital role. Whether this remains to be true in the future depends on neurosurgeons' striving for still improved methods, allowing even improved treatment results for these difficult lesions.

artery, also from the lateral surface of the vessel. The literature is replete with anatomical studies of the proximal intradural ICA and its branches [15, 18, 23, 27, 33, 35, 36].

Dura surrounding the ICA, the so-called "distal (dural) ring" (DR), is the primary landmark for classifying intracranial ICA aneurysms, dividing them into intradural and extradural lesions [26]. The dura, by virtue of its surgical significance, seems superior to the OA for classifying COAs. While a radiographic classification of such lesions by their relationship to the OA may seem simple, it carries little significance for the surgeon, especially in cases where the OA arises extradurally. This is reminiscent of the shortcomings of other classification systems currently in vogue, where the terms, "paraclinoid aneurysms", "carotid cave aneurysms of the ICA", "aneurysms of the ophthalmic segment", etc., result in more confusion than clarity.

Therefore, we propose defining as COAs all aneurysms located on the proximal segment of the intradural ICA, that is, from the DR proximally to an imaginary circle at the level of the origin of the PCom. COAs may be subdivided further with respect to their relationship to the circumference of the ICA (medial, inferior, superior or lateral), their size, also their shapes [12]. Most small COAs arise from the medial half of the ICA and remain lateral and inferior to the ON. Only when they become larger do they project beneath or above the optic apparatus. Such lesions are divided into supraoptochiasmatic and suboptochiasmatic subgroups (Fig. 3.1).

Those cases where the neck of the aneurysm is quite broad, leaving less than one-half of the ICA circumference intact, should be classified as "global" COAs, lesions far more likely to expand beyond the level of the PCom. Although an intradural COA may expand extradurally, such occurrences remain exceptional. In such cases the OA originates from the aneurysm itself and at least the medial part of the DR is incorporated into the lesion. Obviously some COAs defy almost any scheme of classification. It is only pragmatic to elect a system into which the vast majority of these lesions fit.

Pathological anatomy, symptoms and signs, and diagnostic work-up

Typically, unless they rupture and cause subarachnoid hemorrhage (SAH), small COAs remain asymptomatic. In addition to SAH, small unruptured COAs may create visual deficits when the ON is injured by a jet of escaping blood (Fig. 3.2), circumstances to be distinguished from progressive visual deficits resulting from visual pathway compression by large or giant COAs (Figs. 3.3–3.6). The type and extent of visual field deficits generally correlates with the size and location of the aneurysm. Suboptochiasmatic and global aneurysms have a greater effect on visual structures than do supraoptochiasmatic lesions. Depending on the size of a suboptochiasmatic COA, the PSt, optic chiasm, ipsilateral optic tract, hypothalamus, or even the upper brainstem may be affected. Symptoms and signs related to brainstem and pyramidal tract(s) compression are common with giant COAs. Large and giant supraoptochiasmatic COAs may even present with epileptic seizures Fig. 3.7(a, b, c, d). With any partially thrombosed aneurysm, aneurysm-to-artery embolism may occur, resulting in transient ischemic attacks (TIAs), even strokes.

When their mode of presentation is SAH, the diagnostic evaluation of small COAs is no different from other intradural aneurysms. But large and giant COAs, producing symptoms and signs by mass effect, – Figs. 3.3(a and c), 3.4a, 3.7(a and c), 3.8(a–c), and 3.9(a and c) – uniformly require precise ophthalmological assessment as part of their work up. Resulting visual field deficits commonly lead to CTs which typically reveal spherical "tumors" ipsilateral to the deficit. MRI and cross-circulation studies are mandatory in such cases. CTA and DSA are also helpful, characterizing further the nature, location, type, and size of the lesion, as well as its relationship to adjacent structures Fig. 3.8(a–c). Where multiple aneurysms exist, one of which is a COA – whether small or large – and no reliable clue points to the bleeding lesion, it is prudent to design a surgical approach which allows both (all) aneurysms to be clipped at the same operation. This is also true in cases where COAs are bilateral Fig. 3.6(a–d).

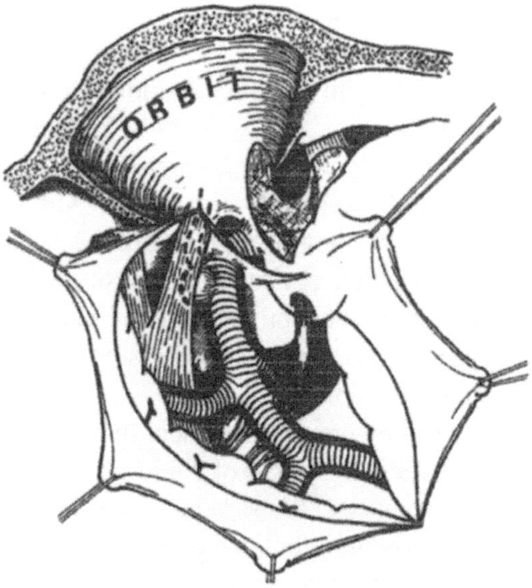

a

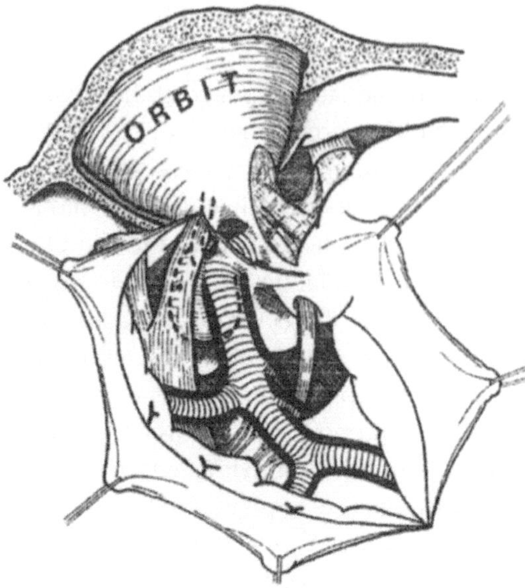

b

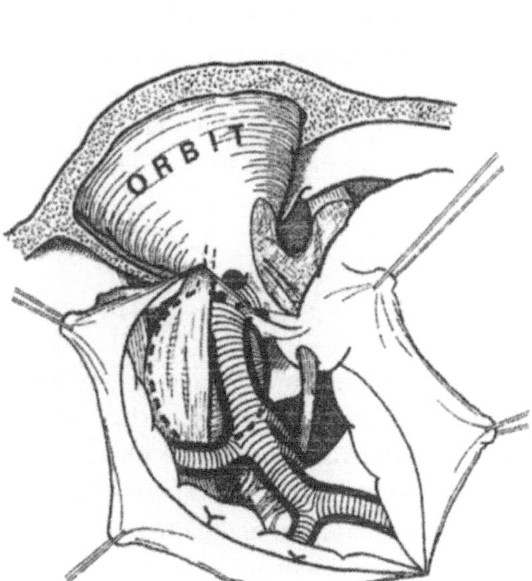

c

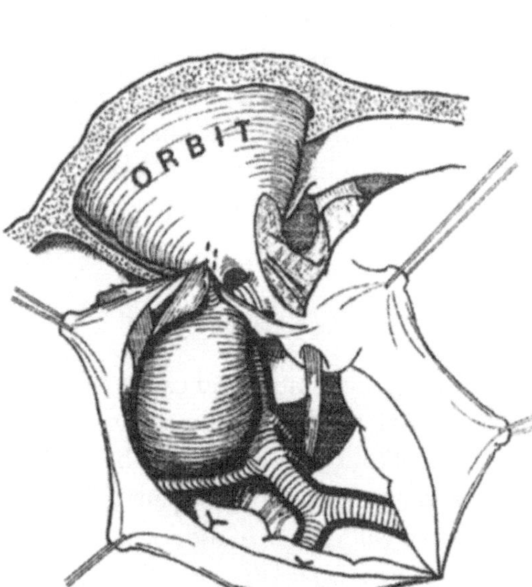

d

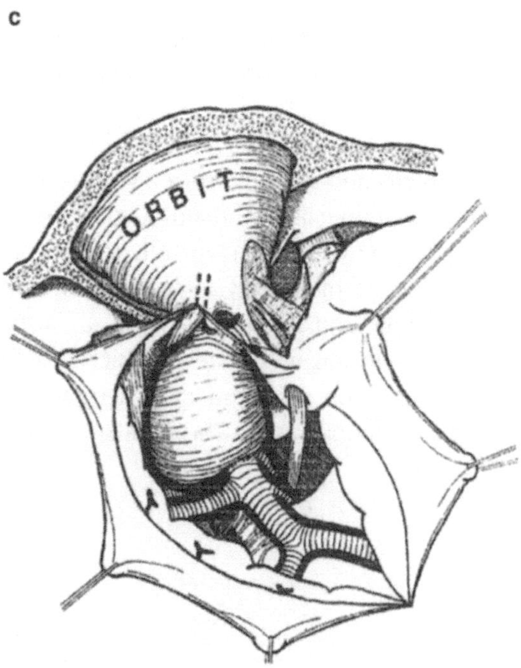

e

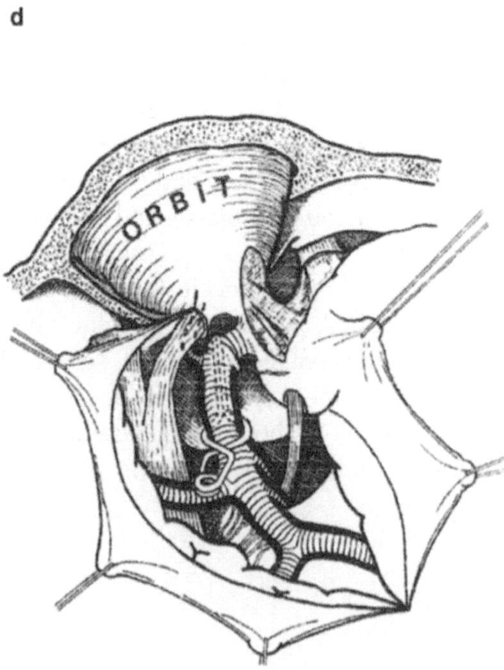

f

Fig. 3.1. Types of carotid ophthalmic aneurysms (*COAs*). (a) A small COA at the origin of the ophthalmic artery (*OA*) is generally inferolateral to the optic nerve (*ON*). Point of origin from the ICA and direction of growth dictate the types of larger aneurysms. (b) A suboptochiasmatic COA, projecting medially and inferiorly, remains above the diaphragm sellae and beneath the ICA. This type is also known as a "carotid cave" aneurysm. (c) Another suboptochiasmatic COA, this one projecting inferiorly and posteriorly, possibly reaching the hypothalamus and upper brainstem causing endocrinological, visual, and long tract deficits. (d) A supraoptochiasmatic COA, projecting posteromedially over the visual apparatus and anterior cerebral artery (*ACA1*), which, if large or giant, may compress the brain. (e) Global COAs (suboptochiasmatic) usually displace the visual apparatus medially and upward, stretching the ipsilateral ON and optic tract. The posterior communicating artery (*PCom*) and anterior choroidal artery may be firmly adherent to its lateral wall, and nerve III may be stretched and displaced laterally. (f) A suboptochiasmatic COA clipped with a fenestrated clip. The distal (dural) ring (*DR*) has been incised circumferentially around the ICA and the OA exposed. The COA has been transected, its now empty sac no longer compressing the visual apparatus and pituitary stalk (*PSt*). Because the ICA has been released by incising the DR and the anterior loop (*AL*) dissected from the lateral wall of the sphenoidal sinus, the position of the ICA as it enters the dura is preserved. (a–f) The completed epidural dissection: the orbit has been unroofed, the sphenoid wing and the ACP resected, along with the superior, lateral, and inferior (optic strut) walls of the optic canal. The anteromedial triangle is wide open. The dural layer of the anterior portion of the lateral wall of the parasellar space (*PS*) has been peeled away, exposing nerves III, IV, V1, V2. The dura has been opened along the longitudinal axis of the sylvian fissure, exposing the MCA, the bifurcation of the ICA (origin of ACA1 and MCA), the anterior choroidal artery, the PCom, nerve III, and the ON. (b–e) Depict four types of COAs

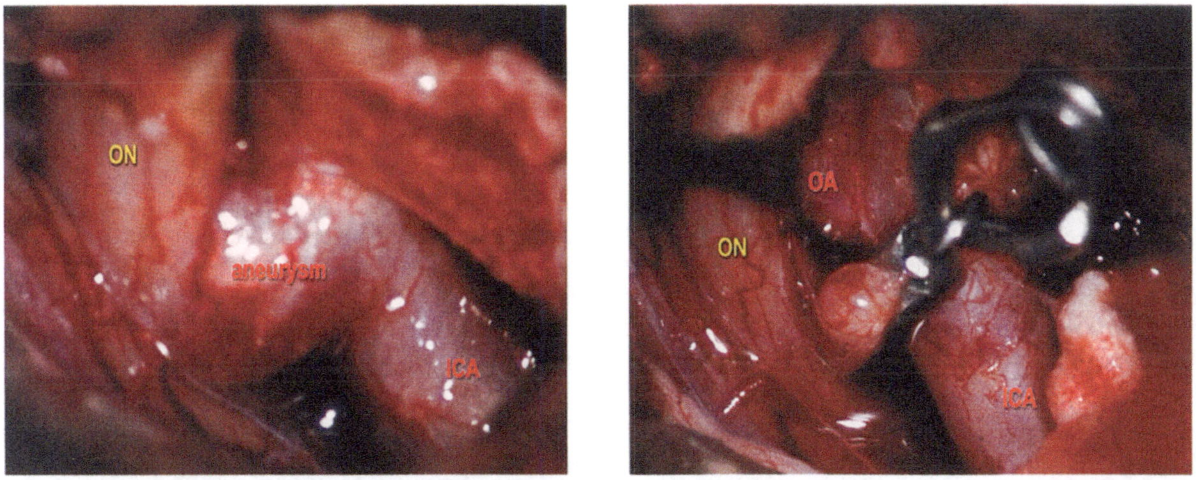

Fig. 3.2. Intraoperative photographs of a small right carotid-ophthalmic aneurysm (*COA*) bulging into the optic nerve (*ON*) causing visual loss in the right eye, (a) before clipping, and (b) after clipping. The patient had sustained a subarachnoid hemorrhage (*SAH*), but DSA failed to determine the cause of bleeding. This ruptured aneurysm was virtually filled with clot, complicating its preoperative detection

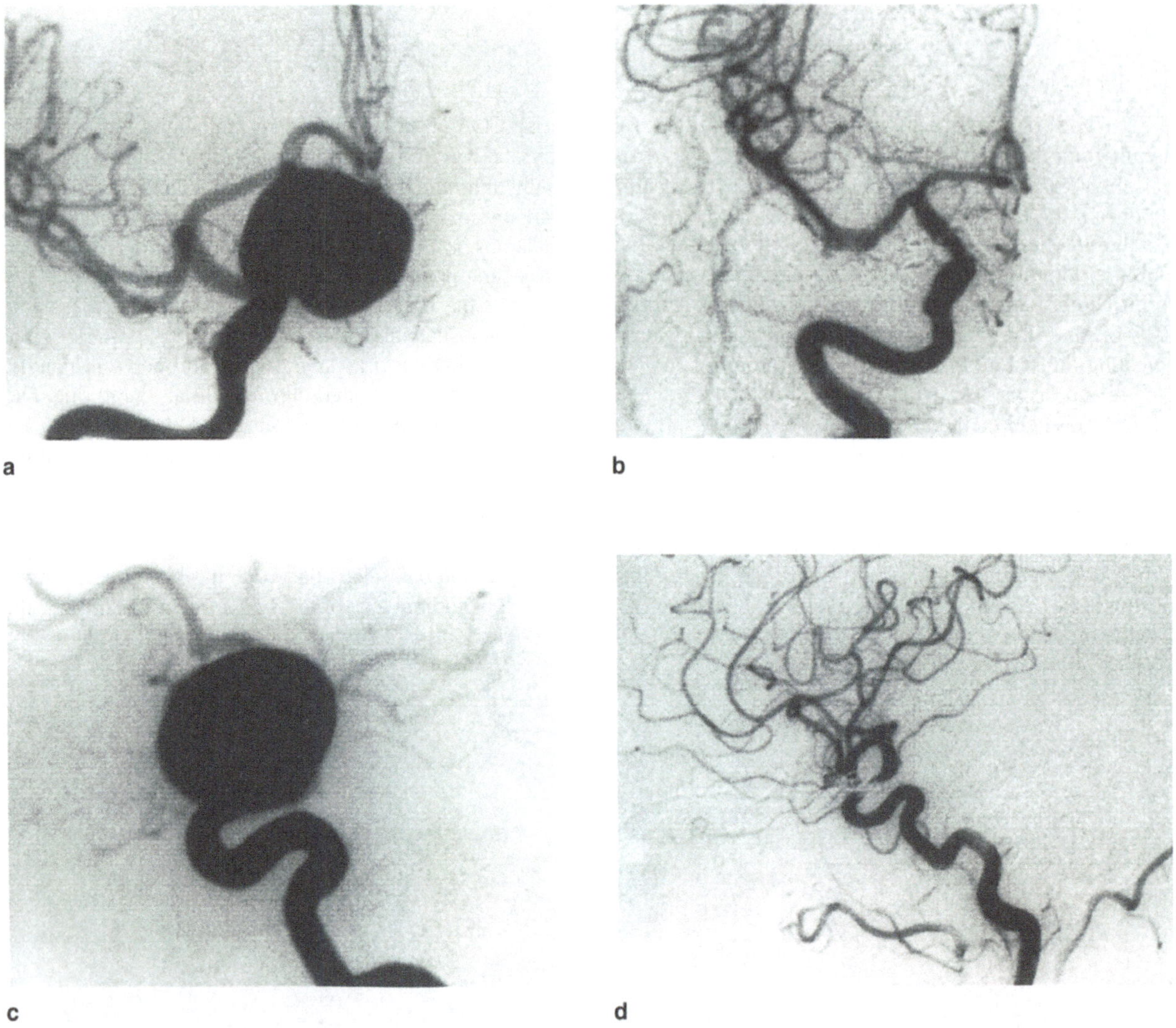

Fig. 3.3. A preoperative right carotid angiogram, AP (a) and lateral (c) views, reveals a large carotid-ophthalmic aneurysm (*COA*) with a relatively small neck. The postoperative study, AP view (b) shows the ICA after the aneurysm has been resected and the ICA reconstructed with sutures. The suture line was reinforced with a fenestrated rectangular clip. The postoperative lateral view (d) reveals relative stenosis of the ophthalmic segment of the (reconstructed) ICA and normal vessels distally. Note that the ophthalmic artery (*OA*) is preserved, and no vasospasm is present

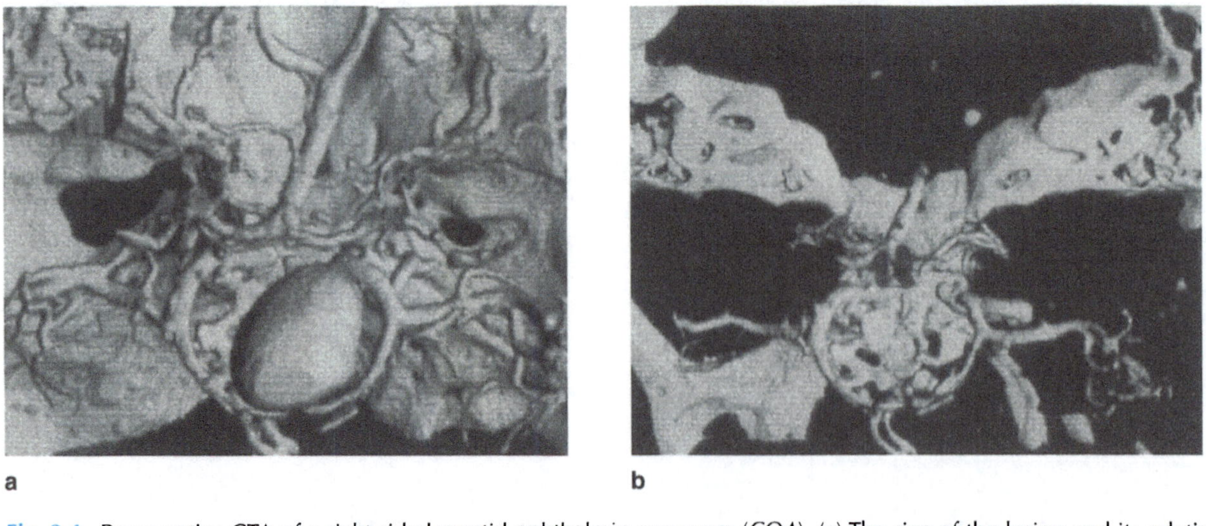

Fig. 3.4. Preoperative CTA of a right-sided carotid-ophthalmic aneurysm (*COA*). (a) The size of the lesion and its relationship to arteries of the circle of Willis and bony structures of the central skull base (*CSB*) are demonstrated. The clinical picture was consistent with displacement of the visual apparatus, pituitary stalk (*PSt*), and the brain. Postoperative CTA (b) confirms the absence of ICA stenosis adjacent to the clip. Note: With large COAs, as presented here, it is quite hazardous to attempt to mobilize the aneurysm before excluding it from the cerebral circulation. It is safer and simpler to expose the anterior loop (*AL*) of the ICA, place temporary clips proximal and distal to the COA, then open the aneurysm, empty it, and reconstruct the ICA wall, either with sutures or by clipping it, or both

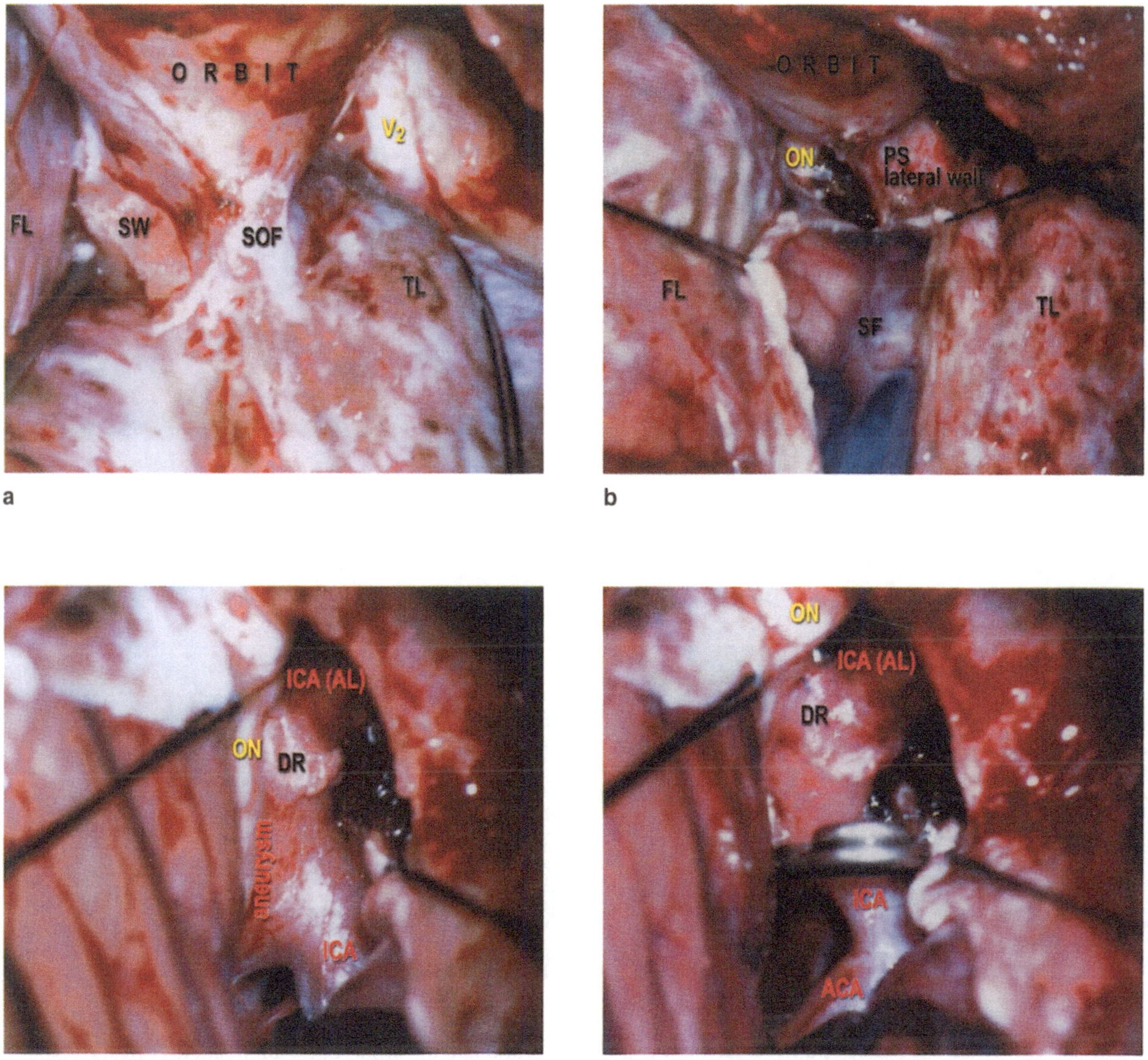

Fig. 3.5. Intraoperative photographs of the approach to the right carotid-ophthalmic aneurysm (*COA*) aneurysm presented in Fig. 3.4. In (a), after resecting the roof of the orbit and much of the sphenoid wing (*SW*), the medial limit of the SW and the anterior clinoid process (*ACP*) remain intact. In (b), after resection of the ACP and dissection of the anterior part of outer layer of the parasellar space (*PS*) lateral wall, the optic canal (*OC*) lies open. Dura has been incised along the longitudinal axis of the sylvian fissure, and the optic nerve (*ON*), the medial border of the anterior part of the PS (nerve III still covered by the PR), and the dura at the base of the anteromedial triangle are seen. Surgicel lines the depths of the triangle. (c) This photograph shows the peripheral part of the anterior loop (*AL*) of the ICA, the distal (dural) ring (*DR*) (incised circumferentially around the ICA), the lateral margin of the right intradural ON, the intradural ICA, and the wide neck of the COA projecting beneath the ON. The proximal ring (*PR*) of the AL covers nerve III, and Surgicel has been placed behind the AL in the anteromedial triangle. (d) A fenestrated clip occludes the neck of the COA, showing also the intradural ICA, including the initial segment of anterior cerebral artery (*ACA1*) and the AL of the extradural ICA, the PR covering nerve III, and the remaining DR proximal to the origin of the aneurysm. Note: The position of the ICA is unchanged and the external diameter of the ICA adjacent to the COA is appropriate

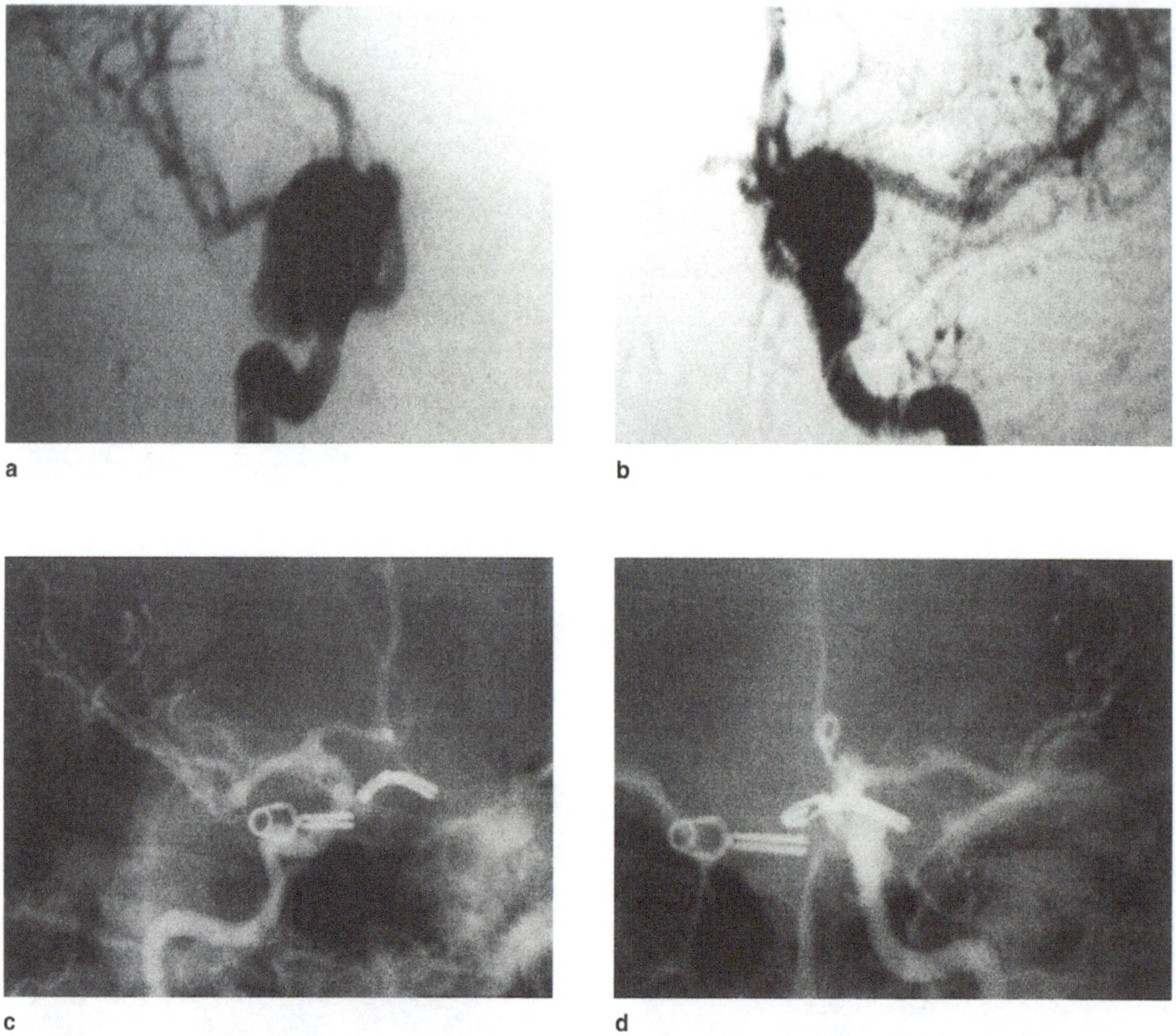

Fig. 3.6. AP views of from a preoperative angiogram (a and b) reveal bilateral carotid-ophthalmic aneurysms (*COAs*). The larger lesion resulted in ipsilateral failing vision, while the COA on the left created no symptoms. Similar views from a postoperative angiogram (c) and (d) confirm total clipping of both lesions. The orientation of the clips indicate that both aneurysms were treated via a right-sided pterional approach

Surgical approach to the carotid-ophthalmic aneurysms

The surgical approach to COAs described initially [10] and subsequently updated [11] proved, in cases of large and giant COAs, to be a demanding procedure, one not generally accepted by neurosurgeons. Some even classified it as "hazardous" [31]. Additional laboratory studies and further surgical experience led to a refinement of the transorbital, transclinoid, and transsylvian approach to COAs [12].

The current approach to COAs is based on the techniques described in detail in Chapter 2, including positioning the patient; performing an osteoplastic craniotomy; unroofing the orbit; transecting the duplicated dura at the lateral end of the superior orbital fissure (SOF), facilitating separation of the outer dural layer of the lateral wall of the PS from its inner layer and nerves III, IV, V1 and V2; and completely resecting the ACP and opening the OC by resecting its superior and lateral walls and the optic strut, thus exposing the peripheral half of the AL of the ICA. It is extremely important that no tension be exerted on dura during the extradural removal of the ACP, and that the ACP is drilled from its infero-lateral surface. It is equally important that, while removing the walls of the OC, all drilling is accomplished with a diamond burr at short intervals beneath continuous saline irrigation and that the thickness of remaining bone near the ON is examined repeatedly. In order to avoid injuring the ON, drilling is performed by gently "rubbing" OC bone rather than boring through it. This technique requires both patience and practice.

Dura is opened at the periphery of the sylvian fissure. It is incised medialward until the aneurysm is visualized. Retracting the frontal or temporal lobes is neither necessary nor recommended since this may result in premature rupture of a large or giant COA as demonstrated in Figs. 3.4(a and b) and 3.5(a–d). The sylvian fissure is split along the entire length of the MCA and the ICA bifurcation is exposed. It is very important to expose the anterior choroidal artery, nerve III and, if possible, the PCom. Only when both the distal ICA – including the anterior choroidal artery – and the extradural ICA (its AL) are exposed sufficiently for temporary clipping that dura should be incised circumferentially around the ICA. If the AL cannot be exposed sufficiently for the placement of a temporary clip proximal to the aneurysm, the ICA should be exposed in the petrous canal [19] or in the neck. The DR should be incised circumferentially around the ICA, providing good exposure of the OA and "mobility" of the ICA. Any venous bleeding from the PS infero-medial to the ICA is controlled with Surgicel. The ON may be lifted, tacking its dura propria with a stitch, in order to gain additional space superior and medial to the COA. The OA is then dissected from the aneurysm and followed into the OC. Lifting the ON also facilitates dissecting the aneurysm from the ON, the PSt, and other adjacent structures.

If the COA is partially thrombosed, handling it, while not likely to result in its rupture, runs the serious risk of releasing an intra-aneurysmal embolus into the bloodstream Figs. 3.6(a–d) and 3.7(a–d). In such cases it is mandatory to temporarily occlude the ICA by applying clips proximally and distally to the aneurysm – but proximal to the anterior choroidal artery. Where the PCom has already been dissected, it should be occluded with an additional small temporary clip placed close to the ICA, taking special care to avoid compromising any perforators arising from the PCom. At this point, the aneurysm may be incised and the thrombus removed, finally irrigating the aneurysmal sac free of debris and fine bits of clot, then clipping the neck of the lesion without constricting the lumen of the ICA and assuring continued patency of the ICA and the OA.

Where the OA arises directly from the lesion, it is clipped together with the aneurysm. Only if it arises from the COA itself can the OA not be preserved. If excessive thickness and/or calcification of the wall at the neck of the aneurysm precludes clipping, the sac is transected and the ICA wall is reconstructed with sutures. If this is necessary, it is extremely important to provide an ample internal diameter of the ICA. The same technique applies when the neck is too broad to be occluded with one or more clips. Temporary occlusion of the ICA should be maintained no longer than is absolutely necessary. Whether the ICA wall is reconstructed by clip(s) or by sutures, adjacent neural structures are inspected critically at this point and dissected further if appropriate. It is crucial that the sacs of large or giant aneurysms not be displaced from the PSt, chiasm, optic tract(s), hypothalamus, and/or the upper brainstem. Mobilizing the fundus of the sac risks damaging important perforators and may result in serious consequences. In such cases the empty sac should be left in place Figs. 3.4(a and b), 3.8(a–d), and 3.11(a–d).

After treating the aneurysm, the dura is closed. Since the DR has been incised circumferentially, a small piece of galea or temporalis muscle or fascia should be placed around the ICA and coated with fibrin glue on its epidural surface. A drain is left in the epidural space, and the osteomuscular flap is replaced, secured with sutures, and the wound closed in layers. In cases where any paranasal sinus may have been entered, continuous lumbar cerebrospinal fluid (CSF) drainage is maintained for at least the first postoperative week.

Our experience with this ipsilateral approach to unilateral COAs leads us to believe that the interhemispheric [2] and contralateral pterional methods [17, 30, 32, 41] are seldom advantageous for most COAs.

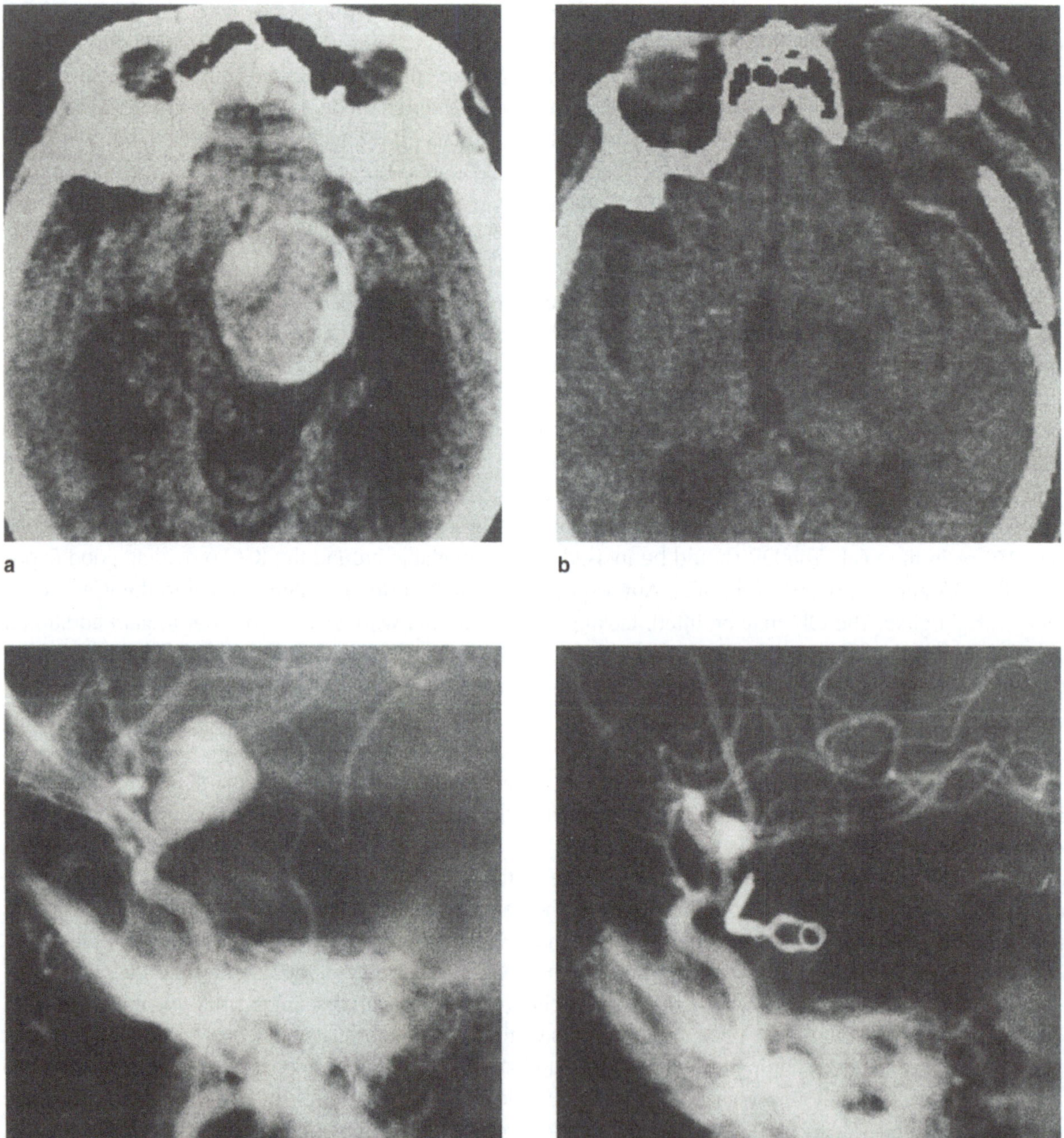

Fig. 3.7. (a) A trans-axial cut from a preoperative unenhanced CT scan reveals a round, marginally calcified lesion on the left side. (b) A corresponding cut from a postoperative unenhanced CT scan reveals a low density lesion in the left frontal lobe which was not seen on the preoperative study. (c) Preoperative left carotid angiography demonstrates an irregularly shaped, medium-sized carotid-ophthalmic aneurysm (*COA*) with a narrow neck arising from the proximal intradural ICA. (d) A corresponding image from the postoperative angiogram confirms ICA patency and absence of the aneurysm

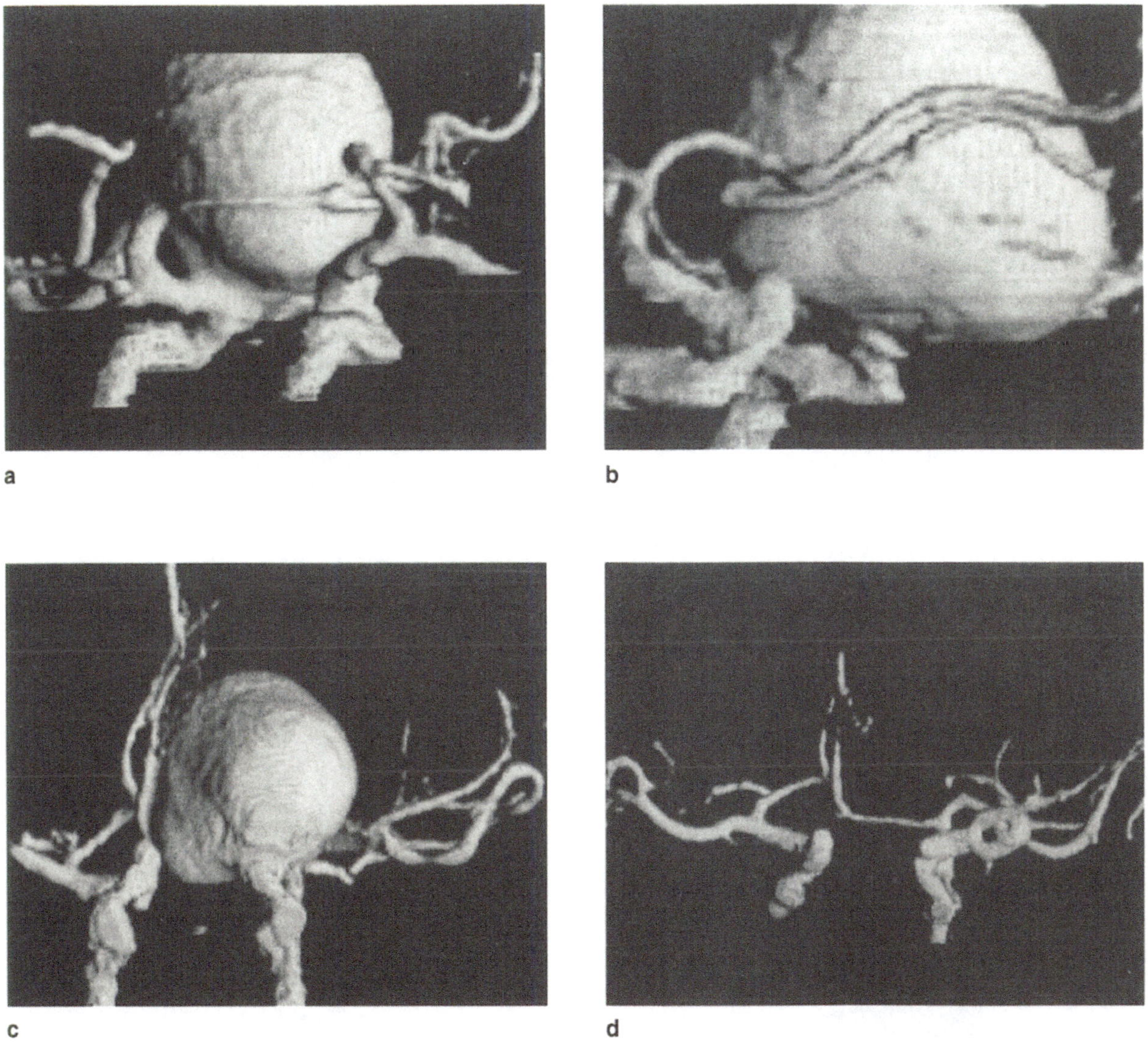

Fig. 3.8. Different views from a preoperative CTA of a left-sided giant supraoptochiasmatic carotid-ophthalmic aneurysm (*COA*), (a–c). Note the size and projection of the lesion and its relationship with the left ICA and both ACAs. The postoperative CTA (d) confirms its complete exclusion from the cerebral circulation while maintaining patency of the ICA. The clip lies adjacent to the lesion's site of origin from the ICA

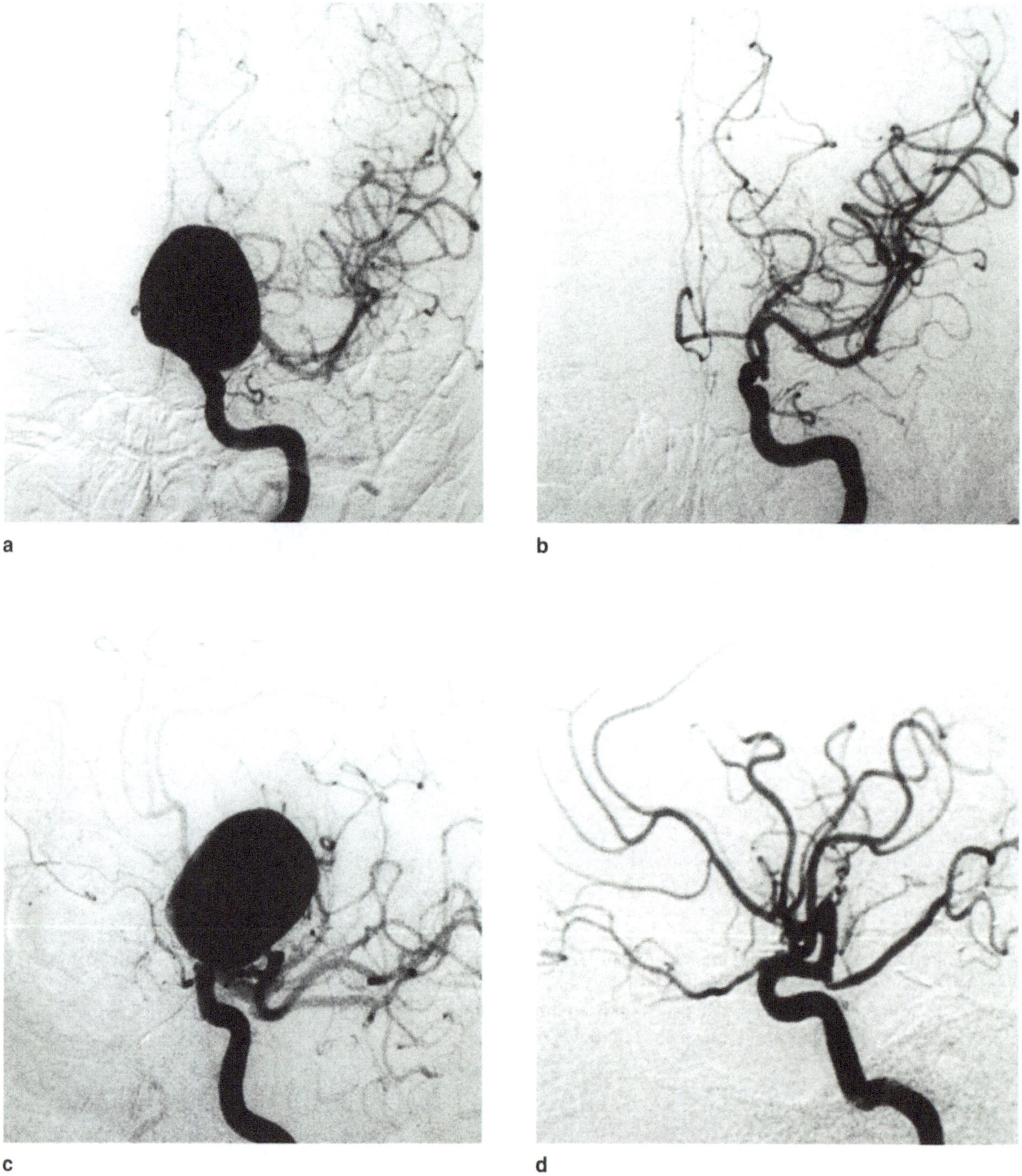

Fig. 3.9. Preoperative left-carotid angiogram, AP (a) and lateral (c) views, of the giant carotid-ophthalmic aneurysm (*COA*) shown in Fig. 3.8. Note: The lateral view confirms that the aneurysm was supraoptochiasmatic and that its extension from the ICA is relatively narrow. The postoperative study, AP view (b), shows the ICA after the aneurysm's neck has been clipped and its sac transsected. A postoperative lateral view (d) reveals a small irregularity of the proximal ICA wall, immediately distal to the ophthalmic artery (*OA*)

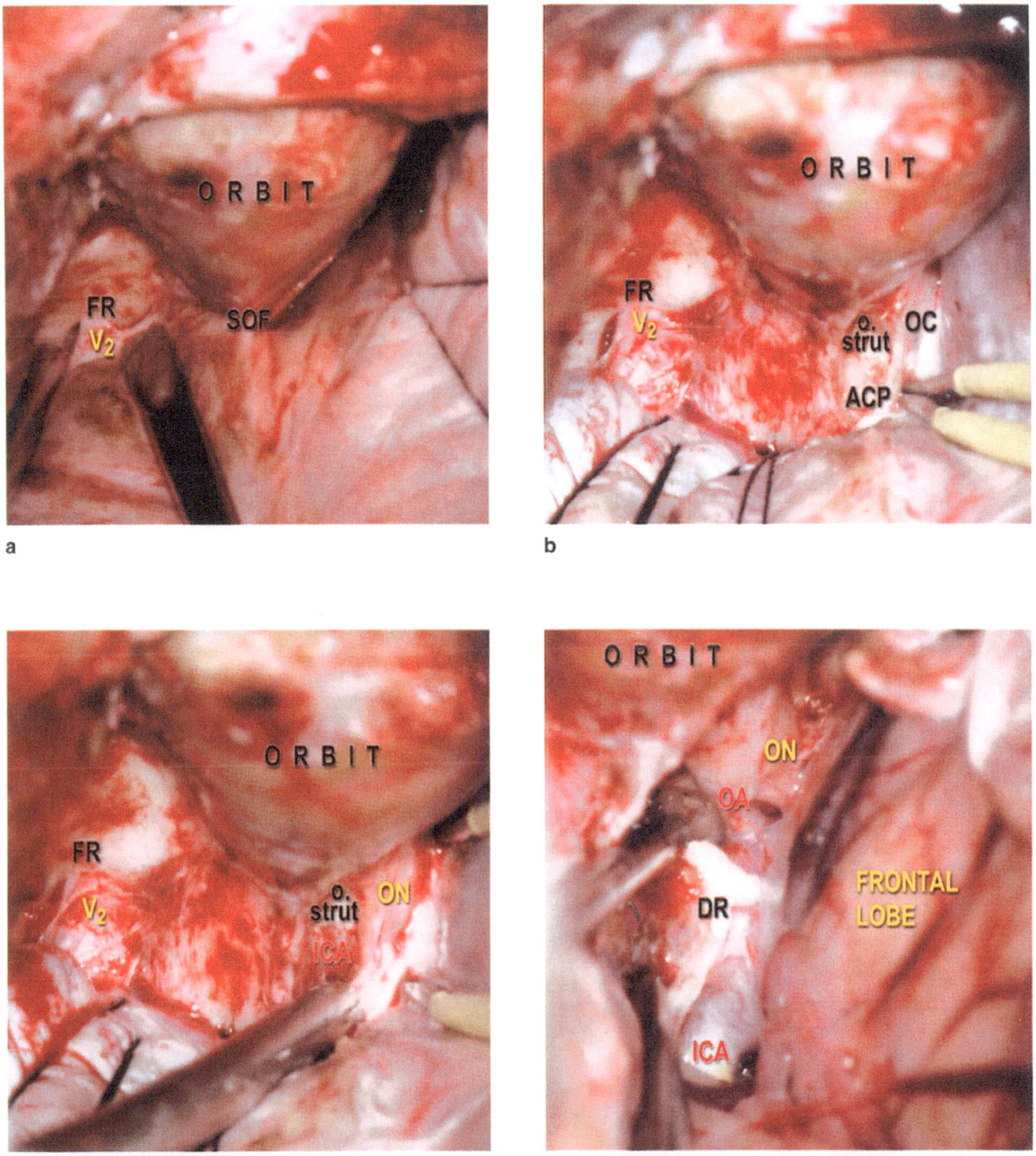

Fig. 3.10. Intraoperative photographs of the extradural approach to the giant carotid-ophthalmic aneurysm demonstrated by CTA (Fig. 3.8) and DSA (Fig. 3.9). (a) After partially unroofing the orbit, dura traversing the superior orbital fissure (*SOF*) has been exposed from its posterior surface. The orbital roof anterior to the SOF remains intact, as does the medial surface of the SOF. The foramen rotundum (*FR*) is apparent, and V2 remains covered by dura in its canal. In (b), the outer (dural) layer of the parasellar space (*PS*) lateral wall has been opened and fixed with stay sutures. The orbital roof anterior to the SOF has been removed to the peripheral limit of the optic canal (*OC*). The sphenoid wing and part of the anterior clinoid process (*ACP*) have been removed while the walls of the OC remain and the lateral PS wall split by opening its dural layer, extending anterolaterally to where V2 nerve enters the foramen rotundum (*FR*). Note: The forceps point to the superomedial surface of the left ACP, where the falciform ligament begins. (c) Further epidural dissection and additional resection of the ACP have been accomplished. The optic strut and the tip of the ACP remain, and the lateral surface of the anterior loop (*AL*) of the ICA is exposed. The posterior and lateral walls of the OC have been removed, and the optic nerve is apparent. (d) Dura lies open along the longitudinal axis of the sylvian fissure, centralward to the distal (dural) ring (*DR*), then medialward over the optic nerve (*ON*). Atrophic left ON and the ICA distal to the aneurysm are shown, with the ophthalmic artery (*OA*) coursing beneath the ON. The base of the neck of the aneurysm has been exposed, but the bulk of the lesion remains beneath the frontal lobe. The optic strut has been removed, and the upper suction tube points to the lateral wall of the sphenoidal sinus

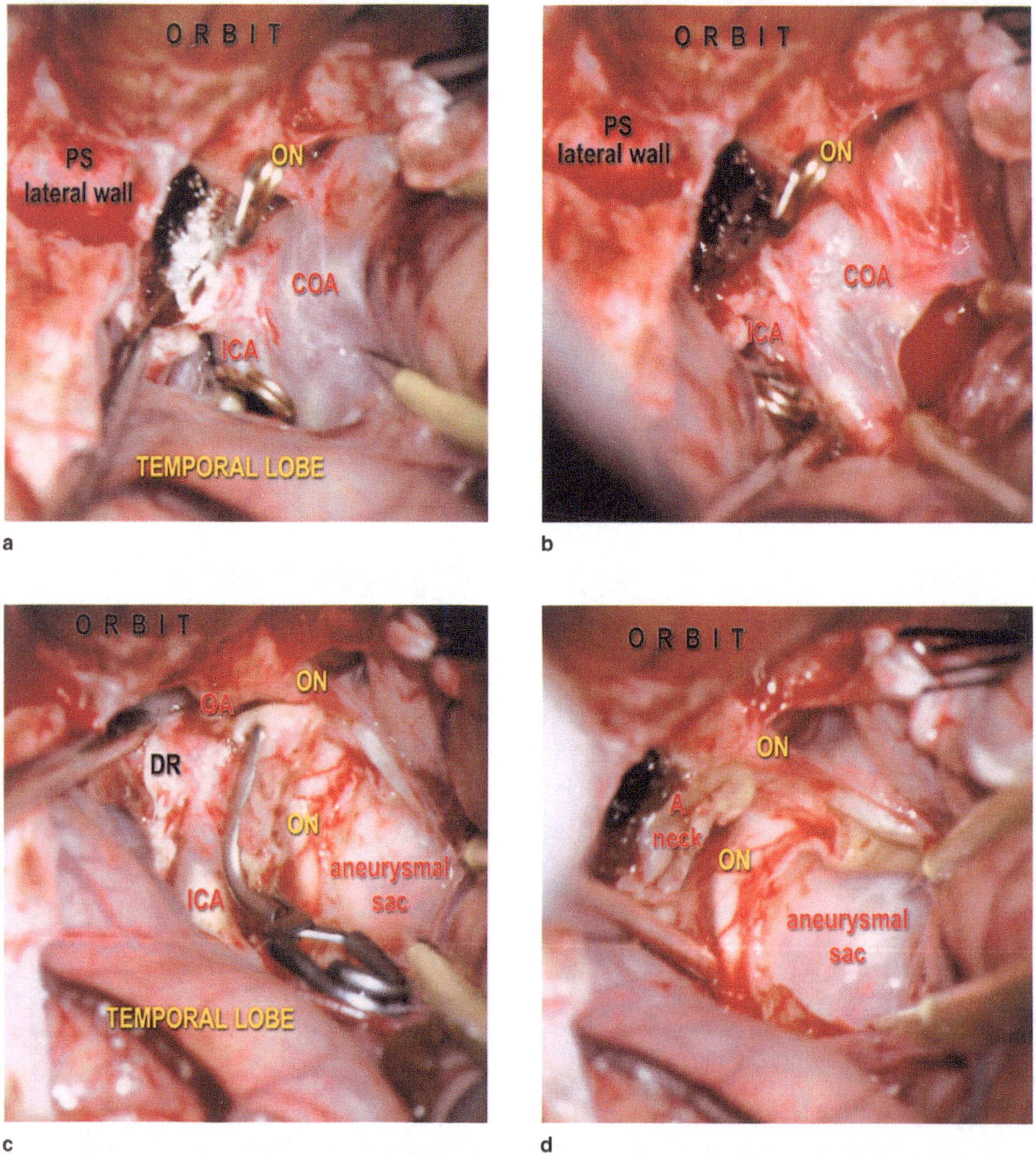

Fig. 3.11. The intradural component of the COA operation presented in Figs. 3.8(a–c) and 3.9(a and c). (a) The anterior loop of the internal carotid artery (*ICA*) lies exposed within the anteromedial triangle, with a temporary clip placed on the extradural ICA proximal to the aneurysm. Surgicel covering the AL of the ICA was placed to control oozing from the parasellar space (*PS*). A second temporary clip has been placed on the ICA distal to the aneurysm but proximal to the anterior choroidal artery. Part of the aneurysm is apparent between the clips. Note: The optic nerve (*ON*) is densely adherent to the anterior surface aneurysm. In (b), the aneurysm has been opened and retrograde ophthalmic artery (*OA*) bleeding complicates dissection of the ON and transsection of the aneurysmal sac. In (c), both temporary clips have been removed and the sac of the aneurysm has been transected and its neck clipped. The ICA remains patent and the OA preserved in a field that is now bloodless. (d) Rotating the clip laterally (away from the aneurysmal sac and the ON) better demonstrates the interior of the sac of the aneurysm. The ON remains adherent to the collapsed aneurysm, its peripheral segment obviously atrophic while proximal elements (prechiasmal and adjacent optic tract) appear more healthy. Notice particularly the vascular supply of the visual apparatus. The sac of a large/giant COA should *never* be dissected from its bed! In this case, the ON was not dissected from the aneurysmal sac, leading to an excellent postoperative result, visual function of the left eye remaining unchanged from prior to surgery

Clinical results

Over a period of two decades 264 consecutive patients with 269 COAs were treated. Results are summarized in Table 1. In five cases, COAs were bilateral, and in 21 cases one or more associated aneurysms were present. Of the 269 COAs, 173 were small, 71 large, and 25 giant. Preoperative bleeding occurred in 161 lesions, while 108 were discovered either incidentally or from neurological deficits resulting from aneurysmal compression of adjacent structures. Visual deficits were present in all patients with large or giant COAs, and in 26 of the patients with small lesions.

Of the entire series, three patients died following surgery. In addition, three had endocrinological deficits, three were hemiparetic, and 57 had visual deficits that were unchanged after surgery in comparison to preoperative deficits. Seventy-two manifest improved vision after surgery. Of all 264 patients, 196 were free of neurological deficits postoperatively, and all 57 with unchanged visual deficits were able to resume their former life-styles. In patients with COA(s) only, vasospasm did not occur. Vasospasm was observed in 20 of 23 patients with multiple aneurysms where SAH had occurred.

Table 1. Presentation of carotid-ophthalmic aneurysms, type of aneurysm exclusion, and postoperative results

Size and presentation of COA			Clipping	Aneurysm resection and reconstruction of the ICA	Patency of the ICA in follow-up angiography	New postoperative deficits	Fatal outcome	Visual function (affected preoperatively)	
								Unchanged postop.	Improved postop.
Small 173	rupture	141	141	–	141	–	–	4	15
	incidental finding	32	32	–	32	–		9	8
Large 71	rupture	14	11	3	13	hemiparesis: 1	1	6	7
	mass effect:	57	36	21	55	endocrinological disorders: 1	–	22	35
Giant 25	rupture	6	3	3	5	hemiparesis: 1 endocrinological disorders: 1	2	4	–
	mass effect	19	4	15	16	hemiparesis: 1 endocrinological disorders: 1	–	12	7
Total 269	161 108		227	42	262	6	3	57	72

Conclusions

Directly treating COAs with a mortality of 1.1%, plus a surgical-morbidity of slightly greater than 2%, would seem to leave little place for indirect treatment therapies such as common carotid artery ligation [25, 39, 40]. Despite controversy regarding the management of incidentally found large and giant aneurysms causing visual impairment, and even endocrinological disturbances, direct surgical occlusion should be considered the treatment of choice for patients whose general medical condition permits craniotomy.

Balloon occlusion or coiling of a large or giant COA – or other lesion with a particularly broad neck – is by no means an ideal alternative to direct surgical treatment. With endovascular techniques it is difficult to preserve ICA patency in cases of COAs with broad necks. Even where this is possible, the mass of the aneurysm remains. Furthermore, unless the lumen of the aneurysm is totally occluded, the risk of repeated emboli remains.

The results of this transorbital-transclinoid-transsylvian (3T) method indicate that surgeons familiar with skull base anatomy and willing to spend more time executing the approach than actually dissecting the aneurysm, can directly exclude COAs from the cerebral circulation – operations once so demanding as to be considered prohibitive – with minimal risk of incurring ischemic and other neurological deficits. Furthermore, preserving and decompressing the visual structures provides conditions for visual improvement over time, which may not be obvious immediately after surgery. It is clear that the risk of postoperative worsening of the visual function should be eliminated.

Contemporary treatment of COAs reflects not only what is possible by microsurgical techniques but also the overwhelming promise of interventional neuroradiology. With COAs there is room for expertise in both disciplines. The opportunity for collaboration is not only theoretical but highly practical.

References

1. Almeida G (1976) Carotid-ophthalmic aneurysms. Surg Neurol 5: 41–45
2. Aoki N (1987) Interhemispheric approach for carotid-ophthalmic artery aneurysm clipping. Case report. J Neurosurg 67: 293–295
3. Bajter HH, Samson DS (1990). Retrograde suction decompression of giant paraclinoidal aneurysms. Technical note. J Neurosurg 73: 305–306
4. Bajter HH Kopitnik Ta, Giller CA, Samson DS (1994) Surgery for paraclinoidal carotid artery aneurysms. J Neurosurg 80: 650–658
5. Benedetti A, Curri D (1977) Direct attack on carotid ophthalmic and large internal carotid aneurysm. Surg Neurol 8: 49–54
6. Dawson BH (1958) The blood vessels of the human optic chiasm and their relation to those of the hyophysis and hypothalamus. Brain 81: 207–217
7. Day AL (1990) Aneurysms of the ophthalmic segment. J Neurosurg 72: 677–691
8. Diraz A, Kyoshima K, Kobayashi S (1993) Dorsal internal carotid artery aneurysm: classification, pathogenesis, and surgical considerations. Neurosurg Rev 16: 197–204
9. Dolenc VV (1983) Direct microsurgical repair of intracavernous vascuar lesions. J Neurosurg 58: 824–831
10. Dolenc VV (1985) A combined epi- and subdural direct approach to carotidophthalmic artery aneurysms. J Neurosurg 62: 667–672
11. Dolenc VV (1989) Anatomy and surgery of the cavernous sinus. Springer Verlag, Wien, New York, pp 1–344
12. Dolenc VV (1995) Carotid-ophthalmic Aneurysms. In: Carter LP, Spetzler RF (eds) Neurovascular surgery. Mc Graw-Hill Inc, New York, pp 673–686
13. Drake CG, Vanderlinded RG, Amacher AL (1968) Carotid-ophthalmic aneurysms. J Neurosurg 29: 24–31
14. Ferguson GG, Drake CG (1981) Carotid-ophthalmic aneurysms: visual abnormalities in 32 patients and the results of treatment. Surg Neurol 16: 1–8
15. Fiore DL, Pardatscher K, Fiore D, Zuccarello M, Iraci G (1981) Persistent dorsal ophthalmic artery. Report of a case with associated fibromuscular hyperplasia of the extracranial internal carotid artery and multiple cerebral aneurysms. Neurochirurgia 24: 106–108
16. Fox JL (1988) Microsurgical treatment of ventral (paraclinoid) internal carotid artery aneurysms. Neurosurgery 22: 32–39

17. Fries G, Perneczky A, Van Lindert E, Bahadori-Martasawi F (1997) Contralateral and ipsilateral microsurgical approaches to carotid-ophthalmic aneurysms. Neurosurgery 41: 333–343
18. Gibo H. Lenkey C, Rhoton AL Jr (1981) Microsurgical anatomy of the supraclinoid portion of the internal carotid artery. J Neurosurg 55: 560–574
19. Glasscock ME (1969) Exposure of the intra-petrous portion of the carotid artery. In: Hamberger CA et al (eds) Disorders of the skull base region. Proceedings of the Tenth Nobel Symposium. Almqvist & Wiksell, Stockholm, pp 135–143
20. Giudetti B, La Torre E (1975) Management of carotid-ophthalmic aneurysms. J Neurosurg 42: 438–442
21. Hieshima GB, Higashida RT, Halbach VV, Cahan L, Goto K (1986) Intravascular balloon embolization of a carotid-ophthalmic artery aneurysm with preservation of the parent vessel. AJNR 7: 916–918
22. Iwabuchi T, Suzuki SH, Sobata E (1978) Intracranial direct operation for carotid-ophthalmic aneurysm by unroofing the optic canal. Acta Neurochir (Wien) 43: 163–169
23. Knosp E, Müller G, Perneczky A (1988) The paraclinoid carotid artery: anatomical aspects of a microneurosurgical approach. Neurosurgery 22: 896–901
24. Kobayashi S, Kyoshima G, Gibo H, Hedge SA, Takemae T, Sugita K (1989) Carotid cave aneurysms of the internal carotid artery. J Neurosurg 70: 216–221
25. Kothandaram P, Dawson BH, Kruyt RC (1971) Carotid-ophthalmic aneurysms. A study of 19 patients. J Neurosurg 34: 544–548
26. Kraemer JL, Schneider FL, Raupp SF, Ferreira NP (1989) Anatomoradiological correlation of the intersection of the carotid siphon with the dura mater. Neuroradiol 31: 408–412
27. Lang J, Kageyama I (1990) Clinical anatomy of the blood spaces and blood vessels surrounding the siphon of the internal carotid artery. Acta Anat 139: 320–325
28. Litofsky NS, Vinuela F, Giannotta SL (1994) Progressive visual loss after electrothrombosis treatment of a giant intracranial aneurysm. Case report. Neurosurgery 34: 548–551
29. Locksley HB (1966) Report of the cooperative study of intracranial aneurysms and subarachnoid hemorrhage. Section V, part I. Natural history of subarachnoid hemorrhage, intracranial aneurysms and arteriovenous malformations: based on 6368 cases in the cooperative study. J Neurosurg 25: 219–239
30. Milenkovic Z, Gopic H, Antovic P, Jovicic V, Petrovic B (1982) Contralateral pterional approach to a carotid-ophthalmic aneurysm ruptured at surgery. J Neurosurg 57: 823–825
31. Mizoi K, Suzuki J, Kinjo T, Yoshimomto T (1988) Bifrontal interhemispheric approach for carotid-ophthalmic aneurysms. Acta Neurochir (Wien) 90: 84–90
32. Nakao S, Kikuchi H, Takahashi N (1981) Successful clipping of carotid-ophthalmic aneurysms through a contralateral pterional approach. Report of two cases. J Neurosurg 54: 532–536
33. Nishio S, Matsushima T, Fukui M, Sawada K, Kitamura K (1985) Microsurgical anatomy around the origin of the ophthalmic artery with reference to contralateral pterional surgical approach to the carotid-ophthalmic aneurysm. Acta Neurochir (Wien) 76: 82–89
34. Nutik S (1978) Carotid paraclinoid aneurysms with intradural origin and intracavernous location. J Neurosurg 48: 526–533
35. Nutik SL (1988) Removal of the anterior clinoid process for exposure of the proximal intracranial carotid artery. J Neurosurg 69: 529–534
36. Ono M, Ono M, Rhoton AL Jr (1984) Microsurgical anatomy of the region of the tentorial incisure. J Neurosurg 60: 365–399
37. Perneczky A, Knosp E, Vorkapic P, Czech T (1985) Direct surgical approach to infraclinoidal aneurysms. Acta Neurochir (Wien) 76: 36–44
38. Punt J (1979) Some observations on aneurysms of the proximal internal carotid artery. J Neurosurg 52: 151–154
39. Sengupta RP, Gryspeerdt GL, Hankinson G (1976) Carotid-ophthalmic aneurysms. J Neurol Neurosurg Psychiatry 39: 837–853
40. Thurel C, Rey A, Thiebaut JB, Chai N, Houdart R (1974) Anevrismes carotido-ophthalmiques. Neurochirurgie 20: 25–39
41. Vajda J, Juhasz J, Pasztor E, Nyari I (1988) Contralateral approach to bilateral and ophthalmic aneurysms. Neurosurgery 22: 662–668
42. Yasargil MG, Gasser JC, Hodosh RM, Rankin TV (1977) Carotid ophthalmic aneurysm: direct microsurgical approach. Surg Neurol 8: 155–165
43. Yasargil MG (1984) Carotid-ophthalmic aneurysms. Microneurosurgery. Georg Thieme Verlag, Stuttgart New York, pp 46–57

3.2

Extradural approach to internal carotid artery aneurysms of the parasellar space

Introduction

Controversy has surrounded the treatment of vascular lesions within the parasellar space (PS) for generations. Initially, negative attitudes were rooted in a basic ignorance of the topography of the region. Even if the problem of venous bleeding could be overcome somehow, how could one establish proximal and distal control of an artery coursing from bone through a venous sinus? And how would one go about exposing and protecting nearby cranial nerves within a sea of venous blood? Indeed, neurosurgeons everywhere viewed the boundaries of the PS as anatomical frontiers impossible to cross.

Parkinson was first to demonstrate that surgery for PS-ICA aneurysms was even possible, solving the problems of exposure and hemostasis with cardiac arrest and extra-corporeal circulation [19–21]. Yet, even with hypothermia, cardiac arrest presented time restrictions, and with extra-corporeal circulation and anticoagulation came a new list of potential complications. Most neurosurgeons considered Parkinson's technique hazardous to the point of having little practical application.

While revolutionary interventional endovascular techniques soon brought hope of a practical method of managing PS-ICA aneurysms, they essentially excluded surgeons from the battlefield [23]. But problems remained. Often endovascular aneurysm occlusion required sacrifice of the ICA. Neurosurgeons around the world continued to re-examine the problems of PS-ICA surgery and improve their skills accordingly. Anatomical concepts had been rooted in preliminary descriptions of the so called "cavernous sinus" available from as early as 1940 [2, 26]. New studies were needed. By 1980, with additional knowledge of PS anatomy, neurosurgeons' attitudes began to change, and ultimately new direct approach to the PS came into being. Slowly the treatment of PS-ICA aneurysms returned to the domain of the surgeon [6]. Still, most neurosurgeons were slow to grasp the new opportunities. Concepts and attitudes honored for generations died slowly. For most, the problems presented by the PS remained intimidating. Over the following 20 years still newer anatomical information [6–11, 19–22] led to the realization that the location of PS-ICA aneurysms indicated that they could be approached safely and treated effectively by an extradural method [7–11]. Cardiac arrest and hypothermia were no longer prerequisites for surgical treatment [19–22].

For a time, spirited competition between surgeons and interventionalists over the treatment of PS vascular lesions generated more heat than light. An extensive literature resulted. But the emotional debate and turf wars were hardly necessary. Over time, discussions between the two specialities became more realistic, more objective, ultimately more productive [1, 3–12, 14–17, 22–27]. A more precise understanding of PS anatomy and intra-arterial and intra-aneurysmal morphology, especially the nature of the ICA wall proximal and distal to ICA aneurysms, led both groups to realize that neither surgical nor endovascular treatment precluded the other, that neither approach provided the ultimate solution for every case [11]. In the interest of good patient care direct surgery and endovascular techniques became more complementary than competitive. Coordinated teamwork became commonplace as diagnostic work-ups were undertaken jointly and all treatment options began to be considered dispassionately. Both groups recognized that in one major respect PS-ICA aneurysms were no different from any other aneurysm, that none should be treated at the expense of parent and daughter arteries. While certain PS-ICA aneurysms could be eliminated by endovascular methods while preserving the ICA, clearly such was not possible in every instance. It became clear that microsurgical resection of the aneurysm and reconstruction of the ICA wall are far preferable to simply occluding the ICA by endovascular methods.

Recently the problem of small, incidentally-found PS-ICA aneurysms has been addressed. With better understanding of PS anatomy and improving treatment options the policy to withhold treatment for such lesions, readily accepted by previous generations of surgeons, is being re-examined. Improving treatment outcomes bring such fatalistic views into sharp focus. Why should treatment be withheld? Although the natural history of small PS-ICA aneurysms has yet to be defined precisely as a group, there is little reason to believe that such lesions are less likely to rupture than those located intradurally. Catastrophic circumstances, even subdural hemorrhage and death, have resulted [14]. Injuries to the PS-ICA resulting in massive epistaxis following craniocerebral trauma have been known for many decades [18] while appropriate treatment has become available only in the past 30 years [6, 23].

Knowledge of PS architecture and the relationship between the ICA and adjacent cranial nerves under normal conditions is essential to understanding the specific relationships between these structures when an aneurysm is present. This allows the neurosurgeon to undertake a direct surgical approach to eliminate these aneurysms, preserving or reconstructing the ICA and restoring relationships between PS structures that approximate normal. Large and giant PS aneurysms, particularly, demand treatment which insures the preservation of ICA patency.

Relevant anatomy of the extradural ICA

At the skull base, the ICA enters the petrous bone and courses from infero-postero-laterally to supero-antero-medially, finally entering the intradural space through the distal (dural) ring (DR) over a total distance of approximately 6 cm as shown in Figs. 3.12, 3.13(a and b), and 3.14. Its proximal (petrous) segment, entirely covered by bone, averages 3 cm in length, compared to the distal (or "free") PS segment (not covered by bone), which is approximately 2.6 mm long. The ICA is again enveloped by bone in its final extradural segment, 3–5 mm in length, at the level of the anterior clinoid process (ACP). As it progresses through the skull base the ICA makes four loops: the posterior (PL), lateral (LL), medial (ML), and anterior (AL) loops (Fig. 3.13a). Several well defined anatomical characteristics are particularly important in describing the configuration and skull base course of the ICA. The PL is entirely within petrous bone and represents the ICA's first turn, where its orientation changes from vertical to horizontal toward the foramen lacerum (FL). From the LL, overlying the FL, the ICA courses superomedially toward the lateral surface of the posterior clinoid process (PCP). Lateral to the PCP, the ICA forms the ML and courses anteriorly, forming the AL inferomedial to the ACP, and finally through the dura at the distal (dural) ring (DR). Each loop lies in a different plane, suggesting that the artery rotates on its longitudinal axis as blood flows within it, possibly resulting in the spiraling, or spinning, of blood elements. The distal part of the PL, coursing almost horizontally toward the LL, is encased by bone in approximately 80% of cases, covered in the remainder by a fibrous membrane, or paper thin bone. Two fairly constant vessels arise from the PS-ICA, the meningohypophyseal trunk (MHT) from the ascending segment of the proximal leg of the ML, and the inferolateral trunk (ILT) from the lateral wall of the distal leg of the ML (the horizontal segment). These two arteries carry important localizing significance since nerve VI courses lateral to the MHT and inferior to the ILT (Fig. 3.14).

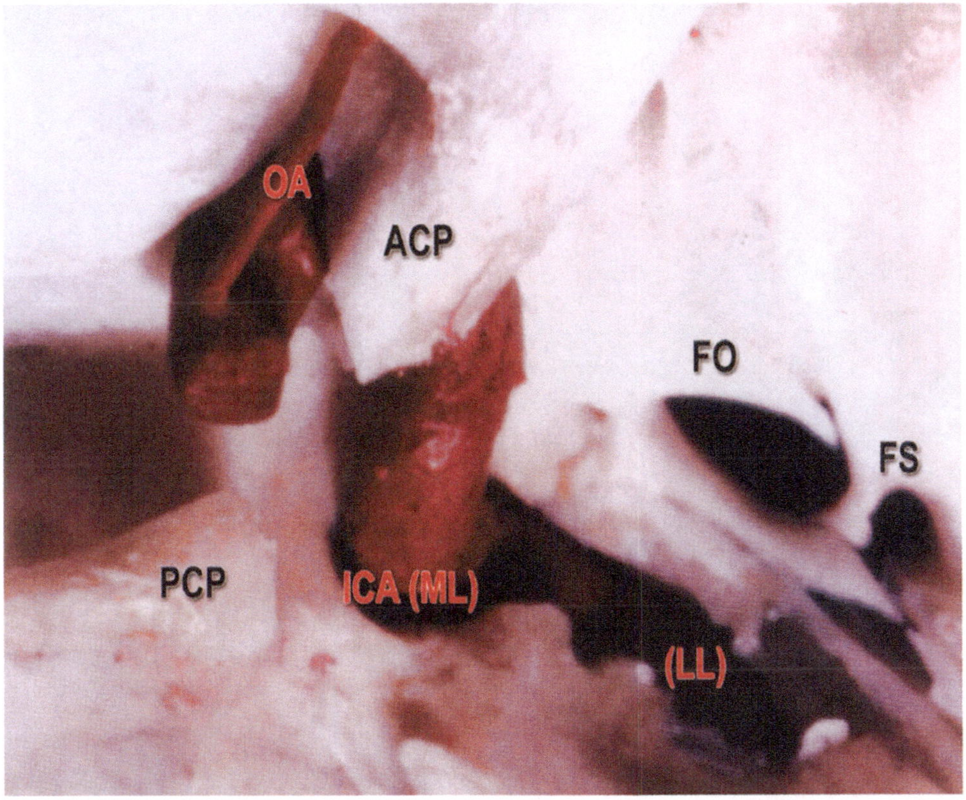

Fig. 3.12. The ICA on the right side of the central skull base. After entering the petrous canal, the ICA courses upward, forming the posterior loop within bone, then a horizontal segment anteromedially. At the foramen lacerum, it turns upward again and medialward, forming the lateral loop (*LL*), then proceeding medially to immediately lateral to the posterior clinoid process (*PCP*) where it turns forward, forming the medial loop (*ML*). From the PCP, the horizontal ICA segment proceeds inferomedial to the anterior clinoid process (*ACP*), where it makes a U-turn around it, forming the anterior loop proximal to the origin of the ophthalmic artery (*OA*). If one were to imagine a cube at the right side of the central skull base, the ICA would enter it (at the petrous canal) from its infero-postero-lateral corner, and exit at its antero-supero-medial corner (where it enters the dura)

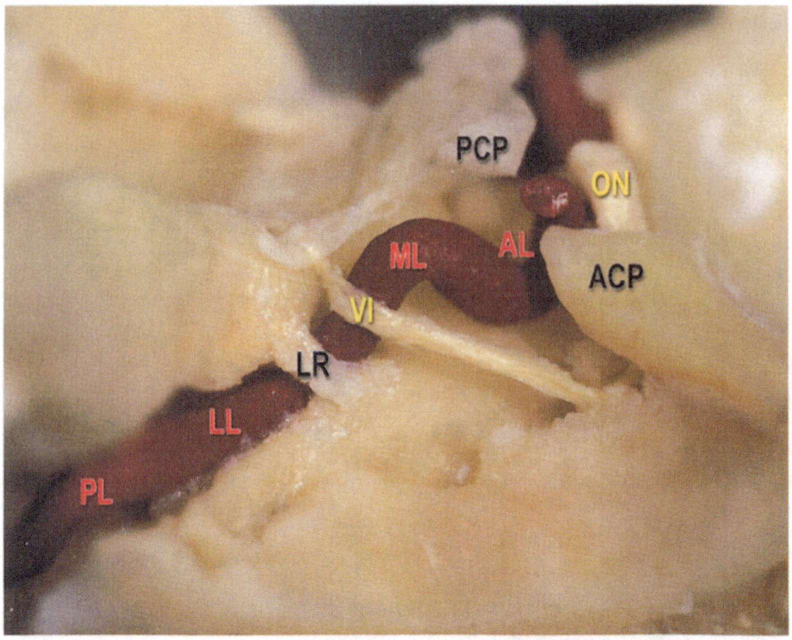

a

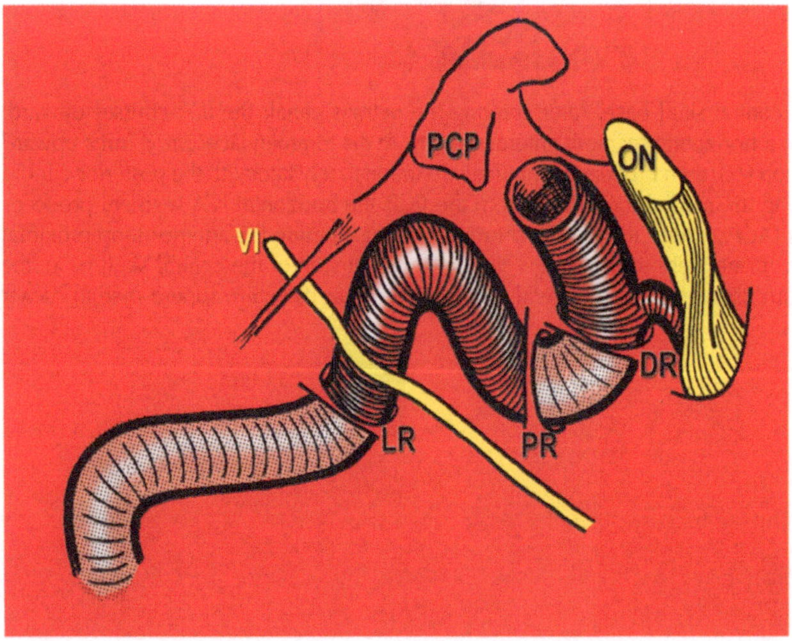

b

Fig. 3.13. (a) The right ICA within a bony central skull base specimen. All four loops – posterior loop (*PL*), lateral loop (*LL*), medial loop (*ML*), and anterior loop (*AL*) – are shown. The lateral ring (*LR*) and nerve VI are apparent, along with the ICA in the parasellar space. The anterior clinoid process (*ACP*) "hangs" superolaterally over the AL, and hence must be resected in all surgical approaches to this ICA segment. (b) An illustration of the different segments of the extradural ICA as it courses through the skull base on the right side, from its entry point into bone to where it pierces the dura. The "covered" segment of the ICA, from its point of entry into the petrous carotid canal to the lateral ring (*LR*), represents the proximal segment of the ICA in the cental skull base. The "free" segment of the ICA is located between the LR and proximal ring (*PR*) and constitutes the site of most parasellar space (*PS*) aneurysms. The "paraclinoid segment," between the PR and DR, is partially covered by bone, completely encased only when the ACP is continuous with the lateral wall of the sphenoidal sinus. Note: The intradural segment of the ICA lies parallel to the horizontal portion of the PS-ICA as it courses from its ML to its AL. The ophthalmic artery (*OA*), optic nerve (*ON*), and posterior clinoid process (*PCP*), are shown schematically. Nerve VI (the only cranial nerve actually inside the PS) overrides the lateral surface of the ascending segment of the "free" ICA

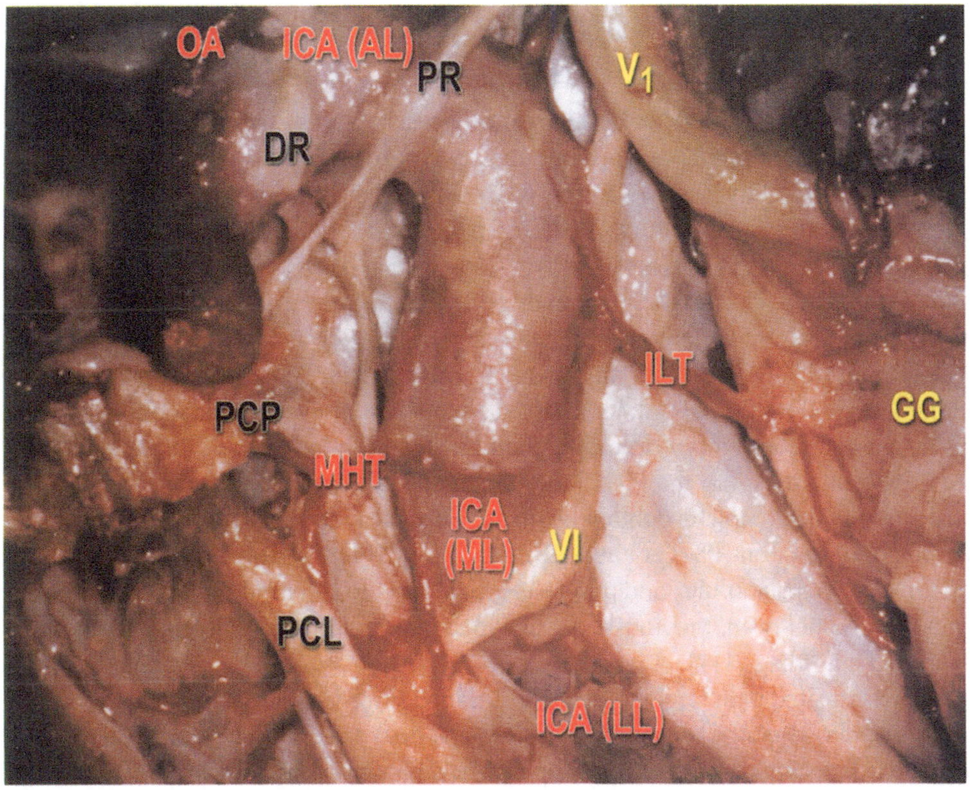

Fig. 3.14. An injected ICA in a fresh cadaver specimen, showing the anterior loop (*AL*) from the proximal ring (*PR*) through the distal (dural) ring (*DR*), after resection of the anterior clinoid process. The inferolateral trunk (*ILT*) originates from the ICA proximal to the PR and is adjacent and inferior to nerve VI, which is oriented parallel to the horizontal segment of the PS-ICA. The meningo-hypophyseal trunk (*MHT*) originates medial to where nerve VI courses lateral to the ascending segment of the ICA

Pathological anatomy, symptoms and signs, and diagnostic work-up

Saccular and fusiform aneurysms of the PS-ICA may be present anywhere from the lateral ring (LR) to the proximal ring (PR). Most, however, are on the proximal or distal legs of the ML, where the MHT and ILT originate. PS-ICA aneurysms vary considerably, not only in location but in size. When they occur bilaterally (occasionally), they generally are not equal in size, the lesion causing symptoms and signs where its mass affects one or several adjacent neural structures (nerves III through VI). Such neurological deficits are particularly likely when the lesion is large or giant. However, since nerve VI is "fixed" by sympathetic fibers originating from sympathetic cords adherent to the ICA wall, it may be compressed by much smaller lesions (Fig. 3.15). Even small ICA aneurysms may be responsible for transient ischemic attack (TIAs) by way of aneurysm-to-artery emboli although this is much more likely with large and giant ICA aneurysms. Small ICA aneurysms may rupture into the PS, resulting in high-flow fistulae. Although typically neither is severe, headaches from large and giant PS-ICA aneurysms are commonly more disturbing than those associated with PS tumors. By contrast, when a small ICA aneurysm ruptures into the PS, causing a large or giant false aneurysm, excruciating pain results, typically associated with complete ophthalmoplegia. Such headaches may persist for several months. More commonly hemorrhages from PS-ICA aneurysms lead to high-flow fistulae, which are associated with only moderate pain, characteristic bruits, and proptosis which may be either mild or severe. In exceptional cases a large or giant PS-ICA aneurysm, previously causing local symptoms and signs, may rupture and result in catastrophic intradural hemorrhage [14].

The most frequent symptom associated with large or giant PS-ICA aneurysms is a fluctuating dull periorbital ache, which is commonly ascribed to migraine by general practitioners, even specialists if not associated with diplopia. Only occasionally is such aching described as true "pain." In addition to such mass-associated symptoms and signs, large and giant lesions result in TIAs and strokes from intra-aneurysmal thrombi or clots. The sudden onset of sharp pain, accompanied by ophthalmoplegia, commonly announces the rupture of a PS-ICA aneurysm of any size. Such pain is typically refractory to most medical therapy. When sudden, spontaneous, intense pain and ophthalmoplegia is associated with a bruit, even pulsating exophthalmos, the diagnosis of aneurysmal rupture causing a high-flow fistula is inescapable.

Small ICA aneurysms in the PS generally are discovered incidentally by carotid angiography following subarachnoid hemorrhage (SAH) from an intradural aneurysm, or patients being studied for symptoms suggesting extracranial carotid artery disease. Large and giant PS-ICA aneurysms typically are diagnosed by CT scans and/or MRI performed in investigation of nerve III, IV, or VI paresis, or unilateral headaches related to nerve V involvement. Diagnosis in such cases is confirmed by CTA and/or DSA as illustrated in Fig. 3.16(a–d). With large and giant aneurysms, CTA is usually adequate for studying the ICA proximal and distal to the aneurysm, as well as their sites of origin from the ICA and the size of their necks. CT, MR, and DSA are also useful with such lesions to determine the quality of the wall of the ICA proximal and distal to the aneurysm, the nature of the aneurysm's wall, the size and location of its neck, and whether or not they contain thrombi or clot.

Once the presence of an PS-ICA aneurysm is confirmed by CT, CTA, MR, or/and DSA, the capacity for cross-circulation must be determined with additional studies, specifically those that define patterns of perfusion and venous return of both hemispheres during temporary balloon occlusion of the ICA giving rise to the aneurysm. Such information enables a rational surgical approach to be planned. It should be clear before surgery whether the ICA will require reconstruction after resecting the aneurysm, even if such reconstruction is possible. Patients not tolerating temporary ICA occlusion should be considered for EC-IC by-pass before attacking the aneurysm.

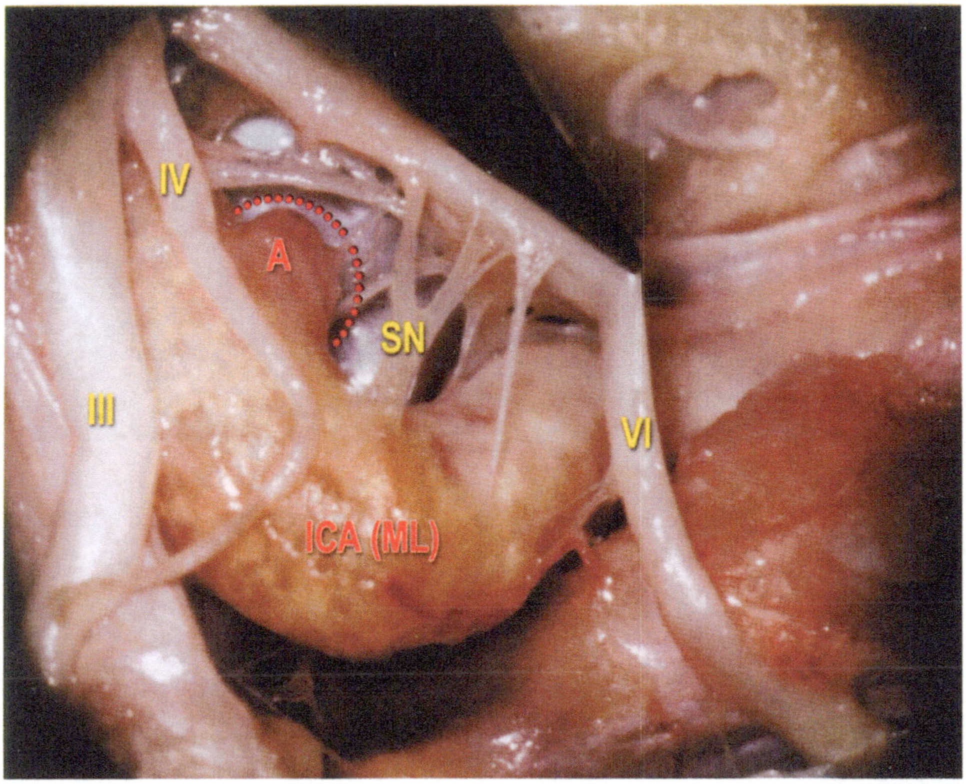

Fig. 3.15. A small parasellar space aneurysm (*A*), arising from the horizontal segment of the ICA in the parasellar space and projecting inferolaterally. In addition to nerves III and IV, a number of sympathetic fibers arising from cords on the surface of the ICA to nerve VI are seen. Aneurysm growth as indicated by the dotted line may stretch these fibers and result in injury to nerve VI. Note: Such nerve VI injuries may explain the "early" abducens palsies which result in temporary diplopia. When such sympathetic fiber(s) are stretched beyond a certain point, they likely rupture, releasing nerve VI from its adherence to the ICA, possibly allowing it to slide away from the sac of the aneurysm to some extent, reducing its compression. This may explain the disappearance or reduction of diplopia that is sometimes seen over a period of time

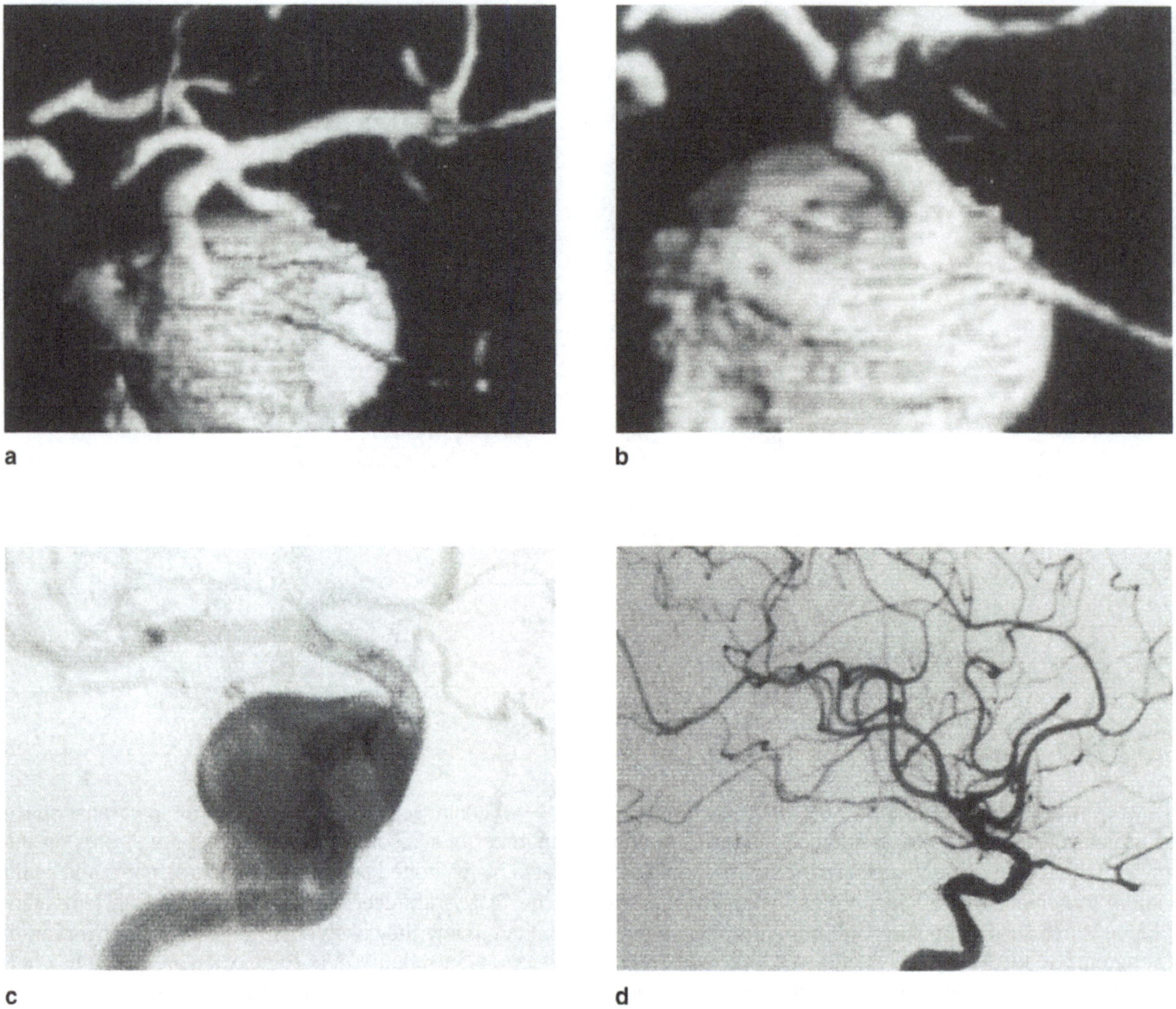

Fig. 3.16. CTA of an ICA aneurysm in the right parasellar space, AP (a) and lateral (b) views. DSA of the same lesion, AP view, preoperative (c) and lateral view, postoperative (d). The ICA, reconstructed with sutures after resection of the aneurysm, has an internal diameter which approaches normal. Note: CTA defines more precisely than DSA the origin of the aneurysm and its involvement with the ICA wall

The surgical approach to the ICA aneurysms in the PS

The procedures described below are based on details described in Chapter 2. 1) The patient is positioned supine, with the head rotated slightly away from the side of the lesion and immobilized in three-point skeletal fixation. 2) In the course of making the scalp incision and reflecting the soft tissue flap laterally, great care must be taken to preserve the superficial temporal artery should an EC-IC bypass become necessary if unforseen ICA injury occurs. 3) After securing the osteomuscular flap with fish hooks, dura is tacked with sutures to the edges of cut bone in the frontal and parieto-temporal regions. 4) The squama of the temporal bone is removed to the floor of the middle cranial fossa. The resection of the lesser sphenoid wing follows unroofing the orbit and exposing the superior orbital fissure (SOF). 5) Dura is peeled from the anterior pole of the temporal lobe, toward the foramen rotundum (FR), the foramen ovale (FO), and the foramen spinosum (FS), and dura lateral to neural structures traversing the SOF is incised, then peeled from the inner layer of the lateral wall of the PS.

With large and giant ICA aneurysms, particularly if they harbor thrombi and/or clotted blood, the FS should be exposed, the middle meningeal artery (MMA) coagulated and divided, and the ICA exposed within the petrous canal [13] before peeling away the outer layer of the lateral wall of the PS overlying the bulging aneurysm. Only when an adequate segment of the ICA in the petrous carotid canal medial to the FS and eustachian tube is dissected and prepared for temporary occlusion, should the anterior clinoid process (ACP) and the walls of the optic canal (OC) be resected. To omit this step is to court disaster in the event of premature rupture of the aneurysm. After removing the ACP and dissecting the AL of the ICA through the anteromedial triangle between the optic nerve (ON) and nerve III, the lateral and the superior walls of the OC and optic strut are removed. When the ICA is prepared for isolating the aneurysm by temporary proximal and distal clipping, the outer layer of the lateral wall of the PS may be gently peeled from its inner layer (covering adjacent nerves and the wall of the aneurysm). When large or giant aneurysms involve the wall of AL of the ICA near the DR, or extend intradurally through the DR, it is necessary to open the dura and to be prepared to temporarily occlude the distal ICA (Figs. 3.17a–d). When the PS-ICA aneurysm is located more proximally, as in Fig. 3.18(a–d), and the AL distal to the aneurysm is fully exposed, allowing it to be temporarily clipped extradurally, opening the dura may be circumvented. The importance of gaining access to the ICA proximal and distal to the aneurysm prior to dissecting the outer layer of the lateral wall of the PS cannot be over emphasized. Nerves III, IV, V1, V2, V3, and the Gasserian ganglion (GG) should be exposed before beginning to dissect the aneurysm. The location, size, and possible presence of thrombi within the aneurysm determines through which "window" (triangle) the aneurysm should be approached. However, in most large and giant aneurysms it is wise first to dissect nerve IV from nerve V1 from the inner layer of the lateral PS wall, and transpose nerve IV medially, thus enlarging Parkinson's triangle. At this point, dissection of the aneurysm may be initiated.

Handling aneurysms containing thrombi should be minimized as much as possible in order to avoid releasing emboli. They must be dissected with great care and delicacy. Whenever possible, the aneurysm, and the ICA proximal and distal to it, should be fully dissected before temporarily trapping the lesion. Most small aneurysms, and some non-thrombosed large and giant lesions, may be clipped without temporarily occluding the ICA. When this is not possible for whatever reason, temporary clips are applied and the aneurysm is incised widely, thrombi and clot washed away, the sac resected, and the ICA reconstructed with clip(s) or sutures, or both. Before considering resecting any aneurysm and reconstructing the arterial wall, the ICA proximal and distal to the aneurysm should be examined carefully to confirm that it is of adequate quality. Depending on the aneurysm's size and the quality of its neck, also the quality of the ICA wall proximal and distal to the lesion, aneurysm resection and ICA reconstruction may be local, direct, or by grafting [9]. With large or giant PS-ICA lesions it is sometimes necessary to interpose a vascular graft from the petrous ICA to its clinoidal segment. If the AL is incorporated into the sac of the aneurysm, or otherwise diseased, the graft is anastomosed to the intradural ICA distal to the origin of the ophthalmic artery (OA). If this becomes necessary, the OA is ligated and transected. When such bypass grafting is anticipated, an adequate length of either saphenous vein or radial artery should be harvested prior to applying proximal and distal clips. A graft should be employed only when direct end-to-end reconstruction of the PS-ICA is not feasible, but a graft should be on hand in questionable cases. To gain additional length for direct anastomosis the proximal and distal stumps of the PS-ICA should be mobilized and placed lateral to nerves III, IV, and VI. Similarly, grafts are placed lateral to nerves III – VI. Following reconstruction of the PS-ICA, its patency should be confirmed with a Doppler probe before closing the wound. Venous bleeding occurring after resecting the aneurysm may be controlled by packing the PS with Surgicel, particularly those sinuses connecting the left and right PSs. When surgery is completed, the entire PS is flooded with fibrin glue in order to prevent postoperative bleeding.

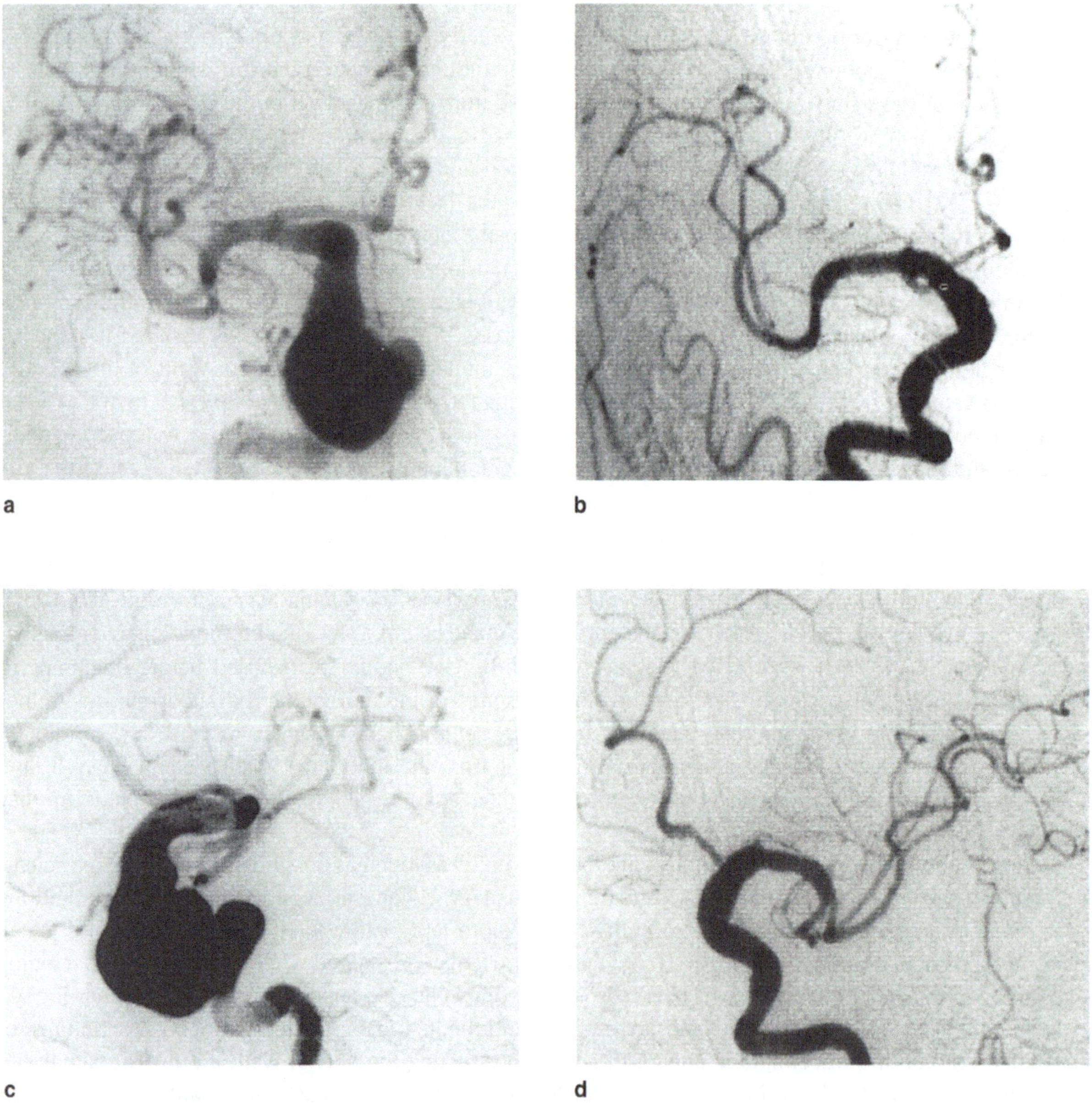

Fig. 3.17. DSA of a giant fusiform ICA aneurysm in the left parasellar space extending into the intradural segment of the ICA, preoperatively (a and c) and postoperatively (b and d). This patient presented with subarachnoid hemorrhage (*SAH*) from rupture of the intradural portion of the ICA aneurysm. At surgery, the ICA was reconstructed with sutures after extensive resection of the sac of the aneurysm

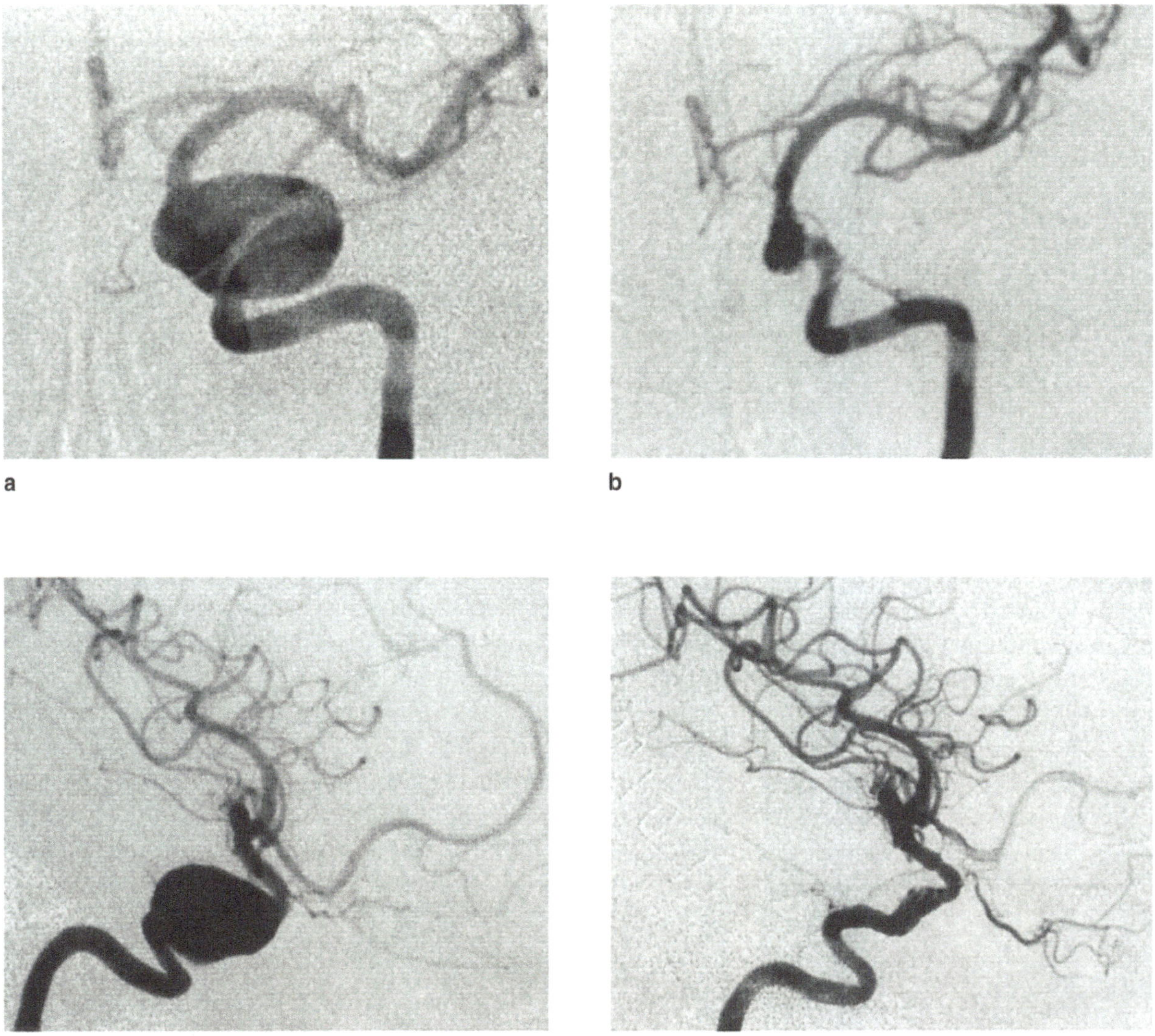

a b c d

Fig. 3.18. DSA of a large ICA aneurysm in the right parasellar space, preoperative (a and c) and postoperative (b and d). The aneurysmal sac was resected and the ICA reconstructed with three clips. Note: Because of this lesion's thin-walled neck, adjacent to a sclerotic ICA wall, suturing was not possible. Repeated repositioning of the clips was required to avoid constricting the ICA

Table 2. ICA aneurysms in the PS, type of exclusion, and results

Size & number of aneurysms		Aneurysm exclusion & type of the ICA reconstruction: Clipping	Local	Direct	Graft	Postoperative ICA patency	Postoperative nerve(s) III, IV, VI deficits after: One week	Six months
Small	119	73	40	6	–	114/114	81	3
Large	49	10	21	7	11	43/47	47	4
Giant	38	4	13	10	11	34/37	38	6
Total	206	87	74	23	22	191/198	166	13

Patients and results

Of 206 PS-ICA aneurysms, 87 were clipped. In 74 cases, the aneurysm was resected and the ICA reconstructed by closing the neck of the aneurysm with interrupted sutures. In 23 patients the ICA was reconstructed by end-to-end anastomosis after resecting the aneurysm and a short segment (5 mm or less) of diseased ICA. Aneurysm resection followed by ICA grafting was necessary in 22 cases. Postoperative angiography was accomplished in 198 cases, confirming ICA patency in 191, occlusion in seven. Although nerve III palsies were apparent in 166 patients immediately following surgery, such deficits cleared within six months in all but 13 cases. All patients with preoperative trigeminal pain experienced gradual relief after surgery. Trigeminal sensory deficits present before surgery usually improved following aneurysm resection, and painful paresthesias did not occur. Four patients demonstrated persistent, although slight, residual contralateral hemiparesis. Postoperative epilepsy did not occur. The overall outcome was considered to be excellent in 186 patients. Postoperative diplopia in the 13 patients with nerve(s) III, IV, VI deficits generated more complaints than among the four with hemiparesis. There were three deaths.

Conclusion

Many PS-ICA aneurysms, particularly large and giant lesions – those likely to present with periorbital pain and neurological symptoms as well as the associated risk of creating spontaneous high-flow fistulae, even devastating SAH – cannot be effectively and completely eliminated from the cerebral circulation by endovascular techniques without sacrificing the ICA. For such lesions the extradural approach provides a safe, efficient, practical surgical method so long as one is willing to adhere compulsively to the techniques described above, emphasizing delicately exposing salient PS structures rather than proceeding quickly to the aneurysm itself. There is no short cut to success. Patience and full knowledge of the anatomy are everything.

References

1. Berenstein A, Ransohoff J, Kupersmith M, Flamm E, Graeb D (1984) Transvascular treatment of giant aneurysms of the cavernous carotid and vertebral arteries. Functional investigation and embolization. Surg Neurol 21: 3–12
2. Dandy WE, Follis RH Jr (1941) On the pathology of carotid-cavernous aneurysms (pulsating exophthalmos). Am J Ophthalmol 24: 365
3. Debrun G, Lancour P, Caron JP, Hurth M, Comoy J, Keravel Y (1978) Detachable balloon and calibrated-leak balloon technique in the treatment of cerebral vascular lesions. J Neurosurg 46: 635–649
4. Debrun GM (1987) Embolization techniques in the treatment of vascular lesions involving the cavernous sinus. In: Dolenc VV (ed) The cavernous sinus: a multidisciplinary approach to vascular and tumorous lesions. Springer, Wien New York, pp 173–181
5. Diaz FG, Ohaegbulam MD, Dujovny M, Ausman JI (1988) Surgical management of aneurysms in the cavernous sinus. Acta Neurochir (Wien) 91: 25–28

6. Dolenc VV (1983) Direct microsurgical repair of intracavernous vascular lesions. J Neurosurg 58: 824–831
7. Dolenc VV (1989) Anatomy and Surgery of the Cavernous Sinus. Springer, Wien New York, pp 1–344
8. Dolenc VV (1990) Surgery of vascular lesions of the cavernous sinus. Clin Neurosurg 36: 240–255
9. Dolenc VV (1991) The necessity for intracavernous ICA reconstruction. In: Sato K (ed) Neurosurgeons 10 (Proceedings of the Japanese congress of neurological surgeons). Sci Med Publications, Tokyo, pp 299–307
10. Dolenc VV (1994) Intracavernous carotid artery aneurysms. In: Carter LP, Spetzler RF (eds) Neurovascular Surgery. Mc-Graw-Hill, New York, pp 659–672
11. Dolenc VV, Valavanis A (1997) Cavernous carotid aneurysm: Direct vs. indirect approaches. In: Bajter HH (ed): Cerebrovascular disease, Lippincott-Raven, Philadelphia, pp 939–955
12. Eguchi T, Nakagomi T, Teraoka A (1982) Treatment of bilateral mycotic intracavernous carotid aneurysms. Case report. J Neurosurg 56: 443–447
13. Glasscock ME (1969) Exposure of the intra-petrous portion of the carotid artery. In: Hamberger CA et al (eds) Disorders of the skull base region. Proceedings of the 10th Nobel Symposium. Almqvist & Wiksell, Stockholm, pp 135–143
14. Hodes JE, Fletcher WA, Goodman DF, Hoyt WF (1988) Rupture of cavernous carotid artery aneurysms causing subdural hematoma and death. J Neurosurg 69: 617–619
15. Johnston I (1979) Direct surgical treatment of bilateral intracavernous internal carotid artery aneurysms. Case report. J Neurosurg 51: 98–102
16. Lasjaunias P, Berenstein A (1987) Aneurysms of the cavernous segment of the internal carotid artery. In: Surgical neuroangiography, vol 2. Endovascular treatment of craniofacial lesions. Springer, Berlin Heidelberg New York Tokyo, pp 248–264
17. Liu MY, Shih CJ, Wang YC, Tsai SH (1985) Traumatic intracavernous carotid aneurysm with massive epistaxis. Neurosurgery 17: 569–573
18. Maurer JJ, Mills M, German WJ (1961) Triad of unilateral blindness, orbital fracture and massive epistaxis after head injury. J Neurosurg 18: 837
19. Parkinson D (1965) A surgical approach to the cavernous portion of the carotid artery. Anatomical studies and case report. J Neurosurg 23: 474–483
20. Parkinson D (1967) Transcavernous repair of carotid cavernous fistula. J Neurosurg 26: 420–424
21. Parkinson D (1973) Carotid cavernous fistula: direct repair with preservation of the carotid artery. J Neurosurg 38: 99–106
22. Sano H (1988) Bilateral giant intracavernous aneurysms. Technique of unilateral operation. Surg Neurol 29: 35–38
23. Serbinenko FA (1974) Balloon catheterization and occlusion of major cerebral vessels. J Neurosurg 41: 125–145
24. Spaziante R (1986) Intracavernous giant fusiform aneurysm of the carotid artery treated with Gianturco coils. Neurochirurgia 29: 31–41
25. Spetzler RF, Fukushima T, Martin N, Zabramski JM (1990) Petrous carotid-to-intradural carotid saphenous vein graft for intracavernous giant aneurysm, tumor, and occlusive cerebrovascular disease. J Neurosurg 73: 496–501
26. Taptas JN (1949) La longe du sinus caverneux, sa constitution et les rapports des éléments vasculaires et nerveux qui la traversent. Sem Hop 25: 1719–1722
27. Vinuela F, Lylyk P (1987) Endovascular therapy of vascular lesions of the cavernous sinus. Experience with 129 cases. In: Dolenc VV (ed) The cavernous sinus: a multidisciplinary approach to vascular and tumorous lesions. Springer, Wien New York, pp 182–197

3.3

Extradural approach to carotid-parasellar fistulae

Introduction

Carotid-parasellar fistulae (CPF) are abnormal communications between the ICA and venous compartments of the parasellar space (PS). Some include external carotid artery (ECA) branches. Generally they follow head trauma, accounting for most high-flow CPFs, but a minority arise spontaneously, often resulting in low-flow lesions. Typically, skull base fractures lacerate the ICA, but penetrating injuries (stab wounds and gun-shot injuries) cause others. In some cases the cause of spontaneous CPFs remains speculative [23, 34, 36], but some are associated with ruptured PS-ICA aneurysms [2, 24, 33]. Arterial hypertension, arteriosclerotic vascular disease, hypertension during pregnancy, minor head trauma, straining, and collagen vascular disease are likely major factors in others [1, 17, 20, 23, 28, 35].

The clinical management of these lesions is dependent on their flow characteristics. High-flow CPFs demand aggressive treatment, while low-flow lesions may be managed effectively more conservatively, sometimes with only close observation over time. No vascular pathology of the central nervous system has generated more therapeutic controversy than CPFs. Currently, endovascular surgeons armed with detachable balloons and coiling techniques play a prime role in the treatment of high-flow lesions [8–11, 30, 36].

Browder was first to describe a direct surgical approach to CPFs in 1936 [5], a method subsequently to be refined by Parkinson [25]. Other surgical options followed, including the injection of various thrombogenic materials into the lesion [21, 22]. Ultimately a safe, relatively "easy," direct microsurgical approach to the PS was described [12, 13, 14]. But endovascular methods continued to evolve as different materials, including various balloons, coils, and liquid embolic agents, were introduced. Detachable balloons and coils remain the best of embolic materials for directly occluding high-flow CPFs. More recently, radiosurgery is being considered for certain low-flow lesions.

Effective treatment results in eliminating the fistula and while maintaining patency of the ICA. If this is achieved, the troublesome bruits disappear and exophthalmos wanes, and, where vision has remained intact, improving visual acuity and clearing diplopia (from improved cranial nerve(s) function) occurs. However vital interrupting the fistula and preserving ICA patency, preserving or improving vision in the affected eye is equally important. Ideally, a full team of specialists is involved with the clinical work-up of such patients – including a neurosurgeon, a neuroradiologist, a radiosurgeon, and an ophthalmologist – each fully cognizant of the therapeutic needs of the specific lesion under consideration, committed to offering the appropriate treatment for each patient rather than falling into the trap of opting for whatever management currently being pushed by the industry.

Relevant anatomy

An understanding of the normal anatomy and function of PS structures is essential for dealing with the changed relationships dictated by these lesions. CPFs acutely destroy the previously normally functioning PS venous network, resulting in pathological shunt circulation and functional consequences.

While relevant characteristics of bone elements of the central skull base (CSB), periosteum, and dural layers – their various folds, clefts, and duplications – are generally accepted, it is the nature of PS venous elements which has been at the root of controversy roiling for years. Proponents of the PS as a network of veins (Fig. 3.19), a venous plexus, have remained at odds with those supporting the concept of cavernous structures constituting a true venous sinus, similar to the corpora cavernosa of the penis or clitoris. But more recent anatomical studies douse cold water on what arguments remain, fully supporting Parkinson's and Taptas' concept of the PS as venous plexus. Nowadays, for practical purposes it is certain that the venous elements of the PS amount to a vastly complicated network of veins, rather than a thin-walled encapsulation of venous blood.

The study of formalin fixed preparations miss that distinction, now confirmed by detailed examination of appropriately injected fresh cadaver specimens.

Not only does the venous anatomy of the involved PS carry important implications for the formulation and execution of CPF treatment, but also veins draining the orbit, the ipsilateral intradural space, and trans-sellar communications with the contralateral PS [13]. The orbit and the PS are unique adjoining spaces, analogous to a large hotel suite, with orbital drainage emptying into the PS, which is connected to the opposite PS through the sella and across the clivus, all draining ultimately into the superior petrosal sinus (SPS) and the inferior petrosal sinus (IPS). Until recently, the venous drainage of the brain itself into the PS had not been studied in detail. Current information confirms that such drainage along the lateral wall of the PS either enters the PS or continues separately between its dural layers into the SPS [16, 27]. Prior to this finding the brain was believed to drain exclusively via the superficial middle cerebral veins through dura into the PS, entering it at the anterolateral triangle [13].

Pathological anatomy, clinical picture, and diagnostic work-up

ICA lacerations resulting in abnormal communications with PS veins occur in the majority of cases between the lateral ring (LR) and the distal (dural) ring (DR). The site of arterial leakage depends on the orientation of the skull base fracture or the trajectory of the penetrating object. Rarely, however, with shearing of the meningohypophyseal trunk (MHT) or the infero-lateral trunk (ILT), the fracture may be relatively distant from the ICA. Similarly, the site of shunting in spontaneously occurring CPFs depends on the location of ruptured PS-ICA aneurysms. The size of the defect in the ICA wall determines a given lesion's flow volume, larger openings resulting in high-flow fistulae. The size of the arterial leak may vary from quite small to complete transsection of the ICA (Fig. 3.20). The size of the escaping stream of blood notwithstanding, the nature and orientation of neighboring structures are important in determining whether a fistula forms or extravasated blood remains localized to the site of leakage. Adjacent dura, periosteum, and bone may dictate the formation of a false aneurysm if escaping blood remains confined locally within space adjacent to the ICA as opposed to CPFs resulting when blood gushes directly into an open venous channel. False aneurysms and low-flow fistulae may coexist [14]. The nature and extent of the fistula varies with the interplay between anatomical and functional factors. Each must be considered if the best possible treatment plan is to be determined and carried out.

Anatomical and radiographic characteristics allow CPFs to be placed into one of four categories as described by Barrow, et al [2, 7]. Type A lesions are direct shunts between the PS-ICA and the PS. Type B consists of dural shunts between PS-ICA meningeal branches and the PS, and type C, dural shunts between ECA meningeal branches and the PS. Type D refers to dural shunts between meningeal branches of both the PS-ICA and ECA and the PS. In addition, direct shunts between the PS-ICA and PS occur from false aneurysms, resulting in a "mixed type" [14].

Most Type A shunts follow trauma or spontaneous rupture of PS-ICA aneurysms. Types B, C and D, dural shunts between meningeal arteries and the PS, involve collateral venous channels. Parkinson's two types amounted to the first attempt to classify CPFs on the basis of anatomical characteristics [25]. Numerous classification systems followed, including the one employed here, which is now widely accepted as the most useful in determining the appropriate treatment for CPFs.

The characteristic clinical presentation of high-flow CPFs includes pulsating exophthalmos with chemosis, a very disturbing subjective bruit (easily confirmed objectively by auscultation) which is synchronous with the heart beat, and ipsilateral visual impairment or total blindness [4, 15, 18, 19]. Visual findings may be bilateral. With spontaneous CPFs, the clinical picture may initially be quite mild, progressing as the fistula increases in size. In such cases, visual deterioration, persistent diplopia, an intolerable bruit, possibly headache, and proptosis with corneal exposure demand urgent treatment [31]. The diagnosis of high-flow CPFs, arising either from trauma or aneurysmal rupture, is obvious and is made on clinical grounds.

Carotid angiography is performed to confirm the presence of a CPF, determine its extent, and characterize attendant changes in regional circulation. When appropriate, diagnostic angiography is followed immediately by endovascular treatment. If direct surgery is to be considered, additional examinations become necessary. Analysis of the opposite carotid artery, the vertebral circulation, and cross-compression studies are important, particularly with high-flow lesions, to determine the adequacy of collateral circulation since temporary interuption of ICA flow must be anticipated in the operating room. Such studies determine the safe length of time ICA flow may be arrested and whether or not a given patient will tolerate permanent carotid occlusion if ICA reconstruction is not possible. Patterns of venous return are also characterized. Balloon occlusion testing should be carried out in supine, semi-sitting, and sitting positions. With low-flow CPFs, complete opthalmological investigations are necessary, including especially the periodic measurement of intraocular pressure. MRI plus CT scans may also be of great value, particularly with traumatic CPFs where skull base bones require detailed study. We prefer that DSA be accomplished prior to surgery, neither MRA nor CTA providing a satisfactory alternative.

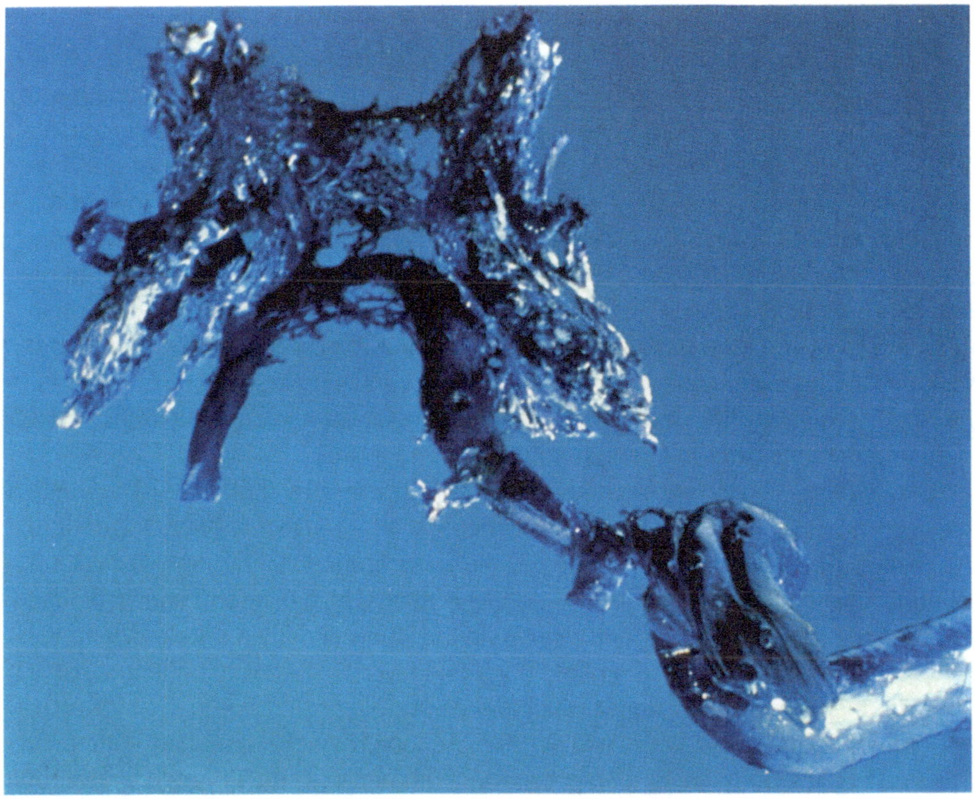

Fig. 3.19. A cast of the central skull base venous circulation demonstrates the rich trans-sellar venous channels connecting the right and left parasellar spaces (*PSs*). These connections also extend posteriorly (into the clival region). Note the drainage of the right PS into the sigmoid sinus via superior petrosal sinus (*SPS*)

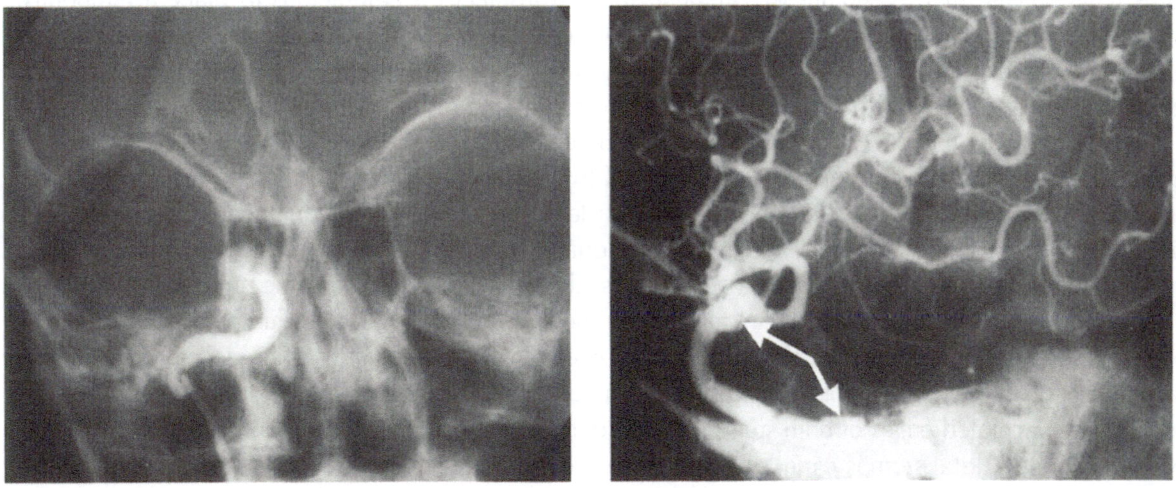

a b

Fig. 3.20. (a) The AP view of a preoperative carotid angiogram reveals a completely transsected ICA in the right parasellar space (*PS*). (b) Postoperative carotid angiography (lateral view) demonstrates a normal arterial tree, with a short high-flow graft connecting proximal and distal stumps of the left PS-ICA. End-to-end anastomoses were constructed proximally and distally, with the suture lines reinforced with Sundt-Kees cuff clips (indicated by *arrows*)

Surgical approach to the carotid-parasellar fistulae

Patient positioning, scalp incision, and the turning of an osteomuscular flap are identical to PS-ICA aneurysm operations, but special care must be made to preserve the superficial temporal artery should extracranial-intracranial bypass surgery become necessary. One must be alert for the presence of fracture lines and free bone fragments as the middle cranial fossa is exposed. Temporal lobe dura is pealed from bone toward the foramen rotundum (FR) and farther laterally toward the foramen ovale (FO) and foramen spinosum (FS) before transsecting dura at the superior orbital fissure (SOF) as described in Chapter 2. Even the slightest traction on dura is to be avoided before exposing the petrous segment of the ICA as a means of establishing proximal control. The middle meningeal artery (MMA) is transected at the FS and the ICA exposed in the petrous carotid canal and prepared for temporary clipping even before resecting the anterior clinoid process (ACP). After exposing the anterior loop (AL) of the ICA through the anteromedial triangle, the lateral and superior walls of the optic canal (OC) are removed and the optic strut thinned, mobilized, and removed.

Exposing the optic nerve (ON) and the AL may be treacherously dangerous when the skull base is fractured, particularly if free bone fragments are present. To remove bone hurriedly with a rongeur is likely to result in massive hemorrhage. In the process of drilling away the ACP and the walls of the OC one must be as conscious of preserving dura as protecting the ICA and the ON. All drilling is done with a diamond burr in limited increments, repeatedly inspecting the relationship between remaining bone and underlying dura. Where the lesion involves the AL, the ACP should not be removed entirely but a thin layer of bone is left intact over the ICA, finally "wiping" it away with the diamond burr under maximum control.

Similarly, the lateral wall of the PS, now tense with swirling arterial blood, must be dissected with great care, beginning at V3 and proceeding anteriorly rather than posteriorly from nerve III. The outer dural layer of the lateral wall of the PS is removed, exposing nerves III, IV, V1, V2, V3, and the Gasserian ganglion (GG), scrupulously avoiding opening the PS.

Once the segments of the ICA within the petrous carotid canal and the anteromedial triangle have been exposed and prepared for temporary clipping, nerve IV is gently mobilized for the purpose of enlarging Parkinson's triangle as described in Chapter 2. Since the inner layer of the lateral PS wall is extremely thin, bulging pulsations may indicate that its rupture is imminent. In such instances it is judicious to place a temporary clip on the ICA in the petrous carotid canal. Often, particularly with low-flow fistulae, this reduces arterial pressure within the PS sufficiently to allow further dissection of the ICA. High-flow CPFs, however, require isolating the ICA between proximal (in the carotid petrous canal) and distal (at the AL, immediately proximal to the ophthalmmic artery) clips. Only then may the PS be entered.

Trapped between clips, the entire PS-ICA, from the LR to the proximal ring (PR), may be exposed through Parkinson's triangle, then fully dissected and the arterial defect repaired. With spontaneous high-flow CPFs from aneurysmal rupture, one or more clips may be adequate. Traumatic lacerations of the ICA wall require closure with sutures or the application of a Sundt-Kees clip or both, reinforcing the suture line with the circumferential clip. The state of the PS-ICA wall dictates the type of repair to be carried out. If the PS-ICA is atherosclerotic or widely lacerated, local reconstruction is extremely difficult. This often necessitates a more extensive dissection of the PS-ICA, possibly constructing an end-to-end anastomosis after transecting the vessel and trimming the lacerated edges of both stumps as appropriate. If the arterial laceration is longer than 5 mm and the proximal and distal stumps cannot be approximated without tension, it becomes necessary to interpose a short graft between the petrous and clinoidal segments Fig. 3.20(a and b). If this is required, it is helpful to reinforce both suture lines with circumferential Sundt-Kees clips. Although we initially employed saphenous vein grafts for this purpose, we currently favor using the radial artery, providing that the ulnar artery and palmar arterial arches are adequate.

Traumatic low-flow CPFs arising from false aneurysms are managed identically to high-flow lesions. In some cases, however, a clip is sufficient to exclude the false aneurysm. If not, the defect is closed with interrupted sutures Fig. 3.21(a and b) and/or reinforced with a Sundt-Kees cuff clip. Either technique is sufficient for lesions of this type.

Low-flow CPFs arising from spontaneous hemorrhage typically may be interrupted by occluding the arterial leak with the bipolar cautery and sealing involved PS veins with Surgicel (Fig. 3.22). Such lesions rarely present a true surgical challenge. Patients harboring them usually present with a reddened ocular globe (conjunctival injection plus subscleral hemorrhage) and/or elevated intraocular pressure rather than full blown exophthalmos. A word of caution, however: with low-flow fistulae the surgeon must be certain he occludes only those venous elements involved with the fistula, preserving overall function of the venous network. To do otherwise is likely to make matters worse than prior to surgery. Only direct surgery allows selective occlusion of the PS, that part involved with the shunt, preserving physiological venous function.

By surgically exposing low-flow CPFs, where arterial blood is shunted to only part of the PS venous network, one is able to observe normally functioning venous structures, once again confirming Parkinson's concept of the anatomy. If the goal of treating the low-flow CPF is *only* to eliminate the shunt, without preserving function of the PS venous network, surgery may prove to be *inferior* to the natural history of the lesion [3, 6, 26, 29, 32, 37]. Excessively packing the PS with Surgicel, occluding normally functioning venous elements, may even result in thrombosis of the ICA.

In treating low-flow CPFs, cooperation between various specialties is essential. Whatever modality is selected (direct surgery, endovascular techniques, or radiosurgery), no invasive treatment should be undertaken without repeated ophthalmological examinations over several months. That excessive therapeutic enthusiasm does not always bode well for the patient certainly holds true in the management of spontaneous low-flow parasellar fistulae.

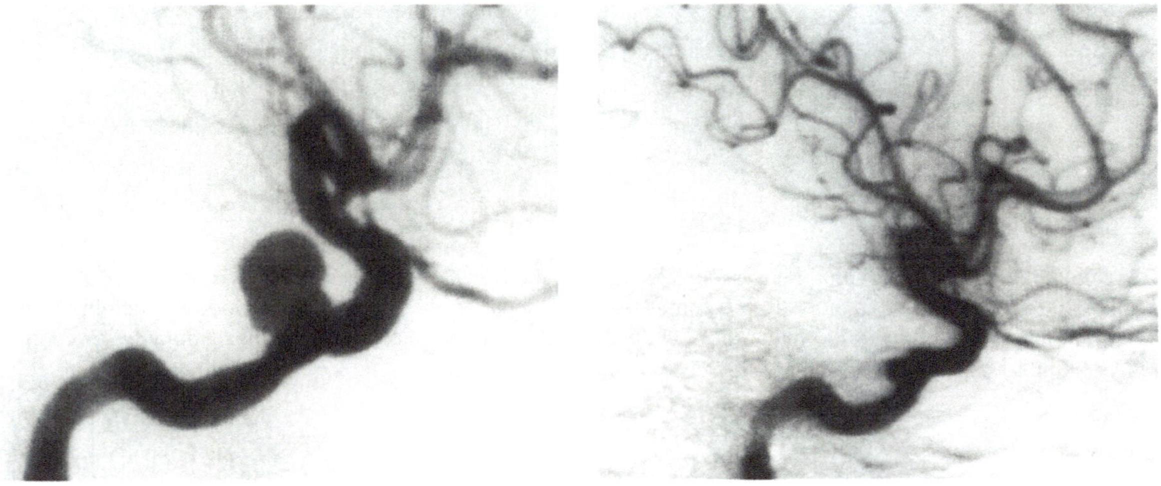

a b

Fig. 3.21. (a) Fig. 3.21 Preoperative (a) left-carotid angiogram showing a false aneurysm as a result of a small laceration of the PS-ICA due to skull base fracture. This patient had a rather light bruit following trauma – indicating an "A" type of parasellar-carotid fistula - which subsided with intermittent compression of the common carotid artery at the neck for several days. However, the nerve VI palsy – initially only subjective – became objective and of progressive character. The postoperative (b) corresponding angiogram showing slightly irregular PS-ICA after repairing the artery with interrupted sutures. Paresis of nerve VI disappeared completely two months after surgery

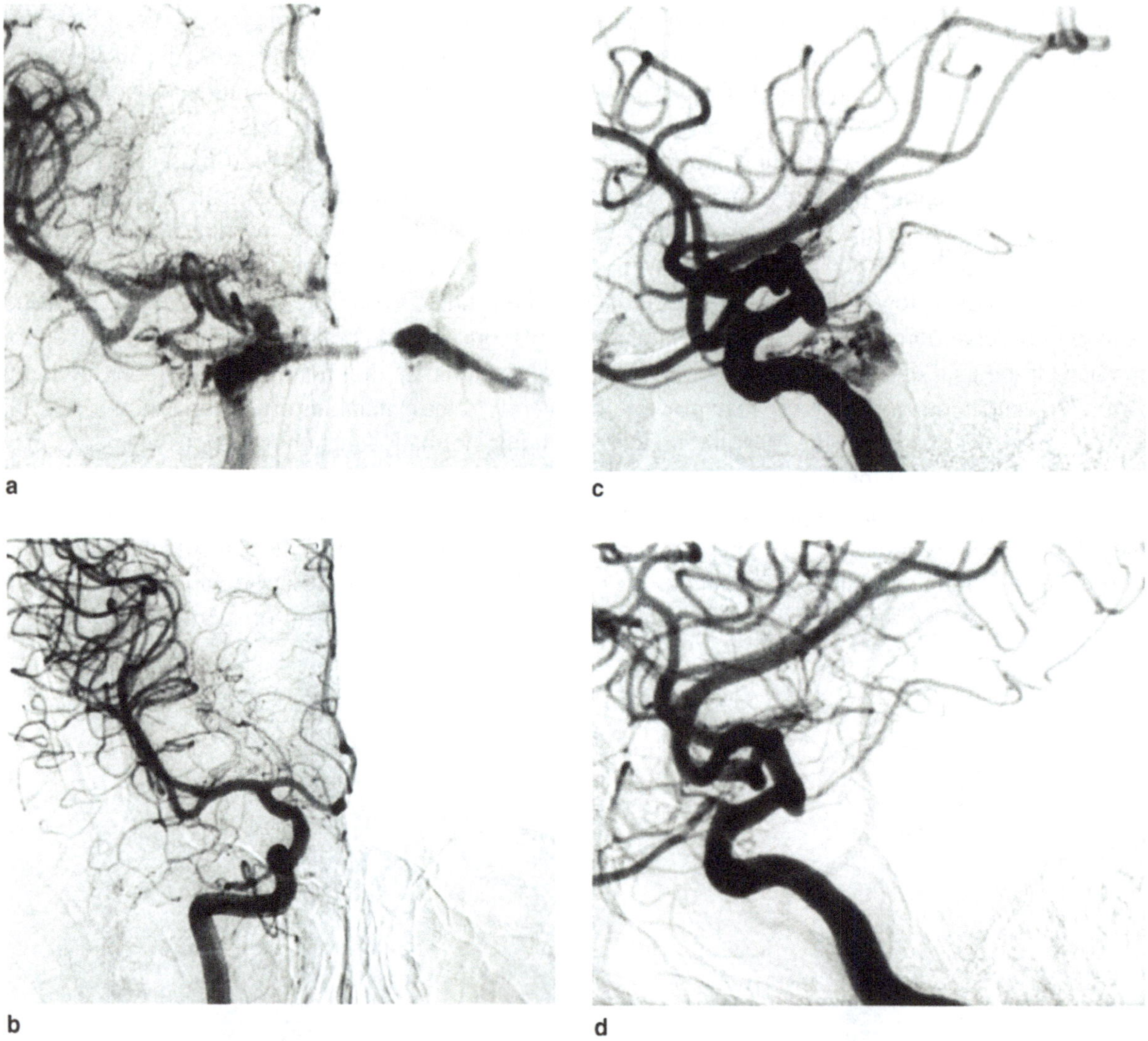

Fig. 3.22. Preoperative (a, c) right carotid angiogram showing a Type B low-flow carotid-parasellar fistula (*CPF*) which is connected by trans-sellar venous channels to the contralateral parasellar space (*PS*) venous network. As seen in (a), high pressure within the lesion results in the filling of a contralateral cortical vein. Postoperative angiography (b, d) confirms the absence of the CPF with preservation of the PS-ICA

Patients and results

Of 66 patients treated for carotid-parasellar fistulae, 52 had high-flow and 14 low-flow lesions (Table 3). Most high-flow CPFs in this series were caused by skull base fractures, resulting in 36 ipsilateral lesions, three contralateral, and two bilateral. Four high-flow CPFs resulted from stab wounds – two ipsilateral lesions, one contralateral, and one bilateral. Three patients had gun shot wounds, and there were four instances of spontaneous rupture of PS-ICA aneurysms. The ICA was reconstructed with one or more clips in 25 patients, in 10 with sutures. Sutures with clip-reinforcement was employed in six. The PS-ICA could not be reconstructed in 11 patients, requiring a grafting procedure. Postoperative angiography demonstrated the repaired PS-ICA to be patent in 46 of 52 patients. Every high-flow CPF was occluded, including those harbored by six of patients with ICAs demonstrated to be occluded following surgery. One patient required re-operation for an incompletely occluded lesion.

Fourteen low-flow CPFs were treated. Two had occurred during transsphenoidal surgery, five from trauma, and seven spontaneously. The types of low-flow CPF and the methods of interuption are presented in Table 4. Endovascular attempts failed in two patients, necessitating open surgery. Of the seven cases with low-flow lesions occurring spontaneously, only five were treated surgically, two remaining untreated since symptoms subsided spontaneously. Of 12 patients operated for low-flow fistulae, the PS-ICA was demonstrated to be patent in 10 after surgery. Prior to surgery all patients with low-flow CPFs had injected conjunctivae, subscleral hemorrhage, slight to moderate exophthalmos, diplopia, and elevated intraocular pressures. Symptoms uniformly cleared completely within four months after surgery.

The preoperative bruits and exophthalmos associated with every high-flow fistula resolved following surgery. Six patients with ipsilateral visual impairment from high-flow lesions did not improve with surgery. By contrast, nerves III – VI impairments resolved considerably if not completely in most cases within three months after surgery. Surgery resulted in no new neurological deficits, and there were no deaths.

Table 3. High-flow CPF: preoperative symptoms and signs, surgical treatment and results

Cause	N° of cases	Exophthalmos			Surgery				ICA patent			
		Ipsilateral	Contralateral	Bilateral	Clip(s)	Suture	Both	Graft	Clip	Suture	Both	Graft
Fracture of the CSB	41	36	3	2	22	5	4	10	19/22	5/5	3/4	8/10
Stab-wound	4	2	1	1	1	3	–	–	1/1	3/3	–	–
Gun-shot	3	3	–	–	1	1	1	–	1/1	1/1	1/1	–
Aneurysm-Rupture	4	4	–	–	1	1	1	1	1/1	1/1	1/1	1/1
Total	52	45	4	3	25	10	6	11	22/25	10/10	5/6	9/11

Table 4. Low-flow CPF: types of the CPFs, types of treatment and results

Cause	N°	Type			Treatment		ICA patency	
		B	C	D	Embolization	Surgery	Yes	No
Iatrogenic	2	2	–	–	–	2	2	–
Trauma	5	5	–	–	2	5	4	1
Spontaneous	7	4	2	1	–	5	4	1
Total	14	11	2	1	2	12	10	2

Conclusion

If vision is preserved following trauma, high-flow CPFs must be treated urgently, either by endovascular methods or direct surgery in order to preserve vision (and save the eye itself) and eliminate the extremely unpleasant bruit. Direct surgery allows preservation of the ICA either by direct repair or by grafting, both associated with good patency rates.

By contrast, low-flow CPFs do not require immediate treatment if intraocular pressures are stable and acceptable. Some of these lesions regress spontaneously. Low-flow CPFs may be managed successfully with direct surgery although venous complications, even PS-ICA occlusion, may occur if venous PS compartments are packed excessively with Surgicel or other material.

References

1. Barrow DL, Krisht A (1993) Cavernous sinus dural arterio-venous malformations. In: Awad IA, Barrow DL (eds) Dural arteriovenous malformations. Illinois, pp 117–130
2. Barrow DL, Spector RH, Braun FI, et al (1985) Classification and treatment of spontaneous carotid-cavernous fistulas. J Neurosurg 62: 284–256
3. Bennett DR, Van Dyk HJL, Davis DO (1973) Carotid-cavernous sinus fistula closure following angiography. JAMA 224: 1637–1639
4. Benoit BG, Wortzman G (1973) Traumatic cerebral aneurysms: clinical features and natural history. J Neurol Neurosurg Psychiatry 36: 127–138
5. Browder J (1937) Treatment of carotid artery-cavernous sinus fistula. Report of a case. Arch Ophthalmol 18: 95–102
6. Castillo M, Silverstein M, Hoffman JC, et al (1989) Spontaneous thrombosis of a direct carotid-cavernous sinus fistula: confimation by Gd-DTPA-enhanced MR. AJNR 10: 75–76
7. Cawley CM, Barrow DL (1996) Cavernous sinus dural arterio-venous malformations. Tech Neurosurg 2(1): 41–52
8. Debrun G, Lacour P, Vinuela F, et al (1981) Treatment of 54 traumatic carotid-cavernous fistulae. J Neurosurg 55: 678–692
9. Debrun G (1983) Treatment of traumatic carotid-cavernous fistula using detachable balloon catheters. AJNR 4: 355–365
10. Debrun G, Vinuela F, Fox A, et al (1988) Indications for treatment and classification of 132 carotid-cavernous sinus fistula. Neurosurgery 22: 285–289
11. Debrun G, Nauta HJ, Miller NR, et al (1989) Combining the detachable balloon technique and surgery in imaging carotid cavernous fistulae. Surg Neurol 32: 3–10
12. Dolenc VV (1983) Direct microsurgical repair of intracavernous vascuar lesions. J Neurosurg 58: 824–831
13. Dolenc V (1989) Anatomy and surgery of the cavernous sinus. Springer, Wien New York, pp 118–136
14. Dolenc VV, Lipovšek M, Slokan S (1999) Traumatic aneurysm and carotid-cavernous fistula following transsphenoidal approach to a pituitary adenoma: treatment by transcranial operation. J Neurosurg 13(2): 185–188
15. Friedmann g, Frowein RA, Luster G (1979) Karotis-Sinus-cavernous-Aneurysmen. Fortschr Neurol Psychiat 38: 57–79
16. Gailloud P, San Millan Ruiz D, Nichols DA, Muster M, Fasel JHG, Dolenc VV, Rufenacht DA (1996) A venous channel in the lateral wall of the cavernous sinus: Anatomy and clinical relevance. Proceedings of the 34th Annual Meeting of the American Society of Neuroradiology, pp 251–251
17. Hamby WB (1966) Carotid-cavernous fistula. Thomas, Springfield, Ill, pp 3–129
18. Harris AE, McMenamin PG (1984) Carotid artery-cavernous sinus fistula. Arch Otolaryngol 110: 618–623
19. Helmke K, Kruger O, Laas R (1994) The direct carotid-cavernous fistula: a clinical, pathoanatomical and physical study. Acta Neurochir (Wien) 127: 1–5
20. Hieshima GB, Cahan LD, Mehringer CM, Bentson JR (1986) Spontaneous arteriovenous fistulas of cerebral vessels in association with fibromuscolar dysplasia. Neurosurgery 18: 454–458
21. Isamat F, Ferer E, Twose J (1986) Direct intracavernous obliteration of high flow carotid-cavernous fistulas. J Neurosurg 65: 770–775
22. Mullan S (1979) Treatment of carotid-cavernous fistulas by cavernous sinus occlusion. J Neurosurg 50: 131–144
23. Newton TH, Hoyt WF (1970) Dural arteriovenous shunts in the region of the cavernous sinus. Neuroradiology 1: 71–81
24. Obrador S, Gomez-Bueno J, Silvela J (1974) Spontaneous carotid-cavernous fistula produced by ruptured aneurysm of the internal carotid artery. Case report. J Neurosurg 40: 539–543
25. Parkinson D (1973) Carotid cavernous fistula: direct repair with preservation of the carotid artery. Technical note. J Neurosurg 38: 99–106
26. Reider-Grosswasser I, Loewenstein A, Gaton DD, Lazar M (1993) Spontaneous thrombosis of a traumatic cavernous sinus fistula. Brain Injury 7(6): 547–550

27. San Millan Ruiz D, Gailloud P, De Miquel Miquel MA, Muster M, Dolenc VV, Rufenacht DA, Fasel HDJ (1999) Laterocavernous Sinus: The anatomical record 254: 7–12
28. Schievink WI, Piepgras DG, Earnest F et al (1991) Spontaneous carotid-cavernous fistulae in Ehlers-Danlos syndrome Tipe IV: case report. J Neurosurg 74: 991–998
29. Seeger JF, Gabrielsen TO, Giannotta SL, et al (1980) Carotid-cavernous sinus fistulas and venous thrombosis. AJNR 1: 141–148
30. Serbinenko FA (1974) Balloon catheterization and occlusion of major cerebral vessels. J Neurosurg 41: 125–145
31. Slusher MM, Lennington BR, Weaver RG (1979) Ophthalmic findings in dural arteriovenous shunts. Ophthalmology 86: 720–731
32. Spinnato S, Talacchi A, Perini S, Dolenc VV, Bricolo A (1997) Conservative Treatment of a Traumatic Direct Low-Flow Carotid-Cavernous Sinus Fistula: A Case Report. Acta Neurochir Wien 139: 1181–1184
33. Taki W, Nakahara I, Nishi SH, et al (1994) Pathogenetic and therapeuric considerations of carotid-cavernous sinus fistulas. Acta Neurochir (Wien) 127: 6–14
34. Taniguchi RM, Goree JA, Odom GL (1971) Spontaneous carotid-cavernous shunts presenting diagnostic problems. J Neurosurg 35: 384–391
35. Toya S, Shiobara R, Izumi J (1981) Spontaneous carotid-cavernous fistula during pregnancy or in the postpartum stage. Report of two cases. J Neurosurg 54: 252–256
36. Vinuela F, Fox AJ, Debrun GM (1984) Spontaneous carotid-cavernous fistulas: clinical, radiological, and therapeutic considerations. Experience with 20 cases. J Neurosurg 60: 976–984
37. Voigt K, Sauer M, Dichgans J (1971) Spontaneous occlusion of a bilateral caroticocavernous fistula studied by serial angiography. Neuroradiology 2: 207–211

3.4

Skull base approaches to basilar artery aneurysms

Introduction

Almost 80 percent of basilar artery (BA) aneuryms are located at or near the basilar bifurcation, or apex, accounting for one-half of all aneurysms of the entire posterior cerebral circulation [6, 7, 16, 32, 36, 39, 40]. Most occur as solitary lesions but some are associated with aneurysms at other intracranial sites, also developmental anomalies such as a persistent primitive hypoglossal artery [1]. Large and giant distal BA aneurysms generally present as manifestations of compression of adjacent structures [5, 8, 12, 19, 20, 27, 30, 34] while smaller lesions typically cause subarachnoid hemorrhage (SAH). Withholding surgery for distal BA aneurysms of any size has resulted in mortality rates exceeding 50% [42], little improvement on the results of direct surgery prior to 1966 [3, 26]. But subsequently the introduction of microsurgical techniques and various new operative approaches has produced outcomes clearly superior to the natural history of these lesions [7, 10, 11, 15, 18, 25, 31, 40, 42, 45]. Hypotension and hypothermia have proved to be useful surgical adjuncts [7, 16]. Similarly, ligating one or both vertebral arteries or the BA itself, combined with by-pass grafting, has provided a promising alternative to directly occluding large and giant lesions [4, 16–18, 24]. The technique of hypothermia coupled with circulatory arrest and appropriate neuroanesthesia offers safe conditions for distal BA lesions not thought to be operable by more standard methods [29, 33]. In addition, with clearer understanding of certain variations of vascular anatomy and specific relationships between the distal BA and the brainstem and skull base, particularly the microsurgical anatomy of the tentorial incisura [9, 16, 18, 35, 37, 40, 44, 45] still further improvements in surgical management have followed [30, 38].

While operations for distal BA aneurysms were met initially with considerable skepticism [26, 39], gradually neurosurgeons have become much more optimistic, presently favoring direct surgical treatment [15–18, 38, 40, 41, 43, 45]. This consensus did not result easily. Not only did the proximity of these lesions to the brainstem (and its blood supply) and the extremely limited working space available made these lesions particularly formidable, their relative rarity precluded a large number of neurosurgeons from gaining sufficient individual experience to become proficient in their treatment. Many felt that these difficult lesions should be attacked only in tertiary centers [43].

Endovascular occlusion techniques [2, 21, 22] represent an important advance, resulting in the belief by many that they constitute the treatment of choice for distal BA aneurysms. Yet certain of these lesions defy complete occlusion by this method, and others present special risk, particularly aneurysms with wide necks where important brainstem perforators likely arise. In such cases a direct surgical approach remains the preferred method of treatment.

The trans-orbital, trans-anterior clinoid process, trans-sellar, trans-parasellar space, trans-sylvian fissure, and trans-posterior clinoid process, approach to the distal basilar artery aneurysms ("6-T" approach)

Our original approach to distal BA lesions, published in 1987 [12], amounted to a "triple trans," or "3-T" method – trans-orbital, trans-anterior clinoid process (ACP), transsylvian. Over the past 15 years expanding the idea of exposing pertinent vascular and neural anatomy at the expense of removing skull base bone, Fig. 3.23(a, b) – compared to the more limited exposures provided by intradural methods, which depend on brain retraction – has resulted in three additional methods, each providing clear advantages [14]. The "6-T" approach expands the 3-T method in that the exposures gained depend on removal of additional skull base bone, hence a trans-posterior clinoid process (PCP), trans-sellar, trans-parasellar space (PS) approach as illustrated in Fig. 3.24(a, b) and Fig. 3.25(a–d). It is of particular value with respect to distal basilar artery (BA) aneurysms, including pure apex lesions and those arising at the origin of the superior cerebellar artery. The 6-T approach allows the skilled surgeon additional operating space [12–14] without retracting brain. It may also be combined with any of the standard approaches described previously.

The "trans-petrous" approach is primarily reserved for small aneurysms of the proximal BA, the putative basilar trunk aneurysms [28]. The "7-T" approach refers to combining the trans-petrous exposure with the 6-T approach, a method of particular value for large and giant lesions arising from the BA trunk.

These approaches offer safe and somewhat easier methods of treating these difficult lesions. However, the details of the individual steps involved, even in the most skilled of hands, require careful and repeated study.

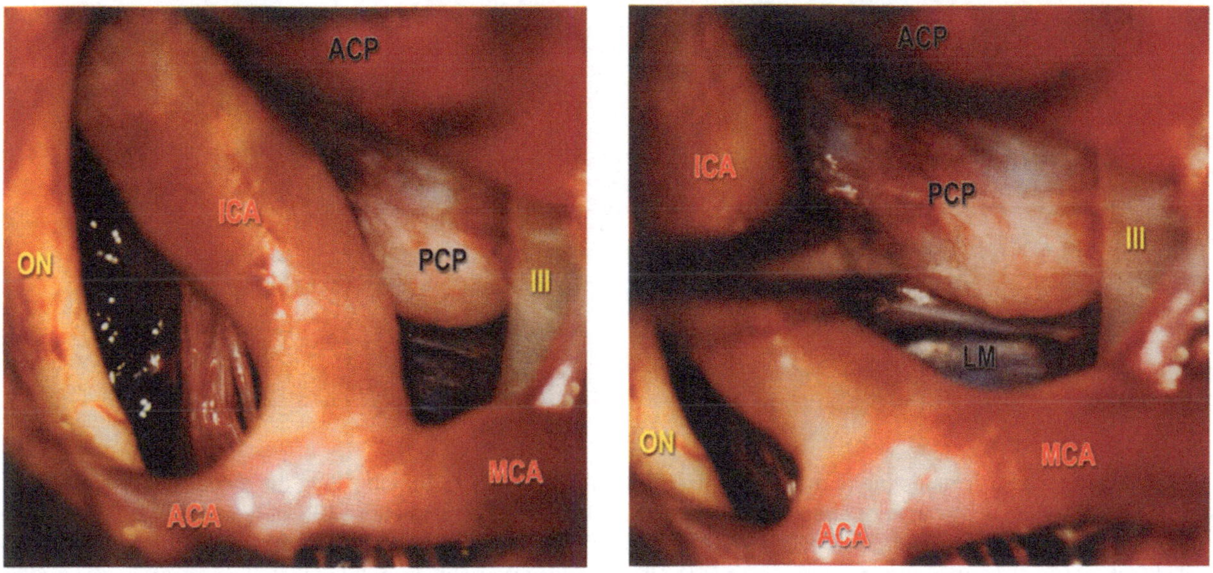

Fig. 3.23. (a) This operative photograph demonstrates the compartments between the right optic nerve (*ON*) and ICA and the ICA and nerve III. Several perforating arteries are visible between the ICA and the ON. These arteries, supplying the optic chiasm, optic tract, pituitary stalk, and hypothalamus, are vital to the normal function of these structures and must be preserved. Displacing the ICA laterally is likely to stretch such vessels, possibly resulting in vasospasm and subsequent occlusion. Lilliequist's membrane is apparent through the quadrangular window lateral to the ICA which is circumscribed by the distal part of the ICA medially, the middle cerebral artery (*MCA*) posteriorly, nerve III laterally, and the posterior clinoid process (*PCP*) anteriorly. Gentle posteromedial displacement of the ICA will not disturb such perforators, but enlarges the quadrangular corridor to Lilliequist's membrane, through which the distal BA may be exposed. (b) This is an intraoperative view of the PCP with the ICA displaced medially. Mobilizing the ICA medially will not affect perforators coursing inferomedially from the ICA, the ophthalmic artery, the ON, the optic chiasm, the anterior cerebral artery (*ACA1*), the MCA, or the lenticulostriate arteries arising from the bifurcation of the ICA or the ACA1. Note: Gentle brief retraction of the ICA does not restrict flow in the ICA. If done intermittently for short periods, this is not dangerous. The quadrangular window described above may be enlarged further only by completely resecting the PCP. Retracting the ICA very gently for short periods with a suction tube does not lead to ICA vasospasm and assures ICA patency (with its perforators) both during surgery and afterward

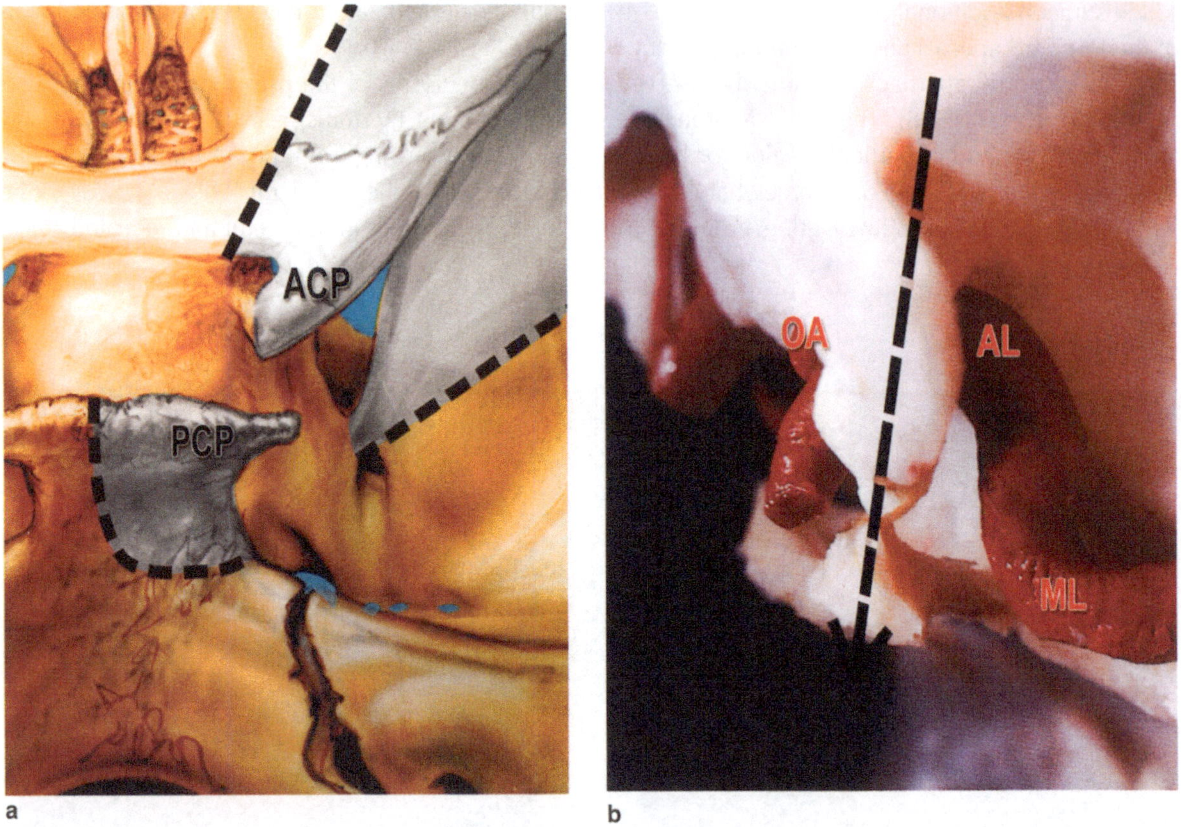

Fig. 3.24. (a) A computer rendering of the skeletonized central skull base (*CSB*), indicating the extent of bone resection: orbital roof; sphenoid wing; anterior clinoid process (*ACP*); dorsal, lateral and inferior walls of the optic canal (*OC*); posterior clinoid process (*PCP*); and dorsum sellae. (b) Skull base specimen with the ICA (filled here with red latex) inferomedial to the ACP where it forms the anterior loop (*AL*). This specimen is oriented similar to that of a patient's head position during surgery. The *arrow* indicates the most direct and safest route to aneurysms of the BA apex and distal BA trunk. Resection of bony structures as indicated in (a) provides ample space for approaching such lesions

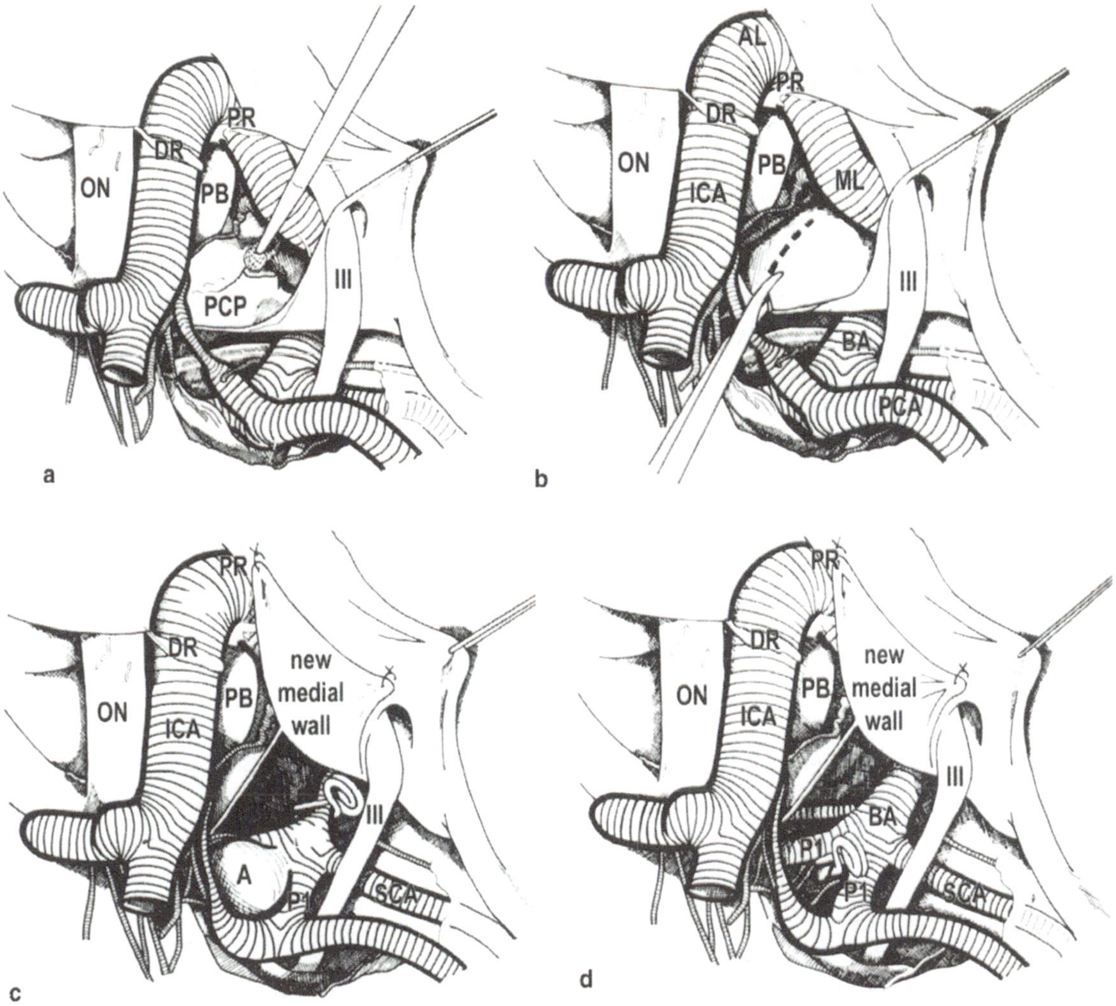

Fig. 3.25. (a) This illustration shows the ICA anterior loop (*AL*), the pituitary body (*PB*), the horizontal segment of the parasellar space (*PS*)-ICA, and the posterior clinoid process (*PCP*). The anterior clinoid process has been removed. Note that the diamond burr is inside the sella, drilling the PCP from its anterolateral surface in order to preserve dura posterior to the PCP, a measure to prevent injury to the basilar artery (*BA*) and/or the aneurysm. Lateral to the PCP, nerve III enters the lateral wall of the PS, coursing parallel to the horizontal segment of the PS-ICA. The PB lies medial to and beneath the AL. (b) Removing the PCP and the ipsilateral half of the dorsum sellae exposes posterior cranial fossa dura, which is marked here with an interrupted line indicating a midline incision. A dural flap is shaped as indicated for "closure" of the ipsilateral PS from its medial surface. (c) The dural flap posterior to the removed PCP and dorsum sellae has been reflected upward, medial to the horizontal segment of the PS-ICA. Its edges are sutured to dura adjacent to nerve III posteriorly, and to the proximal ring (*PR*) anteriorly. After thus fixing the dural flap, "creating" a new medial wall for the ipsilateral PS, the distal segment of the BA trunk is exposed, providing additional space in the vicinity of the BA apex (aneurysm) including the distal BA and its branches. Reflecting the dura in this manner ensures that the lesion and associated vascular structures are easily exposed and that the distal BA trunk is accessible should temporary occlusion become necessary. The space bordered by the intradural ICA (medially), the posterior communicating artery (*PCom*) and posterior cerebral artery (*P1*) (posteriorly), nerve III and the newly fashioned medial wall of the ipsilateral PS (laterally), and the distal leg of the AL of the ICA and PB (anteromedially) becomes much larger after the PCP has been removed. The distal segment of the BA trunk, both superior cerebellar arteries (*SCA*)s, and both posterior cerebral arteries (*PCA*)s are shown. (d) Illustration showing the clipped neck of a BA apex aneurysm the sac of which has been resected. Perforators arising from the BA bifurcation and both P1s have been preserved. Hemostasis is maintained by sealing the connecting venous sinuses adjacent to the PB with Surgicel

Pathological anatomy, presentation of patients, and diagnostic work-up

The apex of the BA, its bifurcation, giving rise to the most important perforators and to both P1 segments, may be located high (well above the dorsum sellae) or low (adjacent to the clivus or the dorsum sellae). Large aneurysms arising from high basilar bifurcations project into the interpeduncular cistern, even higher into ventricle III, hiding and possibly compressing perforating arteries arising from posterior elements of the circle of Willis behind the aneurysm. Such lesions typically compress both pyramidal tracts, resulting in the "heavy legs" described by Lucian Rubinstein, the neuropathologist, even additional motor disturbances, endocrinological deficits, and hypothalamic symptoms and signs. Large lesions arising from lower basilar bifurcations generally compress the brainstem, and, especially, one or both nerve(s) III.

No neurological lesion requires a lengthy surgical approach between vital structures rivaling that necessitated by BA aneurysms, whether arising from the BA bifurcation or along its trunk. That so many different approaches to these lesions have been devised reflects the difficulty of avoiding devastating, life-threatening neurological consequences [23].

Traversing the considerable venous network in the PS is fraught with hazard and may result in a number of untoward effects, ranging from profuse venous bleeding to inadvertent injury to nerves III, IV, V1 and VI. Furthermore, excessive packing of the PS with Surgicel in an effort to control bleeding endangers ICA patency [13]. Resecting the PCP and opening the PS is crucial to gaining sufficient working space adjacent to the aneurysm, allowing its safe dissection from nearby structures and complete clipping while preserving patency of the distal basilar trunk, both posterior cerebral artery (PCA1) segments, and, particularly, all neighboring perforating arteries. To attempt to place a clip, either permanently or for very brief periods, without perfect vision is to risk damaging the contralateral nerve III and/or occluding one or more vital perforators, resulting in death or serious permanent disability for the patient.

Surgical technique

Craniotomy is carried out by the basic central skull base approach described in Chapter 2, including unroofing the orbit, resecting the sphenoid wing (SW) and anterior clinoid process (ACP), and opening the optic canal (OC), removing its superior and lateral walls and the optic strut. Only the anterior and medial half of the lateral wall of the PS need be disturbed, gently stripping away its outer (dural) layer, exposing nerves III, IV and V1. With small distal BA aneurysms the dura may be opened at this point and the sylvian fissure divided, and the aneurysm clipped without necessity of the PCP resection.

With large and giant distal BA aneurysms, trans-PS, transsellar, and trans-PCP dissections are carried out after opening the dura. It is incised from the peripheral end of the sylvian fissure, proceeding along its longitudinal axis toward the distal (dural) ring (DR). The DR is cut circumferentially around the ICA, completely freeing the artery from dura, allowing its subsequent displacement medially without exposing it and its branches to trauma likely to induce vasospasm. Mobilizing the ICA in this manner is essential to providing adequate space for dissecting the aneurysm. It is important to understand that safely incising dura from around the ICA is not possible without first removing the optic strut. Whenever possible, nerve III should be left attached to the inner layer of the lateral wall of the PS. No attempt should be made to dissect it from its canal in the wall of the PS. The membrane posterior to the anterior loop (AL) of the ICA is incised toward the PCP. The diaphragm sellae is then incised transversely (in the coronal plane), parallel to the edge of the dorsum sellae, toward the contralateral PCP. Any venous bleeding encountered from connecting sinuses in the vicinity of the pituitary body may be controlled with Surgicel introduced toward the contralateral side. This affords extradural access to the sella. Venous bleeding may be profuse posterior and lateral to the pituitary body (also anteriorly and caudally) from sinuses connecting two PSs. Bleeding is also likely from the ipsilateral PS medial to the horizontal segment of the PS-ICA. Hemostasis is obtained by placing Surgicel into these venous sinuses as appropriate. Only when all bleeding is stopped can the PCP be drilled away, from its intrasellar surface (from the pituitary body toward dura posterior to the dorsum sellae and the PCP) as shown in Fig. 3.25(a–d). Great care must be taken to protect the ICA and the pituitary body, and particularly to avoid perforating dura overlying the dorsum sellae, PCP, and upper clivus. The PCP should be removed completely, then the dorsum sellae is drilled away, at least to the midline, then finally the lateral surface of the upper clivus, taking care to avoid injuring nerve III as it courses lateral to the PCP. After removing the PCP and examining the parasellar area and ipsilateral PS for absolute hemostasis, the dura overlying the dorsum sellae is transected in the midline and fixed with sutures anterolaterally to the PR as indicated in Fig. 3.25(b, c). This essentially creates a medial wall for the PS. With low lying distal BA aneurysms, the BA proximal to the lesion may not yet be exposed. In such cases the clivus must be resected further, carefully drilling from the floor of the sella. Prior to fixing the dura dorsal to the dorsum sellae to the PR as shown in Fig. 3.25(c, d), Surgicel in the ipsilateral PS should be loosened in order to avoid compression of the ICA. Failing to do this is to risk creating ICA stenosis, or even occluding the vessel.

The sylvian fissure is divided completely, exposing the middle cerebral artery (MCA) and its major branches. Next, the uncus is dissected from arachnoid surrounding the posterior communicating artery (PCom) and its perforators, the anterior choroidal artery, and the proximal segment of nerve III as it passes between the PCA and the superior cerebellar artery (SCA), finally exposing the brainstem. With large and giant aneurysms the PCom may be safely transected so long as its perforators are not disturbed. If they are allowed to remain patent, these perforators will be supplied from stump(s) of the PCom or the PCA or from both.

When sufficient working space around the aneurysm has been achieved, the surgeon may begin dissecting the lesion from adjacent structures. The aneurysm cannot be clipped safely until the BA proximal to it, both SCAs, both nerves III, both PCA1s, and all perforators arising from the BA bifurcation, both PCA1s, and from the neck of the aneurysm have been completely exposed and dissected. When necessary to fully dissect all these structures, the BA may be temporarily occluded for 3-minute intervals by placing a clip proximal to the aneurysm. Temporarily occluding the BA may be repeated safely but not without allowing full circulation in the BA and its branches for periods of at least five minutes. Such a maneuver allows the aneurysm's wall tension to be reduced, thereby allowing safe dissection of the sac from adjacent structures.

Definitive clip(s) should be adjusted and readjusted until the entire aneurysm has been occluded and the proximal BA, both PCA1s, and all perforators are demonstrated to be patent Fig. 3.26(a–d). No clip should be permitted to contact or compress nerve(s) III or the ICA or any other important structure. After clipping the aneurysm, flow within both PCA1s should be confirmed with a Doppler probe. Once this is done and all perforators are judged to be patent, the sac of the aneurysm may be opened and emptied of its contents. Peeling the sac of the aneurysm from the brainstem and adjacent vessels is not only extremely dangerous but entirely unnecessary. It should not be attempted.

Incising the DR precludes watertight dural closure unless a piece of temporalis muscle or fascia is placed around the ICA and covered with fibrin glue.

The transpetrous approach to proximal basilar trunk aneurysms

This approach is appropriate for small aneurysms of the BA trunk, arising from any point between the confluence of the vertebral arteries and the origin of the SCAs Fig. 3.27(a–d). The standard craniotomy for approaching the CSB is performed as described in Chapter 2. The squama of the temporal bone is trimmed sufficiently to allow exposure of the foramen ovale (FO), the foramen rotundum (FR), and the foramen spinosum (FS). The middle meningeal artery (MMA) is coagulated and divided and dura is peeled from the anterior wall of the petrous bone. The geniculate ganglion, when not covered by bone, is exposed along with the greater petrosal nerve, finally the dorsal border of Meckel's cave and the nearby V3 and Gasserian Ganglion (GG). Dura lateral to V3, in front of the FO, is incised perpendicular to the longitudinal axis of V3, then dura is peeled from V3 and the GG. Dura is incised farther anteriorly over the base of the lateral triangle (circumscribed by V2, V3, and middle fossa bone) across V2 into the anterolateral triangle (defined by V1 and V2). The outer layer of the lateral wall of the PS is then peeled from V2 and the anterolateral triangle, accessing the lateral border of V1, exposing the GG completely and the entry to Parkinson's triangle (between nerves IV and V1). The apex of the petrous pyramid is then drilled away as described in Chapter 2, the interior auditory canal (IAC) opened, preserving dura covering nerves VII and VIII within the IAC. Drilling petrous bone medial to the carotid canal allows removal of the lateral part of the clivus. Care must be taken to avoid damaging nerve VI in Dorello's space. If the petrous apex cannot be removed posterior and medial to the plexiform segment of the nerve V root, and from beneath and medial to GG, dura behind nerve V, along the longitudinal axis of nerve V, is incised additionally. This allows complete exposure of nerve V from the brainstem to the GG. Additional mobilization of the GG is possible, allowing access to the space anterior to nerve V and beneath the GG, facilitating the removal of the remainder of the petrous apex. When possible, the superior petrosal sinus (SPS) should be left intact. Resecting additional dura beneath the GG and toward Dorello's space exposes nerve VI (as it courses from the brainstem to its point of entering the dura).

Opening dura posterior and anterior to nerve V, lateral to nerve VI, is sufficient for dissecting and clipping most small BA trunk aneurysms, those located from the confluence of the vertebral arteries proximally to the origin of SCAs distally. As with all aneurysms, such dissection includes exposing the BA proximally and distally and the entire neck of the lesion. If the BA distal to the aneurysm cannot be adequately visualized, typically in cases where the BA is higher and closer to the clivus than usual, additional drilling of the clivus and trimming of adjacent dura becomes necessary. In such cases this allows entry into PS and exposure of nerve VI throughout its course in Dorello's space and beyond. When the posterior compartment of the PS has been occluded with Surgicel, nerve VI should be dissected from its dural attachment where it enters Dorello's space, as well as throughout Dorello's space. Only when it is free from dura, can nerve VI be gently displaced, making further drilling of the clivus, medial and ventral to Dorello's space, safe. Gently lifting and retracting the medial loop (ML) of the ICA affords sufficient space for clipping the aneurysm under direct vision. When employing this approach, skeletonizing the PS anteromedially beyond Parkinson's triangle is not recommended. Neither is dissecting nerve IV. Isolating nerve IV is far more difficult and dangerous than combining this method with the 6-T approach. Generally, clips should be placed on the necks of proximal BA trunk aneurysms parallel to the longitudinal axis of the BA (Fig. 3.27b) rather than perpendicular to it. Once again, after final clip placement, patency of the BA at the neck of the aneurysm should be confirmed with a Doppler probe. This is particularly important when the aneurysm's neck is broad. Once this is done, and when all brainstem perforators and nerves IV, V, and VI are demonstrated to be free of the clip, the sac of the aneurysm may be opened. Dural defects medial to the GG and over the apex of the petrous pyramid should be covered with muscle and coated with fibrin glue.

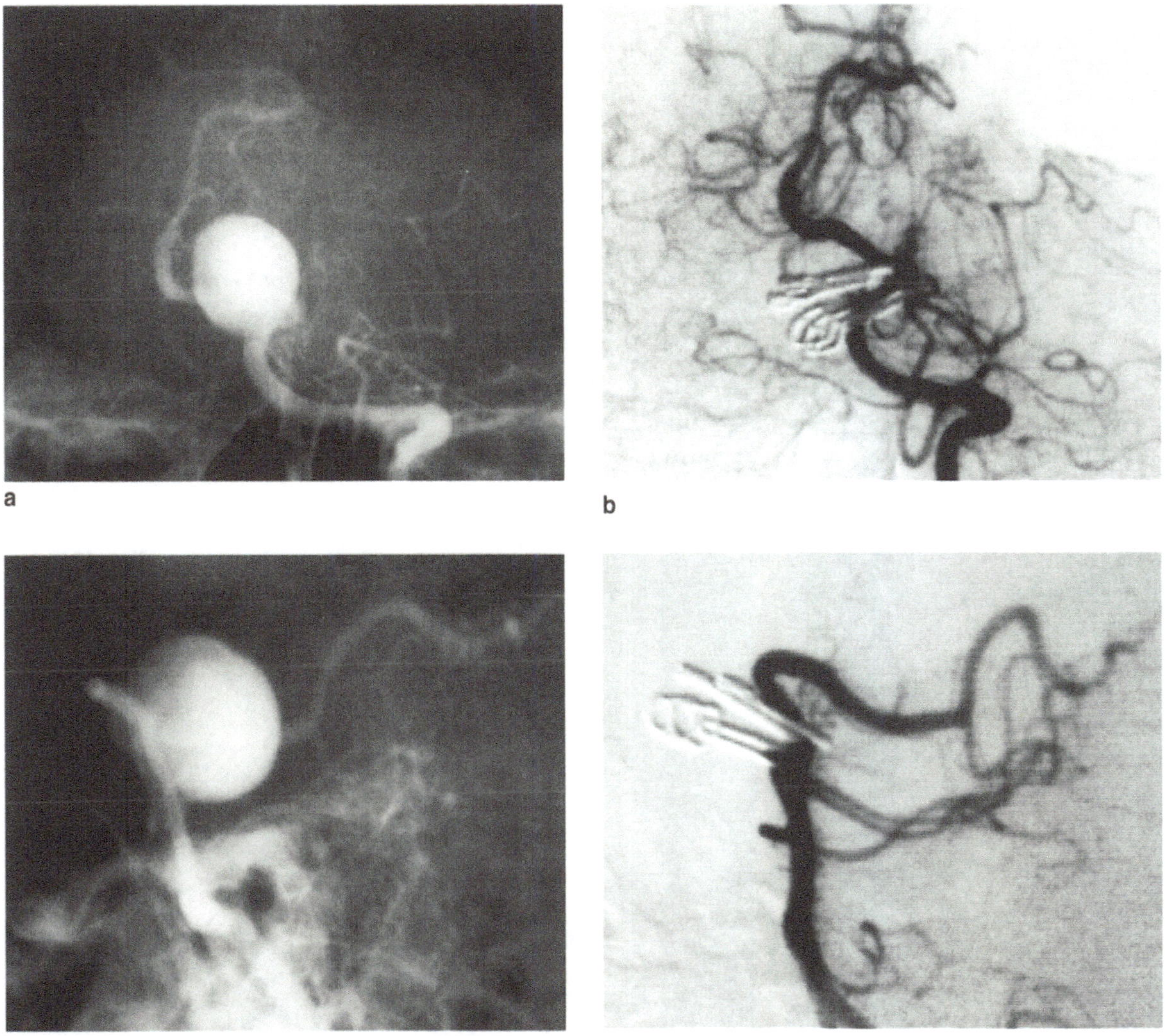

a b c d

Fig. 3.26. AP (a) and lateral (c) views of a preoperative left vertebral angiogram demonstrating a large aneurysm of the basilar artery (*BA*) apex and distal BA trunk. Note: The left posterior cerebral artery (*PCA*) does not arise from the BA. AP (b) and lateral (d) views of a postoperative angiogram confirm satisfactory clipping of the aneurysm with right PCA remaining patent. In (a), note the broad and indistinct neck of the aneurysm, characteristics which preclude endovascular treatment. The aneurysm has been occluded with several clips (b, d), and perforators arising from the basilar apex and the right P1 remain patent (d). Dissection of the aneurysm from nerves III on either side, the right P1, and perforators to the upper brainstem required intermittent temporary clipping of the basilar trunk. Temporary interruption of flow in the distal BA for 3-minute periods (alternating with 5-minute intervals of perfusion), was a crucial part of this procedure. Total occlusion time was 56 minutes. The patient was discharged from the hospital one week after surgery in good condition. His only neurological deficit, an incomplete ipsilateral nerve III paresis, was not present two months following surgery

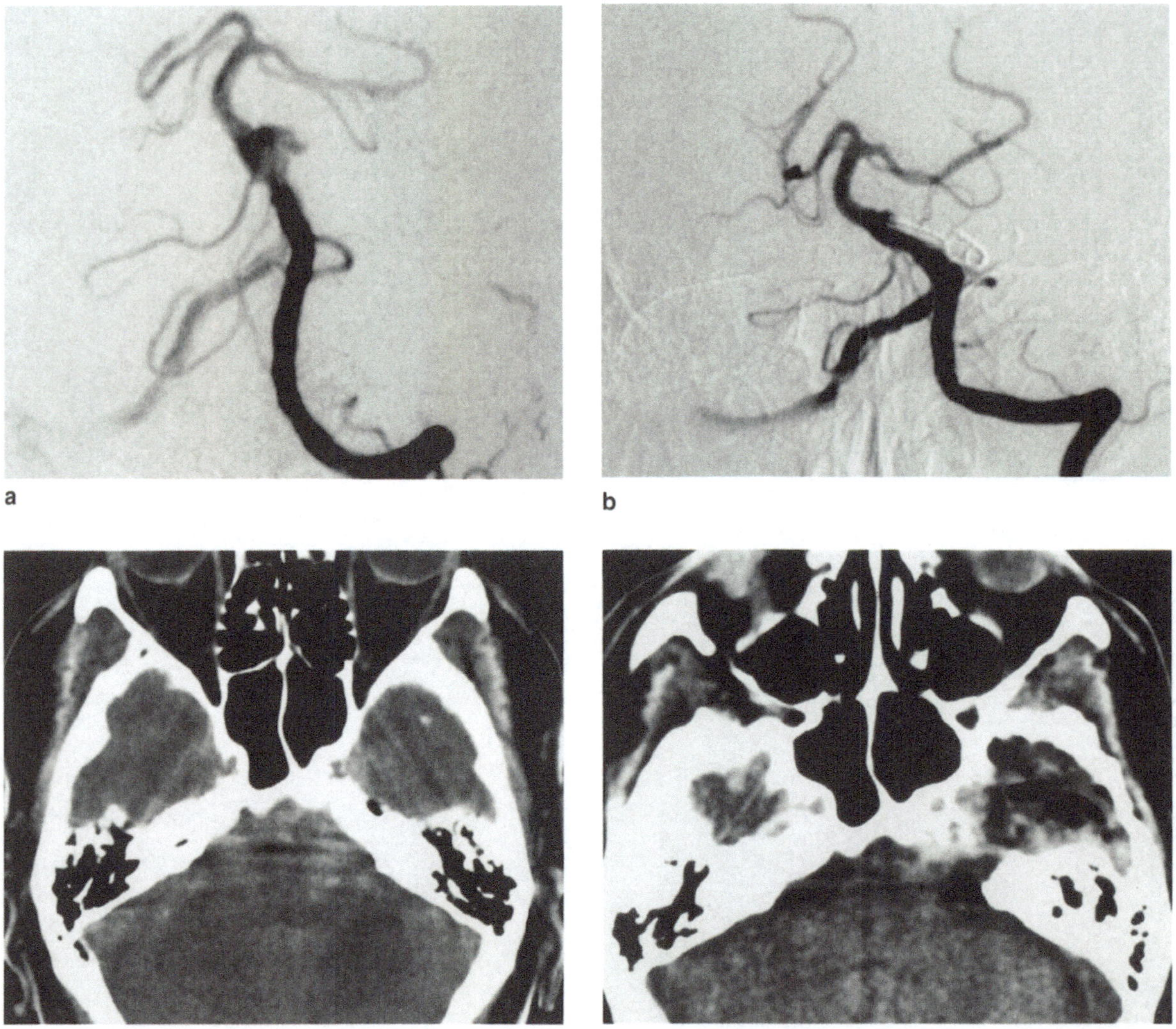

a b c d

Fig. 3.27. A left vertebral angiogram shows a small basilar trunk aneurysm preoperatively (a) and postoperatively (b) a small residual aneurysm (with two clips placed parallel to the BA trunk). In (a), the aneurysm's irregular shape and heterogeneous filling with contrast material suggest the presence of a thrombus and possibly a sclerotic, calcified wall (which was confirmed at surgery). The thick, calcified BA wall extending into the aneurysm's neck precluded the ideal positioning of clips. (c) A preoperative unenhanced CT of the same patient demonstrates normal petrous apices bilaterally. An unenhanced CT following the transpetrous approach to this aneurysm (d) reveals a surgical defect in the left petrous apex

The 7-T approach to large and giant basilar artery aneurysms

While the 6-T approach is ideal for exposing the distal BA trunk, it does not allow exposure and dissection of the BA proximal to large and giant aneurysms. Similarly, the transpetrous approach to such lesions arising from the proximal or middle segments of the BA trunk precludes visualization of the distal BA without injuring the PS-ICA and nerves IV and III. Combining these approaches – adding the trans-petrous approach to the 6-T approach – amounts to the "7-T approach," which essentially results in skeletonizing the PS, affording exposure of the entire BA, from the confluence of the vertebral arteries to the BA bifurcation. Since this approach carries additional risk for nerves III and IV, it should be employed only in situations where no better alternative exists. It is designed to treat primarily large and giant BA trunk aneurysms.

Patients and results

Over the past 15 years 153 consecutive patients have been treated for BA aneurysms – 79 female, 74 male. This includes 115 small lesions, 31 large, and seven giant. Of the 115 patients with small aneurysms, 108 had bled, and seven experienced SAH from the rupture of an anterior circulation aneurysm, the basilar lesions being discovered incidentally by angiography. Of 31 large aneurysms, five had bled and 26 presented as space occupying lesions. One of the seven giant aneurysms had bled, and six presented with symptoms and signs of mass effect.

Of the seven patients with distal BA aneurysms found incidental to remote ruptured lesions, five were treated at the time the leaking aneurysm was clipped and two after three months, allowing all manifestations of the SAH to clear prior to recommending another craniotomy. Both patients did well at surgery, and the postoperative results were good.

Of the 115 small aneurysms, 93 arose from the distal BA and 22 from the basilar trunk. All were clipped without technical problems. Of 31 large aneurysms, 25 were located at the distal BA and six arose from the BA trunk. Twenty-five of them were clipped, four wrapped, and two defied attempts at treatment. Of seven giant aneurysms, two arose at the distal BA and five from the BA trunk. While three of these lesions were clipped successfully and two were wrapped, two could not be treated. Ninety-eight of the 115 small aneurysms were clipped without temporary interuption of BA flow. Each of the remaining 17 required temporary intermittent proximal clipping of the BA proximal to the lesion.

In the group of 31 large lesions, the BA proximal to the aneurysm was prepared for temporary occlusion, but interuption of flow proved to be necessary in only 25, facilitating successful clipping in all 25. Four others were wrapped, and two could not be treated. Of the seven giant aneurysms, three required multiple clips, two were wrapped, and two could not be treated. Among six large BA aneurysms and four smaller lesions, calcific atherosclerosis precluded safe temporary clipping of the BA proximal to the lesion and this maneuver was not attempted. Of the entire series, six large and four giant aneurysms defied occlusion by clipping. One hundred four were excluded from the circulation without disturbing BA flow, while temporary intermittent BA occlusion was required in 49 cases.

One hundred six patients recovered completely, 24 were moderately disabled, and 15 had severe disabilities. There were eight postoperative deaths, five with large and three with giant aneurysms.

Table 5. Basilar artery aneurysms: size, location, presentation, surgical approach, type of the aneurysm exclusion, and results

Size	Distal BA / Proximal BA	Presentation: SAH	Tumor effect	Incidental finding	Approach: 3T or 6T	Trans-petrous	7T	Aneurysm exclusion: Clipping	Wraping	None	Postoperative results: Exellent	Good	Defficits	Died
Small	93	89	–	4	93	–	–	93	–	–	91	2	–	–
	22	19	–	3	–	22	–	22	–	–	10	10	2	–
Large	25	3	22	–	25	–	–	22	2	1	3	11	8	3
	6	2	4	–	–	–	6	3	2	1	2	1	1	2
Giant	2	1	1	–	–	–	2	1	–	1	–	–	1	1
	5	–	5	–	–	–	5	2	2	1	–	–	3	2
Total	153	114	32	7	118	22	13	143	6	4	106	24	15	8

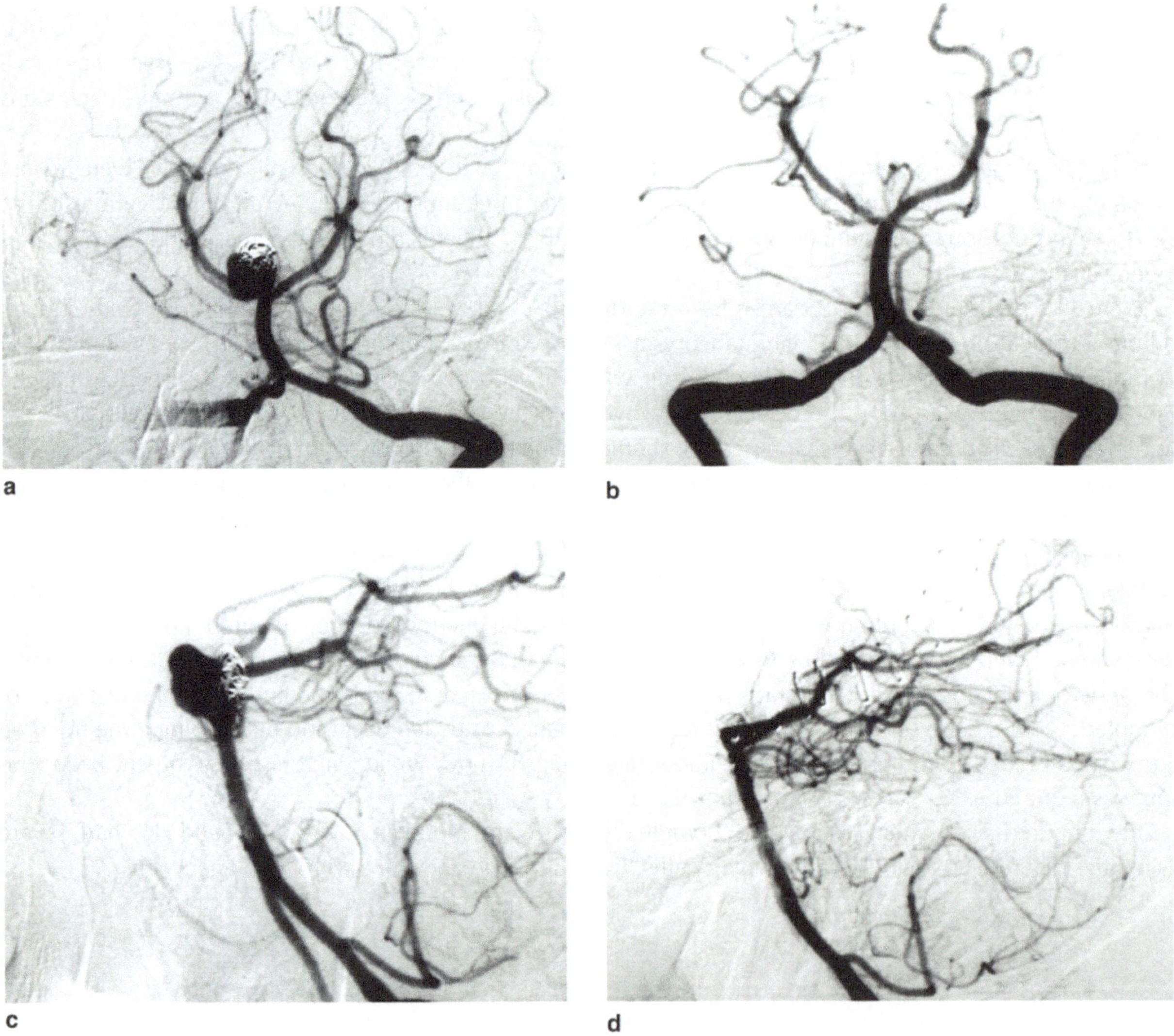

Fig. 3.28. Preoperative vertebral angiograms (a, c) demonstrate a basilar apex aneurysm treated previously with coils. Postoperative (b, d) angiography confirms that the aneurysm has been clipped and both P1 segments remain patent. Complete occlusion of the aneurysm was confirmed immediately following endovascular treatment after the original hemorrhage. Two years later, after a second SAH, angiography demonstrated coils posteriorly but clear filling of the ventral part of the aneurysm (a, c). At surgery the aneurysm was clipped, perforators preserved, the aneurysmal sac opened, and the coils removed. The patient left the hospital the eighth postoperative day, free of neurological deficit

3.5

Evolution of the "6-T" and "7-T" approaches to basilar artery aneurysms

Devising the 6-T and 7-T approaches seemed desirable in order to maximize exposure of important perforators "behind" BA aneurysms and to gain additional space for clip(s) placement (Fig. 3.26). In our experience, such limitations are especially acute with BA apex lesions larger than 1 cm in diameter whatever more standard approach is employed – pterional (lateral or medial to the ICA), transsylvian, or sub-temporal [16, 18, 40, 45]. The transoral-transclival approach to the distal BA trunk and BA apex aneurysms does not seem to be a satisfactory alternative because of its high risk of postoperative CSF leak [23]. After dissecting fresh cadaver specimens for years and performing a large number of CSB operations, particularly surgery for COAs and various tumors, it seemed that approaching BA apex aneurysms by removing more skull base bone than for COAs operations described previously [12] would result in clear advantages. Our current approach to BA apex and distal BA trunk aneurysms evolved in phases from the original 3-T approach [12, 13, 14]. Realizing that resecting the sphenoid wing, the ACP, and the PCP provided the shortest route to the BA apex and distal BA trunk amounted to a major breakthrough (Figs. 3.23–3.25). Opening the OC, gently displacing the ICA medially (possible only if the DR has been incised circumferentially), and preserving vital perforators and nerve III (while working along the longitudinal axis of nerve III) are also crucial components of this method (Fig. 3.23).

For years we wrestled with the problem of over-packing the PS with Surgicel, therefore risking critical ICA compression. We had overlooked the possibility of incising dura posterior to the PCP in such a way as to "create" a medial wall for the PS (Fig. 3.25). Creating such a safe PS medial wall is a major element of the approach. Not only does this minimize the likelihood of troublesome bleeding from the PS, it circumvents the potential problem of restricting ICA flow.

The transpetrous approach to the BA trunk is a logical extension of numerous operations performed for CSB tumors. Furthermore, where the size of the aneurysm prevents placing a temporary clip proximal to the lesion, the sub-temporal [15, 18] or transpetrous approach [28] should be combined with the 6-T approach. Combining the 6-T approach with the transpetrous approach results in the 7-T approach. The 7-T approach includes also all previously described methods – the pterional [45], transsylvian [40], and subtemporal approaches [16, 18].

Conclusions

Compared to the transoral-transclival approach [23], the 6-T and 7-T approaches obviate the necessity of an ENT or oral surgical team and their special instrumentation and other requirements, while carrying little risk of postoperative CSF leakage and meningitis. The three skull base approaches ("6-T", transpetrous, and "7-T") provide safe, effective methods of completely excluding these difficult lesions from the cerebral circulation. They complement other approaches by providing maximum exposure of the BA proximal and distal to such aneurysms, important particularly with large and giant lesions.

Sometimes the nature of the BA wall, occasionally quite fragile or excessively firm and brittle from calcification, requires that additional measures be considered. Compressing calcified, atherosclerotic plaques risks dislodging debris into the lumen of the BA, leading to life-threatening stroke. In many such cases temporarily occluding the BA internally by endovascular techniques would seem to be much preferred.

Large and giant BA aneurysms, also those with very broad necks from which perforators may originate, typically preclude satisfactory endovascular treatment (Fig. 3.26). Such lesions provide opportunities for surgeons and neuroradiologists to work together. Collaboration offers the advantage of temporarily occluding the BA internally, obviating the risk of releasing debris into the lumen of an atherosclerotic vessel by external clip-compression. The surgeon may use his time more efficiently, dissecting the aneurysm for three minutes while the neuroradiologist temporarily arrests BA flow by inflating a balloon proximal to the lesion, restoring normal BA flow at intervals of at least five minutes between periods of occlusion. Not only would such teamwork result in less trauma to the vessel wall, but the periods of direct dissection of the aneurysm are lengthened by at least 30 seconds since the surgeon does not have to worry himself with the sometimes tedious business of removing temporary clips placed in awkward positions in limited working spaces.

The technique of extra-corporal circulation and complete cardiac standstill for limited periods of time enables the surgeon to dissect large and giant aneurysms circumferentially with safety, allowing the preservation of all perforators and precise, complete clipping of the lesion. Yet dissecting time is limited, and the necessary anticoagulation and cardiovascular manipulations introduce the potential for new complications. While the controlled hypotension afforded by extra-corporal circulation makes temporary intermittent BA clipping unnecessary and may be friendly to the BA, what is the overall effect on the patient's brain? Or on the patient as a whole [29, 33]?

We believe strongly that neuroradiologists and neurosurgeons should combine their efforts in determining the ideal treatment modality for BA aneurysms and work together in its execution. The capacity to simply introduce coils into a BA apex lesion may fool one into losing sight of the fact that incomplete occlusion may result in continued growth of the aneurysm while total occlusion may compromise important perforators. Figs. 3.28(a–d) and 3.29(a and b). Furthermore, flow defects adjacent to the orifice of the aneurysm may lead to BA occlusion. We have observed that the number of aneurysms requiring surgery following repeated attempts at coil occlusion is increasing. One of our patients, referred for surgery after six coiling procedures (with three episodes of bleeding during the course of treatment), asked the simple question, "Why didn't you guys cooperate in the first place?" How many did not survive to voice such poignant advice? Neurosurgeons and neuroradiologists must work together from the very outset with patients harboring lesions not entirely meeting criteria for either surgery or endovascular treatment. Every questionable lesion should be exposed in the operating room, then treated by endovascular techniques, giving direct inspection a role in determining when treatment is optimal and complete. Only then will the highest percentage of safe, certain BA aneurysm occlusions be realized, and the number of catastrophic events reduced.

Some BA aneurysms defy satisfactory treatment by any method currently available – surgery and endovascular treatment, either alone or in combination Fig. 3.30(a–d). Pioneers paving the road to treating seemingly impossible lesions in the past danced on the cutting edge of what was possible, sometimes breaking every rule of accepted management, even at times the law (Fig. 3.31). Advancement from this point will depend on teamwork if further understanding is to be realized. In the future, the size and location of BA aneurysms will no longer be limiting factors in their treatment. What will remain is the limited capacity of local circulation to compensate the loss of blood supply contributed by diseased BA segments. Severe atherosclerosis, often leading to fusiform dilatation, severely restricts brainstem circulation (Fig. 3.30), commonly insurmountable by current treatment methods. The solution lies in some form of revascularization, either partial or total.

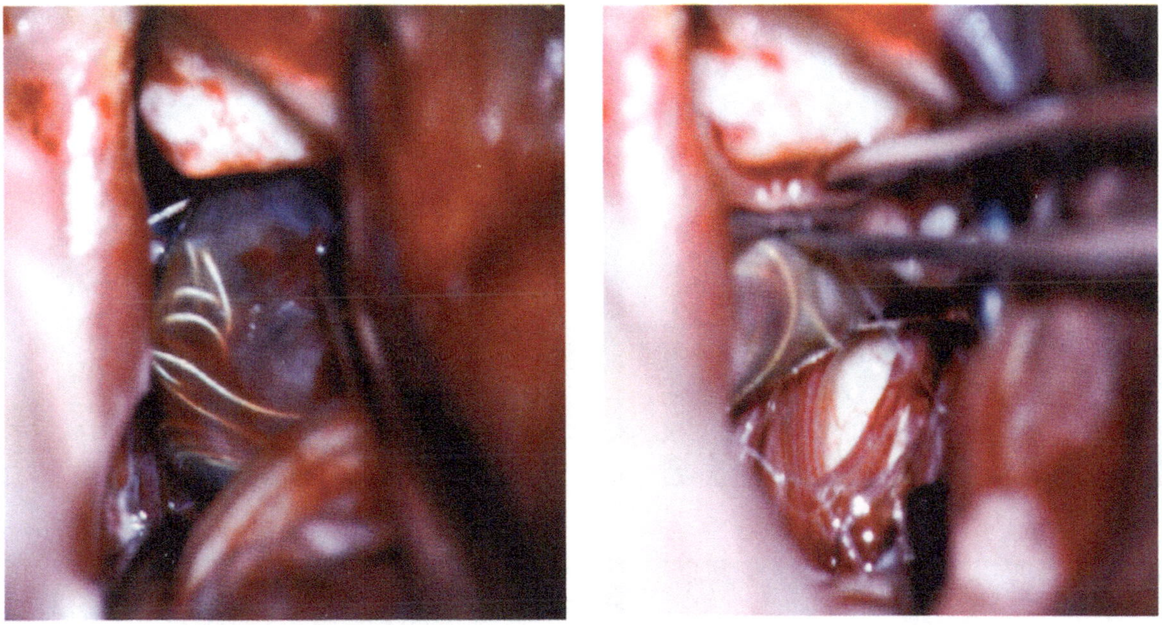

a b

Fig. 3.29. An intraoperative photograph of the aneurysm presented in Fig. 3.28 before clipping (a), the coils visible through the transparent wall of lesion. Prior to final clipping, all perforators originating from the BA apex and both P1s were identified and preserved (b). The aneurysm was incised, thrombus and coils removed, and the aneurysm's sac partially resected

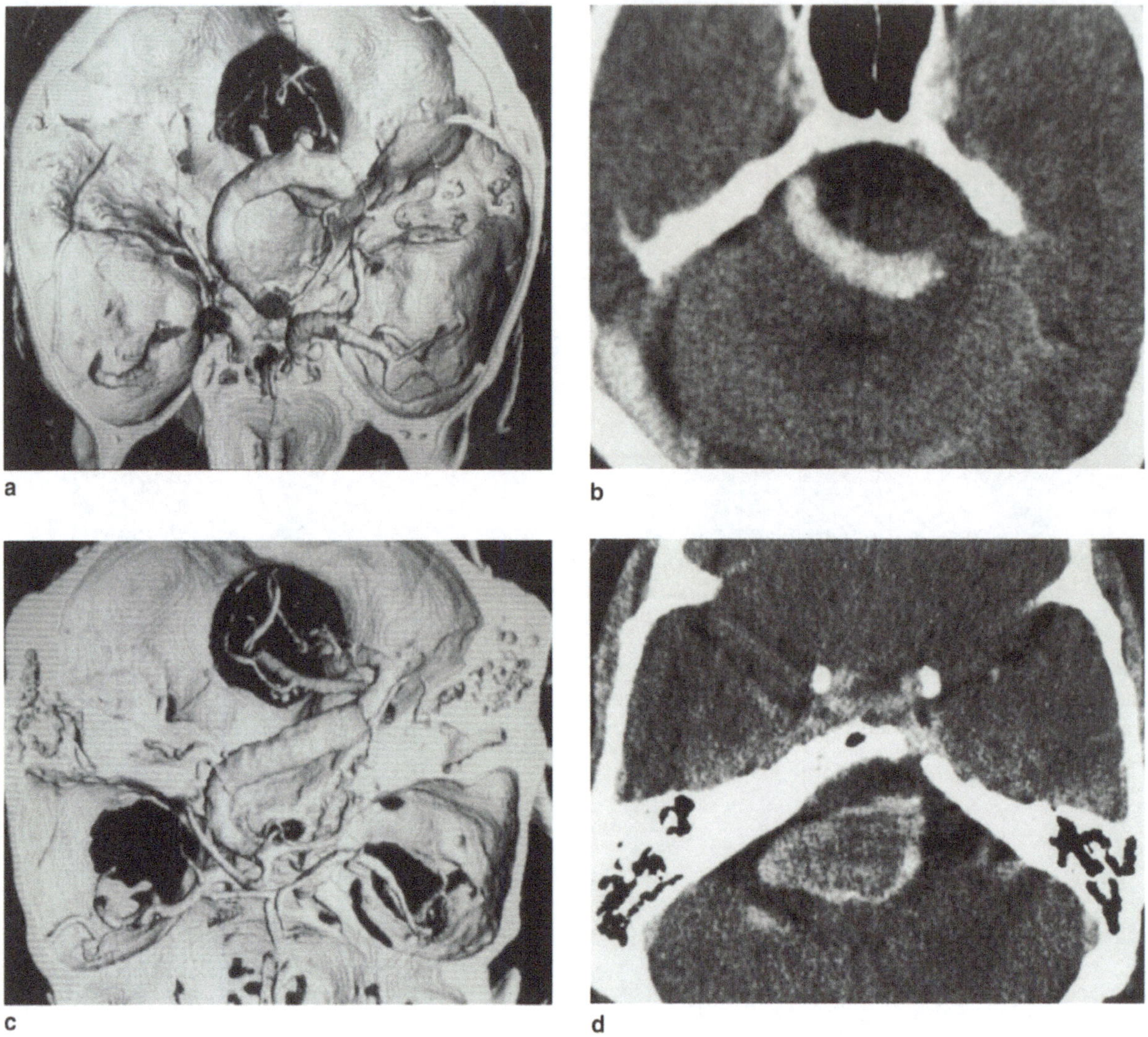

Fig. 3.30. CTA (a) and CT (b) of a patient with diplopia from a sclerotic, calcified, elongated basilar artery with fusiform dilatation. At surgical exploration, the rock hard basilar artery could not be mobilized from nerve VI. Any displacement risked compromising the brainstem's blood supply by stretching vital perforators. CTA (c) and CT (d) of another 71-year old patient hospitalized with symptoms and signs identical to the patient presented in (a) and (b). At craniotomy, the calcified fusiformly-dilated basilar artery could not be mobilized from nerve VI and the brainstem decompressed. After consultation with neurosurgeons around the world, we are convinced that no solution to this problem currently exists

Fig. 3.31. A personal snapshot of Dr. Charles Drake, shown here at a 1981 meeting in Europe where he chaired a discussion on basilar artery aneurysm surgery. Without his pioneer work in vertebrobasilar aneurysm surgery our art would remain far from its present state. Drake was a true giant, one whose vast skill and experience allowed him to walk with grace and relative safety on the cutting edge of BA aneurysm surgery. His genius compelled him to break rules established by the unenlightened, advancing vascular neurosurgery to a higher orbit of understanding, making it possible for his followers to avoid many of the disastrous results attendant to these frightening lesions

References

1. Anderson M (1976) Persistent primitive hypoglossal artery with basilar aneurysm. J Neurol 213: 377–381
2. Aymard A, Gobin YP, Hodes JE, Bien S, Rufenacht D, Reizine D, George B, Merland JJ (1991) Endovascular occlusion of vertebral arteries in the treatment of unclippable vertebrobasilar aneurysms. J Neurosurg 74: 393–398
3. Bartal A, Schiffer J, Winstein J (1968) A successful case of basilar bifurcation aneurysm surgery. Acta Neurochir (Wien) 19: 163–170
4. Brushan C, Hodges FJ, Posey J (1978) Successful surgical treatment of giant aneurysm of the basilar artery: case report. J Neurosurg 49: 124–128
5. Bugiani O, Piola P, Tabaton M (1983) Nontraumatic dissecting aneurysm of the basilar artery. Eur Neurol 2: 256–260
6. Bull JWD (1962) Contributions of radiology to the study of intracranial aneurysm. BMJ 2: 1701–1708
7. Chou SN, Ortiz-Suarez HJ (1974) Surgical treatment of arterial aneurysms of the vertebrobasilar circulation. J Neurosurg 41: 671–680
8. Constans JP, Visot A, Fredy D, Dorland P (1976) Aneurysme géant du tronc basilaire revele par une nevralgie faciale essentielle. Neurochirurgie 22: 493–502
9. Dawson BH (1958) The blood vessels of the human optic chiasm and their relation to those of the hyophysis and hypothalamus. Brain 81: 207–217
10. Dolenc VV (1983) Direct microsurgical repair of intracavernous vascular lesions. J Neurosurg 58: 824–831
11. Dolenc VV (1985) A combined epi- and subdural approach to carotidophtalmic artery aneurysms. J Neurosurg 62: 667
12. Dolenc VV, Škrap M, Šušteršič J, Škrbec M, Morina A (1987) A transcavernous-transsellar approach to the basilar tip aneurysms. BJN 1: 251–259
13. Dolenc VV (1989) Anatomy and surgery of the cavernous sinus. Springer, Berlin Heidelberg New York Tokyo
14. Dolenc VV, Prestor BP, Šušteršič J, Pregelj R (1994) Transclinoid-Transsellar–Transcavernous Approach to Basilar Tip Aneurysms. In: Pasqualin A et al (eds) New Trends in Management of Cerebro-Vascular Malformations. Springer, Wien New York, pp 231–237
15. Drake CG (1961) Bleeding aneurysms of the basilar artery. J Neurosurg 18: 230–238
16. Drake CG (1973) Management of aneurysms of posterior circulation. In: Youmans JR (ed) Neurological surgery, vol 2. Saunders, Philadelphia, pp 707–806
17. Drake CG (1975) Ligation of the vertebral (unilateral or bilateral) or basilar artery in the treatment of large intracranial aneurysms. J Neurosurg 43: 255–274
18. Drake CG (1978) Treatment of aneurysms of the posterior cranial fossa. In: Progress in neurological surgery, vol 9. Karger, Basel, pp 122–194
19. Emanuele MA, Dorsch TR, Scarff TB, Lawrence AM (1981) Basilar artery aneurysm simulating pheochromocytoma. Neurology (Ny) 31: 1560
20. Gale AN, Crockard HA (1982) Transient unilateral mydriasis in basilar aneurysm. J Neurol Neurosurg Psychiatry 45: 565
21. Guglielmi G (1990) Balloon embolization of a basilar bifurcation aneurysm. AJNR 11: 653–655
22. Guglielmi G. Vinuela F, Duckwiler G, Dion J, Lylyk P, Berenstein A, Strother C, Graves V, Halbach, V, Nichols D, Hopkins N, Ferguson R, Sepetka I (1992) Endovascular treatment of posterior circulation aneurysms by electrothrombosis using electrically detachable coils. J Neurosurg 77: 497–500
23. Hayakawa T, Kamikawa K, Ohnishi T, Yoshimine T (1981) Prevention of postoperative complications after a transoral-transclival approach to basilar aneurysms. Technical note. J Neurosurg 54: 699–703
24. Hopkins LN, Eudny JL, Castellani D (1983) Extracranial-intracranial arterial bypass and basilar artery ligation in the treatment of giant basilar artery aneurysms. Neurosurgery 13: 189–194
25. Ikeda K, Yamashita J, Hashimoto M, Futami K (1991) Orbito-zygomatic temporopolar approach for a high basilar tip aneurysm associated with a short intracranial internal carotid artery: a new surgical approach. Neurosurgery: 28: 105–110
26. Jamieson KG (1968) Aneurysms of the vertebrovascular system. Further experience with nine cases. J Neurosurg 28: 544–555
27. Katakura R, Yashimoto T, Suzuki J (1979) A case of giant aneurysm of the basilar artery. Acta Neurochir (Wien) 49: 87–93
28. Kawase T, Toya S, Shiobara S, Mine S (1985) Transpetrosal approach for aneurysms of the lower basilar artery. J Neurosurg 63: 857–861
29. Khambatta HJ, Matteo RS, MCMurtry JG, Bowman FO Jr (1974) Anaesthesia for basilar arterial aneurysms with elective circulatory arrest and moderate hypothermia. Anaesthesiology 41: 512–516
30. Krayenbuhl HA, Yasargil MG (1968) Cerebral angiography. Lippincott, Philadelphia, p 74

31. Litvak J, Sumners TC, Barron JL, Fisher LS (1981) A successful approach to vertebrobasilar aneurysm. Technical note. J Neurosurg 55: 491–494
32. Logue V (1964) Posterior fossa aneurysms. Clin Neurosurg 11: 183–207
33. McMurty JG, Housepian BM, Bowman FO, Jr, Matteo RS (1974) Surgical treatment of basilar artery aneurysms. Elective circulatory arrest with thoracotomy in 12 cases. J Neurosurg 40: 486–494
34. Nijensohn DE, Saez TJ, Reagan TJ (1974) Clinical significance of basilar artery aneurysms. Neurology 4: 301–305
35. Ono M, Ono M, Rhoton AL Jr (1984) Microsurgical anatomy of the region of the tentorial incisure. J Neurosurg 60: 365–399
36. Pia HW (1979) Classification of vertebro-basilar aneurysms. Acta Neurochir (Wien) 47: 1–30
37. Regli L, de Tribolet N (1991) Tuberothalamic infarct after division of a hypoplastic posterior communicating artery for clipping of a basilar tip aneurysm: case report. Neurosurgery 28: 456–459
38. Samson DS, Hodosh RM, Clark K (1978) Microsurgical evaluation of the pterional approach to aneurysms of the distal basilar circulation. Neurosurgery 3: 135–141
39. Sharr MM, Kelvim FM (1973) Vertebrobasilar aneurysms. Experience with 27 cases. Eur Neurol 10: 129–143
40. Sugita K, Kobayashi S, Shintani A, Mutsuga N (1979) Microneurosurgery for aneurysms of the basilar artery. J Neurosurg 51: 615–620
41. Tiyaworabum S, Wanis A, Schirmer M, Bock WJ (1982) Aneurysms of the vertebro-basilar system: clinical analysis and follow-up result. Acta Neurochir (Wien) 63: 221–229
42. Troupp H (1971) The natural history of aneurysms of the basilar bifurcation. Acta Neurol Scand 47: 350–356
43. Wilson CB , Sang UN (1976) Surgical treatment for aneurysms of the upper basilar artery. J Neurosurg 44: 537–543
44. Yasągil MG, Fox JL, Ray MW (1975) The operative approach to aneurysms of the anterior communicating artery. In: Krayenbuhl H (ed) Advances and technical standards in neurosurgery, vol. 2. Springer, Wien New York, pp 113–170
45. Yasąrgil MG, Anič J, Laciga R, Jain KK, Hodosh RM, Smiths RD (1976) Microsurgical pterional approach to aneurysms of the basilar bifurcation. Surg Neurol 6: 83–119

4

Tumors originating in the central skull base

Tumors of the parasellar space (PS) and adjacent regions in the central skull base (CSB) occur most frequently among the middle–aged, but they are also common in the elderly. They also appear in children. Approximately one-half of all histologically benign CSB tumors are meningiomas. Females with meningiomas outnumber their male counterparts by nearly 2:1. Other benign tumors common at the CSB are neoplasms of the pituitary gland, which may invade one or the other or both PS(s). Craniopharyngiomas, the third most frequent benign CSB tumor, generally occur in the midline, growing upward and posteriorly, only occasionally lateralward into the PS(s). Trigeminal neurinomas, fourth in incidence among benign lesions, are most often unilateral, while chordomas, chondrosarcomas and even chondromas may be unilateral but more commonly extend across the midline. Parasellar tumors in children most commonly are neurinomas of nerve V. Malignant CSB tumors include direct invasion by sarcomas and carcinomas arising from adjacent structures. Even distant metastases appear within the CSB.

Symptoms and signs generated by CSB tumors include headache, cranial nerve(s) paresis, exophthalmos, ophthalmoplegia, and neuralgic pain (predominantly in the V1 distribution, less frequently in V2, only occasionally in V3). Headaches are usually mild at the onset of disease and generally remain unchanged throughout the course of illness, although subsequently they become associated with additional symptoms and signs. Headache may be heralded by, or accompanied quite early by, diplopia related to cranial nerve(s) paresis. Exophthalmos occurs among approximately one fifth of these patients while ophthalmoplegia is present only occasionally prior to initial hospitalization. Severe neuralgic (trigeminal) pain is more likely to indicate Tolusa-Hunt syndrome or a malignant PS tumor than a benign tumor within the superior orbital fissure (SOF) or elsewhere in the PS. Severe unilateral trigeminal pain, frequently associated with sudden paresis of one or more nerve(s) III, IV, VI, is extremely rare among patients with benign PS tumors, but this type of pain occurs occasionally with trigeminal neurinomas presenting with intratumoral hemorrhage.

Since different pathological processes occurring in the PS may result in similar or identical symptoms and signs, the clinical pattern typical to various PS lesions will be described in detail. The advent of CT and MRI has facilitated considerably the diagnosis of CSB tumors. While the clinical presentation of various CSB lesions often do not identify a specific diagnosis, but they always justify requesting a CT scan and/or an MRI. Imaging studies performed without contrast enhancement may easily fail to provide a diagnosis if the lesion is quite small or isodense to brain. Imaging studies may not predict a histological diagnosis, especially if examination is incomplete. While the scan diagnosis of inflammation in the PS or at the apex of the orbit may be highly probable, needle biopsies remain useful in Institutions where endoscopic transsphenoidal biopsies are not yet available.

Digital subtraction angiography (DSA) clearly differentiates tumors from vascular lesions and provides important information about changes in ICA caliber and position within the PS where tumors are present, as well as changes in other intracranial arteries. However, MRI, magnetic resonance angiography (MRA), and CT-angiography may provide accurate surgical information, and often may be substituted for DSA. But if the PS-ICA is displaced and/or stenotic, cross-circulation and ICA-occlusion studies, including examination of brain perfusion and venous return, become mandatory if surgery is to be considered.

The precise characterization of changes of an ICA likely to be encased by tumor within the PS is crucial to formulating a strategy for resecting tumor from the ICA. Narrowing of the PS ICA in the PS may not only indicate vascular stenosis or circumferential compression by tumor, but direct invasion of the ICA wall. In such instances it is extremely important to obtain complete cross-circulation studies before undertaking surgery. Such information is required to determine if temporary occlusion and reconstruction of the ICA (either by local suturing, patching, or resecting and reconstructing a short segment of the vessel) is possible should the ICA rupture during surgery. If cross-circulation studies indicate that collateral flow is inadequate for such an attempt, a cuff of tumor must be left in place around the ICA in order to prevent such a life-threatening event. By determining the presence or absence of ACom and PCom patency and sufficient flow through them with ICA occlusion, one may accurately predict if temporary or permanent ipsilateral ICA occlusion will be tolerated during surgery. If cross-circulation studies indicate that direct ICA reconstruction is not possible, or even questionable, the surgeon must accept the fact that total resection may not be possible. It is always preferable to combine a subtotal microsurgical tumor resection with postoperative focused beam (Gamma Knife) radiation (for meningiomas) or Proton Beam radiation (for chondromas and chondrosarcomas) than to risk rupturing the ICA by aggressively peeling tumor from a diseased arterial wall.

EMG examination of periocular muscles and muscles enervated by nerve V are quite useful prior to surgery. Such studies provide a valuable estimate of the state of cranial nerves and corresponding muscles, especially where clinical paresis is not yet apparent. Perioperative monitoring may also be useful but is certainly no substitute for precise knowledge of surgical anatomy. For patients with meningiomas located anteriorly in the PS – adjacent to the ACP, inside the sella, crossing the tuberculum sellae, and those possibly invading the optic canal – objective visual investigations, including perimetry, assessment of visual acuity and optic fundi, and visual evoked potentials, are helpful both prior to and following surgery.

Clinical, neurophysiological, and neuroradiological information gained prior to surgery may allow the surgeon to more accurately predict functional deterioration likely to occur even with the most gentle dissection of nerves III–VI and the ICA, also the rate of improvement likely to be expected if temporary impairment occurs, ultimately the final outcome of treatment.

From a practical standpoint, the nature of the tumor and its location play the primary role in planning surgery and determining to a great extent if it is possible to totally excise a given lesion with minimal risk of permanent neurological impairment. The large number of cases treated in this series for tumors residual or recurring after surgery at other institutions (Table 6) confirms the complexity of operations for tumors occurring at the CSB.

Table 6. Central skull base tumors and their resectability

Tumor	N° of cases		Virgin cases		Residual and/or recurrent tumors		Resection					
							Complete		Gross-total		Sub-total	
Meningiomas	817	47%	512	63%	305	37%	321	39%	238	29%	258	32%
Large pituitary adenomas	411	24%	190	46%	221	54%	359	87%	41	10%	11	3%
Craniopharyngiomas	275	16%	183	67%	92	33%	161	59%	91	33%	23	8%
Trigeminal neurinomas	85	5%	80	94%	5	6%	82	96%	3	4%	–	–
Chordomas, chondromas and chondrosarcomas	71	4%	50	70%	21	30%	–	–	29	41%	42	59%
Cavernomas	19	1%	15	79%	4	21%	15	79%	3	16%	1	5%
Other benign lesions	12	1%	11	92%	1	8%	–	–	8*	67%	–	–
Malignant lesions	31	2%	23	74%	8	26%	16	52%	8	26%	7	22%
Total	1721	100%	1064	62%	657	(38%)	954	55%	421	24%	342	20%

*Of eight cases in which gross-total resection was achieved, four were cholesteatomas, two were venous thromboses occurring within the PS (operated unnecessarily), and two cases of Tolusa-Hunt syndrome in which granulomatous tissue in the PS was found. Two additional patients with Tolusa-Hunt syndrome and two with venous thrombosis within the PS were not operated upon.

4.1

Central skull base meningiomas

At the turn of the 20th century intracranial neoplasms arising from the brain's meninges carried a confusing array of names, including "dural endothelioma," "arachnoid fibroblastoma," and "meningeal fibroblastoma," leading Cushing to coin the word, "meningioma," which has survived to the modern lexicon [1, 2]. Initially thought to arise from dura mater itself, these tumors were subsequently found to form from arachnoidal cap cells, which are particularly plentiful near groups of arachnoidal villi at dural sinuses and veins draining into them. While pathologists and others have preferred to classify meningiomas by histological type, neurosurgeons have emphasized the importance of their location and the physical nature of their mass, characteristics of primary importance with respect to their treatment. Although the incidence of meningiomas may vary depending on referral patterns specific to institutions called on to treat them, it is generally accepted that meningiomas represent approximately 20% of all primary intracranial tumors [9, 12, 16]. They are primarily tumors of adult life [5], generally detected most commonly between the fourth and sixth decades and accounting for no more than 3% of childhood brain tumors [3, 17]. Meningiomas occur more frequently in women, particularly in middle age [8, 15]. Meningiomas have been found associated with other disease entities, e.g. multiple rheumatoid nodules [10], also as incidental findings during life as well as in autopsy studies [14, 20]. A clear familial incidence exists with the central form of neurofibromatosis (NF-2), but other families have been confirmed to harbor unexpectedly high numbers [18]. Head trauma has long been considered to play a role in the development of meningiomas, occurring in 33% of the Cushing and Eisenhardt series [2]. Previous irradiation treatment may also be an etiological factor [7, 13].

A massive literature confirms that meningiomas continue to pose unsolved treatment problems [5, 6]. Approximately half occur at the cerebral convexity, either adjacent to the superior sagittal sinus or at varying distances from it. While the resection of many of these lesions is essentially straightforward, those occurring at the central skull base (CSB) are particularly demanding [4, 19] owing to neural and vascular structures coursing through and adjacent to bone and problems presented by neighboring paranasal sinuses. As with convexity lesions, histological type constitutes an obvious factor with respect to the total resectability of CSB meningiomas. Firm, fibroblastic lesions are clearly more difficult to excise than softer angioblastic lesions. Yet, notwithstanding the current World Health Organization's [11] listing 11 histological types of meningiomas, it is the anatomical relationship of CSB lesions to vital neighboring structures which determines their resectability.

CSB meningiomas described here are divided into 14 groups with respect to their specific location, the likelihood of complete surgical resection, and the outlook toward preserving adjacent neural and vascular structures. This classification system and list of resectability rates are summarized in Table 7. These data emphasize the role the complexity of local anatomy plays in surgical success. The importance of the proportion of virgin lesions treated versus those residual from previous surgical attempts speaks of varying complexities in different anatomical locations.

The extradural approach to CSB meningiomas offers two major advantages. First, by interrupting the blood supply of adjacent dura first, these lesions are in large part devascularized, minimizing surgical blood loss and essentially converting the complexity of their removal to that of convexity lesions. Second, by initiating tumor resection from its epidural surface, important neural structures and their blood supply are positioned best to be preserved. These two principles amount to the keys to success in the surgery of these difficult lesions.

Table 7. Location(s) of central skull base meningiomas and their resectability in virgin and residual cases

Location of meningioma	N° of cases		Virgin cases		Residual cases		Tumor-resection Complete		Gross-total		Sub-total	
Sphenoid wing	51	6%	47	92%	4	8%	46	90%	5	10%	–	–
Anterior clinoid process	67	8%	54	81%	13	19%	62	92%	5	8%	–	–
Middle cranial fossa	12	2%	12	100%	–	–	12	100%	–	–	–	–
Parasellar space	123	15%	89	72%	34	28%	38	31%	52	42%	33	27%
Parasellar space + adjacent regions	388	47%	172	44%	216	56%	51	13%	127	33%	210	54%
Superior orbital fissure	7	1%	1	14%	6	86%	–	–	2	29%	5	71%
Dorello's space	5	1%	4	80%	1	20%	3	60%	2	40%	–	–
Petroclival	41	5%	36	88%	5	12%	35	85%	6	15%	–	–
Tuberculum sellae	44	5%	35	80%	9	20%	42	95%	2	5%	–	–
Diaphragm sellae and optic canal(s)	8	1%	7	88%	1	0%	6	75%	2	25%	–	–
PCP, dorsum sellae, and upper clival meningiomas	6	1%	6	100%	–	–	6	100%	–	–	–	–
Clival	22	3%	18	82%	4	18%	9	41%	13	59%	–	–
Large CSB	27	3%	15	55%	12	44%	–	–	21	78%	6	22%
Multiple: parasellar + distant	16	2%	15	94%	1	6%	4	25%	8	50%	4	25%
Total	817	100%	511	63%	306	37%	314	38%	245	30%	258	32%

Sphenoid wing meningiomas

Pathological anatomy and clinical presentation

Meningiomas involving the lesser wing of the sphenoid bone may easily be confused with anterior clinoid process (ACP) meningiomas. Such lesions originate from dura overlying the lesser sphenoid wing, lateral to the ACP, then grow medially, even across the ACP into the superior orbital fissure (SOF) and/or parasellar space (PS). Meningiomas arising somewhat more laterally may spread in three directions: laterally toward the calvarium, medially along the lesser wing, and inferiorly along the greater sphenoid wing. Typically they become very large prior to diagnosis. "En plaque" sphenoid wing meningiomas spread over the roof of the orbit, along the lesser and greater wings of the sphenoid, and, in many cases, laterally and posteriorly, creating hyperostosis along the way (Fig. 4.1) which acts as a space-occupying lesion within the orbit and the middle cranial fossa, also narrowing the SOF. "En plaque" lesions may actually invade the optic canal (OC) and/or the SOF, resulting in unilateral visual deterioration and/or diplopia as early clinical manifestations.

Sphenoid wing meningiomas may present with retro-orbital discomfort (even frank pain), a visible mass lateral to the orbit, or moderate exophthalmos. Visual loss may occur, but it is not the leading symptom. Such symptoms and signs typically lead to CT and/or MR imaging, either of which is diagnostic.

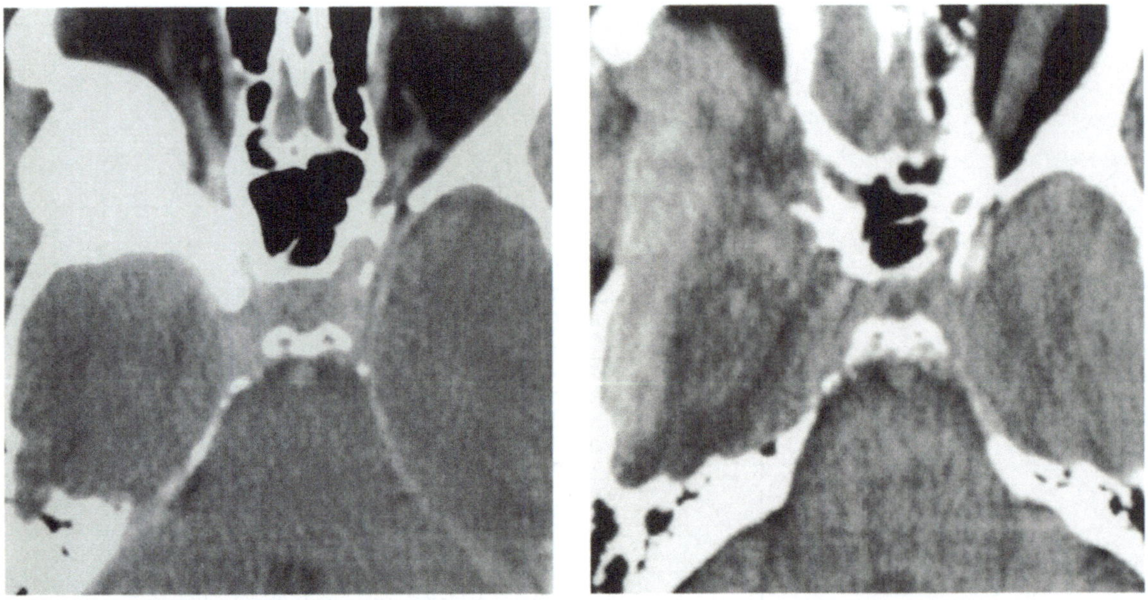

a b

Fig. 4.1. A preoperative trans-axial CT scan (a) shows a large hyperostotic right sphenoid wing meningioma which involved the anterior clinoid process and the orbital roof. The study obtained three days after surgery (b) demonstrates complete resection of tumor and all thickened bone

Surgical approach to sphenoid wing meningiomas

The treatment of choice for all sphenoid wing meningiomas is total microsurgical resection.

Positioning the patient, three-point fixation of the skull, skin incision, and osteoplastic craniotomy are carried out as described in Chapter 2. Unroofing the orbit may be more demanding when the orbital roof and sphenoid wing are hyperostotic. To avoid damaging periorbita, the orbital roof should be drilled away quite carefully, continuously irrigating the diamond burr with saline. Dura overlying the roof of the orbit must be reflected with care, and a plane between periorbita and drill-thinned hyperostotic bone must be established before the remaining orbital roof is removed with a small rongeur. The surgeon must be patient and meticulous at this point. Resecting the orbital roof with a rongeur can be extremely dangerous since fractures easily occur, posing serous risk of injury to adjacent structures. It must be drilled to a very thin remnant before a rongeur is used. Drilling thickened bone, even with a high-speed drill equipped with a diamond burr, is typically very tedious work, resulting in the surgeon's becoming fatigued and impatient before all hyperostotic bone has been removed. One must resist the urge to employ a cutting burr, which, especially in inexperienced hands, is extremely dangerous. Irreversible injury to vital structures may occur before the tumor is even exposed.

After removing the roof of the orbit and sphenoid wing laterally to the foramen ovale (FO), one may drill the medial orbital wall from the anteromedial side of the SOF. Dura is carefully peeled from the lesser wing of the sphenoid, the ACP, and over the OC, finally even more medially from the tuberculum sellae. The sphenoid wing, the ACP, the lateral and proximal walls of the OC (also the optic strut), and the lateral surface of the tuberculum sellae are resected extradurally in order to devascularize both dura and underlying tumor.

Following totally resecting all hyperostotic bone, the dura is incised along the sylvian fissure anteriorly until tumor-infiltrated dura is encountered. Normally the dura is infiltrated by tumor whether or not the intradural tumor is globular or spread out, carpet-like, as with "en plaque" meningiomas. The intradural meningioma and infiltrated dura are resected completely, finally reaching healthy dura and coagulating its edges to assure absolute hemostasis. Then the tumor is carefully dissected from brain and vascular and neural structures. Dura is reconstructed by suturing an appropriately tailored graft of commercially available dural substitute in place, then sealed with fibrin glue from the epidural surface. After watertight closure, an epidural drain is left in place, and the osteomuscular flap is replaced and secured with sutures. The soft tissue flap is closed in layers.

Patients with sphenoid wing meningiomas rarely require postoperative lumbar drainage or antibiotics, but they should be given a prophylactic anticonvulsant for at least three months.

Patients and results

The surgical results for sphenoid wing meningiomas are summarized in Table 8. Of 51 patients, tumor resection was considered to be complete in 46. In five cases, the resection was only gross-total. Forty-one tumors were larger than 5 cm in diameter, ten somewhat smaller. Despite 14 cases with "en plaque" lesions, the overall postoperative surgical outcome was good in 48 patients. Three did not recover completely. In one, preoperative unilateral visual loss did not return fully, and in another (with tumor involving the SOF) diplopia (from partial paresis of nerve III) did not clear after six months. In the third, one who had a huge tumor, anticonvulsant drugs are still required since the EEG has not yet returned to normal. He has had no epileptic seizures, however.

Table 8. Surgical resection of sphenoid wing meningiomas: results with respect to location, size and type

Location	N° of cases		Type				Size				Resection				Recovery			
			Globular		En plaque		<5 cm		>5 cm		Complete		Gross-total		Complete		Incomplete	
Medial SW	22	43%	19	86%	3	14%	7	32%	15	68%	20	91%	2	9%	21	95%	1	5%
Lateral SW	14	27%	12	86%	2	14%	3	21%	11	79%	14	100%	–	–	14	100%	–	–
Entire SW	9	18%	2	22%	7	78%	–	–	9	100%	8	89%	1	11%	8	89%	1	11%
Beyond SW	6	12%	4	67%	2	33%	–	–	6	100%	4	67%	2	33%	5	83%	1	17%
Total	51	100%	37	73%	14	27%	10	20%	41	80%	46	90%	5	10%	48	94%	3	6%

Anterior clinoid process meningiomas

Relevant anatomy of the anterior clinoid process

The size, location, and anatomical relationships of the anterior clinoid process (ACP) are described in detail in Chapter 1. It is important to emphasize that the sphenoidal sinus may extend into the ACP Fig. 4.2(a–d). Detailed anatomy of the ICA, optic canal (OC), optic nerve (ON), and nerve III is also described in Chapter 1. A new growth (meningioma) at this site alters the normal relationships with adjacent anatomical structures by displacing, compressing, invading, or at least by irritating them, any of which may affect their function.

Pathological anatomy, clinical presentation, and diagnostic work-up

Anterior clinoid process (ACP) meningiomas constitute a well defined group. The dural layers in the region of the ACP are manifold, continuous with dural walls of adjacent central skull base (CSB) compartments as follows: the superior orbital fissure (SOF), the parasellar space (PS), the sella, the optic canal (OC), and the canal of nerve III, the most important of which are the lateral wall of the PS, the tentorial edge, and the diaphragm sellae. Meningioma growth patterns are important. Globular lesions, those likely to result in localized masses, are far more common, but "en plaque" lesions occur as well.

The nearby optic nerve (ON) and nerve III, medial and lateral to the ACP respectively, plus the internal carotid artery (ICA), are the most likely important structures to be affected by these lesions and are generally the sources of early symptoms and signs, also the structures most likely to be associated with potential post-operative morbidity.

Owing to the restricted nature of their locations, involvement of the ON and nerve III by tumor spread are similar in several ways. Both are sheltered within compartments with well defined borders, the ON in the bony OC and nerve III in the occulomotor canal, its own sheath of dura in the lateral portion of the anterior petro-clinoid fold adjacent to the posterior clinoid process (PCP) and extending anteriorly within the lateral wall of the PS. Since both canals enclose very limited space, each nerve is particularly vulnerable to "en plaque" lesions spreading along adjacent dura, and are likely to result in clinical manifestations before significant tumor mass is apparent. Fortunately, such tumors are rare. Globular lesions invading the OC, even if small, are likely to cause rapid deterioration of vision. Such lesions frequently displace and stretch the ON medially, but some may even encase the ON, compressing it circumferentially. Meningiomas in this group typically compress the ON from its lateral and inferior surfaces, wedging it against the rigid falciform ligament at the superomedial limit of the dural rim at its entry into the OC. It is important to keep this tendency in mind when assessing unilateral visual loss patterns in such patients.

Although nerve III is not surrounded by bone until it enters the SOF, its dural canal isolates it similar to the way the OC houses the ON. Nerve III is vulnerable to "en plaque" lesions invading the occulomotor canal, also to encasement and circumferential compression. Similar also to the ON, though occurring less frequently, meningiomas of any shape or growth type may trap the intradural segment of nerve III against the sharp edge of the tentorium.

Hyperostosis is commonly associated with ACP tumors, resulting in intense bony hypertrophy not only of the ACP but affecting the walls of the optic canal and occasionally the tuberculum sellae. Hyperostosis narrows the OC further, making the ON even more vulnerable to tumor growth.

ACP meningiomas generally displace the ICA caudally and medially to some extent. However, this vessel, which is fixed by the distal (dural) ring (DR), is relatively immobile at its point of entering cranial vault dura where it may be encased by tumor. Even its more peripheral branches, the posterior communicating artery (PCom) and the anterior choroidal artery, may be affected. If the meningioma is large, even the ICA bifurcation, including the proximal anterior cerebral artery (ACA1), the lenticulostriate arteries, and primary elements of the middle cerebral artery (MCA), may be encased by tumor. Elderly patients, with rigid, atherosclerotic ICAs (and branches), are at particular risk of vascular compromise from tumor growth. But even in the young, once an ACP meningioma invades the arachnoidal membrane surrounding the ICA, its wall is rendered defenseless and may be invaded (Figs. 4.3–4.6). The same is true for any other artery in the region.

Large ACP meningiomas may also compress the ipsilateral optic tract, the optic chiasm, even the pituitary stalk (PSt) on occasion (Figs. 4.5 and 4.6). These nasty lesions have the potential to grow in all directions, having no respect for any anatomical boundary, not infrequently invading paranasal sinuses and the orbit.

Symptoms and signs associated with ACP meningiomas are easily recognized and often provide important clues to the tumor's unique characteristics of growth. Unilateral visual deterioration is the most common clinical manifestation, sometimes associated with diplopia. Visual loss and double vision are followed in frequency by ipsilateral retro-orbital discomfort (occasionally even frank pain) and epileptic seizures. Since unilateral deterioration of vision, especially if combined with other symptoms or signs, necessitates radiographic imaging (CT and/or MR), ACP meningiomas are easily detected. Depending on the CT and/or MRI findings, whether or not to include angiography in the work-up is determined on an individual case basis.

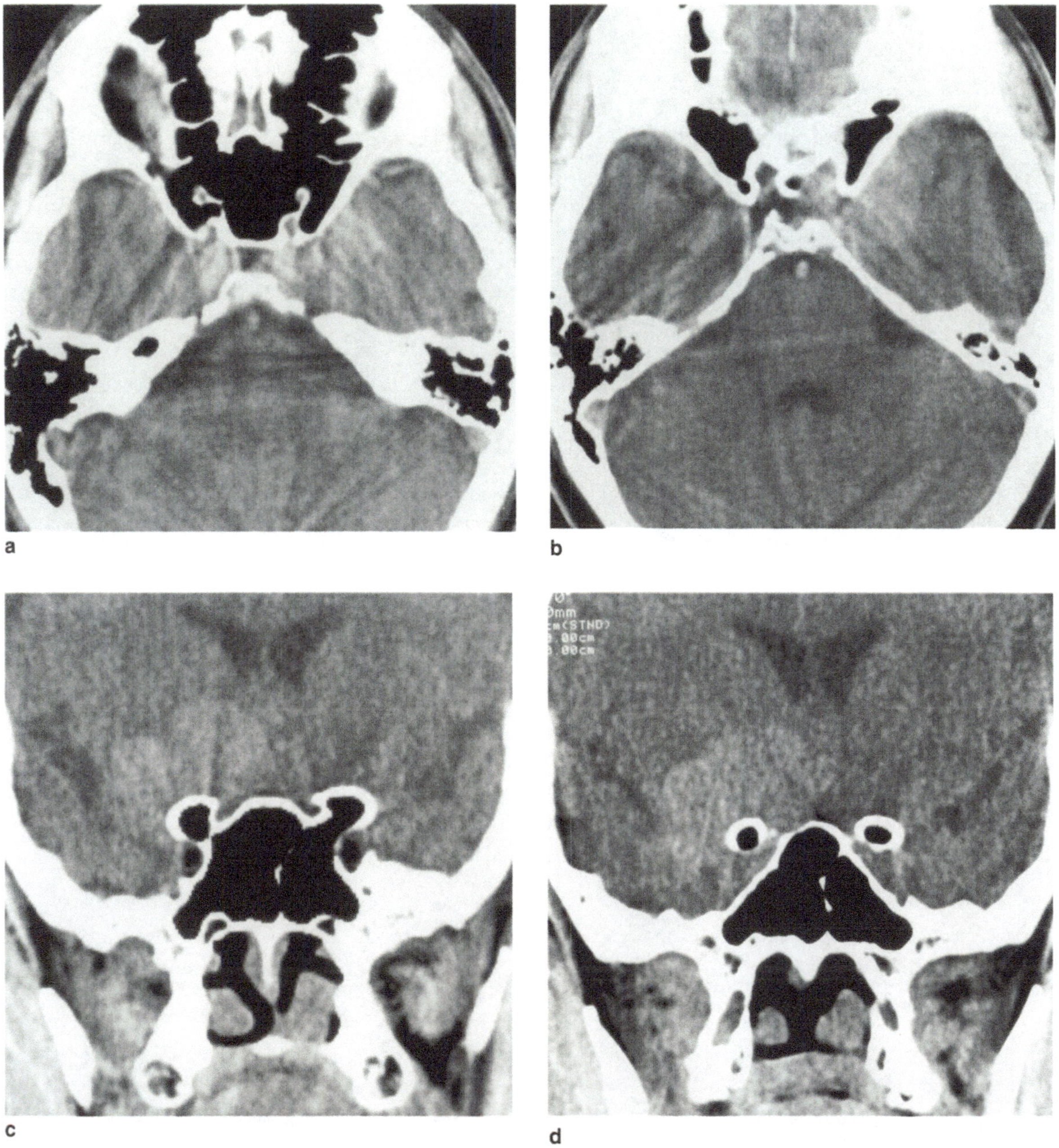

Fig. 4.2. Contrast-enhanced transaxial (a, b), and coronal (c, d) CT images demonstrate pneumatization of both anterior clinoid processes (*ACPs*) in different sections. Note the continuity between ACP and sphenoidal sinus. In cases such as this one, complete resection of the ACP would create a communication between the subarachnoid space and the sphenoidal sinus, making a cerebrospinal fluid (*CSF*) leak all but inevitable

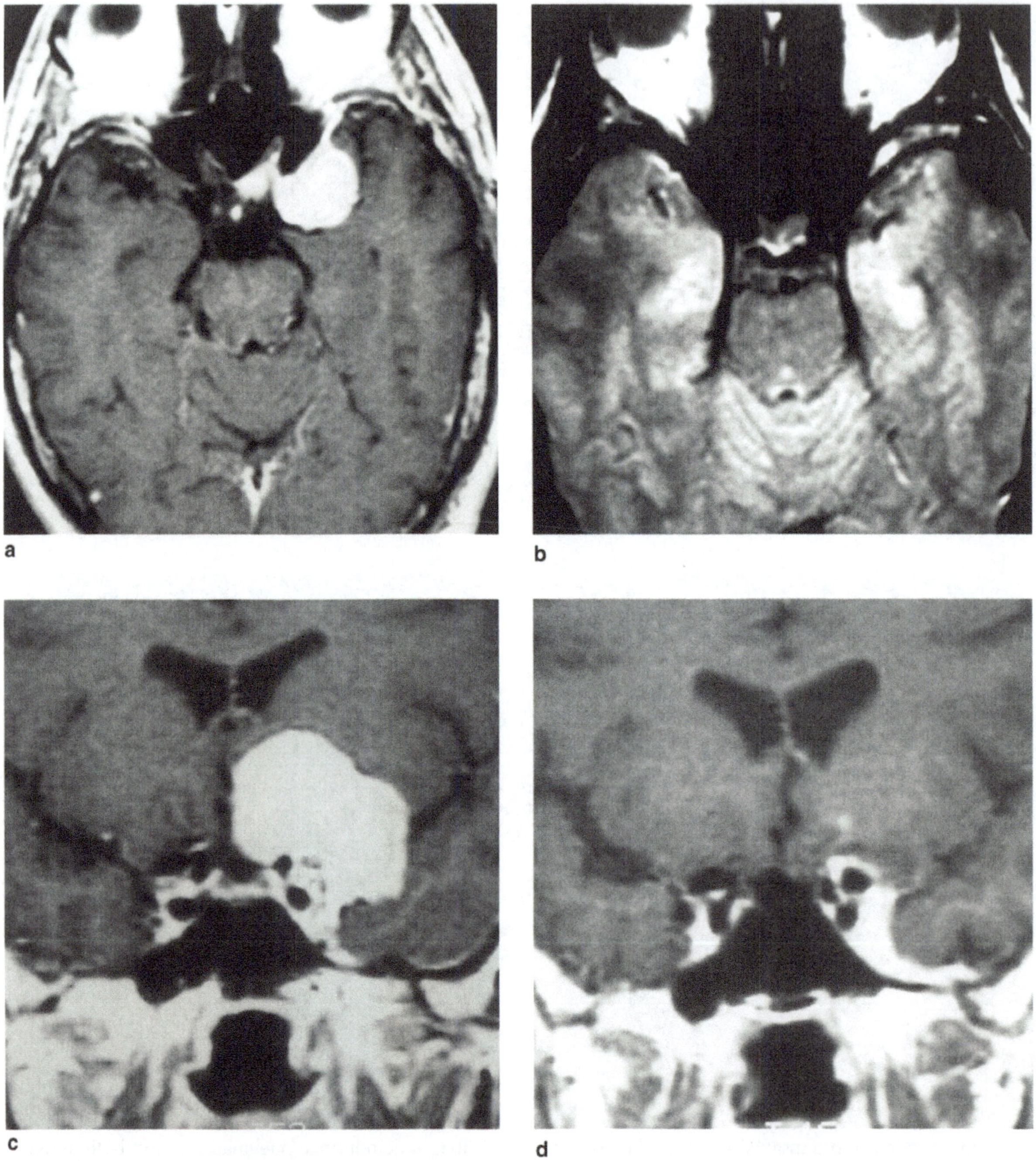

Fig. 4.3. Preoperative contrast-enhanced transaxial (a) and coronal (c) MR images demonstrate a rather large meningioma surrounding a pneumatized anterior clinoid process (*ACP*). Corresponding postoperative contrast-enhanced MR images (b and d) demonstrate total resection of the tumor. Note that extension of the sphenoidal sinus into the ACP is better demonstrated by CT than by MRI (Fig. 4.2). On the other hand, identification of the limits of the tumor is superior by MRI (Fig. 4.2)

Fig. 4.4. Preoperative transaxial (a), coronal (c), and sagittal (e) MR images show a rather large anterior clinoid process (*ACP*) meningioma with adjacent (arachnoidal) cysts (a) incasing the ICA, the ACA, and the MCA, as well as surrounding cerebral edema (c). ACP hyperostosis, encasement of the ICA, and surrounding edema are apparent in (e). Corresponding postoperative contrast-enhanced MR images demonstrate total tumor resection without evidence of injury to surrounding brain (b, d, f)

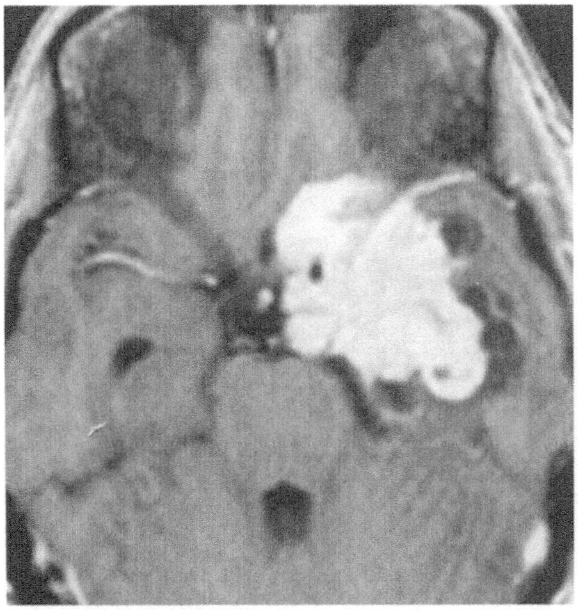

a

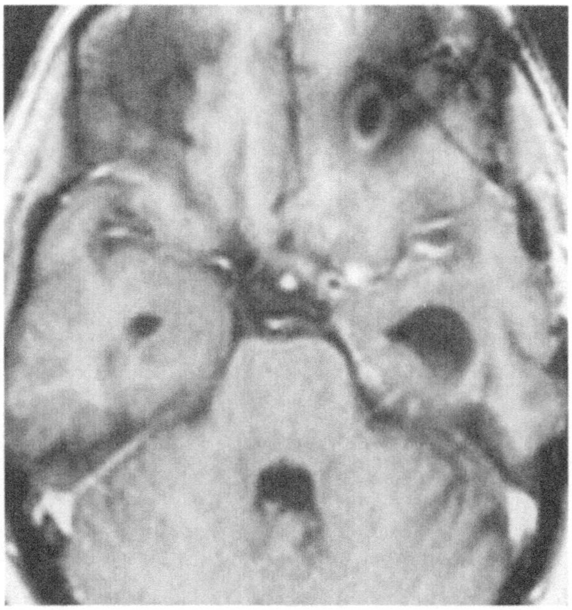

b

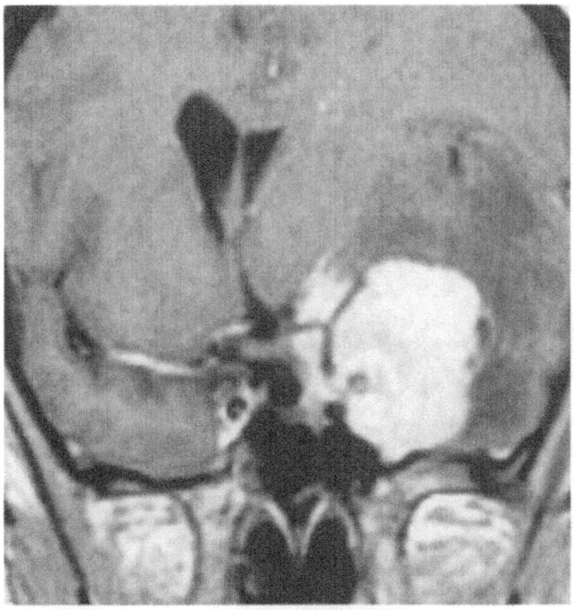

c

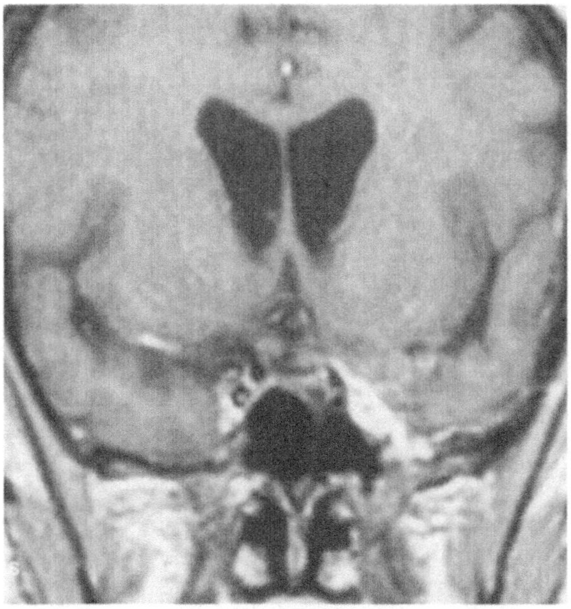

d

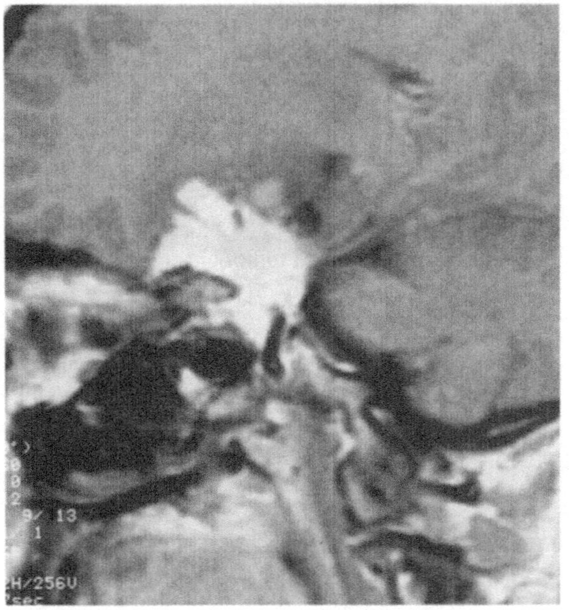

e

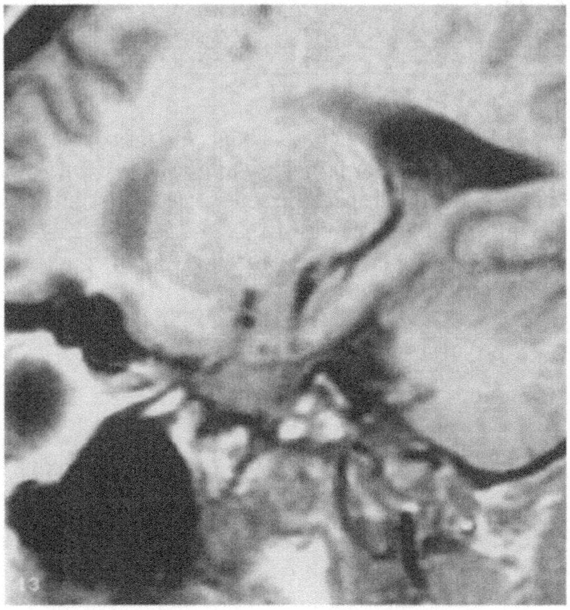

f

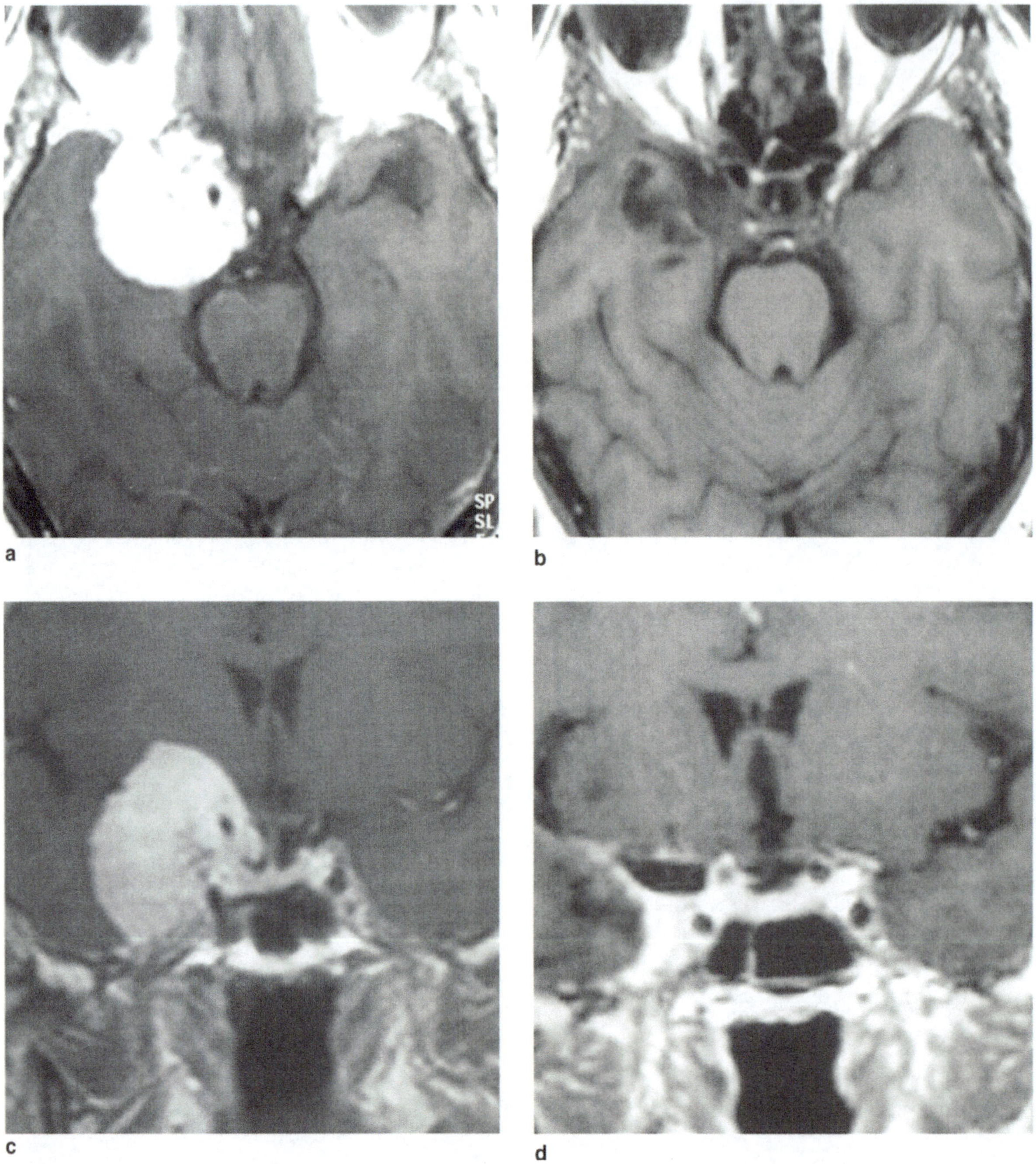

Fig. 4.5. Preoperative contrast-enhanced transaxial (a) and coronal (c) MR images show a large anterior clinoid process (*ACP*) meningioma which incases and invades the ICA wall, compressing the optic nerve, chiasm, and nerve III. Corresponding contrast-enhanced postoperative images, transaxial (b) and coronal (d) show gross-total resection of the lesion. Note: The lateral wall of the right parasellar space (*PS*) is clearly enhanced by the injected contrast material. This may indicate only persisting hypervascularization, but the possibility of a small amount of residual tumor (invading the PS) cannot be excluded

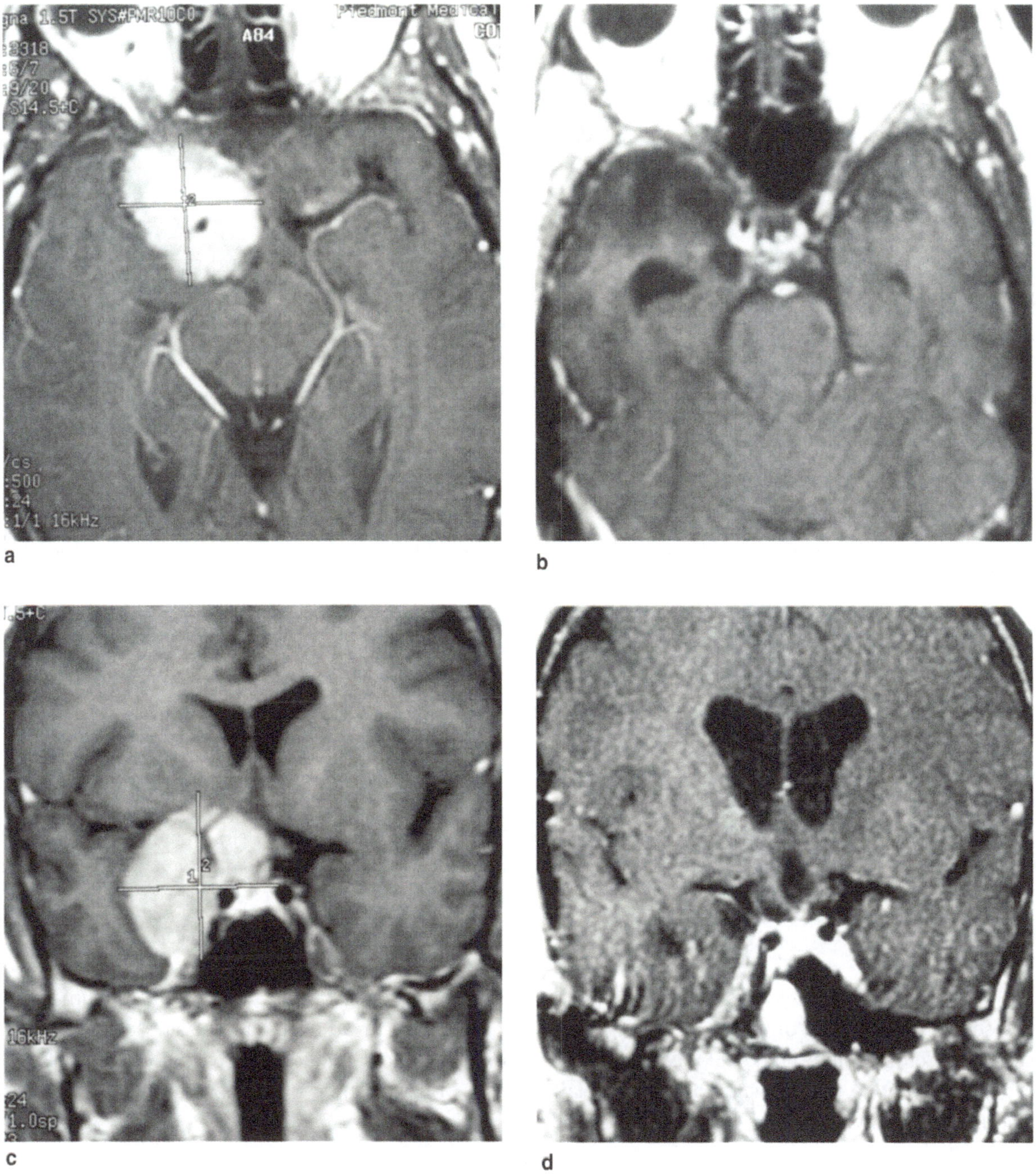

Fig. 4.6. Preoperative contrast-enhanced transaxial (a) and coronal (c) MR images show a large anterior clinoid process (*ACP*) meningioma encasing the intradural ICA and initial segments of the ACA and MCA, compressing the visual apparatus and extending into the parasellar space (*PS*) and across the midline of the dorsum sellae. Corresponding postoperative contrast-enhanced MR images reveal complete resection of the lesion, transaxial (b) and coronal (d), with significant atrophy of the anterior pole of the right temporal lobe

Surgical classification of ACP meningiomas

ACP meningiomas may be divided into three groups, depending on size and growth pattern (whether globular or "en plaque"). A fourth group contains tumors of all sizes, shapes, and growth patterns, but those whose bony substrates carry special surgical significance.

Type I

These ACP meningiomas originate from dura rostral to the ACP, spreading medial and superior to the ON, lateral and superior to the ICA, and superior to nerve III. They may displace the ON medially but the OC remains free of tumor. Similarly, nerve III may be displaced caudally and posteriorly to some extent, but there is no extension into the oculomotor trigone or the nerve III canal. The ICA, tethered by the DR as it enters the dura, is not displaced but may be in contact with the meningioma. At least three quarters of its circumference is likely to be free of tumor. Dura inferolateral to the ICA is not involved with meningioma.

Type II

ACP meningiomas of this type extend beyond the region occupied by Type I lesions, entering the oculomotor trigone and possibly the nerve III canal. Nerve III is displaced even more posteriorly and may be encased by tumor. More medial extension results in compressing the ON against the falciform ligament or encasing it circumferentially, even invading the OC. Extending along the dural surface, these tumors encase the ICA, and, in many cases, the PCom and more peripheral ICA branches. Large Type II lesions are particularly likely to stretch or displace the ACA1 and MCA and their branches.

Type III

These tumors are similar to Type II lesions, but grow transdurally into the paranasal sinus(es) and/or the SOF, the PS, and/or the sella. In rare cases, they invade and destroy associated bony structures.

Type IV

Type IV refers to ACP meningiomas otherwise classified as Type I, II or III but associated with pneumatized bone as seen in Figs. 4.2(a–d) and 4.3a. Along with Type III lesions invading paranasal sinuses, they pose serious risk of cerebrospinal fluid (CSF) leakage following total resection (including diseased dura and hyperostotic bone), requiring that the surgeon be especially certain of a watertight dural closure and maintaining continuous lumbar drainage for at least one week following surgery.

Surgical approach to ACP meningiomas

The primary goal is total microsurgical resection of the lesion. Patient positioning, three-point skeletal fixation of the head, osteoplastic craniotomy, and unroofing the orbit are accomplished as described in Chapter 2. Resecting the hypertrophic, highly vascularized sphenoid wing and further unroofing the orbit medially should be performed sequentially, using a drill equipped with a diamond burr in order to devascularize dura in the vicinity of the ACP, a measure which will minimize blood loss. The diamond burr must be irrigated continuously with saline. After exposing the dural sleeve at the SOF covering neural structures entering the orbit from the middle cranial fossa, this dura is transected laterally and any resulting bleeding is controlled by meticulous bipolar coagulation. Then the outer layer of the lateral wall of the PS is carefully stripped away, exposing the inner layer of the PS wall and nerves III, IV, V1 and V2 within it. Only by exposing the lateral wall of the PS and these cranial nerves may one determine whether or not the meningioma has invaded the PS. Most ACP meningiomas do not enter the PS.

The ACP is resected by drilling it extradurally from its lateral surface with a diamond burr. The importance of continuously irrigating the burr to avoid thermal injury to ON and/or nerve III cannot be emphasized too much. Drilling should also be interrupted at frequent intervals. Dull burrs should be replaced immediately since they produce far more heat than when sharp. Furthermore, frequent inspection of bone yet to be removed minimizes the risk of plunging through the final bony barrier, injuring the ON, rupturing the ICA, or entering the lateral wall of the sphenoidal sinus.

Removing the ACP, the lateral and proximal walls of the OC, and the optic strut devascularizes ACP meningiomas significantly. Exposed dura must be handled quite gently to avoid transmitting mechanical pressure to the underlying tumor and adjacent vascular and neural structures. Separating the layers of the anterior half of the lateral wall of the PS also interrupts much of the intradural tumor's blood supply. Tumor devascularization is enhanced medially by peeling dura from the tuberculum selle and planum sphenoidale then drilling these structures, but this must be done very carefully to avoid entering the sphenoidal sinus.

Following the resection of bone as appropriate, dura is incised along the sylvian fissure, beginning posteriorly and proceeding anteriorly until the lesion is encountered. The sylvian fissure is split and branches of the MCA are exposed. It is important to resect all involved and abnormal dura with the meningioma. The neoplasm is dissected from its periphery, at its dural attachment. Depending on its size, the meningioma is resected with attached dura laterally over the PS, incising infiltrated dura until healthy dura is encountered. The lesion plus infiltrated dura are then resected medially, toward the ON. Debulking the now devascularized meningioma facilitates further dissection of the MCA and resection of tumor in its vicinity. It is crucially important to avoid exerting traction on any artery during tumor removal since the tumor may be densely attached to the ICA wall. After removing tumor from the MCA near the ICA bifurcation and the ACA1, the most basal part of the lesion may be resected with the DR surrounding the ICA. Finally, tumor is dissected from the tentorial edge, nerve III, the PCP, the ON, the PCom, the anterior choroidal artery, and from other ICA perforating arteries. Resecting tumors from their epidural surface, rather than peeling brain from the lesion, has the very clear advantage of protecting adherent vessels from stretch injury. If the diaphragm sellae has been invaded, the DR must be resected circumferentially around the ICA, and the meningioma-infiltrated diaphragm sellae is resected with the tumor, finally coagulating its free edges. This is particularly important since the DR is actually an extension of the diaphragm sellae.

The OC should be opened along its entire length sufficiently widely to remove meningioma extending into it, along with the ON dura propria. Dura propria must be resected until it becomes clearly normal peripherally in the OC. Where meningioma has invaded the ICA, it should never be peeled forcefully from the arterial wall but resected sharply, coagulating the film of residual tumor circumferentially around the vessel. This is important. It is far better to leave a coat of tumor attached to the arterial wall than to risk rupturing the ICA.

After totally resecting the meningioma and preserving all neural and vascular structures, the dura is closed in a watertight fashion. Since dura has been excised together with the lesion, an appropriate dural graft should be tailored and secured with interrupted sutures to healthy, dura surrounding the ICA, the ON, and over the lateral wall of the PS, then coated with a film of fibrin glue. An epidural drain provides additional insurance against an epidural clot forming soon after surgery.

After total removal of very aggressive meningiomas occupying extradural and intradural compartments extensively destroying dural and bony structures, dural closure should be achieved with fascia lata or temporalis muscle and fascia, which is reinforced with fibrin glue. This is superior to employing any of the commercially available dural substitutes. If lumbar drainage is required after surgery, preoperative and postoperative antibiotics should be given.

Patients and results

Sixty-seven ACP meningiomas were treated. Preoperative symptoms and signs were generally mild despite the fact that the tumor was greater than 3 cm in diameter in more than 2/3 of cases. Two patients died after surgery following intraoperative ICA rupture. After these tragic events, all patients with ACP meningiomas of Types II, III and IV had DSA and cross-circulation studies prior to surgery, and efforts to carefully and gently dissect tumor from the intradural ICA were redoubled. Where tumor was densely adherent to the ICA, we have become satisfied with incomplete, although gross-total, resections. Complete tumor resection was achieved in 62 of 67 cases, including operations to remove virgin and residual or recurrent meningiomas.

Sixty-one patients recovered fully intact neurologically. Four manifest continuing neurological deficits. Two continue to have a visual deficit in the ipsilateral eye, which was aggravated by surgery. The third patient still has diplopia from stretch-injury to nerve III. The nerve was observed to be quite attenuated at surgery, but the injury was clearly compounded by dissection. The fourth patient failing to recover completely following surgery is quite interesting and instructive. It was a woman with a Type III ACP meningioma who postponed surgery for three years even while experiencing impaired monocular vision and diplopia from ipsilateral ON and nerve III involvement. Her tumor resection was complete, and the entire surgical procedure, including anesthesia, was uneventful. She was extubated immediately following surgery and was fully awake and conscious. Suddenly she had a generalized seizure which led quickly to status epilepticus, requiring reintubation and mechanical ventilation while she was treated with anticonvulsants in very high doses. Only after three days was it possible to control her seizures with medications in more standard doses and allow her to regain consciousness. Since then, we have insisted on preoperative EEGs for every patient with a large ACP meningioma. Preoperative anticonvulsants are now employed where irritating EEG foci corresponding to the tumor location are demonstrated.

Conclusions

Total resection of even large ACP meningiomas is possible in the majority of cases by the measures described here. Occasionally incomplete tumor resection is related to tumor invasion of the ICA wall. It seems only prudent to accept less than complete tumor removal when this is the case.

Table 9. ACP meningioma type, size, extent of resection, and postoperative outcome

			Type				Tumor size				Resection				Recovery					
Location	N° of cases		Globular		En plaque		<3 cm		>3 cm		Complete		Gross-total		Complete		Incomplete		Died	
Type I	26	39%	22	85%	4	15%	12	46%	14	54%	26	100%	–	–	26	100%	–	–	–	–
Type II	23	34%	22	96%	1	4%	6	26%	17	74%	22	96%	1	4%	20	87%	2	9%	1	4%
Type III	12	18%	12	100%	–	–	1	8%	11	92%	9	75%	3	25%	9	75%	2	17%	1	8%
Type IV	6	9%	4	67%	2	33%	2	33%	4	67%	1	17%	1	17%	6	100%	–	–	–	–
Total	67	100%	60	90%	7	10%	21	31%	46	69%	62	93%	5	7%	61	91%	4	6%	2	3%

Middle cranial fossa meningiomas

Pathological anatomy, clinical presentation, and diagnostic work-up

Meningiomas may arise from various middle cranial fossa sites, most commonly near the sphenoid wing. Globular lesions most frequently are attached to dura of the lateral wall of the parasellar space (PS) as illustrated in Figs. 4.7(a, c) and 4.8(a, c, e) while "en plaque" lesions typically advance along dura adjacent to the sphenoid wing and neighboring areas, most commonly spreading medially toward the anterior clinoid process (ACP), even across it to the diaphragm sellae and into the optic canal (OC). Meningiomas penetrating dura render adjacent bone hyperostotic, particularly "en plaque" lesions. Their capacity to infiltrate and traverse dura precludes any adjacent region from establishing a safe haven from them, including the PS and the orbit.

Globular meningiomas compete with brain for intracranial space as demonstrated in Figs. 4.7(a, c) and 4.8(a, c, e), resulting in the generalized headaches of increased intracranial pressure, often in settings of persistent retro-orbital discomfort lasting for years or unilateral headaches previously misinterpreted as "migraine". "En plaque" lesions, on the other hand, generate headaches infrequently. They are far more likely to be heralded by a vague retro-orbital discomfort. Double vision and/or ipsilateral visual deterioration are also common complaints. Sometimes CT or MRI lead to diagnosis in the course of evaluating a seizure disorder, often where no other complaints exist. Once diagnosis is established, the treatment of choice is total microsurgical removal when possible.

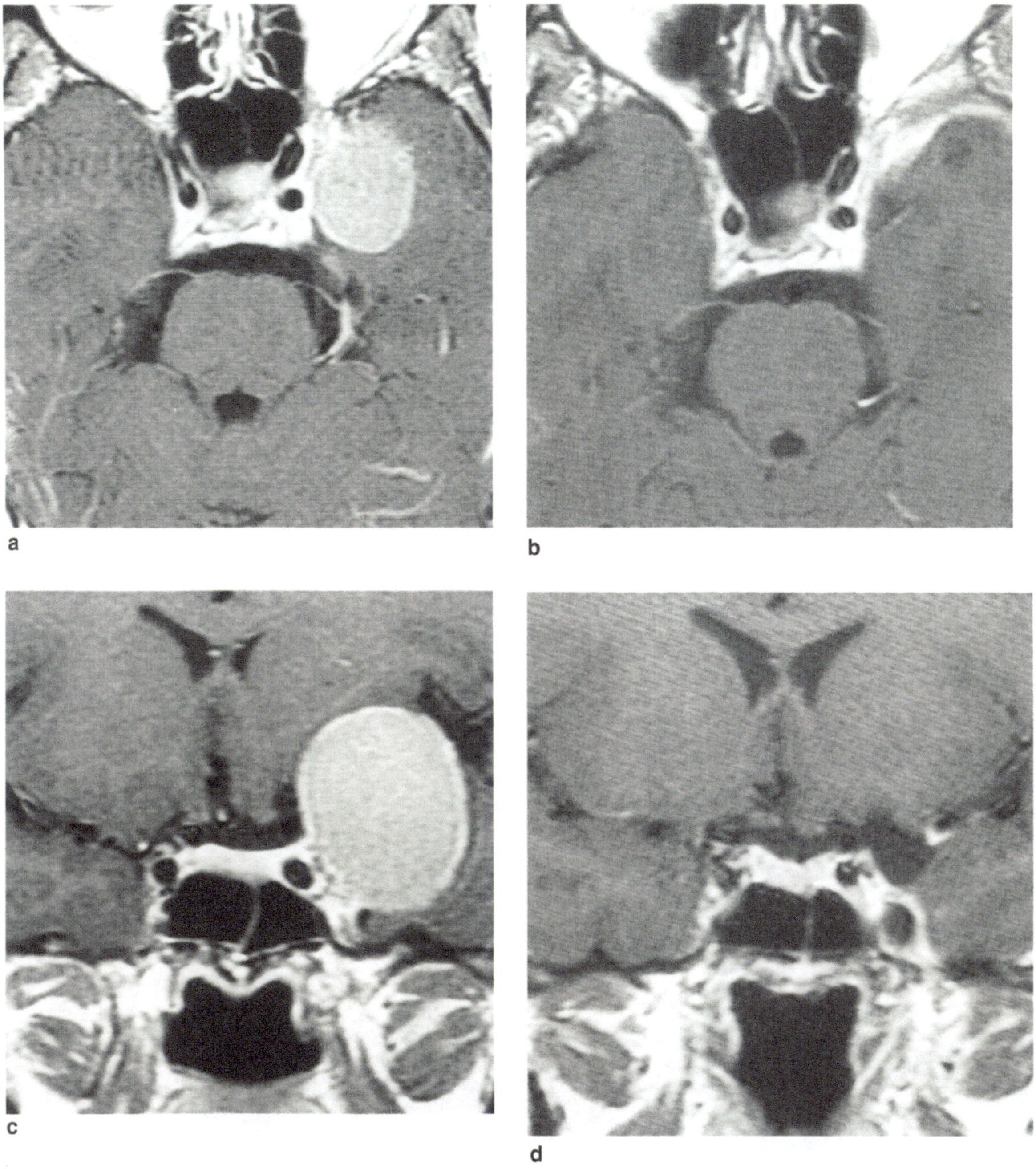

Fig. 4.7. Trans-axial (a) and coronal (c) contrast-enhanced MR images demonstrate a left middle cranial fossa meningioma which is fixed to the lateral PS wall. Corresponding postoperative contrast-enhanced MR images, trans-axial (b) and coronal (d), confirm complete resection of the lesion

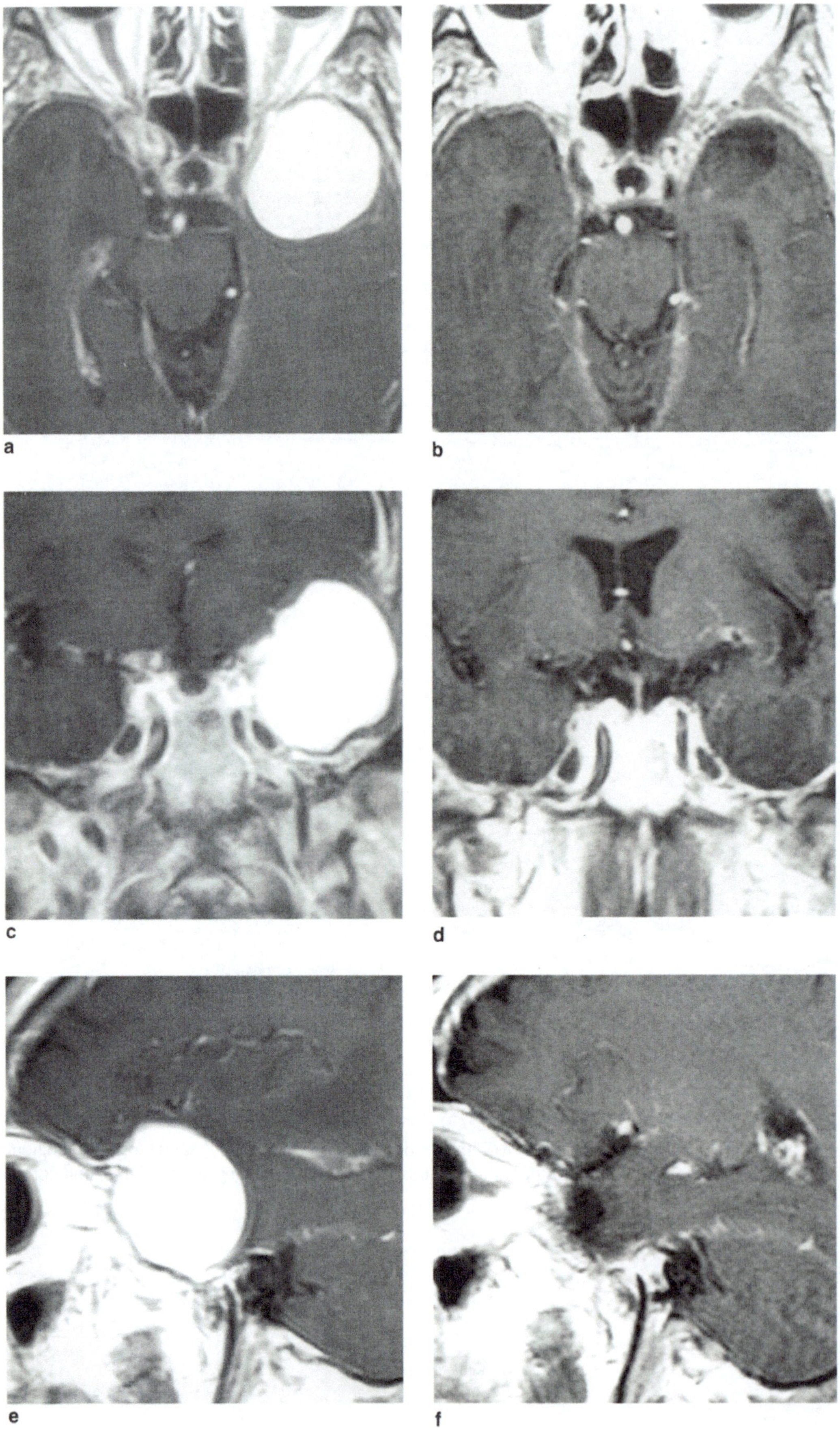

Fig. 4.8. Contrast-enhanced MR images, trans-axial (a), coronal (c), and sagittal (e), reveal a large left middle cranial fossa meningioma. Corresponding postoperative contrast-enhanced images, trans-axial (b), coronal (d), and sagittal (f), demonstrate its total removal

Surgical approach to the middle cranial fossa meningiomas

Patient positioning, three-point skeletal head fixation in the Mayfield device, osteoplastic craniotomy, unroofing the orbit, and resecting the sphenoid wing and ACP are accomplished as described in Chapter 2.

The roof of the orbit, whether hyperostotic or normal, and the lateral and posterolateral sphenoid wing are resected sufficiently to visualize V2 entering the foramen rotundum (FR). Dura at the pole of the temporal lobe and postero-laterally is peeled from the lateral wall of the PS. In addition, all tumor-infiltrated dura throughout the middle cranial fossa is stripped from bone until normal dura and bone are encountered. Bone bleeding is controlled with a drill equipped with a diamond burr. After exposing the dural sleeve traversing the superior orbital fissure (SOF) posteriorly and anteriorly, the central part of the sphenoid wing and adjacent orbital roof, anterior and medial to the SOF, are resected with a diamond burr until encountering normal bone. Then the SOF dura is incised laterally, and the outer layer of the lateral wall of the PS is gently stripped from its inner lining, exposing nerves III, IV, V1, V2. Bleeding from branches of the PS-ICA is controlled with the bipolar cautery, scrupulously avoiding nerves III, IV, V1, V2 and V3. Dura is then elevated from the lateral surface of the ACP. If the ACP is hyperostotic or otherwise involved with tumor, it should be removed entirely. The same is true for the walls of the OC. If the ACP and dura posterior and lateral to it appear normal, the ACP and the OC may be left intact. After the outer layer of the lateral wall of the PS has been peeled medially to the tentorial edge, and posteriorly to the Gasserian ganglion (GG) and V3, even farther if necessary to identify normal dura, the extradural dissection is complete.

By thus dividing the skull base dura and interrupting arterial branches supplying the tumor, most middle fossa intradural meningiomas are largely devascularized, essentially "converting" them into convexity lesions, greatly simplifying their total resection. Dura is incised along the sylvian fissure toward the ACP, ultimately exposing the tumor and allowing its resection, including all infiltrated dura, in a stepwise manner by the piecemeal technique. Tumor resection is initiated from its dural surface, never from the surface adjacent to brain. Cerebral tissue, especially with large globular meningiomas, is gradually displaced from the underlying tumor in a posterior direction. The venous drainage of this cap of brain, dislocated and compressed by tumor, has been compromised, leaving it attenuated and edematous. An arachnoidal plane may no longer exist around it, and its pial covering may have been destroyed. Such brain has lost its normal shape and consistency and may be damaged further if not handled with great care. This is especially important when the meningioma involves the dominant hemisphere. Resecting infiltrated dura, then debulking the tumor from its dural surface results in only minimal disturbance of such eloquent brain and avoids stretching or otherwise damaging adherent vessels.

Globular meningiomas in this region rarely present problems freeing the MCA and its branches and/or the posterior cerebral artery (PCA) since these vessels generally remain confined to the layer of compressed brain tissue overlying the tumor. Problems may occur, however, with respect to veins draining into the vein of Labbé, where arachnoid and pia have been disrupted. Such veins must be preserved at whatever cost in time and effort, even if it means removing a dominant hemisphere tumor incompletely.

Even "en plaque" lesions traversing the dura in this region may be resected without great difficulty so long as nerves III, IV, V1, V2, V3 are exposed and protected extradurally, as well as the intradural optic nerve (ON), nerves III and IV, the ICA (and its branches), and the PCA. When tumors of this type spread medially, all diseased dura over the ACP and the ipsilateral surface of the diaphragm sellae and ON dura propria within the OC must be fully resected, leaving only totally normal dura.

Following tumor resection and establishing complete hemostasis, the dura may be reconstructed with fascia lata or commercially available dural substitute which is tailored appropriately and secured with interrupted sutures to healthy dura. Fibrin glue should be applied to the epidural surface and an epidural drain left in place prior to replacing the osteomuscular flap, securing it with sutures, and closing the wound in layers.

Patients and results

Among 12 large and very large middle cranial fossa meningiomas, 11 were globular, only one of the "en plaque" type. Ten were over 5 cm in diameter. Eleven were completely resected, only a single case relegated to only gross-total removal. That particular tumor partially invaded the PS, and the patient experienced unpleasant postoperative dysesthesias in the distribution of V1 which have persisted, likely the result of thermal injury to V1 while dealing with particularly troublesome bleeding with the bipolar coagulator. Eleven of the 12 patients recovered completely although two still require anticonvulsant medications because of persisting foci of epileptiform discharges (EEG) ipsilateral to their operations. Neither has had a clinical seizure.

Table 10. Middle cranial fossa meningiomas: lesion type and size, extent of resection, and postoperative results

Origin of meningioma	N° of cases		Type				Size				Resection				Recovery			
			Globular		En plaque		<5 cm		>5 cm		Complete		Gross-total		Complete		Incomplete	
Lateral wall of PS	7	58%	6	86%	1	14%	1	14%	6	86%	7	100%	–	–	7	100%	–	–
Sphenoid wing & Lateral wall of PS	2	17%	2	100%	–	–	1	50%	1	50%	2	100%	–	–	2	100%	–	–
Petrous bone & Lateral wall of PS	1	8%	1	100%	–	–	–	–	1	100%	1	100%	–	–	1	100%	–	–
Transdural & complex	2	17%	2	100%	–	–	–	–	2	100%	1	50%	1	50%	1	50%	1	50%
Total	12	100%	11	92%	1	8%	2	17%	10	83%	11	92%	1	50%	11	92%	1	8%

Parasellar space meningiomas

Pathological anatomy, clinical presentation, and diagnostic work-up

Strictly defined, parasellar space (PS) meningiomas are confined entirely to the extradural PS compartment Figs. 4.9a, and 4.10(a, c and e). Only those lesions will be discussed here. Once a meningioma extends through one or several PS apertures, or directly through its walls into an adjacent anatomical compartment (the orbit, the sella, the posterior cranial fossa, or Dorello's space, for instance), it is no longer classified as a pure "PS meningioma," becoming instead a "parasellar space and adjacent regions meningioma" the bulk of which is either within or outside the PS. Such lesions will be discussed in the following section. Similarly, meningiomas arising in neighboring regions – the superior orbital fissure (SOF), the diaphragm sellae, the anterior clinoid process (ACP), the tuberculum sellae, the petroclival region, etc. – may invade the PS. Such lesions are classified according to their primary sites of origin and the location of the bulk of their mass. They too are discussed elsewhere. The distinction is not entirely arbitrary since it carries important practical implications in designing a surgical approach to various lesions. It becomes even more important in the course of such operations.

Meningiomas originating within the PS most probably arise from arachnoidal cells accompanying cranial nerves coursing through the PS (nerve VI) or within its lateral wall (nerves III, IV, V1, and V2). Such lesions may remain confined to the PS or spread to adjacent compartments.

Experience and anatomical study confirms that certain segments of nerves III, IV, V1, V2, VI are more vulnerable than other segments to extrinsic compression and therefore more likely to result in deterioration of function when affected by tumor. Direct proof of this lies in the not infrequent incidental finding of even large PS meningiomas which do not cause neurological deficits while much smaller lesions, located between the posterior PS wall and the ICA, commonly result in diplopia from nerve VI compression. Similar circumstances occur with meningiomas in the anterior corner of the PS, where compression of one or more branches of nerve III, nerve VI, or nerve IV result in symptoms and signs. *Size* of the PS lesion is much less important to the generation of clinical phenomena than *locations* where nerve segments are fixed, and therefore vulnerable to compression by an expanding mass. *Histological tumor-type* also plays an important role in the generation of neurological deficits. A fibroblastic meningioma, a relatively firm lesion, typically results in neurological manifestations much earlier than a hemangioblastic lesion, regardless of its size and location.

Position and size of the ICA as determined by preoperative study are also important since displacement and variations in internal diameter may indicate tumor involvement of its wall. An ICA with a normal diameter, even though possibly displaced, is a good sign compared to a clearly non-displaced "stenotic" PS-ICA. The latter typically indicates tumor-encasement, with tumor possibly even protruding into the arterial wall where adventitia has been destroyed. This frequently occurs with incidentally found PS meningiomas. Usually such patients are advised by their doctors not to consider treatment, specifically surgical treatment. After repeated MR imaging over several years, comparing each to the original study, such meningiomas may be declared to be "stable," implying the absence of further tumor growth. So long as no additional symptoms occur, such patients may believe their tumors to be quiescent. The possibility of temporarily arrested growth notwithstanding, this is often not the case! It is true that some meningiomas grow more slowly than others, but they all grow. If not, how have such incidentally-found lesions achieved sufficient size for diagnosis in the first place?

The trap is that, for years, such PS tumors remained confined to the PS, since free space afforded them room for growth. Even when all available space has become consumed, PS tumors may continue grow at the expense of the PS-ICA volume, hence leading to PS-ICA "stenosis." Pressure inside such tumors then competes with systemic arterial pressure. In other words, such tumors, confined between bone and rigid dura, continue to grow "inward". It would be incorrect to describe a follow-up MRI, performed one year after another study, as revealing no tumor change if the diameter of the ICA is less. Such patients, desperate to hear that their PS tumors are "stable," may not become symptomatic until their PS-ICA "stenoses" reach the point of reducing blood flow, complicating its removal at the very least, possibly even precluding total resection. Unfortunately, such patients do not become aware of continued tumor growth until one or more nerves within the PS begin to deteriorate. There is little question that the likelihood of invasion of the ICA wall increases as the PS becomes more filled with tumor.

Most PS meningiomas present with diplopia or pain from injury to nerves III, IV, VI or V1, V2 respectively. Often the basis for such diplopia is not understood prior to CT or MRI. Some patients with nerve VI impairment have undergone surgical procedures to correct their "squints" many years prior to diagnostic imaging studies. Typically they assume that their continued symptoms are the result of incomplete surgical correction of their extraocular "muscle(s) imbalance" problems. Compensating or partially compensating such signal symptoms and signs with extraocular muscle surgery may only delay the diagnosis of a PS meningioma until it has spread well beyond the frontiers of the PS, affecting other cranial nerves, the ICA, or causing exophthalmos.

Unfortunately, in the great majority of cases, even establishing the correct diagnosis provides little relief. Operations either poorly planned or poorly executed result in doubling or tripling tumor morbidity. Commonly the initial attempt to resect the tumor is not only unsuccessful, but additional neurological deficits appear following surgery, many of which are not only destined to become permanent but could have been avoided had a surgical plan been correctly formulated and executed. Furthermore, the residual tumor continues to grow, creating still additional symptoms and signs and compounding the problems facing surgeons equipped for definitive resection. Meningiomas demand surgical excision. Other means of therapy are ineffective so far.

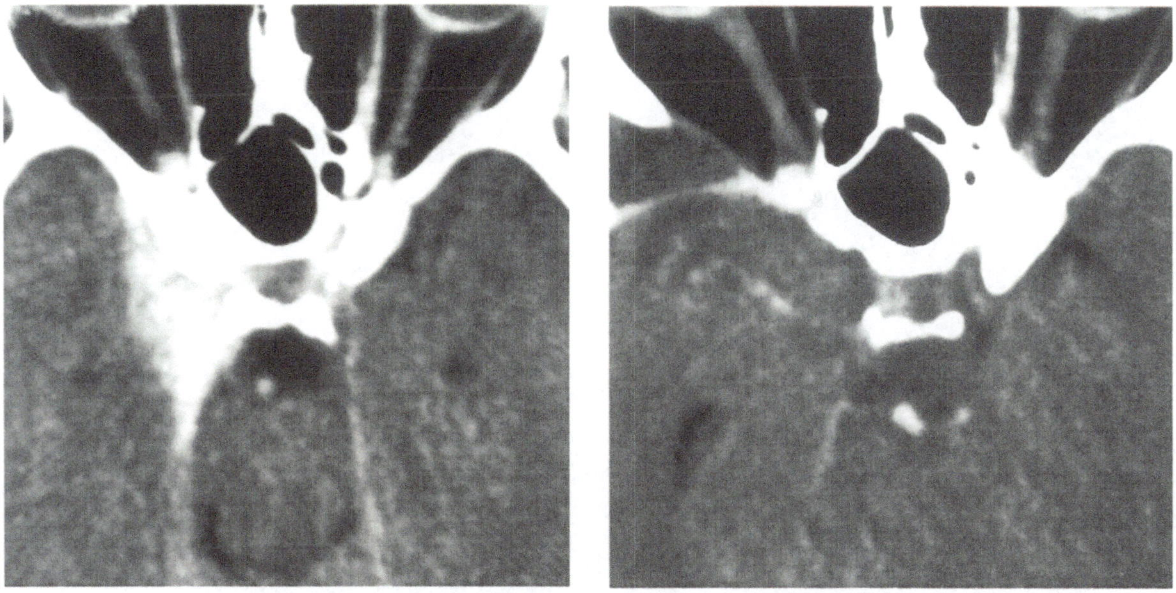

a b

Fig. 4.9. Preoperative contrast-enhanced, transaxial (a), CT image shows a clear-cut PS meningioma with a slight hypervascularization along the tentorial edge. The corresponding postoperative contrast-enhanced CT image demonstrates complete resection of the tumor as well as resection of the sphenoid wing and anterior clinoid process

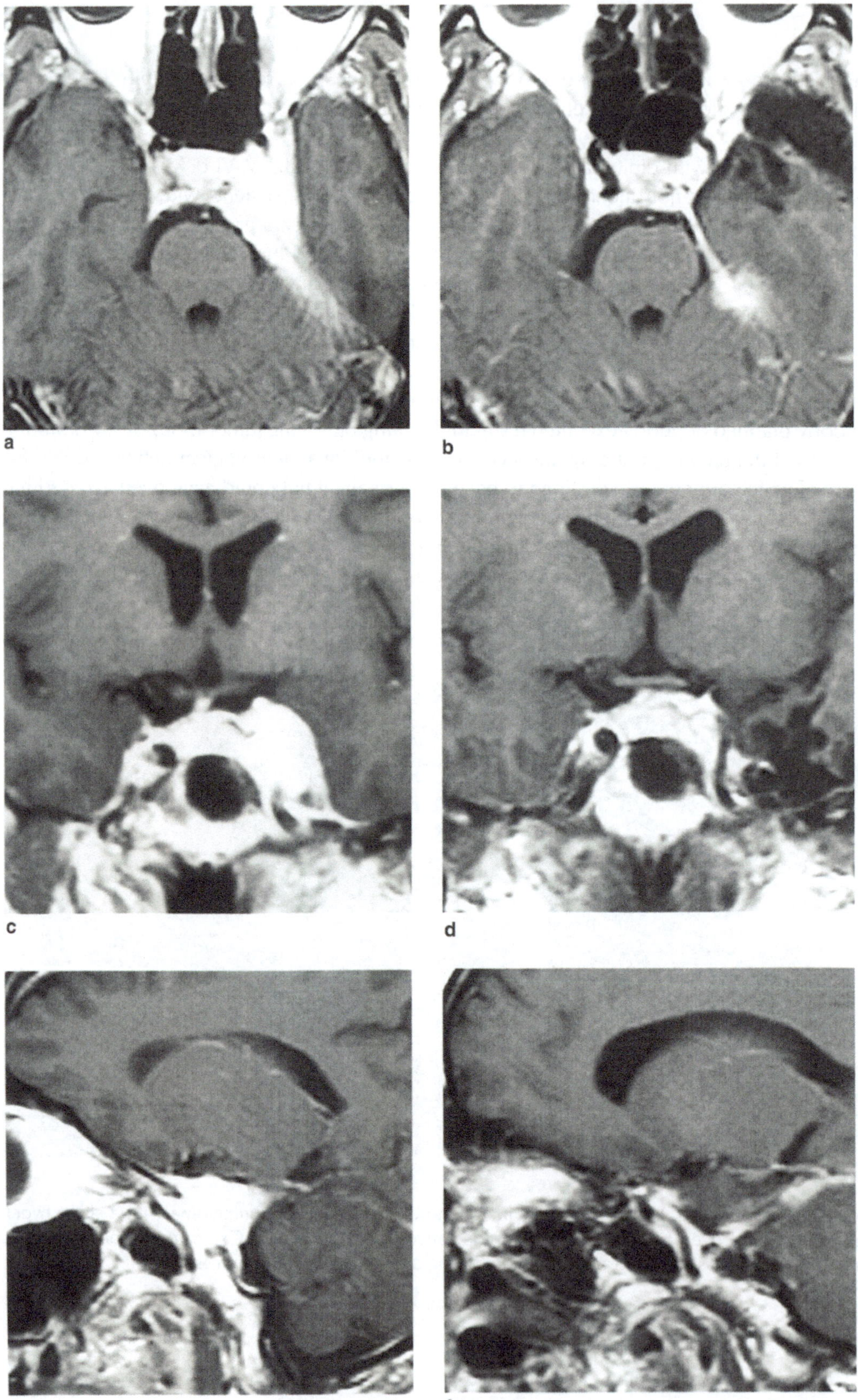

Fig. 4.10. Preoperative contrast-enhanced, transaxial (a), coronal (c) and sagittal (e), MR images demonstrate a PS meningioma with the characteristic "meningeomatous tail" along the tentorial edge and infiltration of the postero-medial wall of the PS (a). Corresponding postoperative contrast-enhanced images (b, d, f) show gross-total resection (b). Note: Narrowing and displacement of the PS-ICA and slight increased contrast-enhancement along the tent (b)

Surgical approach to the meningiomas inside the parasellar space

The initial surgical approach to meningiomas confined to the PS is identical to that described in Chapter 2, including patient positioning; scalp incision; osteoplastic craniotomy; unroofing the orbit; resecting the sphenoid wing; transecting the dural sleeve traversing the SOF and splitting the layers of the lateral wall of the PS; resecting the ACP and the superior and lateral walls of the optic canal (OC) together with the optic strut; exposing the foramen rotundum (FR), foramen ovale (FO), and foramen spinosum (FS); transecting the middle meningeal artery (MMA); dissecting the greater petrosal nerve (GPN); and exposing the ICA in the petrous canal. It is crucial that each of these steps is carried out completely and precisely. Only when the general extradural approach is executed as described may one begin to resect a PS meningioma through one or more of the triangular windows circumscribed by nerves traversing the PS.

It is especially important that the surgeon isolate the ICA in the petrous canal for temporary occlusion if necessary. Dissecting a meningioma encasing the PS-ICA may result in life-threatening bleeding. *Repairing the vessel is not possible without proximal control.* In addition, since retrograde bleeding from the distal ICA may be significant, it is equally important to fully expose the ICA through the anteromedial triangle for placement of a second clip on the anterior loop (AL) of the ICA, distal to the lesion. Accurate repair of the ruptured segment is greatly facilitated by a relatively bloodless field. Nevertheless, no attempt to isolate the PS-ICA between clips should be made until absolutely necessary, when the PS-ICA is bleeding. Bilateral carotid angiography and cross-circulation studies are vitally important prior to surgery to assure that any given patient will tolerate temporary ICA occlusion sufficiently long to allow its repair.

Parkinson's triangle provides the largest window allowing relatively safe entry into the PS in order to debulk tumors that are not particularly firm or fibrous. This triangle may be greatly enlarged by dissecting nerve IV from the medial surface of V1 throughout its course on the outer surface of the tumor, mobilizing it over nerve III, near the SOF. When nerve IV is displaced toward nerve III and over it medially, the entire medial loop (ML) of the ICA and the horizontal segment of the PS-ICA may be dissected from the tumor. Using the piece-meal technique one may resect tumor from the horizontal PS-ICA and the ML, until the meningo-hypophyseal trunk (MHT) is encountered. If the lesion is soft, the MHT may be preserved intact. In most cases, however, this is not possible, requiring that the MHT be coagulated and transected, then its central stump coagulated a second time. It is important to be aware that nerve VI lies adjacent to the ascending PS-ICA near the MHT, always inferolateral to its origin. Locating the MHT facilitates the identification of nerve VI, and avoiding confusing it with sympathetic cords lying on the PS-ICA wall. When nerve VI is identified, it is dissected from the inferior surface of V1 and from the lateral surface of the ICA throughout its course through the PS toward the SOF. Adjacent to the horizontal segment of the PS-ICA, nerve VI is also inferior to the infero-lateral trunk (ILT), which is usually coagulated, transected, then clipped or coagulated again as appropriate. The MHT and the ILT are the two most reliable landmarks for locating nerve VI within the PS.

When the meningioma is resected from the lateral surface of the horizontal PS-ICA (between its ML and AL) and when nerves VI and V1 are freed from it, remaining tumor (anterior to the ICA toward the floor of the PS) may be resected. This may be accomplished through either Parkinson's triangle or the anteromedial triangle. Tumor is also resected from the posterior compartment of the PS, adjacent to the ascending PS-ICA and Dorello's space (extending from the site of nerve VI's entry into clival dura from the prepontine cistern to the PS), finally from the PS-ICA medially (the ML and the horizontal segment).

Meningiomas confined to the PS are resected entirely by the extradural approach, without opening the dura. Once this is done, precise hemostasis must be established. Straining and coughing when patients are extubated following anesthesia sometimes results in retrograde bleeding from the venous sinuses connecting the two PSs. This may be prevented by placing Surgicel within the sella at this site and toward the inferior petrosal sinus (IPS) and superior petrosal sinus (SPS). Then the PS is filled with small pieces of muscle or fat, which should be reinforced with fibrin glue.

At this point, the dura is incised parallel to the longitudinal axis of the sylvian fissure to determine if the meningioma has extended intradurally. If intradural tumor is encountered, it should be resected with any involved dura. After thorough inspection, confirming adequate resection, a watertight dural closure is achieved in the usual manner. An epidural drain is left in place, the osteomuscular flap is replaced and secured with sutures, and the wound is closed in layers.

Patients and results

Surgery was carried out in 123 patients with meningiomas confined to the PS. They may be placed into one of four groups, depending on the amount of tumor within the PS and the status of the PS-ICA by preoperative study. Group 1 (six patients) included only those tumors approximating 50% of the PS volume or less. Twenty-two patients were in Group 2, where the PS was up to 90% filled, the ICA remaining normal. Forty-nine patients composed group 3, where tumor completely filled the PS, with minimal changes in ICA diameter or position. In Group 4, consisting of 46 patients, the entire PS was occupied by meningioma as evidenced by bulging of its lateral wall, plus severe changes in ICA diameter or position. In no patient in Group 4 did tumor significantly invade an adjacent compartment or extend though the lateral PS wall. The lateral wall of the PS was infiltrated by tumor in almost all Group 4 cases.

Eighty-nine patients underwent surgery for the first time, whereas 34 patients had been operated on previously elsewhere. Of the 89 virgin cases, the lesion was resected completely in 38, with gross-total resections in 37 and sub-total resections in 14. In none of 34 patients operated on for residual lesions was complete resection possible. Gross-total resection was achieved in 15, sub-total in 19.

No patient died after surgery, but six require continued anticonvulsant treatment, five because of EEG abnormalities related to previous operations, only one from among those having surgery for the first time. CSF leaks occurred in two patients operated for residual tumor, both of which recovered on conservative treatment (continuous lumbar drainage for one week).

No new cranial nerve deficits resulted from surgery, but five of the 71 virgin cases with preoperative cranial nerve deficits experienced only incomplete recovery after six months. Eighteen of the 34 patients treated for residual tumors who had cranial nerve deficits prior to surgery had continued deficits at the end of a six month follow-up period.

Table 11. Virgin and residual cases of parasellar space meningiomas treated: resectability and outcome

% of PS occupied by Tumor and ICA status	N° of cases		Resectability						Postoperative deficits					
	Virgin / Residual		Complete		Gross-total		Sub-total		Cranial nerve(s) after six months		CSF leak		Other	
M ≤ 50%, ICA = normal	6	5%	6	100%	–	–	–	–	1	17%	–	–	–	–
	–	–	–	–	–	–	–	–	–	–	–	–	–	–
M ≤ 90%, ICA = normal	17	14%	13	76%	4	24%	–	–	1	6%	–	–	–	–
	5	4%	–	–	4	80%	1	20%	2	40%	–	–	1	20%
M = 100%, ICA = changes	39	32%	10	26%	21	54%	8	20%	1	3%	–	–	–	–
	10	8%	–	–	6	60%	4	40%	6	60%	1	10%	2	20%
M ≥ 100%, ICA – severely changed	27	22%	9	33%	12	45%	6	22%	2	7%	–	–	1	4%
	19	15%	–	–	5	26%	–	–	10	53%	1	5%	2	11%
Collective	89	72%	38	43%	37	41%	14	16%	5	6%	–	–	1	1%
	34	28%	–	–	15	44%	19	56%	18	53%	2	6%	5	15%
Total	123	100%	38	31%	52	42%	33	27%	23	19%	2	2%	6	5%

Meningiomas of the parasellar space and adjacent region(s)

Pathological anatomy, clinical presentation, and diagnostic work-up

Meningiomas adjacent to the sella may involve the parasellar space (PS) plus one or more neighboring areas, generally originating outside the PS, then invading it. The opposite occurs but is less common. Typically, any parasellar meningioma invades the sella and displaces the pituitary gland contralaterally, sometimes even extending across the clivus into the posterior cranial fossa as seen in Figs. 4.12a, 4.13a, and 4.14a. Meckel's cave and the entire petroclival region may be involved (Fig. 4.15a). Dura at the superior orbital fissure (SOF) is commonly infiltrated, and significant "tails" of tumor along the edge of the tentorium are not infrequent Figs. 4.14a and 4.15a. Intradural tumors arising beneath the temporal lobe commonly invade the dural layer of the lateral PS wall, varying from a thin carpet of tumor (Fig. 4.15a) to a true globular mass Figs. 4.11a, 4.12a and 4.13a. Although extension of such tumors in any direction may vary both with respect to the volume of the mass and the extent of dural involvement, most predominantly infiltrate dura rather than spread as large masses. Untreated lesions in late stages of growth are obvious exceptions.

The surgeon must understand how much tumor is extradural Figs. 4.11a, 4.12(a and c), and 4.13a and how much is intradural, also the extent of bone involvement. Bone may be hyperostotic, directly infiltrated by tumor, even destroyed, allowing the extradural component to invade paranasal sinuses caudally and/or extracranial compartments laterally. Tumors may also encase the ICA intradurally as well as extradurally, displacing the vessel, invading its wall, or creating stenosis as illustrated in Figs. 4.11b and 4.13(a and b).

Clinical manifestations are more dependent on the site of tumor origin, its growth and histological type, and its potential to spread than its total volume. In most cases cranial nerve injury sufficient to cause symptoms and signs does not occur until the latter stages of growth. Only when arising in the vicinity of a nerve the mobility of which is restricted by its relationship to adjacent bone and/or rigid dura are small tumors likely to result in symptoms and signs. Such sites of particular cranial nerve vulnerability are easily identified. Before the onset of diplopia (from deterioration of nerves III, IV, and VI) the only manifestation of the presence of a tumor may be headaches (frequently confused with "migraine" disease) or retro-orbital discomfort (from irritation of skull base dura or a branch of nerve V). Specific eye movement signs typically do not occur until later. Exophthalmos also occurs late, likely related to the re-routing of venous channels draining the orbit. Another late symptom is unilateral visual loss, occurring as transient blurring alone or in combination with detectable deficits. With significant tumor extension medially, compressing the optic tract and/or the chiasm, homonymous hemianopsia may be present with or without visual symptoms. A large tumor mass at the tentorial edge, compressing the temporal lobe from beneath or medially, may result in epileptic seizures. The sequence of presenting symptoms and signs varies from patient to patient and often fails to reflect the precise relationship between the parasellar meningioma and adjacent neural structures.

Any of the symptoms and signs mentioned above require investigation with CT and MRI, either of which will lead to diagnosis and precise definition of the mass, extent of spread, and involvement of adjacent dura. CT is particularly sensitive to hyperostosis and/or bone destruction, also possible extension of tumor into paranasal sinuses or other extracranial bony compartments. CT is also likely to detect evidence of calcification within the tumor, indicating that it is likely very hard and may be separated from surrounding neural and vascular structures only with difficulty. While hyperostosis may also be apparent on MRI Fig. 4.15a.

On the other hand, MRI provides information clearly superior to CT, particularly with respect to the position of the ICA and possible narrowing of its internal diameter at the site of tumor (Fig. 4.13a). MRI also defines the relationship of the tumor to the ICA wall, both extradurally and intradurally, and the interface between tumor and brain, reflecting the existence or destruction of arachnoid and/or pia mater. Both MRI and CT are indispensable to precise preoperative evaluation of tumors in this group. Both must be performed prior to and following the injection of contrast material. Although the changes between pre-contrast and post-contrast images are similar by both techniques, the amount of brain edema adjacent to the tumor mass is more accurately determined by MRI.

Angiography is mandatory in every case, including detailed evaluation of the ipsilateral carotid artery (Fig. 4.13b) plus cross-circulation studies (Fig. 4.13c, d). If ICA stenosis or significant displacement is demonstrated, a balloon occlusion test with simultaneous comparison of the venous drainage patterns of both hemispheres must be obtained. Neurophysiological testing of nerves III, IV, VI and optic nerve (ON) function, plus classical ophthalmological evaluation of visual acuity and visual fields should be carried out. These studies are important in cases without obvious clinical involvement of associated structures, especially if there is reason to suspect dysfunction. Such information will serve as an objective baseline for subsequent studies, whether or not the patient considers immediate surgical treatment.

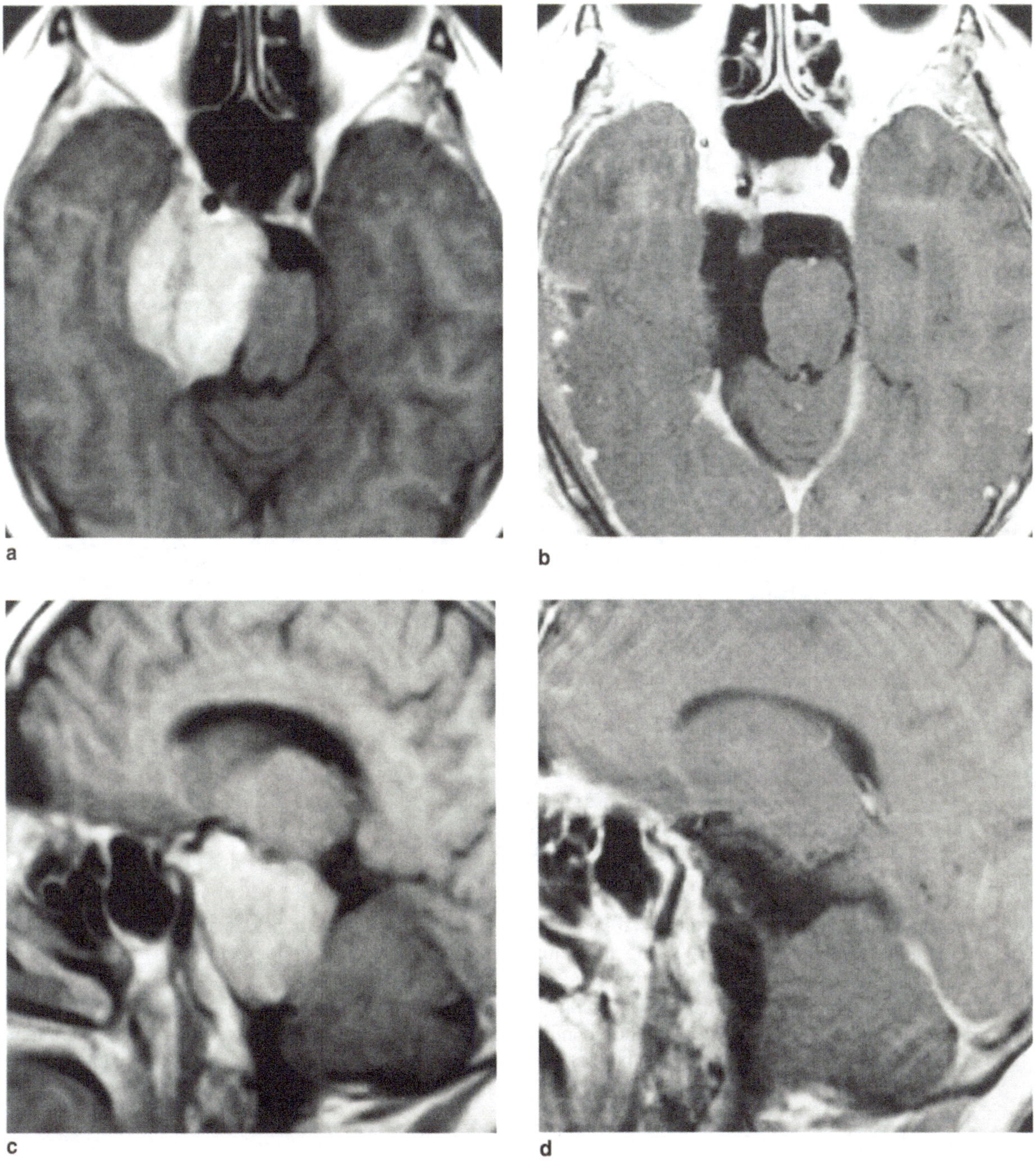

Fig. 4.11. Preoperative trans-axial (a) and sagittal (c) contrast-enhanced MR images show a medium-sized right-sided meningioma occupying the parasellar space (*PS*), middle cranial fossa, and posterior cranial fossa. Note the brainstem compression (a). Corresponding postoperative contrast-enhanced MR images, transaxial (b) and sagittal (d). In (b), a small residual tumor posterior to the PS-ICA is shown. Radiosurgery for such residual tumors may be effective if carried out in a timely manner. Moderate PS-ICA stenosis as demonstrated in (d)

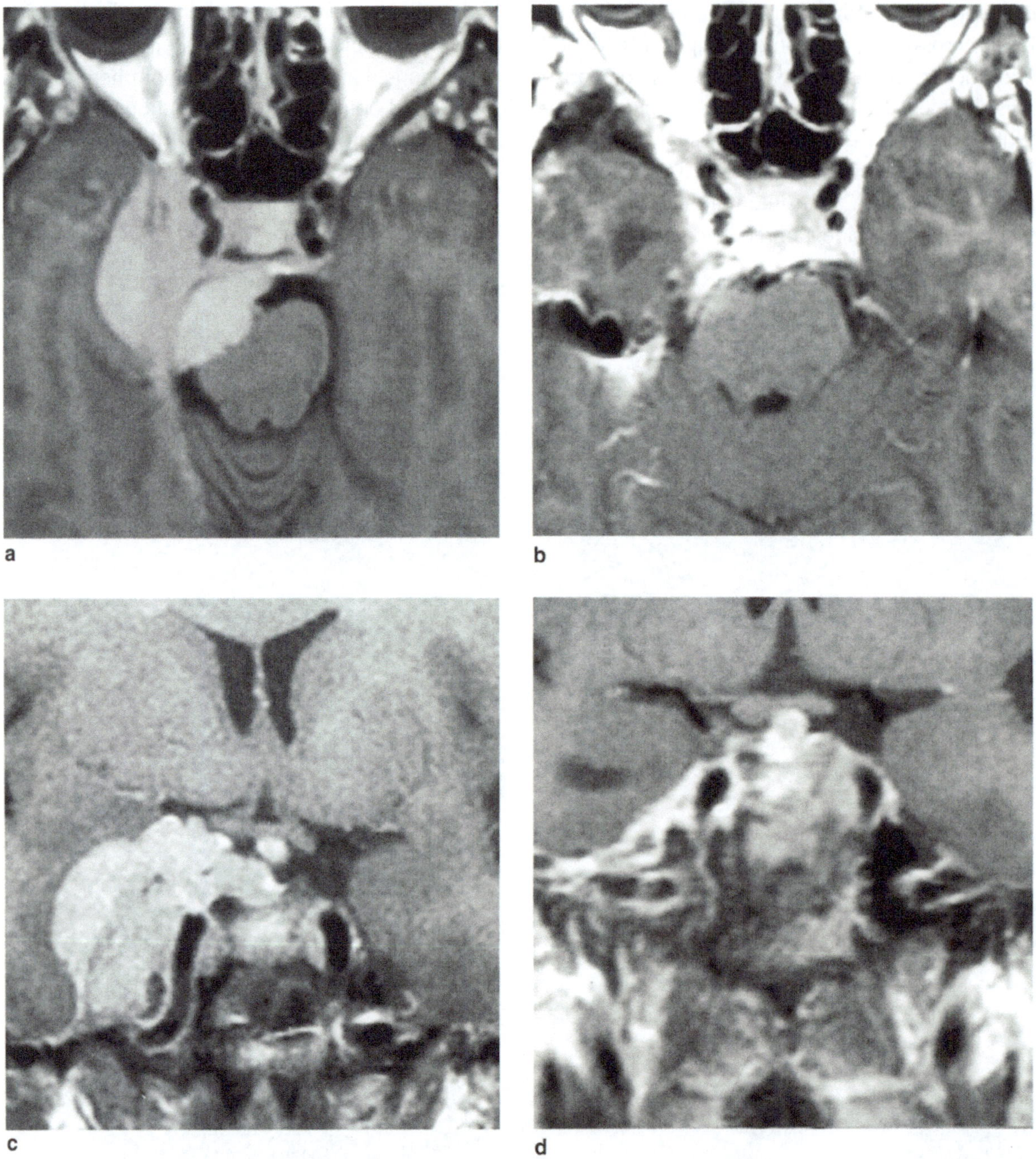

Fig. 4.12. Preoperative contrast-enhanced transaxial (a) and coronal (c) MR images demonstrate a meningioma involving the right PS and middle and posterior cranial fossae. In (a), tumor extension into the posterior cranial fossa compresses the brainstem. Both cuts (a, c) reveal the diameters of the left PS-ICA and the intradural ICA to be almost normal even though they are encased by tumor. This is typical of soft, richly vascularized lesions. Corresponding postoperative contrast-enhanced MR images, trans-axial (b) and coronal (d), confirm total resection of the tumor. Note the enlarged right temporal horn and the hypodensity in right temporal lobe (b). In (c), the lateral PS wall is delineated by a linear low density

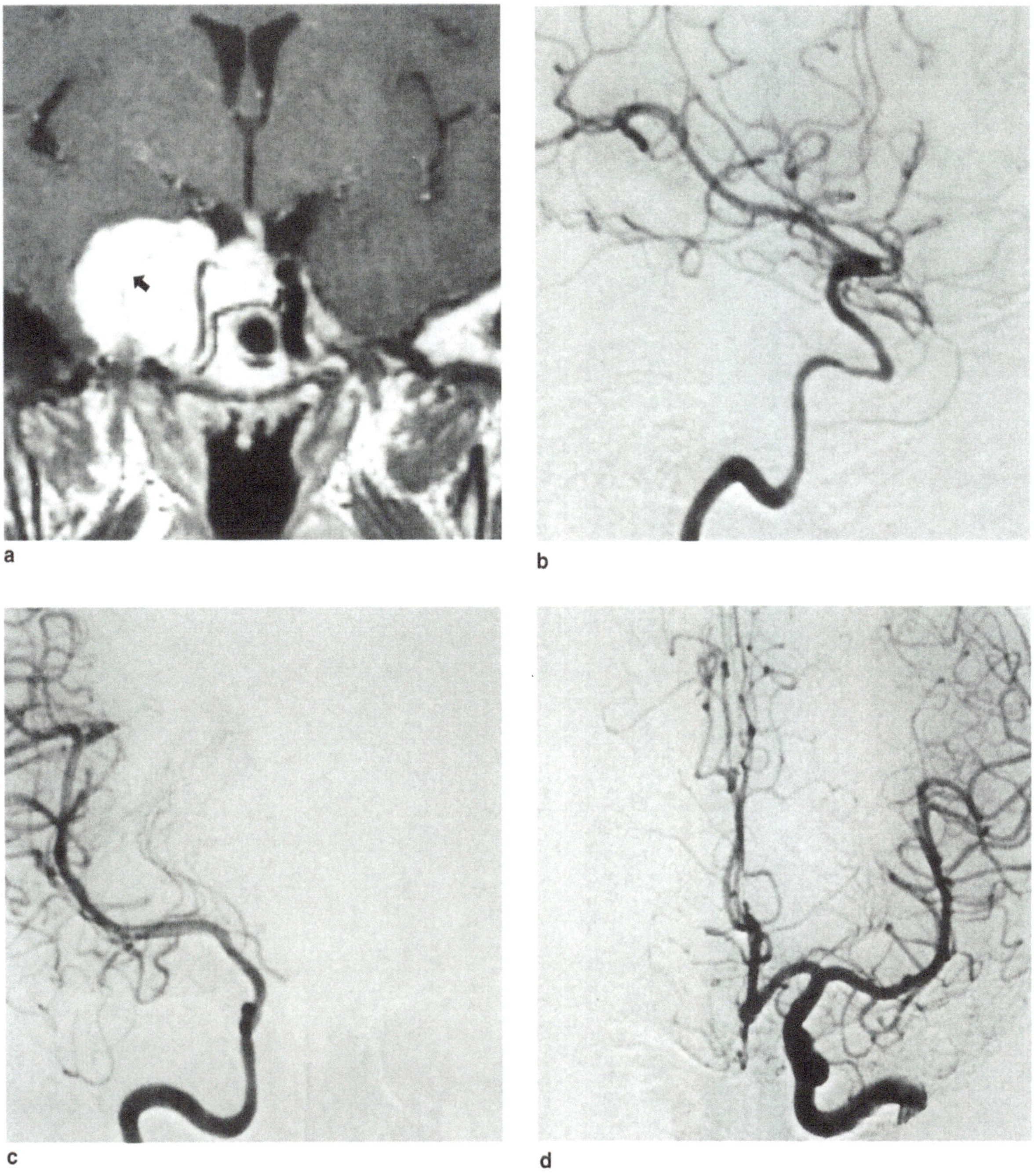

Fig. 4.13. Preoperative contrast-enhanced MRI (a) reveals a meningioma involving the right parasellar space (*PS*) and middle cranial fossa. Compare the severely stenotic PS-ICA, found at surgery to be encased by tumor, to the contralateral (left) PS-ICA, which is normal. Note the hypodense arcuate line indicating the PS lateral wall, "dividing" the lesion into intra-PS and extra-PS components (*arrow*). In (b), carotid angiography, lateral projection, confirms the PS-ICA stenosis shown by MRI in (a). reoperative AP films from bilateral carotid angiography (c, d) demonstrate the absence of the right ACA1, indicating limited collateral circulation to the right ICA territory. Note: Meningiomas with preoperative evidence of severe PS-ICA stenosis likely invade the arterial wall, making total resection impossible without serious risk of arterial rupture. In such cases, a small cuff of tumor should be left in place around the artery, to be treated subsequently by radiosurgery. The absence of collateral circulation from the opposite carotid tree precludes en block resection of tumor and artery and bypassing them with a high-flow graft

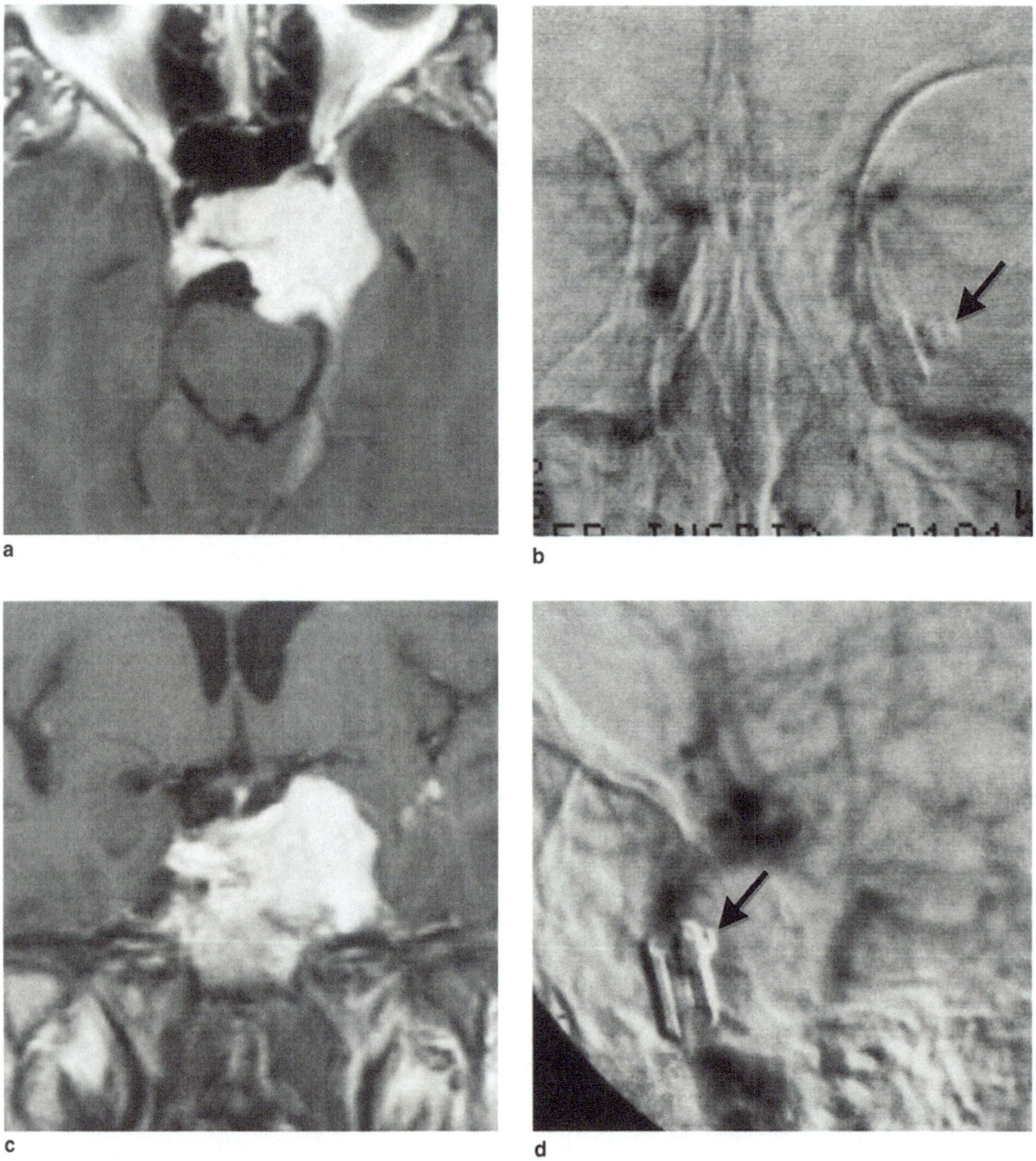

Fig. 4.14. Preoperative trans-axial (a) and coronal (c) contrast-enhanced MR images demonstrate a meningioma occupying the left parasellar space (*PS*) and the sella, extending into the middle and posterior cranial fossae. Note the left temporal lobe hypodensity which resulted from previous surgery (a). AP (b) and lateral (d) postoperative MRA demonstrate a functional carotid artery and middle cerebral artery distal to the cuff-clip placed on the left PS-ICA (b, d). Note the "shadow" of the basilar artery in (d). This tumor was totally removed, but the ICA ruptured. After resecting the diseased segment (less than 5 mm), the ICA was reconstructed by end-to-end anastomosis and the suture-line reinforced with a Sundt-Keys clip *arrow* (b, d)

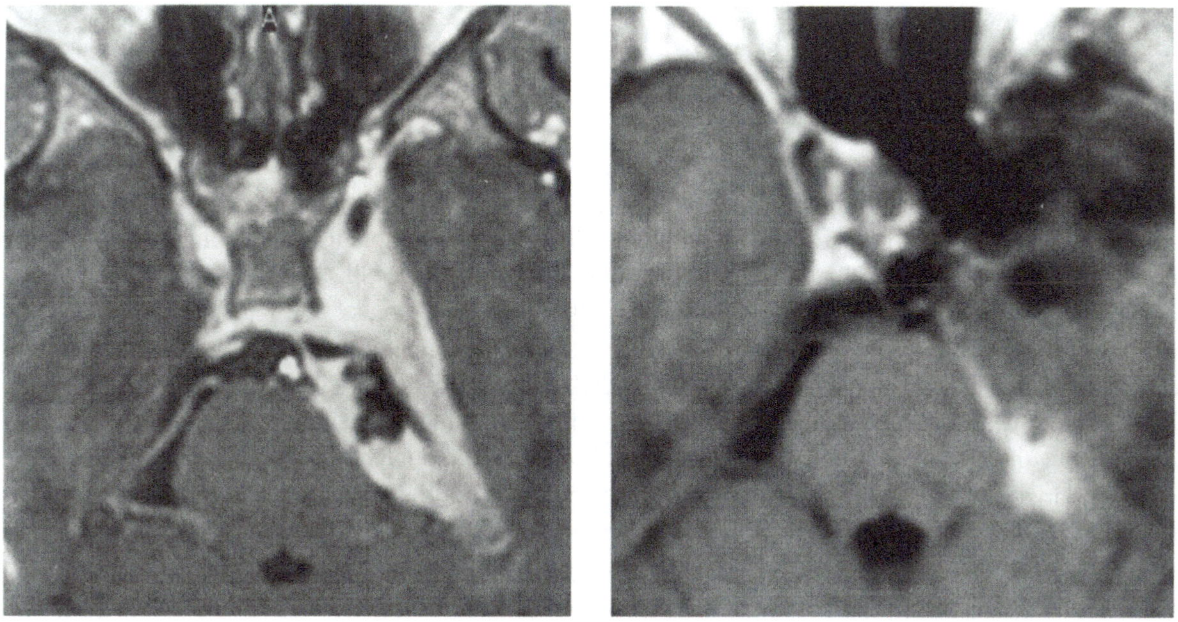

Fig. 4.15. Preoperative trans-axial (a) contrast-enhanced MR image demonstrates a meningioma extending from the left parasellar space (*PS*) along the tentorium into the posterior cranial fossa. Notice the ossified inclusions within tumor lining the tentorium. Corresponding postoperative contrast-enhanced MR image one year later confirms gross-total removal of the lesion (b). The small hyperdense zone well posterior to the resected tumor probably represents a vascular change that likely indicates the site of a new meningioma. Such cases require subsequent MR imaging at intervals of no longer than six months. Any sign of progression of this apparent lesion will demand additional therapy

Surgical approach to meningiomas of the parasellar space and adjacent regions

Basic steps of the general extradural approach to central skull base (CSB) lesions are described in Chapter 2, including patient positioning; scalp incision; osteoplastic craniotomy; unroofing the orbit; resecting the sphenoid wing, the anterior clinoid process (ACP), and the walls of the optic canal (OC) including the optic strut; transecting dura traversing the SOF and splitting the dural layer from nerves III, IV, V1, V2, V3, the Gasserian ganglion (GG), and the inner layer of the PS; transecting the middle meningeal artery (MMA) at the foramen spinosum (FS) and unroofing the intrapetrous ICA; plus dissecting the greater petrosal nerve (GPN), resecting the petrous apex, and exposing posterior fossa dura medial to it and dura within internal auditory canal (IAC). Each step must be carried out completely and precisely with all meningiomas in this category.

By completing the extradural approach as described above, the intradural component of the tumor is effectively devascularized and brain is fully protected since its dura has not been violated. Exposing the ICA in the petrous canal provides proximal control if necessary during tumor resection. The ICA is exposed for distal control by dissecting its anterior loop (AL) circumferentially through the anteromedial triangle. Once avenues for proximal and distal control have been established, attention may be turned to tumor within the PS.

Ideally, tumors within the PS are most directly removed through Parkinson's triangle. By dissecting nerve IV and transposing it medially, this window is effectively enlarged at the expense of the paramedial triangle. To avoid damaging nerve IV distally (at the SOF), it should be dissected from the inner layer of the lateral wall of the PS throughout its course, with particular care at its distal segment, where it crosses nerve III.

Although it provides excellent access to most PS meningiomas, there are clear disadvantages to beginning to resect certain of them through Parkinson's triangle, particularly lesions likely to invade the wall of the PS-ICA. By progressing through especially firm tumors or lesions containing calcifications to the medial loop (ML) of the ICA the surgeon courts real danger. Since there are no objective landmarks to the ICA medial to the tumor, he is left with only his instinct and experience as allies. Even modern monitoring technology will fail him. The very best "neuro-navigation" systems presently available, broadcasting data from every nook and cranny of intracranial space, will not protect him from disaster. Similarly, since nerves III, IV, V1, V2, V3 have been manipulated by splitting the layers of the lateral wall of the PS, somatosensory evoked potential monitoring will not shield him from the cruel realities of this complex anatomy, or absolve him from responsibility when things go wrong.

Where its bulging lateral wall indicates that the PS is fully packed with tumor (increasing the likelihood of ICA wall involvement) or PS-ICA stenosis is present on preoperative MRI or/and angiography, it is more reasonable to begin resecting the PS component of the tumor from Meckel's cave. (This is also true of certain of the lesions limited to the PS discussed previously.) Dura on the anterior surface of the petrous apex is incised, giving access to the intradural space. The root of nerve V is exposed and freed from tumor by additionally incising dura posterior to nerve V, then lateral and anterior to it, exposing the medial corner of the GG. When the root of nerve V and the GG are exposed and separated from tumor, nerve V may be mobilized posteriorly and laterally, and any tumor present resected from Meckel's cave and its environs, also intradurally at the apex of the petrous pyramid. If the petrous apex has not been destroyed by tumor, it may be drilled away anteromedial to the nerve V root and the GG. Dura is then coagulated and resected toward the base of the petrous pyramid, posterior and lateral to the nerve V root and the GG, finally removing bone and dura toward Dorello's space. After exposing the anterior inferior cerebellar artery (AICA) and nerves VII and VIII, nerve VI is identified, and followed to its point of entry into Dorello's space.

At this point the surgeon should direct his attention from posterior and lateral to nerve V, to its anteromedial side. Nerve VI is identified and followed from the posterior cranial fossa throughout Dorello's space, from where the petroclinoid ligament (PCL) is attached to the petrous apex. The PCL is resected and tumor is removed from the medial surface of the ICA as it exits the petrous canal, through its lateral ring (LR). The surgeon may now proceed along the lateral surface of the ICA as it ascends from its lateral loop (LL) toward its medial loop (ML), where it is primarily encased by tumor.

The next step is to dissect V1 and V2 from the tumor, then nerve VI over the ICA, finally along the PS-ICA (horizontal segment) toward the SOF. The tumor may now be dissected from the PS-ICA at the ML. Separating tumor from the PS-ICA is particularly dangerous. It may be prudent to postpone such work until later, instead returning to nerve III, which is dissected from tumor along the horizontal segment of the PS-ICA.

Tumor still remaining within the PS is resected from the dorsal surface of the anterior loop (AL) of the ICA (anterior to the posterior clinoid process (PCP), adjacent to the sella). Intrasellar tumor may be resected at this point, until the pituitary gland is exposed. When the ICA is exposed distal to the LL (including the distal part of the horizontal segment of the PS-ICA), tumor may now be resected from the medial surface of the ML, close to the medial dural wall, through Parkinson's triangle and the paramedial triangle. Following dura on its outside surface (within the PS), the ICA will not be damaged since the floor of the PS is entirely exposed.

If the ML of the ICA remains firmly encased by tumor, it is almost certain that the arterial wall has been invaded. In such cases, the surgeon faces three options. a) To coagulate the "coat" of tumor surrounding the ICA and leave it in place Figs. 4.11d and 4.13(a an b). b) To proceed with dissecting tumor from the arterial wall, risking its rupture, either during surgery Fig. 4.14(b and d) or afterward. Or, c) To resect the involved ICA segment and reconstruct the vessel either with an end-to-end anastomosis or by interposing a short vascular graft between the proximal and distal ICA stumps.

Even though dura making up the medial wall of the PS is quite nicely exposed at this point, it should not be resected. This is important. Nerve IV and nerve III are not yet visible intradurally. Instead, dura is incised at the medialmost corner of the PS, at the clivus, where nerve VI enters Dorello's space medially and anteriorly.

Only after the lesion has been resected from Meckel's cave, Dorello's space, the PS, the SOF, and the ipsilateral half of the sella, should its remaining intradural component be addressed. So long as the edge of tumor abutting the clivus has been exposed, removing intradural tumor should not be difficult. If tumor actually invades the clivus, diseased bone should be drilled away through the posterior part of Parkinson's window. This is not necessary if normal dura is already visible, allowing resection of all diseased dura.

To expose any remaining intradural tumor, dura is incised along the sylvian fissure medialward, toward the distal (dural) ring (DR). Upon reaching tumor the dural incision is extended laterally and posteriorly in order to expose the lesion from its periphery, allowing its straightforward resection, together with infiltrated dura. Since this component of the tumor has been devascularized by the extradural dissection, its resection should present no major problems, especially if it is debulked from its dural surface.

After resecting dura to the point where nerve III exits, it is very important to identify nerve III both extradurally and intradurally. The intradural segment may be encased by tumor. The safest technique is to follow the nerve extradurally back toward the dural opening into its canal in the lateral wall of the PS, then through tumor into the intradural space. Only after isolating a sufficient segment of the nerve intradurally should dura lateral to nerve III be incised and resected. Resecting tumor toward the brainstem along nerve III rarely poses the difficulty and risk of dissecting tumor from nerve III at the extradural-intradural interface. Infiltrated dura and tumor are now resected posteriorly, toward where nerve IV passes through the dural opening into its "canal" between the layers of the lateral wall of the PS.

Finally, the medial wall of the PS may be resected between nerves III and IV, along with tumor bulging toward the brainstem, allowing the superior cerebellar artery (SCA) and the posterior cerebral artery (PCA) to be exposed and dissected. The intradural segment of nerve IV is freed from tumor toward the ambiens cistern, and any infiltrated dura along the tentorial edge is removed. The superior petrosal sinus (SPS) should be occluded with Surgicel, sutures, or titanium clips. The petrous vein must be preserved as it enters the SPS.

When tumor and infiltrated dura have been resected from the PCP, along the clivus, and toward the petrous area, any tumor remaining adjacent to the PCP, the DR of the ICA, the OC, the dorsum sellae, and upper clivus may be identified and removed. If necessary, the PCP and dorsum sellae may be drilled away until clearly normal dura is exposed. Finally, tumor is resected from the ON, the chiasm, the pituitary stalk, the ICA, the posterior communicating artery (PCom), the anterior choroidal artery, the anterior cerebral artery (ACA1), the middle cerebral artery (MCA), and adjacent brain.

After complete excision of all tumor and infiltrated dura, precise hemostasis is achieved and the dura may be reapproximated. If a large dural defect exists, it is best closed with a patch of fascia lata draped over nerves III, IV, VI, V1, V2, V3 and the ON and secured with interrupted sutures to edges of normal dura and periorbita. Then the epidural surface of the dural graft (including suture lines) is coated with fibrin glue. An epidural drain is left in place, the osteo-muscular flap replaced and secured with sutures, and the wound is closed in layers.

Patients and results

Parasellar meningiomas represent the largest group of meningiomas originating at the CSB, 388 operated on in this series. Depending on the direction and extent of tumor spread, patients with these lesions were divided into four groups.

Patients in Group I had meningiomas in the parasellar space and also the SOF and orbit. Forty-eight were treated as virgin cases, while 56 had residual tumor from operations performed previously elsewhere. In no patient with residual tumor was complete excision possible. In only five of the patients with virgin tumors was complete resection possible. Twenty-seven of the virgin cases had gross-total resections, 16 subtotal resections. Among patients treated for residual tumors in this group, 12 had gross-total and 44 subtotal resections.

Group II is composed of patients with PS meningiomas extending into the sella and/or Dorello's space. Of 31 virgin cases, resections were complete in 18, gross-total in 12, and subtotal in one. Of 42 patients undergoing surgery for residual lesions, complete excision was not possible in any case. Gross-total excision was achieved in 18 patients with residual tumors, and subtotal resections in 24.

Group III included patients with meningiomas in the PS, middle cranial fossa, and/or anterior cranial fossa, and/or posterior cranial fossa. Among 54 patients with virgin lesions, complete resection was achieved in 26, and gross-total and subtotal removal in 19 and 9 respectively. In none of the 71 patients treated for residual tumors in this group was complete resection possible, while gross-total resection was achieved in 17, and subtotal in 54.

Group IV includes patients with even broader tumor involvement of the skull base, including patients with PS meningiomas that also involved any other intracranial compartment (intradural and/or extradural) plus any ipsilateral extracranial compartment. Thirty-nine patients in this group were operated as virgin cases, two with complete resections, 15 with gross-total resections, and 22 with tumors subtotally removed. Of 47 patients treated for residual lesions, not one had a complete resection, and seven and 40 had gross-total and subtotal resections respectively.

Nine of 388 patients were treated for cerebrospinal fluid (CSF) leaks following surgery, seven for epilepsy, and five patients suffered hemispheral symptoms, motor and/or sensory. Cranial nerve deficits persisted longer than six months in 34 cases. Three patients died, all three with extremely difficult postoperative courses, including CSF leakage.

Table 12. Cases with virgin and recurrent meningiomas treated: resectability and results

Extension of meningioma	N° of cases Virgin Residual		Resectability Complete		Gross-total		Sub-total		Postoperative deficits Cranial nerve after Six months		CSF leak		Other		Died	
PS + SOF + intraorbital	48	12%	5	10%	27	56%	16	33%	2	4%	–	–	–	–	–	–
	56	14%	–	–	12	21%	44	79%	7	13%	2	4%	1	2%	–	–
PS + intrasellar and/or	31	8%	18	58%	12	39%	1	3%	1	3%	–	–	–	–	–	–
Dorello's Space	42	11%	–	–	18	43%	24	57%	3	7%	1	2%	2	5%	–	–
PS + MCF and/or ACP,	54	14%	26	48%	19	35%	9	17%	1	2%	1	2%	1	2%	–	–
and/or PCF	71	18%	–	–	17	24%	54	76%	8	11%	2	3%	3	4%	2	3%
PS + any intradural, and/or	39	10%	2	5%	15	38%	22	56%	3	8%	1	3%	1	3%	–	–
extradural, and/or extra neurocranial compartments	47	12%	–	–	7	15%	40	85%	9	19%	2	4%	2	4%	1	2%
Collective	172	44%	51	30%	73	42%	48	28%	7	4%	2	1%	2	1%	–	–
	216	56%	–	–	54	25%	162	75%	27	13%	7	3%	8	4%	3	1%
Total	388	100%	51	13%	127	33%	210	54%	34	9%	9	2%	10	3%	3	1%

Superior orbital fissure meningiomas

Pathological anatomy, presentation of the lesion, and diagnostic work-up

The superior orbital fissure (SOF) is the bony opening between the intracranial cavity and the orbit. Anatomical descriptions here will assume its longitudinal axis to be oriented as dictated by head positioning as described in Chapter 2, essentially perpendicular to the floor of the operating room.

The SOF is formed by the lesser sphenoid wing (medially) and the greater sphenoid wing (laterally), with the optic strut making up its base medially and the body of the sphenoid bone composing the remainder (majority) of its base. The lesser sphenoid wing medial to the SOF continues centrally into the anterior clinoid process (ACP). There are few sites in the human body where so many neural structures reside in such a small compartment, making operations in this area particularly demanding and complex.

It is necessary to master not only the extradural intracranial approach along the longitudinal axes of nerves III, IV, V1 and VI (where orientation of the nerves is simpler), but also surgical techniques affording dissection farther anteriorly and downward, into the orbit. Understanding the anatomy from both intradural and orbital perspectives is important. Although the position of each individual nerve is not altered, frequently all nerves traversing the SOF are encased by tumor at some point in their course from the parasellar space (PS) into the apex of the orbit. Meningiomas "en plaque", typically associated with hyperostosis of adjacent bone, are especially prone to narrow the superior orbital fissure (SOF). Some such lesions arise adjacent to the medial sphenoid wing and remain intradural yet without penetrating the dural sleeve passing through the SOF. Other lesions, particularly those likely to cause surgical problems, remain confined to the SOF interdural space, intermingling with nerves III, IV, V1 and VI. Such lesions may represent anterior extension of meningiomas of the PS, either arising from within the PS or lesions extending to it from the sphenoid wing, the ACP, the middle cranial fossa, the tentorial edge, or from any other central skull base (CSB) site. Meningiomas actually arising at the SOF are extremely rare.

Symptoms and signs of lesions arising from or extending into the SOF are related to dysfunction of the cranial nerves within it. The manifold functions of nerve III make its deterioration most likely to become manifest initially, but nerves IV and VI may be involved as well. Diplopia is often preceded by a vague retro-orbital discomfort or periorbital or forehead pain. Pupillary enlargement and sluggish reflexes may occur. Hyperostosis narrowing the SOF may impede orbital venous drainage into the PS, resulting in exophthalmos. Although exophthalmos is not a leading symptom of SOF meningiomas (owing to collateral pathways of venous drainage), occasionally it is the first sign of tumor presence.

Gradually progressive diplopia is quite disturbing for both patient and his or her doctor, even more so if its onset is sudden. With large ACP meningiomas or meningiomas in the PS, unilateral visual loss accompanied by orbital or facial pain plus diplopia all but establish the diagnosis prior to imaging study. Nevertheless, CT and MRI are necessary to confirm the diagnosis, determine the effect of tumor on adjacent bone (CT), characterize in detail the size, shape and extent of spread of the lesion, and determine its effect on the ICA. From MR imaging one may determine if preoperative angiography and cross-circulation studies are necessary. With small SOF meningiomas, angiography is seldom required, but when they occur as extensions of tumor within the PS or anterolaterally from medial sphenoid wing and/or ACP lesions, preoperative carotid angiography and cross-studies become extremely important.

CT and MRI, with and without contrast-enhancement, provide important clues to tumor resectability and whether or not postoperative deficits (with respect to nerves III, IV, VI and V1) will likely result or possibly become permanent. Because of the complexity of the anatomy of the SOF and the apex of the orbit and their propensity to adhere to individual nerves, meningiomas involving the SOF are least likely to be totally resected among all CSB meningiomas. This is true even in the most experienced and skilled of hands, unless the surgeon is prepared to risk almost certain permanent cranial nerve deficits.

Surgical approach to the SOF meningiomas

Patient positioning, scalp incision, osteoplastic craniotomy, unroofing the orbit, and resecting the lesser and greater wings of the sphenoid bone are currently carried out as described in Chapter 2. Unroofing the orbit and resecting the lesser and greater sphenoid wings may be more demanding when hyperostosis is present. These resections must include the removal of all abnormal bone. Often this necessitates drilling far posterolaterally, even opening foramen rotundum (FR) and the anterior and medial wall of the V2 canal. Bone anterior and medial to the SOF frequently must be removed well medially, requiring that one remain especially alert to avoid entering a paranasal sinus. Resecting the ACP with these tumors must be performed with extreme care since nerves III and IV commonly are compressed against its inferolateral surface. The proximity of these nerves makes them particularly vulnerable to any overheating of the diamond burr. Permanent cranial nerve damage may occur. Hollowing out the ACP remains the safest method of resecting this structure in order to preserve function of nerves III and IV (inferolaterally) and the ON (anteromedially).

The lateral and superior walls of the optic canal (OC) and the optic strut must be removed. Transecting dura covering the neural structures traversing the SOF is carried out as described previously, along with separating the outer layer of the lateral wall of the PS from its inner layer, exposing nerves III, IV, V1 and V2 in its anterior half. These nerves must be exposed throughout their entire courses. Achieving hemostasis near the SOF is particularly dangerous since nerves III, IV, V1 and VI converge at this site, making them vulnerable to the extent of coagulation dictated by the blood supply of most meningiomas. Nerve IV should be dissected from the inner layer of the PS wall first, particularly its anterior segment, quite near the SOF, where it crosses over nerve III from its lateral to its medial surface. Then nerve III is dissected along its common trunk anteriorly, to the site of its branching toward individual extraocular muscles. Finally nerve VI may be dissected, beneath V1 and lateral to nerves III and IV. Only when nerves III, IV, V1 and VI are freed from the inner layer of the lateral PS wall and from tumor posterior to the annulus of Zinn, allowing their displacement away from bleeding sites at the anterior corner of the PS, may hemostasis be achieved with reasonable safety. Venous bleeding from the PS is controlled with Surgicel. It is supremely important that the floor of the anterior PS, the vicinity of the AL of the ICA, be examined for possible tumor extension. Meningioma is then resected very carefully at the apex of the orbit, insuring that individual branches of nerves III and V1 are freed, also nerves IV and VI. To achieve this it is necessary to incise the annulus of Zinn, to separate tumor from it, the extraocular muscles, nerves III, IV, V1, VI, and the optic nerve (ON). Care must be taken to avoid damaging the ophthalmic artery (OA) and the ciliary ganglion. This is only possible if the surgeon is prepared to spend considerable time dissecting these individual structures from the lesion, then removing it piecemeal.

The individual pieces of tumor removed at any one time must be very small, using minimal force to peel them from individual nerves. These tumors must be removed slowly and carefully, bit by bit. Once the meningioma has been resected from the anterior corner of the PS, from the SOF, and from the apex of the orbit, the dura should be opened to determine if intradural tumor is present. With intradural extension, the tumor's dural attachment must be identified and excised with the lesion. After excising intradural tumor and infiltrated dura and opening the canal of nerve III, dura where nerve III enters the PS (surrounding the oculomotor "slit") is coagulated or, if infiltrated by tumor, resected. If tumor extends along the tentorial edge, toward the site of nerve IV's exit from the intradural space, nerve IV is dissected free from its dural canal, and tumor-infiltrated dura is resected.

After completely excising the meningioma and establishing hemostasis, the dura is closed with commercially available dural substitute tailored to the size of the defect and fixed to normal dura with interrupted sutures. The periorbital defect is closed with a small piece of temporalis fascia or galea which is also secured with interrupted sutures. Water-tight closure is insured by coating both grafts and suture lines with fibrin glue. An epidural drain is placed, the osteomuscular flap replaced, and the wound closed in layers. Where particularly large dural defects exist, continuous external lumbar CSF drainage for 5–7 days postoperatively is useful to reduce intradural pressure and excessive tension on orbital contents, thus minimizing the risk of postoperative CSF leak or a troublesome epidural hygroma.

Patients and results

Although meningiomas arising at the SOF are rare, the great majority of CSB meningiomas extend to this area. In our series only seven lesions were classified as primary SOF meningiomas. One patient was operated as a virgin case, while six had been operated previously elsewhere. In actual fact, only a single case was truly demonstrated to arise at the SOF. Residual tumor at the SOF from previous operations speaks to the difficulty of complete resection at this site. In no case was total resection possible, and only two gross-total resections were achieved. All seven patients had preoperative cranial nerve deficits which remained six months following surgery. In every case the surgery was demanding, but no additional neurological deficits occurred.

Dorello's space meningiomas

Pathological anatomy, clinical presentation, and diagnostic work-up

Dorello's space begins where nerve VI pierces clival dura and extends lateral to the clivus beneath the petroclinoid ligament, finally ending where the nerve enters the PS lateral to the ICA coursing from the lateral (LL) to the medial loop (ML). Meningiomas very rarely originate within Dorello's space and remain confined to it, but those extending into it along posterior fossa dura (petroclival lesions) or from the PS are not uncommon. The abducens nerve, the neural structure traversing, essentially defining, this compartment, provides signal symptoms and signs of Dorello's space lesions and is also the key to their successful dissection and excision. Nerve VI is vulnerable to injury from meningiomas primarily at three specific anatomical sites at it courses from the prepontine cistern to the parasellar space (PS): first, where the nerve enters Dorello's space; second, where it exits Dorello's space (into the PS); and third, on the lateral surface of the PS-ICA between its LL and ML, where it is tethered by sympathetic fibers to the surface of the ICA.

Large meningiomas, whether within the PS or in the petroclival region, may spread to one of these three critical locations, resulting in impairment or loss of nerve VI function along with additional symptoms and signs. Small meningiomas at one of these three sites may cause slowly progressive diplopia from nerve VI involvement. Diplopia initially is very subjective and evidence of nerve VI impairment may not yet be detectable. Commonly, general practitioners and even neurologists may explain its presence as a metabolic derangement, most frequently diabetic neuropathy. In many cases only when an abducens palsy becomes full blown are diagnostic studies obtained.

The site of nerve VI's entry into Dorello's space is a round dural ring which is rigid and cannot be enlarged, making the nerve likely to be trapped, and compressed, by a meningioma in the immediate vicinity. Dorello's space, formerly called a "canal," in reality is not a canal at all. It provides little restriction to nerve VI displacement by lesions within it. Nerve VI compression sufficient to cause deteriorating function by a very small meningioma within Dorello's space is uncommon if it exists at all. But meningiomas within Dorello's space are free to spread in various directions, including along clival dura or periosteum on the floor of the PS or into the PS via venous channels. Peripherally, where nerve VI exits Dorello's space into the PS, it passes between the clivus and the petroclinoid ligament (PCL). Within this narrow channel it may easily be compressed to the point of dysfunction. Meningiomas at this site, typically those filling the entire posterior compartment of the PS, may compress the nerve either against the PCL (laterally), against bone (medially), into the narrow angle between bone and the PCL (posteriorly), or against the ICA (anteriorly), resulting in loss of function. As nerve VI ascends across the ICA (between the LL and the ML), sympathetic fibers in the lateral ICA wall actually join it and bind it to the ICA wall. Lesions growing between the ICA wall (anteromedially) and sympathetic fibers and nerve VI (laterally) may compress or stretch nerve VI to a point beyond which it can no longer compensate.

Unenhanced CT and/or MR often do not reveal small lesions hidden within the skull base. Sometimes even studies performed with contrast also fail to detect such neoplasms initially. Patients scheduled for follow-up examinations after six to eight months frequently are doomed to the untimely diagnosis of such tumors, particularly those rare lesions confined to Dorello's space. Typically it is MRI which finally determines the cause of diplopia as demonstrated in Fig. 4.16(a, c, e).

Surgical intervention for meningiomas confined to Dorello's space or those fixed to posterior cranial fossa dura or to the membrane separating Meckel's cave and the posterior PS compartment is recommended in order to prevent further deterioration of nerve VI function. Dorello's space lesions or those in dorsal PS compartments (caudal to the LL and ML of the ICA) cause slow but steady nerve VI deterioration. They may have remained small for years, unnoticed on CT scans and/or MRI. Some suggest that such lesions be treated with focused beam radiation, the "gamma knife" or "linac". But radiosurgery is no solution for such tumors since nerve VI continues to be compressed. Tumor growth is destined to progress within a few short years. Furthermore, radiosurgery results in scar tissue, even relative devascularization of nerve VI, further compromising the results of definitive surgical treatment when it is finally offered.

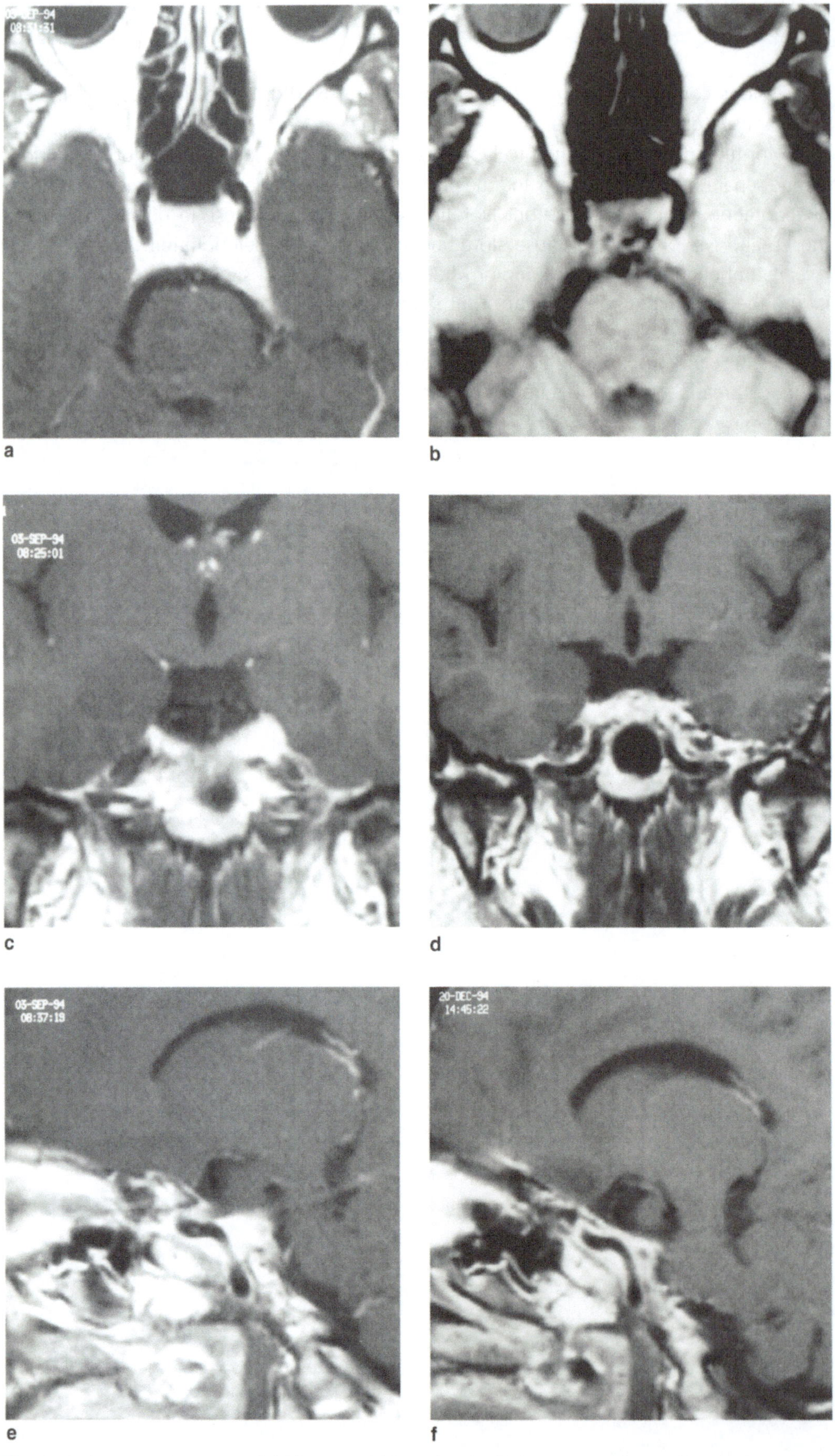

Fig. 4.16. Preoperative trans-axial (a), coronal (c), and sagittal (e) contrast-enhanced MR images reveal a rather small hypedense meningioma posterior to the left PS. Corresponding postoperative contrast-enhanced MR images, trans-axial (b), coronal (d), and sagittal (f), demonstrate resection of the lesion. Notice that the slight preoperative stenosis of the ascending PS-ICA (e) is less following surgery (f). This patient had a "diabetic neuropathy" and an abducens palsy which progressed slowly and steadily for two years prior to diagnosis. Nerve VI function improved within six months after surgery although a slight deficit remains

Surgical approach to meningiomas in Dorello's space

Patient positioning, scalp incision, and osteoplastic craniotomy are carried out as described in Chapter 2. It is *not* necessary to perform the following steps of the general central skull base (CSB) approach: unroofing the orbit, resecting the sphenoid wing and anterior clinoid process (ACP), opening the optic canal (OC), and removing the optic strut. Dura should be peeled from medial middle fossa bone laterally to the foramen rotundum (FR), the foramen ovale (FO), and the foramen spinosum (FS), then posteriorly from the anterior surface of the petrous bone, finally along the petrous bone to the superior petrosal sinus (SPS). After exposing bony structures in the middle cranial fossa laterally and posteriorly, the middle meningeal artery (MMA) is coagulated and transected at the FS, subsequently filling the FS with Surgicel and sealing it with bone wax. Dura overlying V3 is incised in a rectangular fashion, parallel to the longitudinal axis of the nerve, fully exposing it. Dura is then incised anteriorly along the base of the lateral triangle (from V3 to V2). Posterior to V3 and the Gasserion ganglion (GG), dura is peeled from the petrous apex (from the floor of Meckel's cave).

As described in Chapter 2, the apex of the petrous pyramid is drilled away (from posterior and lateral to anterior and medial) rostral to the internal auditory canal (IAC), taking care to avoid injury to the geniculate ganglion, the greater petrosal nerve (GPN), and the SPS. These structures must be preserved. In order to create sufficient posterior cranial fossa exposure, bone should be drilled toward the carotid canal, beneath the GPN. When the petrous apex has been transected, exposing dura on its posterior surface, additional petrous bone may be drilled caudally in order to expose dura within the IAC. Bone between the IAC and the SPS is drilled without opening the SPS or damaging dura medial and superior to nerves VII and VIII. The remainder of the petrous apex, medial and caudal to the GG and the plexiform segment of the nerve V root, is then drilled away.

After exposing the ICA within the petrous canal (from the PL, anteriorly toward the LL), the GPN (where it leaves the geniculate ganglion), the IAC dura, and dura medial to the GG and the plexiform segment of the nerve V root, dura on the posterior surface of the petrous apex may be incised perpendicular to the SPS and medially from it, anterior to the IAC, then further medially toward the clivus, caudal to where nerve VI enters Dorello's space. By transecting dura medial to the resected petrous apex, the posterior cranial fossa is opened and cerebrospinal fluid (CSF) released. Dura is then examined for tumor beneath it and its free margins coagulated. Nerves VII and VIII and the anterior inferior cerebellar artery (AICA) are identified. Anterior and medial to these structures, nerve VI becomes visible on the surface of the brainstem. If nerve VI is not yet apparent, dura may be transected farther medially and its edges carefully cauterized to prevent bleeding from venous structures between its layers. If necessary, the lateral part of the clivus may be drilled additionally, and dura sectioned medially until nerve VI comes into view. When nerve VI is exposed, the ipsilateral vertebral artery and the proximal trunk of the basilar artery (BA) will be seen. Dura is then incised anteriorly, peripheral to the edge of the meningioma. The size of the tumor determines how far cranially dura must be incised.

In order to follow nerve VI from the posterior cranial fossa into Dorello's space and throughout the tumor into the posterior compartment of the PS along the posterolateral surface of the ICA, it is mandatory to free the entire nerve V root and the GG and retract them posterolaterally, exposing the anterior half of Meckel's cave and the dural membrane making up its anterior wall (facing the posterior compartment of the PS).

If tumor is found in Meckel's cave, invading its anterior wall, or between Meckel's cave and the PS, it should be resected along with all infiltrated dura. If necessary, additional petrous apex may be drilled safely through the space gained anterior to the nerve V root. By incising the dural wall between Meckel's cave and the PS, nerve VI's point of exit from Dorello's space is exposed within the angle circumscribed by the PCL and bone. After resecting tumor from the posterior compartment of the PS where nerve VI exits Dorello's space, the posterior border of the PS lies open. After completely removing tumor from this compartment, the PS should be lined with Surgicel. With large tumors considerable dura from the floor and medial side of the PS must be resected and cauterized. Much venous bleeding may occur, requiring control with Surgicel.

The relationship of tumor within the PS to the ICA wall is particularly important. In some cases, rather than spreading circumferentially around the ICA, the meningioma becomes adherent to the ICA wall and protrudes into it. Such cases require that tumor be sharply resected from the ICA wall if total resection is to be achieved, obviously a measure taxing the surgeon's best judgement.

Only when the dural margins surrounding the meningioma have been safely and precisely identified in the posterior cranial fossa, and when nerve VI is exposed where it enters (from the posterior cranial fossa) and exits the tumor (into the posterior compartment of the PS), and when the lesion is freed from the PS-ICA, may nerve VI be dissected completely from the tumor, including the segment in Dorello's space. Uncommonly it becomes necessary to drill the clivus medial to nerve VI, adjacent to Dorello's space. Only when the entire segment of nerve VI traversing the meningioma in Dorello's space has been completely freed from the tumor, and healthy dura exposed medial to Dorello's space, has its dissection been completed.

Meningiomas located in the PS, regardless of size and degree of extension into Dorello's space, cannot be eradicated completely until all tumor has been resected from Dorello's space. It is not possible to remove the Dorello's space component from the PS surface only. Therefore, such lesions must be approached transpetrously from the middle cranial fossa and Dorello's space explored throughout, exposing nerve VI completely. To resect a meningioma from the PS, exposing nerve VI as it exits Dorello's space, and believe that Dorello's space at this point and caudally is free of tumor is only wishful thinking! It is naive to assume that meningioma left at this site will not continue to grow. This is never the case. Neither is it satisfactory, as discussed above, to subject a patient to radiosurgery for meningioma left within Dorello's space. The best time to treat the tumor is when it is first exposed. Sooner or later, another attempt at direct microsurgical removal of the remaining meningioma will be required. Years later, notwithstanding radiosurgery, prayer, and the best intentions of all concerned, residual tumor will not only have become larger, but virgin bony landmarks will have been destroyed and scar tissue formed. Nerve VI will have been subjected to additional trauma and will be more difficult to separate from recurrent tumor. Surgeons familiar with the transpetrous approach will better serve their patients and save trouble for themselves to spend whatever additional time is necessary at the first operation. It is essential, prior to operating on meningiomas extending from the PS into Dorello's space, to formulate an approach that allows the lesion to be resected completely, obviating the necessity for additional treatment.

Patients and results

Meningiomas originating within Dorello's space are very few, while patients with skull base meningiomas invading Dorello's space are numerous. Of 817 CSB meningiomas, five were lesions confined to Dorello's space only. In each case diagnosis was made with difficulty since isolated nerve VI deficits are commonly ascribed to diabetic neuropathy. Four of five patients with Dorello's space meningiomas were operated as virgin cases, with one additional patient having been operated on elsewhere and tumor left behind. In three patients, complete tumor resection was achieved, the remaining two only gross-total removal. Six months following surgery, two of the four patients operated as virgin cases and the one treated for residual tumor had improved but persisting partial nerve VI deficits. No other postoperative deficits occurred.

Petroclival meningiomas

Pathological anatomy, clinical presentation, and diagnostic work-up

Depending on growth type, meningiomas spread differently, varying from carpet-like lesions extending over wide areas but occupying little three-dimensional space, to large space-occupying lesions which remain confined to their sites of origination and dural attachment. Figures 4.17(a, c). Dural anatomy is also important, particularly in areas where it is duplicated. Where layered dura at the petrous apex joins the tentorium, the lateral wall of the parasellar space (PS), and, especially, dural layers in the vicinity of Meckel's cave, constitutes the richest, most complex junction of intracranial dural structures. Important and well defined spaces between leaves of dura in this region are as follows: the superior (SPS) and inferior (IPS) petrosal sinuses, venous channels draining the posterior compartment of the PS into the SPS and IPS, the walls of Meckel's cave, and venous sinuses within the broad expanse of tentorium laterally and along its free edge. Given the complexity of these spaces within dura and their anatomical connections, meningiomas originating at any site in the petroclival region may spread easily into one or several adjacent areas. Lesions originating medial and posterior to the petrous apex may grow not only medially to the clivus, but also upward and laterally, reaching Meckel's cave. In addition to these primary avenues of spread, petroclival meningiomas may extend along spaces within the tentorium and along its edge. On several occasions meningiomas invading venous sinuses (SPS, IPS) have spread more quickly within other potential spaces between layers of dura than within venous sinuses, particularly along the tentorium and its free edge.

Symptoms with little localizing value in the absence of clear clinical findings result in most petroclival tumors defying diagnosis until they have become quite large. The most frequent symptom is described more as a "discomfort" than true headache. Some patients may experience severe but quite atypical trigeminal pain, generally continuous pain in the distribution of one or all sensory branches, V1, V2, or/and V3. Although not at all characteristic of trigeminal neuralgia, this pain is sometimes so severe that it is confused with that condition, typically by doctors with limited experience with facial pain syndromes. Rather than direct compression, the basis of this type of pain may very well be related to relative ischemia of one or more nerve V branches.

Diplopia from compression of nerve(s) VI and/or IV is never an early symptom. Neither is brainstem compression nor displacement of nerves VII and VIII, all of which may be tolerated for long periods, allowing the tumor to become quite large and compression of adjacent structures severe. When the predominant symptom is continuous pain (from local irritation of nerve V at the petrous apex or within Meckel's cave), CT and/or MR may establish the diagnosis of a petroclival meningioma while it is still small. In most cases, however, such imaging is delayed until symptoms and signs occur which indicate marked compression of nerves IV, VI, VII and VIII, and/or the brainstem. Additional studies, such as DSA and neurophysiological testing, are rarely required since they offer little to the planning of surgery.

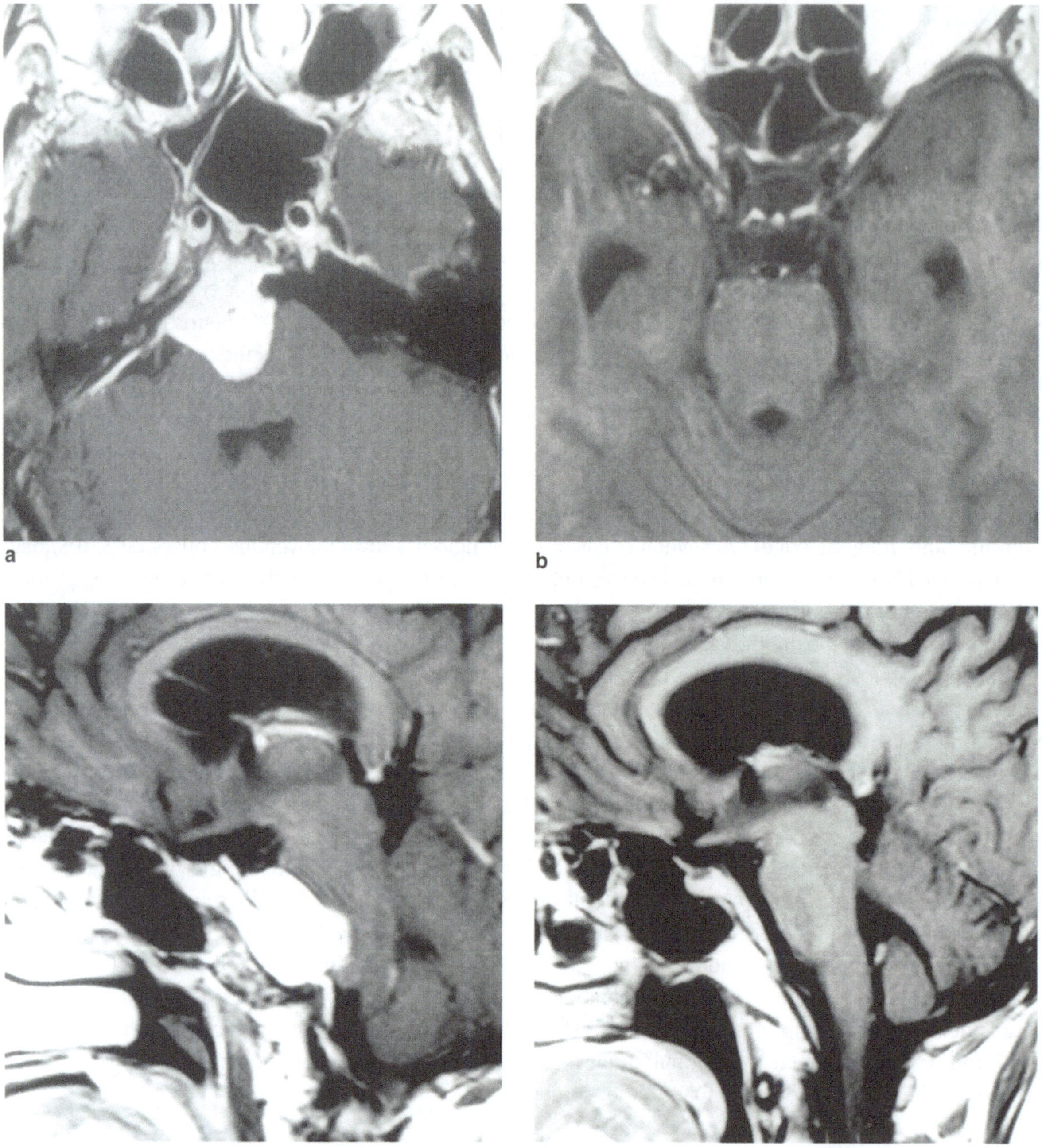
a b c d

Fig. 4.17. Preoperative contrast-enhanced transaxial (a) and sagittal (c) MR images demonstrating a broad based meningioma which is fixed to the right petrous apex and upper and middle part of the right clival region. Note the bulging of tumor into the brainstem. Corresponding postoperative transaxial (b) and sagittal (d) images demonstrate complete resection of the lesion. Notice the slightly enlarged temporal horn of the right lateral ventricle and the hypodense zone at the apex of the right temporal lobe. Following tumor resection the brainstem "re-expanded" to its normal size and shape (b and d)

Surgical approaches to petroclival meningiomas

Petroclival meningiomas may be resected via one of several routes, including the suboccipital retrosigmoid approach, the occipito-suboccipital transtentorial approach, the subtemporal approach, the presigmoid approach, and the transpetrous approach. While the majority of lesions limited to the posterior cranial fossa lend themselves to resection by the transpetrous approach, the converse is not true. It is clear that lesions extending partially into the middle cranial fossa (across and lateral to the petrous apex) cannot be safely removed by way of the suboccipital approach. Positives and negatives are associated with each of these two methods. A major advantage of the transpetrous approach is that it allows the surgeon to work anterior and medial to nerves VII and VIII, structures across which the suboccipital approach requires one to work continuously, increasing the risk of their injury.

In this series the transpetrous approach was chosen for all petroclival meningiomas predominantly in the region of Meckel's cave, Dorello's space, the posterior compartment of the PS, plus a few in the posterior cranial fossa. This approach allows petroclival lesions not extending caudal to the line between the internal auditory canal (IAC) and the posterior limit of Dorello's space to be successfully and completely removed. Although tumors within Dorello's space and the IAC may easily be accessed by the transpetrous approach, elements of meningiomas encasing nerves VI, VII, and VIII (involving their inferior and posterior surfaces) commonly exceed the limits of what can be achieved by this method. Therefore, the suboccipital retrosigmoid approach was employed for meningiomas smaller than 3 cm in diameter located exclusively within the posterior cranial fossa.

In our experience, most petroclival meningiomas may be resected via the transpetrous approach, only a small number of cases requiring the suboccipital retrosigmoid method. We feel that the risk of devastating consequences of temporal lobe retraction precludes the use of the subtemporal intradural approach while the occipital-suboccipital transtentorial and suboccipital presigmoid approaches offer nothing that cannot be achieved by the transpetrous and suboccipital retrosigmoid approaches when selected as indicated.

Transpetrous approach

This method allows most petroclival meningiomas, those located in the middle and posterior cranial fossae (in the vicinity of the petrous apex) to be completely resected with minimal risk. Patient positioning, scalp incision, and craniotomy are carried out as described in Chapter 2. After completing the osteoplastic craniotomy, dura is peeled from middle cranial fossa bone toward the foramen rotundum (FR), foramen ovale (FO), and foramen spinosum (FS). Then dura is dissected from the anterior surface of the petrous pyramid, taking great care to preserve the greater petrosal nerve (GPN), exposing petrous bone to its apex, the site of the SPS. Epidural meningioma in Meckel's cave may be debulked at this point. In all other cases where the dura remains intact, the petrous apex is drilled away, exposing posterior fossa dura. Drilling is begun posterior and lateral to the Gasserian ganglion (GG), proceeding beneath it, toward the anterior and medial petrous surface, removing bone at the floor of Meckel's cave. Depending on the size of the tumor, petrous bone may be resected farther posteriorly and medially, making it possible to open the IAC and expose dura laterally and superiorly within it.

Dura medial to the petrous apex is incised, and the root of nerve V is identified and dissected from the meningioma along its entire length, from the pons, including its plexiform segment, to the GG. This done, the meningioma is debulked, then the peripheral end of the nerve V root and the GG are dissected from tumor and gently retracted caudally, allowing exposure of tumor-infiltrated dura rostrally, which is resected. The petrous apex is drilled away anteromedial to nerve V, and meningioma is removed from Meckel's cave and from its anterior dural wall (facing the PS).

The tentorium is then incised and the SPS opened. The SPS should be stuffed with Surgicel or clipped, permanently occluding it peripheral to the site of entry of the petrosal vein. Patency of the petrosal vein must be preserved. The tentorium and its free edge are resected farther, beyond where dural layers are infiltrated by tumor, until healthy dura is encountered.

Nerve IV is identified within the ambiens cistern and dissected free toward its entry point into the duplicated dura beneath the tentorial edge. If the meningioma invades the posteromedial compartment of the PS medial to the medial loop (ML) of the ICA, this compartment should be opened and totally evacuated of tumor through Parkinson's triangle (Fig. 4.18d). Although petroclival meningiomas have a tendency to spread posteriorly along the tentorial edge, they may also extend anteriorly, requiring additional dissection of nerve IV from the tumor, between layers of the tentorial edge anteriorly.

After resecting all tumor from the posterior compartment of the PS, venous bleeding is controlled with Surgicel. If the lesion infiltrates the membrane between Meckel's cave and the PS, it too must be resected and free edges of normal dura cauterized. After totally resecting the meningioma and completing hemostasis, the dural defect at the petrous apex, in the vicinity of the nerve V root and in the posterior part of Parkinson's triangle, should be closed with a piece of muscle, which is then covered, along with the resulting suture lines, with fibrin glue from the extradural surface. Following dural closure, an epidural drain is installed and the osteomuscular flap is replaced and secured with sutures, and galea and scalp edges reapproximated in the standard manner.

Suboccipital retrosigmoid approach

The patient is placed in the lateral park-bench position with the head immobilized in the Mayfield device as for other CPA tumors. In contrast to the general extradural approach to the CSB, this approach is entirely intradural as demonstrated in Fig. 4.19(a, b). Following evacuation of cerebrospinal fluid (CSF), hyperemic dura beneath the tentorium and on the posteromedial surface of the petrous apex (rostral to the IAC) is coagulated. Nerves VII and VIII, typically displaced caudally and anteriorly and hidden beneath the meningioma, must be identified and dissected with great care. Similarly, the anterior inferior cerebellar artery (AICA) and the internal auditory artery are identified and preserved.

After exposing nerve V and freeing it from the tumor, it is possible to detach the lesion from the petrous apex by cauterizing the extremely vascular dura. The petrous vein must be preserved intact. Continued detachment of tumor from the tentorium is achieved by coagulating the tentorium and removing the thus-devascularized meningioma piece by piece.

When the ambiens cistern is reached, nerve IV is identified and dissected from the tumor. Finally, the most anterior and medial portion of the meningioma is separated from nerve VI and from dura surrounding its entrance into Dorello's space. The meningioma is then dissected from the cerebellum and brainstem, preserving all veins and arteries. All tumor-infiltrated tentorium must be resected, including occluded dural venous sinuses. However, the SPS must be preserved at all cost, insuring especially inflow from the petrous vein.

If the lesion invades the medial part of Meckel's cave, enough of the petrous apex may be drilled away to allow resection of tumor from beneath the plexiform segment of the nerve V root. This approach is not suitable for petroclival meningiomas traversing the membrane between Meckel's cave and the PS, invading the posterior compartment of the PS. Such tumors should be resected via the transpetrous approach.

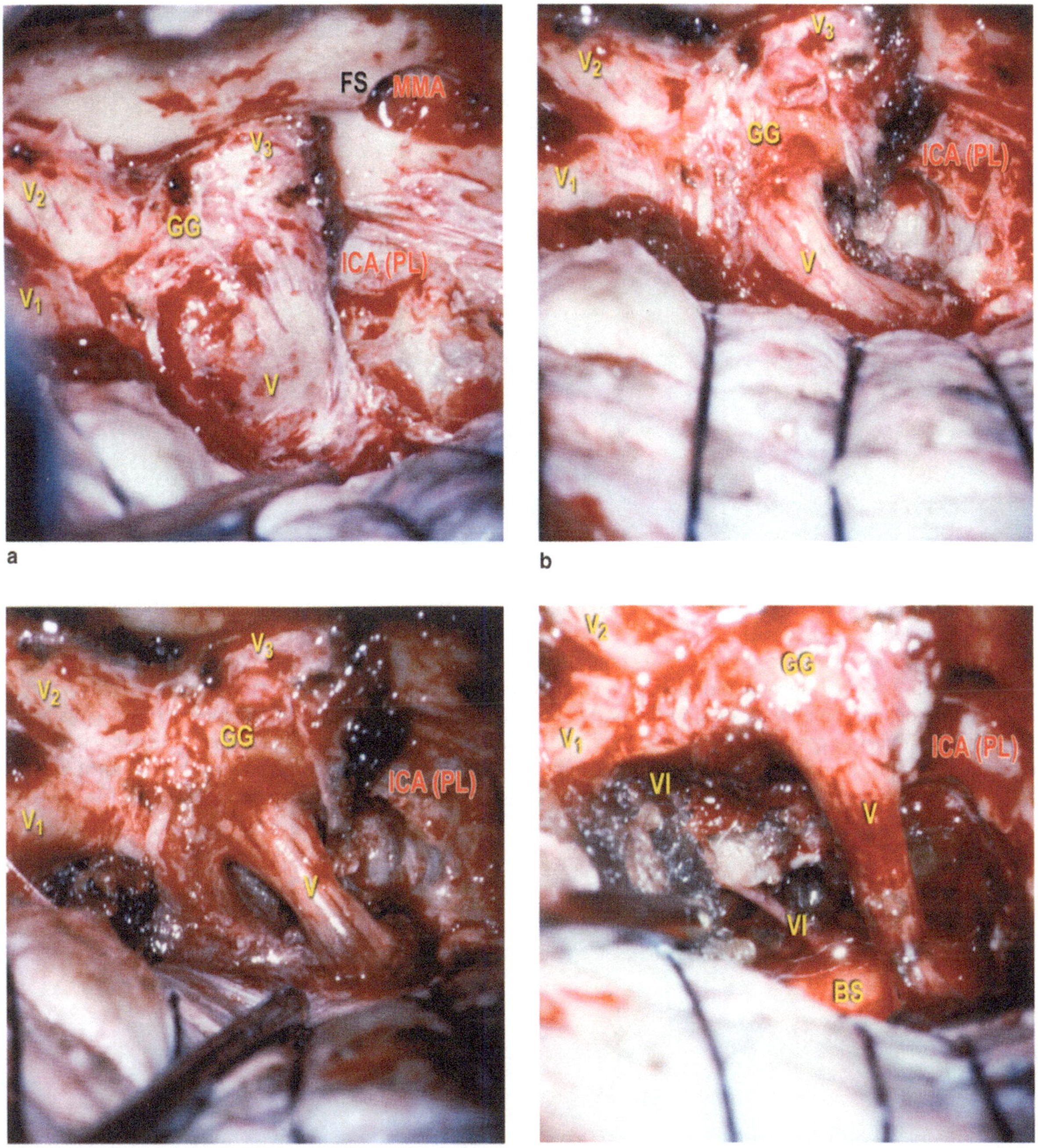

a b c d

Fig. 4.18. Photographs made during resection of the meningioma shown in Fig. 4.17. (a) The outer dural layer of the lateral parasellar space (*PS*) wall has been reflected, exposing V3, the Gasserian ganglion (*GG*), the nerve V root, and the anterior surface of the temporal bone. The middle meningeal artery (MMA) has been coagulated and sectioned, its proximal stump inverted into the foramen spinosum (*FS*), which has been sealed with Surgicel and bone-wax. The petrous carotid canal has been opened, exposing the posterior loop (*PL*) of the ICA. Notice that Nerve V seems stretched superolaterally. In (b), the meningioma has been partially resected from posterior to and beneath nerve V, which remains elevated because of residual meningioma medial to it, extending anteriorly. (c) Resection of the meningioma from to anterior nerve V (now straight) is not yet complete. The posterior corner of Parkinson's triangle, now lined with Surgicel, is free of tumor. Meningioma remains medial to the GG, adjacent to the clivus. In (d), the tumor has been resected completely. The brainstem is visible posterior and anterior to nerve V. Nerve VI courses from the brainstem anterolaterally to and through Dorello's space, ultimately entering the parasellar space (*PS*) lateral to the PS-ICA. Note the wide space between the brainstem medially and the GG laterally. Nerve V is no longer displaced upward, and the GG, V1, and VI are free of tumor. Note: No brain retraction was required at surgery. Sufficient working room was obtained by periodic repositioning of the operating table and the microscope, also by lifting dura fixed by stay sutures as appropriate (b, c, d)

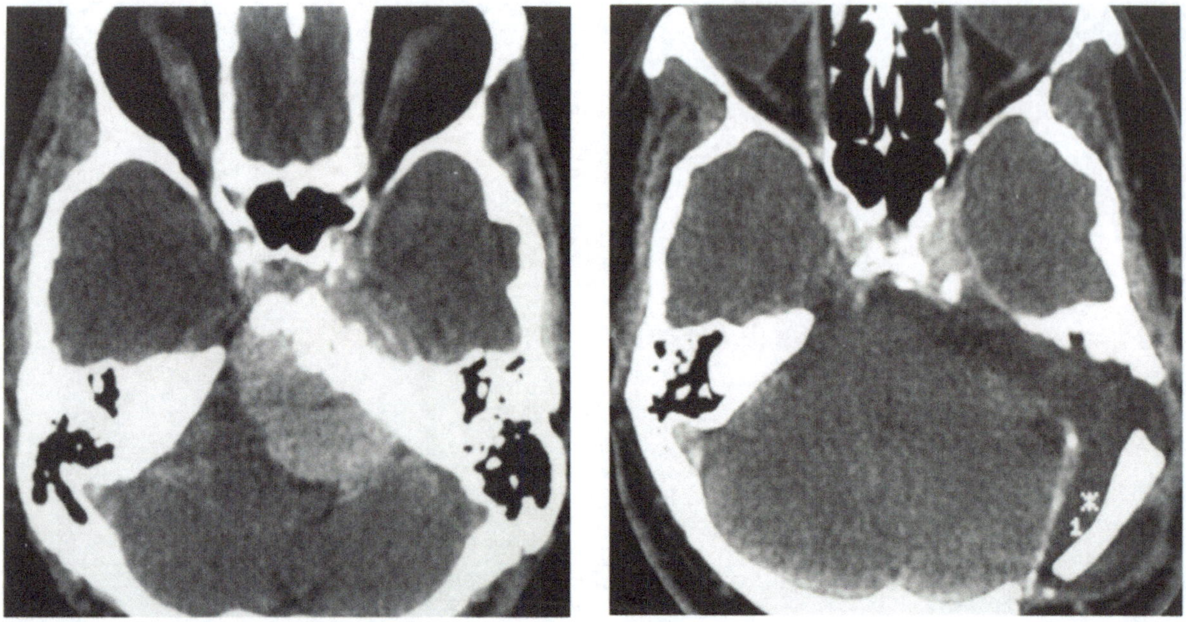

a b

Fig. 4.19. Preoperative contrast-enhanced trans-axial CT (a) reveals a meningioma lying anterolateral and posteromedial to an extremely hyperostotic petrous apex which extends across the clivus, almost reaching the contralateral (right) posterior clinoid process. Corresponding postoperative contrast-enhanced transaxial CT (b) confirms complete resection of the hyperostotic petrous apex and subtotal resection of the meningioma, a significant remnant remaining in the PS. (With the bulk of the tumor behind the petrous apex, in the posterior fossa, a retrosigmoid suboccipital approach was employed.) Note: Now one year after surgery and free of neurological deficit and the severe trigeminal pain she had experienced prior to surgery, this young woman is quite content. But her surgeon is not! He understands that the residual tumor in the left parasellar space (*PS*) is destined to cause more trouble and must be removed via the transpetrous approach

Patients and results

Forty-one patients with petroclival meningiomas were treated surgically. Five had been operated previously elsewhere, with 36 managed as virgin cases. The transpetrous approach was employed in 36 cases, the remainder resected via the suboccipital retrosigmoid approach. Tumor resectability and neurological outcomes are summarized in Table 13.

Complete tumor resection was achieved in 21, gross-total and subtotal removals in 12 and eight respectively. None of the five residual tumors were completely excised. Immediate neurological loss following surgery was limited to the cranial nerves, nerve IV deficits in two patients, nerves V and VI impairment in three each, and one patient with facial paresis and hearing loss. All cranial nerve deficits were transient with the exception of two patients with persisting nerve VI deficits. Patients with complete and/or gross-total resections receive annual MRI studies, and those having had subtotal resections are followed with clinical and MRI examinations at six-month intervals. To date no patient has required additional treatment.

Table 13. Petroclival meningomas treated: surgical approach, tumor resectability, and results

Location & type of meningioma	N° of cases		Approach				Resection						Postoperative deficits one week post-op							
	Virgin		Transpetrous		Suboccipital		Complete		Gross-total		Subtotal		IV		V		VI		VII, VIII	
	Redue																			
Medial	6	15%	4	67%	2	33%	4	67%	1	17%	1	17%	–	–	–	–	1	17%	–	–
	1	2%	1	100%	–	–	–	–	1	100%	–	–	–	–	–	–	1	100%	–	–
Lateral	14	34%	12	86%	2	14%	12	86%	2	14%	–	–	–	–	1	7%	–	–	–	–
	1	2%	1	100%	–	–	–	–	1	100%	–	–	–	–	1	100%	–	–	–	–
Combined	8	20%	8	100%	–	–	5	63%	2	25%	1	13%	1	13%	–	–	–	–	–	–
	2	5%	2	100%	–	–	–	–	1	50%	1	50%	–	–	1	50%	–	–	–	–
En-plaque	8	20%	7	88%	1	13%	–	–	4	50%	4	50%	–	–	–	–	1	13%	–	–
	1	2%	1	100%	–	–	–	–	–	–	1	100%	–	–	–	–	–	–	1	100%
Collective	36	88%	31	86%	5	14%	21	58%	9	25%	6	14%	1	3%	1	3%	2	6%	–	–
	5	12%	5	100%	–	–	–	–	3	60%	2	40%	–	–	2	40%	1	20%	1	20%
Total	41	100%	36	88%	5	12%	21	51%	12	29%	8	17%	2	5%	3	7%	3	7%	1	2%

Tuberculum sellae meningiomas

Pathological anatomy, clinical presentation, and diagnostic work-up

Meningiomas of the tuberculum sellae constitute a well-defined pathological entity, one recognized for generations. Their point of origin plus patterns of extension determine whether the symptoms and signs they produce are related to encroachment of the visual apparatus, the pituitary stalk (PSt), or the frontal lobes. From their sites of origination these meningiomas spread into the anterior cranial fossa and/or posteriorly, displacing the optic chiasm posteriorly and one or both optic nerves (ONs) laterally Figs. 4.20–4.26.

Impaired vision normally leads to clinical evaluation. Although tuberculum sellae meningiomas may displace the PSt, endocrinological disturbances are only occasionally present at the time of diagnosis. Globular lesions in the anterior cranial fossa may become quite large before frightening visual symptoms send the patient in search of medical assistance Figs. 4.20–4.26. But mental changes from frontal lobe involvement may impair a patient's capacity to report early visual symptoms, and his or her family and general practitioner may be so distracted by distressing personality changes as to overlook visual changes until the patient is totally blind (Fig. 4.26). Among the elderly, visual loss may be misinterpreted as being related to cataract formation. Continued visual dysfunction, despite cataract surgery or repeated corrections of vision with glasses, typically leads to a CT and/or MRI and the correct diagnosis. "En plaque" lesions, however, by virtue of their minimal mass may escape CT or MRI diagnosis, growing primarily into the optic canal, resulting in ON dysfunction both inside and outside the optic canal (OC).

Tuberculum sellae meningiomas which displace an ON laterally and upward, trapping it against the sharp falciform ligament above it, create clear-cut visual field deficits Fig. 4.24c. Further deterioration of vision may be rapid. Being aware of this sometimes leads neuroradiologists and neuro-ophthalmologists focused on ONs within the OC to ultimately achieve a diagnosis by sequential MR scanning.

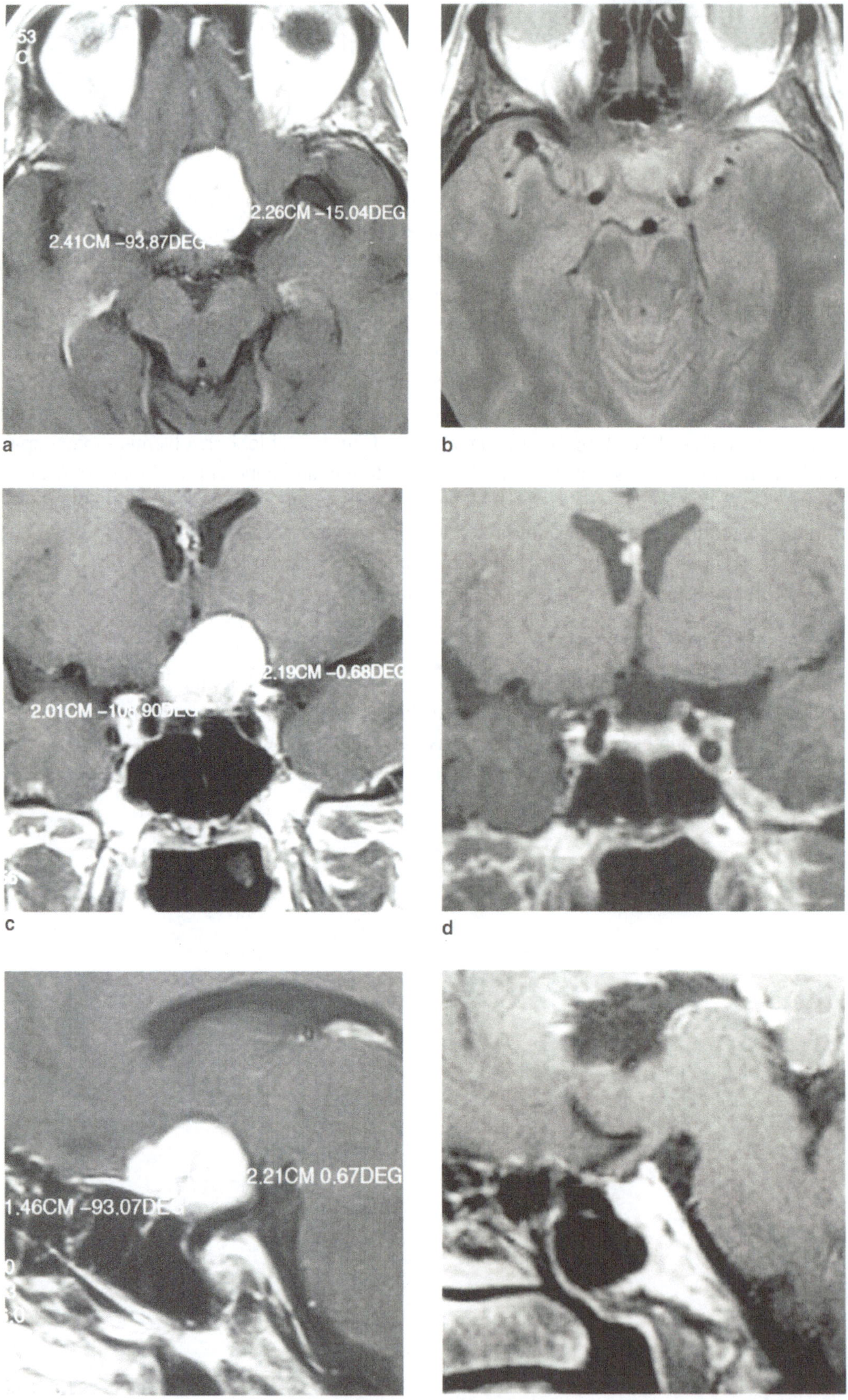

Fig. 4.20. Preoperative contrast-enhanced transaxial (a), coronal (c), and sagittal (e) MR images of a medium-sized tuberculum sellae meningioma. Note: This 62-year old lady was in excellent health with the exception of slightly reduced visual acuity in her left eye, a condition present for almost one year. Repeated glasses changes did not improve her vision. In the three weeks immediately prior to surgery vision in her left eye became increasingly blurred, and she began to experience obscurations of vision on the right side as well. Postoperative contrast-enhanced MR images – one month postoperatively – demonstrate complete resection of the tumor (b, d, f) with no frontal or temporal lobe abnormality. During the first month after surgery her vision on the right eye become normal. On the left eye the vision improved to that extent that she can read with the left eye, while the right eye is covered with her hand

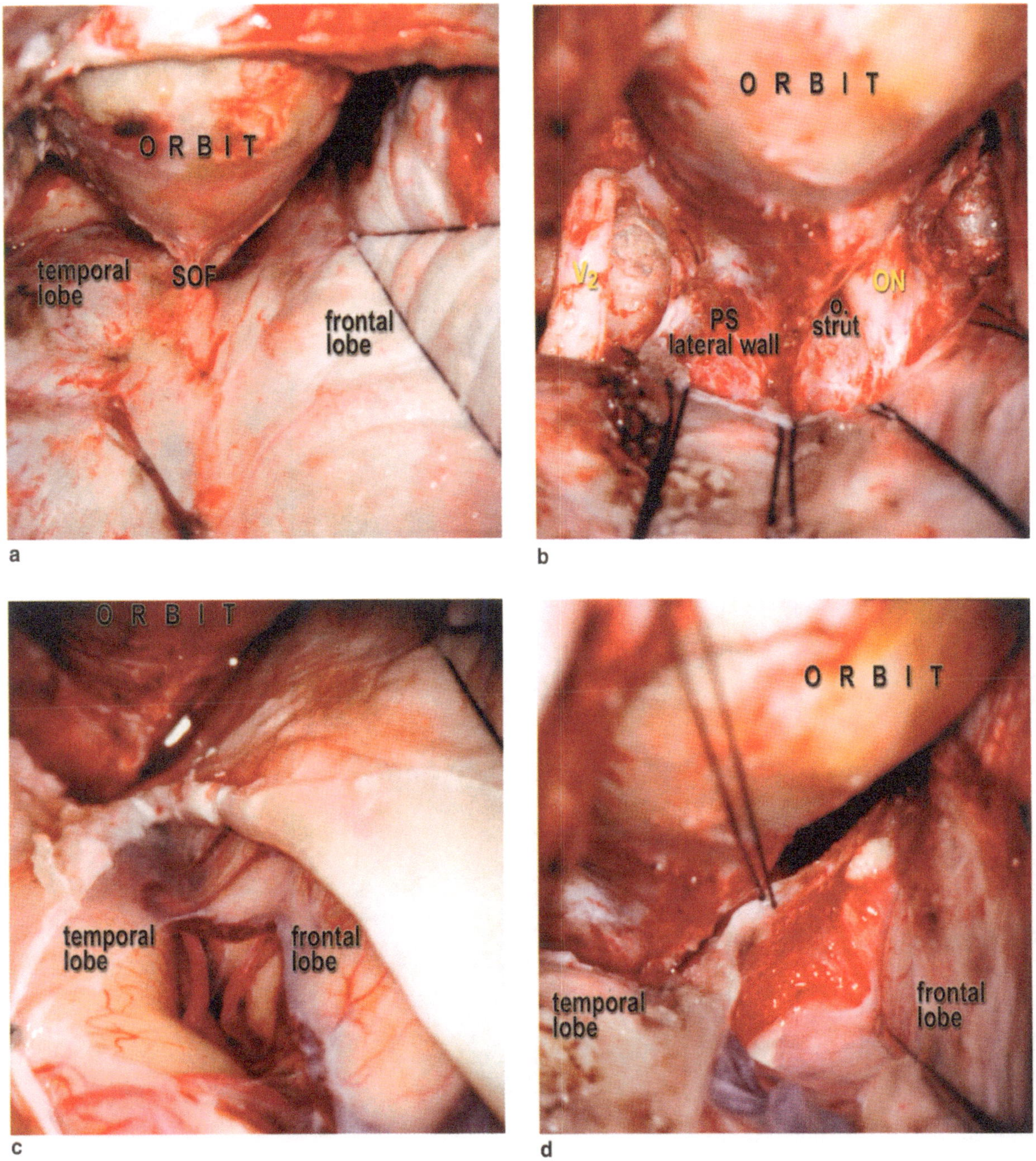

Fig. 4.21. The patient whose tumor is shown in Fig. 4.20 underwent surgery via the approach described in Chapter 2. (a) The orbital roof has been opened and the lateral surface of the superior orbital fissure dura (*SOF*) exposed. (b) Orbital roof bone has been removed further, the sphenoid wing resected, and the optic canal (*OC*) opened (its dorsal and lateral walls removed). The optic strut remains in place, but the anterior clinoid process (*ACP*) has been resected. V2 has been skeletonized and the parasellar space (*PS*) lateral wall has been divided, its outer (dural) layer fixed with stay sutures. The optic nerve (*ON*) remains covered with dura propria, and the ICA (*AL*) traversing the anteromedial triangle is exposed. (c) After incising dura along the sylvian fissure, temporal and frontal lobes are exposed and the sylvian fissure has been partially opened. (d) Continuing the dural incision centralward, then turning medialward approximately 5 mm proximal to the distal (dural) ring (DR) exposes the falciform ligament. Note the hyperemic dura and tumor adherent to the inner surface of the dura

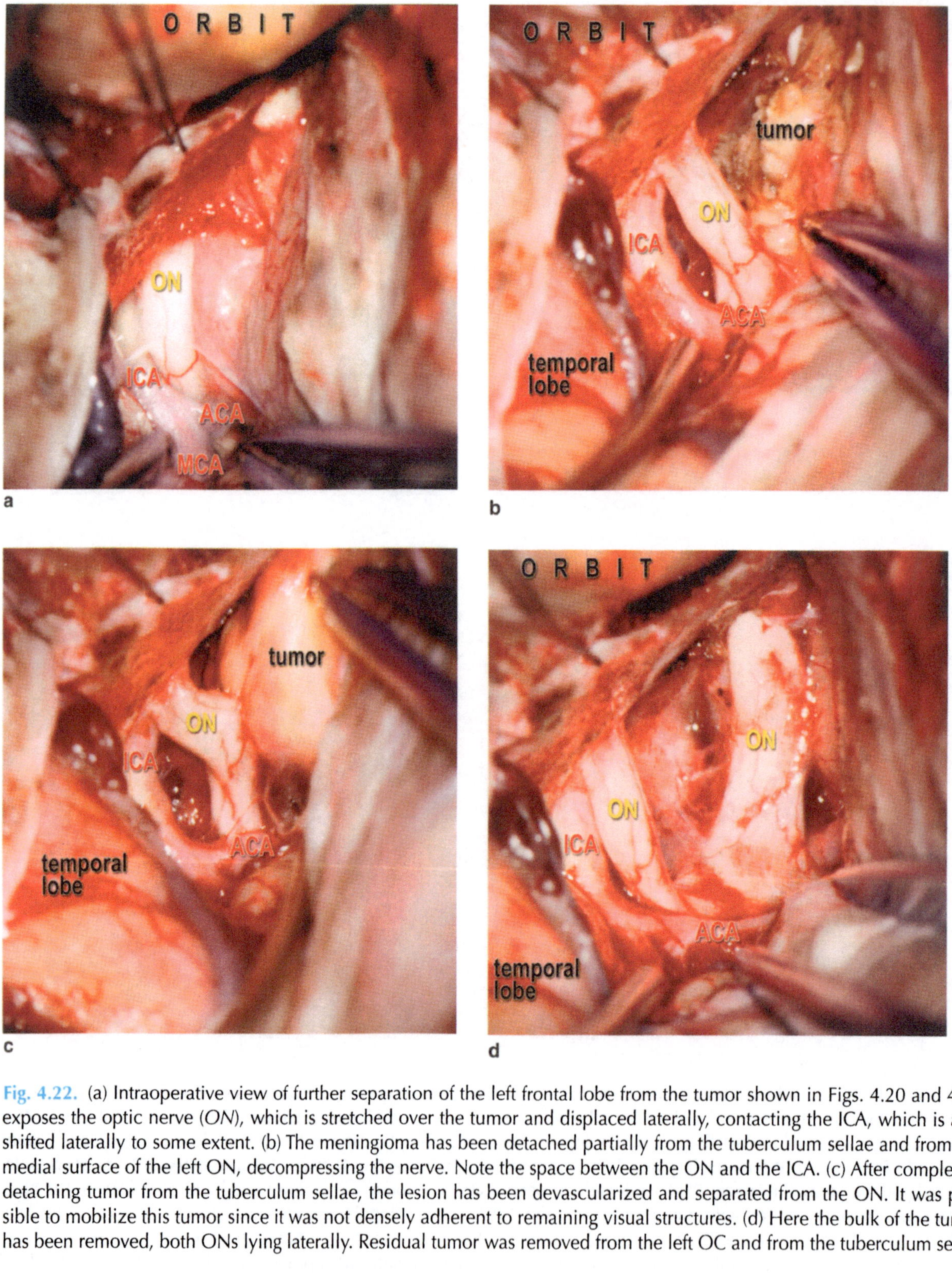

Fig. 4.22. (a) Intraoperative view of further separation of the left frontal lobe from the tumor shown in Figs. 4.20 and 4.21 exposes the optic nerve (*ON*), which is stretched over the tumor and displaced laterally, contacting the ICA, which is also shifted laterally to some extent. (b) The meningioma has been detached partially from the tuberculum sellae and from the medial surface of the left ON, decompressing the nerve. Note the space between the ON and the ICA. (c) After completely detaching tumor from the tuberculum sellae, the lesion has been devascularized and separated from the ON. It was possible to mobilize this tumor since it was not densely adherent to remaining visual structures. (d) Here the bulk of the tumor has been removed, both ONs lying laterally. Residual tumor was removed from the left OC and from the tuberculum sellae

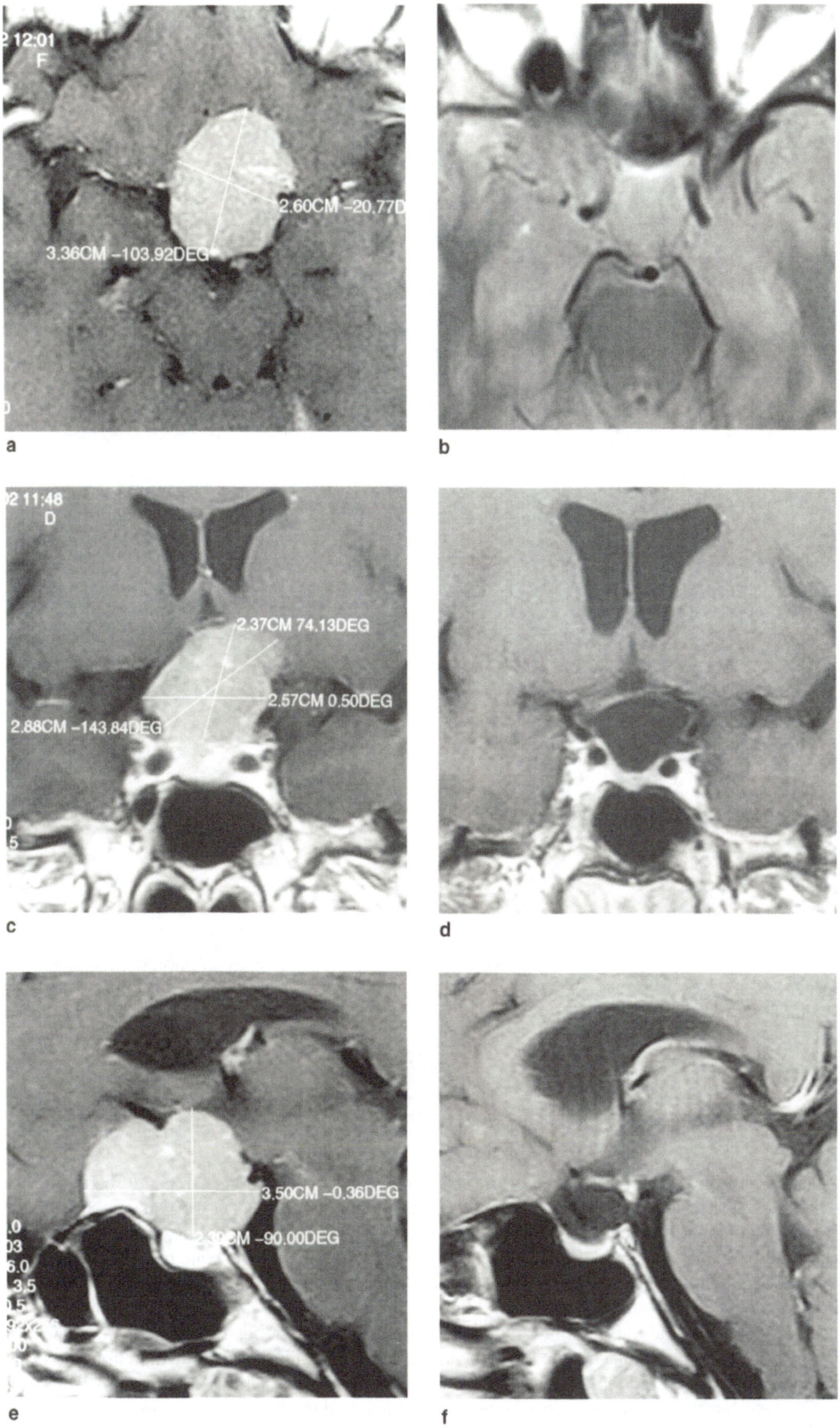

Fig. 4.23. Preoperative contrast-enhanced, transaxial (a), coronal (c), and sagittal (e), MR images show a rather large tuberculum sellae meningioma. Postoperative corresponding contrast-enhanced MR images (b, d, f) confirm complete resection of the tumor

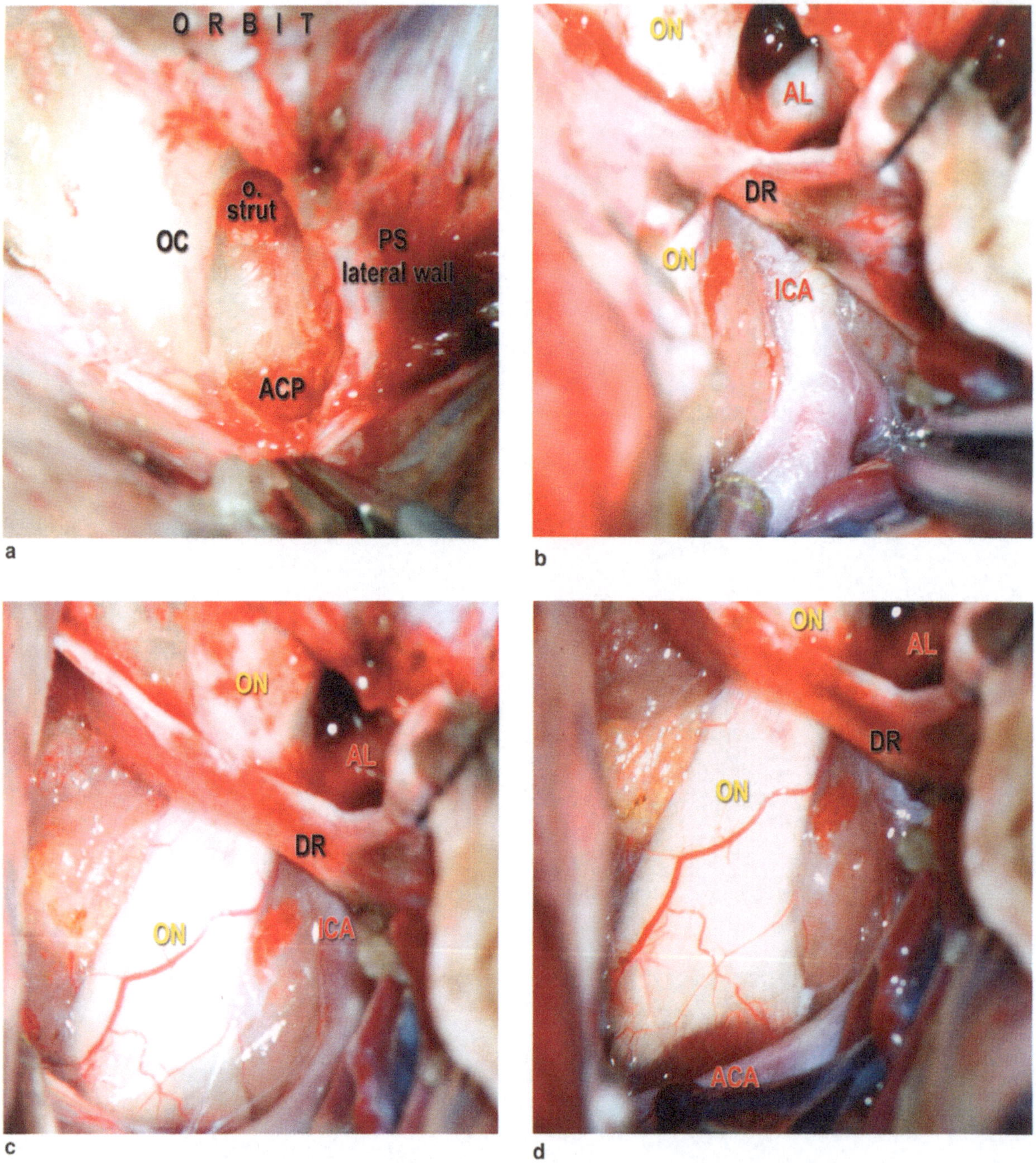

Fig. 4.24. The intraoperative view of resection of the tuberculum sellae meningioma shown in Fig. 4.23. (a) The lesion was approached from the right side as described in Chapter 2. The bulk of the anterior clinoid process (*ACP*) has been removed although the anterior loop (*AL*) of the ICA in the anteromedial triangle remains covered by bone. The apex of the ACP and optic strut remain, contiguous with the lateral and caudal walls of the optic canal (*OC*). (b) The optic nerve (*ON*) has been exposed after completely removing the walls of the OC. The AL of the ICA is apparent within the anteromedial triangle. Dura has been divided along the sylvian fissure toward the dural rim, exposing the intradural ICA and tumor on either side of it. The forceps on the right side point to the PCom. In the left half of the photograph, beneath the dura, a segment of the intradural ON has been stretched and displaced proximally by the tumor. (c) Additionally incising dura from the distal (dural) ring (*DR*) medialward better exposes the displaced ON, tumor on both sides of it. The (*AL*) of the ICA lies to the right of the extradural ON, which remains covered by dura propria, and the intradural ICA is adjacent to the right side of the tumor. Note: The falciform ligament, where the ON enters the OC, compresses the ON from above, compounding the tension applied ventrally and rostrally by the tumor (d). The ON is also sandwiched more centrally, between the ACA1 and tumor beneath it, possibly even more crucially over time because of the artery's pulsations. Note the discoloration of the ON adjacent to the ACA

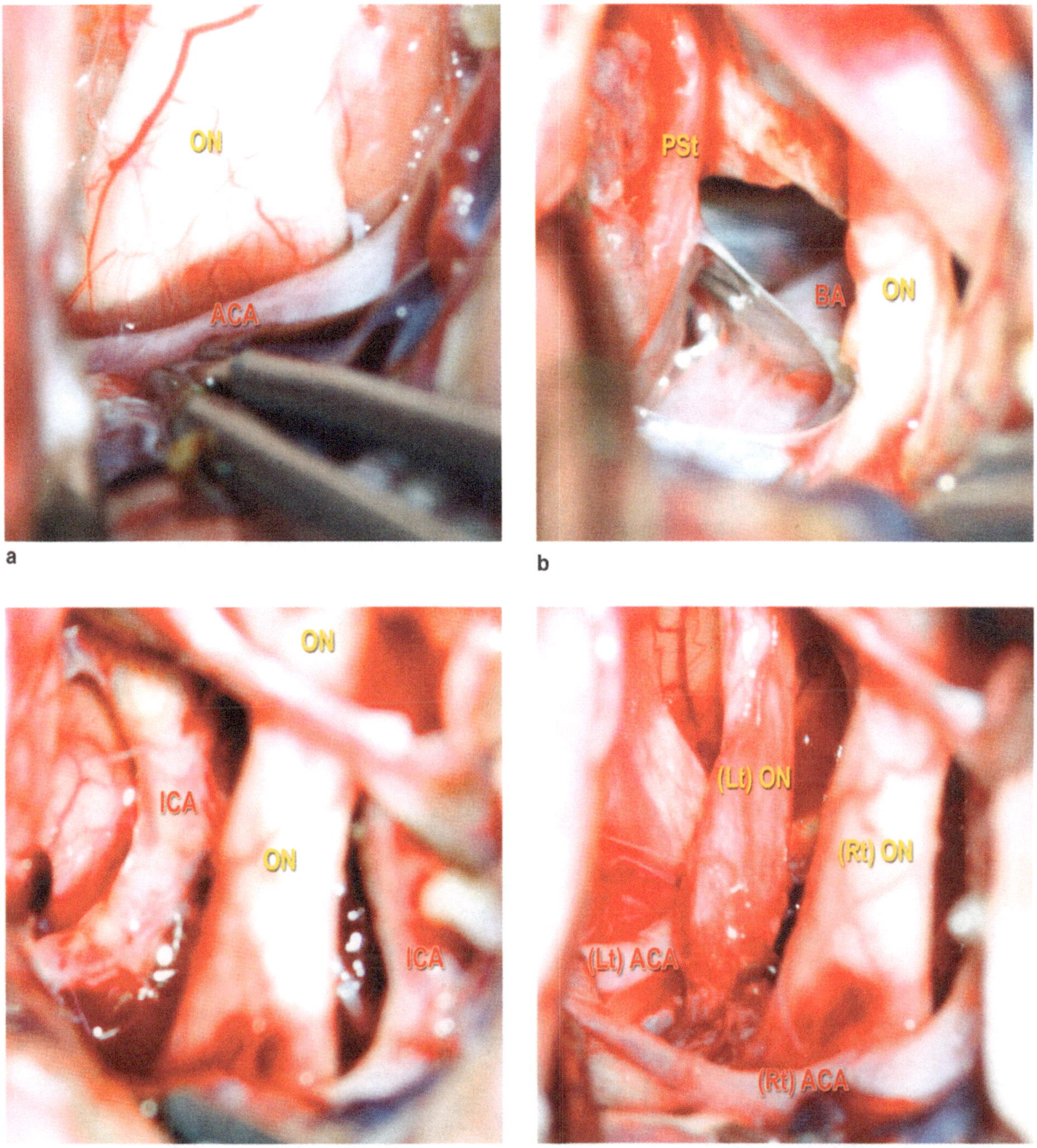

a b c d

Fig. 4.25. (a) The entire width of the optic nerve (*ON*) is demonstrated where it is crossed by the left ACA. This segment of the nerve is trapped between the ACA (above) and the meningioma (below). Note the ON's discoloration. (b) Additionally, tumor has been removed piece by piece, decompressing the right ON. The basilar artery (*BA*) is apparent deep in the exposure, also the pituitary stalk (*PSt*) to the left of which meningioma remains. (c) Complete resection has been achieved, demonstrating both ICAs and the right ON between them. (d) Here both ONs are shown after total resection of the tumor. The visual apparatus had been compressed bilaterally, particularly by both ACAs from above and tumor from below. Both ONs and optic chiasm were sandwiched between the ACAs and tumor. From this angle the PSt lies hidden behind the right ON

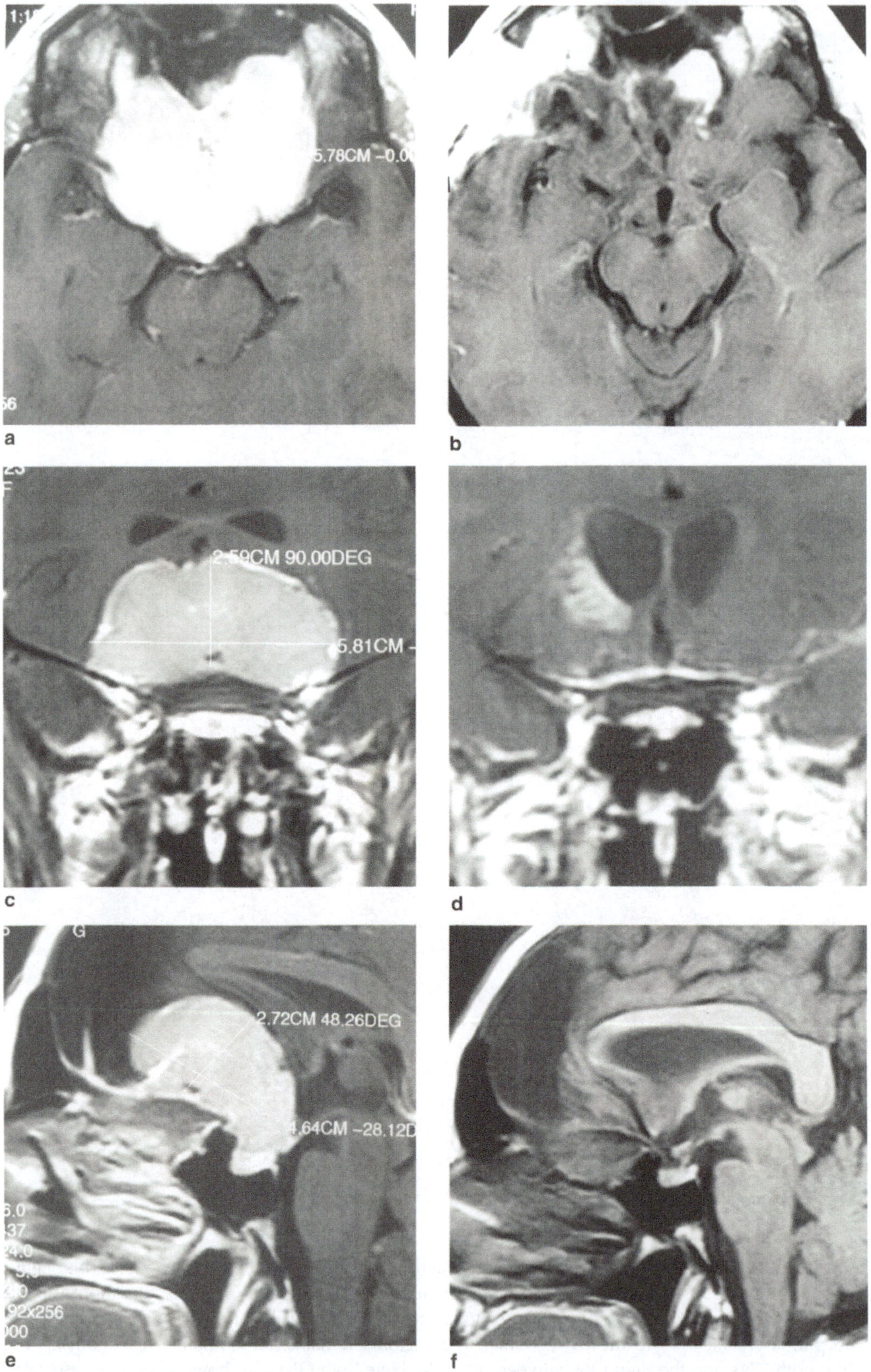

Fig. 4.26. Preoperative contrast-enhanced transaxial (a), coronal (c), and sagittal (e) MR images demonstrate a large (rest) tuberculum sellae meningioma (operated somewhere else). Note the considerable destruction of both frontal lobes above and anterior to the tumor (a and e). Preoperatively the patient's vision had deteriorated to the point that she was not capable of independent living, seeing large objects only as shadows. At the first operation, via a bifrontal approach (in an other horizontal), the tumor was only biopsied, surgery amounting to little more than a continuous struggle with swollen brain and barely controllable tumor bleeding (due to the wrong approach). Subsequently, a definitive operation was carried out from the right side by the approach described in Chapter 2, requiring no brain retraction at all. The tumor was devascularized initially, detaching it from the tuberculum sellae and the floor of the anterior cranial fossae, then resecting it small piece by small piece from adjacent structures, including both ONs, the optic chiasm, the pituitary stalk, and both ICAs. Postoperative contrast-enhanced transaxial (b), coronal (d), and sagittal (f), MR images confirm total resection of the tumor. Visual function improved considerably (already during two months period following second surgery), allowing the patient to resume a completely independent life

Surgical approach to tuberculum sellae meningiomas

Patient positioning, skin incision, and craniotomy are performed as described in Chapter 2. Unroofing the orbit and resecting the sphenoid wing are unnecessary. Resecting the anterior clinoid process (ACP) and the walls of the OC are not required in most cases, but "en plaque" lesions spreading into one or both OCs must be resected with all involved dura propria. With such cases it is necessary to resect at least the lesser sphenoid wing, the ACP, the superior and lateral walls of the OC, and the optic strut in order to expose the ON and its dura propria throughout the OC.

With both "en plaque" and globular meningiomas it is wise to peel dura from the tuberculum sellae and devascularize the lesion and infiltrated dura by coagulating dura and drilling away all hyperostotic tuberculum sellae. While drilling the tuberculum sellae, one must take special care to avoid entering the sphenoidal sinus in order to avoid a cerebrospinal fluid (CSF) leak postoperatively. The entire width of the tuberculum sellae, from one OC to the other, is drilled away with care, ultimately allowing dura to be detached posteriorly and caudally from the attachment of the diaphragm sellae to bone.

When tumor infiltrates the diaphragm sellae, the involved portion must be resected, not simply coagulated on its intradural surface. When necessary, the contralateral OC should be opened, at least proximally, and inspected for possible tumor between dura and the ON. Only when tumor-involved dura has been completely detached from bone and normal dura beyond the lesion exposed, may the meningioma be resected piece by piece from its dural surface.

Resecting tumor from the ipsilateral ON, the optic chiasm, brain, and the contralateral ON may be achieved atraumatically and with safety only as the final step of the procedure, after the bulk of the tumor has been removed. The PSt must be dissected from tumor without exerting pressure on the ON or the chiasm, either rostrally or caudally. It is crucially important to preserve all Dawson's arteries, arising from each ICA. These vessels supply the entire visual apparatus and the PSt, structures already made vulnerable by tumor growth.

Generally, the lesion may be dissected from the frontal lobe(s), the olfactory nerve(s), the first segment of the anterior cerebral arteries (ACA1s), the anterior communicating artery (ACom), and the second segment of the anterior cerebral arteries (ACA2s) with minimal difficulty. Following total resection of the tuberculum sellae meningioma and establishing precise hemostasis, the dural defect may be bridged with commercially available dural substitute. The graft should be fixed with interrupted sutures to healthy dura and covered with a film of fibrin glue. An epidural drain is installed, the osteomuscular flap replaced and secured with sutures, and the wound is closed in layers.

Where CSF leakage may seem imminent, patients should be maintained on continuous lumbar drainage for one week. Typically such drains are installed within the first six or eight hours following surgery. In general, it is simpler to prevent a leak than to be forced to treat one later.

Patients and results

Of 817 patients harboring central skull base meningiomas, 44 were classified as tuberculum sellae meningiomas. Tumor size, resectability, and neurological outcomes are summarized in Table 14. Complete tumor resection was achieved in 35 virgin cases and in seven of nine patients operated on for residual tumors. Gross total resections were possible in two recurrent cases of both groups. Forty-one of the 44 had preoperative visual deficits, 29 of them also manifesting mental disturbances and endocrinological deficits. Thirty experienced improved vision after surgery, and in 11 patients visual function remained unchanged. Twenty-two had improved mental function following surgery, seven remaining unchanged with respect to frontal lobe function.

Table 14. Cases of virgin and residual tuberculum sellae meningiomas operated: resectability and outcome

	N° of cases										Postoperative deficits							
	Virgin		Resectability				Preoperative deficits				Visual				Other			
Meningioma	Residual		Complete		Gross-total		Visual		Other		Improved		Static		Improved		Static	
Tumor < 3 cm	9	20%	9	100%	–	–	17	189%	–	–	6	67%	1	11%	–	–	–	–
	2	5%	2	100%	–	–	2	100%	–	–	1	50%	1	50%	–	–	–	–
3 cm < tumor < 5 cm	14	32%	14	100%	–	–	13	93%	11	79%	11	79%	2	14%	8	57%	3	21%
	3	7%	2	67%	1	33%	3	100%	2	67%	1	33%	2	67%	1	33%	1	33%
Tumor > 5 cm	12	27%	12	100%	–	–	12	100%	12	100%	10	83%	2	17%	10	83%	2	17%
	4	9%	3	75%	1	25%	4	100%	4	100%	1	25%	3	75%	3	75%	1	25%
Collective	35	80%	35	100%	–	–	32	91%	23	66%	27	77%	5	14%	18	51%	5	14%
	9	20%	7	78%	2	22%	9	100%	6	67%	3	33%	6	67%	4	44%	2	22%
Total	44	100%	42	95%	2	5%	41	93%	29	66%	30	68%	11	25%	2	50%	7	16%

Diaphragm sellae and optic canal(s) meningiomas

Pathological anatomy, clinical presentation, and diagnostic work-up

Meningiomas attached exclusively to the diaphragm sellae are very rare. Such lesions originate beneath and posterior to the tuberculum sellae, invading the diaphragm sellae posteriorly and laterally. The majority of diaphragm sellae meningiomas occur as extensions of other meningiomas, mainly those arising at the tuberculum sellae but also those arising from the anterior clinoid process (ACP), the posterior clinoid process (PCP), or the dorsum sellae. Diaphragm sellae meningiomas initially involve the distal segment of the pituitary stalk (PSt), later extending along this structure to compress the optic chiasm – displacing it upward and posteriorly, stretching the optic nerves (ONs) – and hypothalamus. Most displace the ON laterally, often against the falciform ligament, at the entry into the optic canal (OC).

"En plaque" lesions may invade either or both OCs, creating the frightening danger of rapid bilateral visual loss, even total blindness. On the other hand, globular lesions typically invade only one OC (if at all), and then only partially, most often medially at the site of the ON entry.

Even though these lesions spread along the PSt early in their growth, endocrinological derangements are only rarely clinically prominent, probably for two reasons. First, the PSt's attachment to the diaphragm sellae is commonly eccentric, providing space for tumor growth without distorting the PSt. The entire compartment between diaphragm sellae and the visual apparatus must be occupied with tumor to result in either endocrine or visual symtoms and signs. Second, the PSt, whether compressed from one side or circumferentially, seems inherently better equipped to compensate mechanical injury than the chiasm or the ONs. The result is that diaphragm sellae meningiomas most frequently present with visual symptoms and signs but normal endocrinological function at the time of diagnosis. Diplopia from nerve III compression also may be an associated symptom where such lesions extend by way of oculomotor trigone dura into the canal of nerve III. In some cases retro-orbital discomfort or pain is not accompanied by visual manifestations until the meningioma is relatively large, fully filling the space between the diaphragm sellae and the visual apparatus.

It is worrisome that generalized visual deterioration, particularly among the elderly, may be ascribed to ocular problems rather than diaphragm sella neoplasms when they are present. Only when diplopia (from occulomotor palsy) or other symptoms appear does the diagnosis become apparent.

CT and the MRI establish the diagnosis. Carotid angiography is seldom required although one must always be aware of the possibility of a carotid-ophthalmic aneurysm, a diagnosis generally made by comparing contrasted and non-contrasted studies.

Surgical approach to the diaphragm sellae, and/or optic canal(s) meningiomas

Patient positioning, scalp incision, osteoplastic craniotomy, resecting the lesser sphenoid wing, the ACP, and the superior, lateral, and inferior walls of the OC are carried out as described in Chapter 2. Dura at the superior orbital fissure (SOF) is transected laterally and the layers of the anterior part of the parasellar space (PS) lateral wall are separated, which enables the ACP to be resected with maximum safety from surrounding structures. Dura is then peeled from the tuberculum sellae, and the edge of the tuberculum sellae is drilled medial to the origin of the ipsilateral OC toward the medial wall of the contralateral OC. This allows exposure of the space beneath the tuberculum sellae and the most anterior part of the diaphragm sellae. The hyperostotic edge of the tuberculum sellae is carefully "brushed" away with the diamond burr, leaving only a thin transparent layer of bone along the entire width of the tuberculum sellae covering the mucus membrane of the sphenoidal sinus. Generally, it is possible to identify the anterior attachment of the diaphragm sellae at the posterior surface of the tuberculum sellae.

Only after completing the necessary extradural bone removal should dura be incised along the sylvian fissure centrally toward the distal (dural) ring (DR) of the ICA. Before reaching the DR, the dural incision is angled medially over the optic nerve (ON). The diaphragm sellae is coagulated anterior to the posterior clinoid process

(PCP) and medially, beneath the ICA. Next, dura in the very narrow space between the ICA and the ON is coagulated with great caution in order to avoid damaging the ophthalmic artery (OA) and/or Dawson's arteries. Then the tumor is detached from the central portion of the anterior half of the diaphragm sellae, all the way to the contralateral side, until encountering normal diaphragm. The thus-devascularized lateral and anterior portion of the tumor may then be debulked.

While resecting tumor, it is extremely important to avoid stretching Dawson's arteries and exerting tension on the ipsilateral ON. Perfect hemostasis should be maintained throughout the procedure. When the optic tract is short and space between the ICA and the ON does not allow sufficient working room, the DR around the ICA may be incised circumferentially and the OA dissected from the ICA to and into the OC by transecting dura propria along the longitudinal axis of the ON, finally resecting dura infiltrated by meningioma. This will allow the ICA to be mobilized gently laterally, enlarging the working space between the ICA and the OA (anterolaterally) and the ON and optic tract (medially and posteriorly), affording good exposure of the ipsilateral half of the diaphragm sellae. By carefully debulking the meningioma, tension on the ipsilateral ON and the optic tract will be reduced, thus providing additional working space beneath and above the ON. It is important to identify the PSt and dissect it free.

Changing the angle of the approach, the contralateral half of the diaphragm sellae may be accessed from above, between the ONs. When all visible meningioma has been removed, the diaphragm sellae should be examined carefully for evidence of tumor infiltration and possible tumor extension into the sella. If the diaphragm sellae is infiltrated by tumor, it should be resected and any tumor within the sella evacuated.

Tumor in the ipsilateral OC is removed along with infiltrated dura propria. Meningioma in the contralateral OC may be removed medial to the ON. It is not safe to open the contralateral OC circumferentially throughout its entire length. Forcing the resection of tumor from the walls of the contralateral OC may result in injury to the ipsilateral ON or entering the sphenoidal sinus, violating its mucous membrane, which may result in a cerebrospinal fluid (CSF) leak. When the contralateral OC is truly packed with tumor, surgery should be staged, a second operation planned for after a week or so.

After completely resecting all tumor and infiltrated dura and establishing precise hemostasis, the dura is closed. Any dural defect at the skull base should be bridged with a small piece of galea or temporalis muscle or fascia. The graft should be sutured to healthy dura throughout, then covered with a film of fibrin glue. Following watertight dural closure the osteomuscular flap is replaced and secured with sutures. An extradural drain is left in place and the wound is closed in layers. When the sphenoidal sinus may have been violated, the risk of a CSF leak is minimized by maintaining continuous lumbar drainage (initiated up to six hours following surgery) for one week.

Patients and results

Eight patients underwent surgery for diaphragm sellae and OC meningiomas. Seven were operated as virgin cases, and one had been operated on previously elsewhere. Two of the virgin lesions involved both OCs in addition to the diaphragm sellae, one patient experiencing rapid deterioration of vision on the left, the right eye remaining unaffected. At her first operation meningioma was resected from the diaphragm and the left OC (Figs. 4.27 and 4.28). Although vision was preserved on the left, she declined surgery to remove the remaining lesion since vision in her right eye remained normal. After two months, however, she decided to go ahead with a second operation when vision rapidly deteriorated in her right eye. Prior to that procedure, she actually had better vision on the left (previously operated) side. Following surgery to remove residual meningioma from the right OC, vision in the right eye remained unchanged. The second patient, who had experienced bilateral visual symptoms for two years, suddenly developed deteriorating vision in both eyes, progressing quickly to total blindness on the right (Fig. 4.29). His left-sided vision was preserved unchanged following the operation to remove the meningioma in his left OC (Fig. 4.30). Two months later he consented to have surgery on the right side (Fig. 4.31). Despite complete resection of the tumor, blindness in his right eye (then present for almost a full year) persisted.

Two patients had relatively small meningiomas of the diaphragm sellae with extension into one OC. Vision was preserved in both by total tumor removal.

Two additional patients had diaphragm sellae tumors, both essentially globular in shape, each compressing one ON at its entrance into the OC. Both patients had experienced ipsilateral visual loss prior to surgery. In each, the affected ON was found to be displaced upward, sandwiched between tumor medial to it and the falciform ligament above, tumor extending between the ICAs. Following meningioma resection, one lost what slight vision had been present prior to surgery, and vision was preserved unchanged in the other.

The seventh patient operated as a virgin case had experienced bilateral severe symmetrical visual impairment from a globular diaphragm sellae meningioma (Fig. 4.32). The tumor was found to be extremely hard at surgery, but gross-total resection was achieved, leaving the patient blind postoperatively, likely from ON dissection and vasospasm of Dawson's arteries.

The eighth patient had been operated on for a residual lesion from an operation at another institution. (That surgery was for an ACP meningioma although the ACP had not been resected and the OC had not been opened.) We encountered a meningioma which extended from the diaphragm sellae medial to the origin of the right OA, completely filling the OC. The OC was widely opened and all tumor and infiltrated dura propria were resected, but vision remained unchanged after surgery.

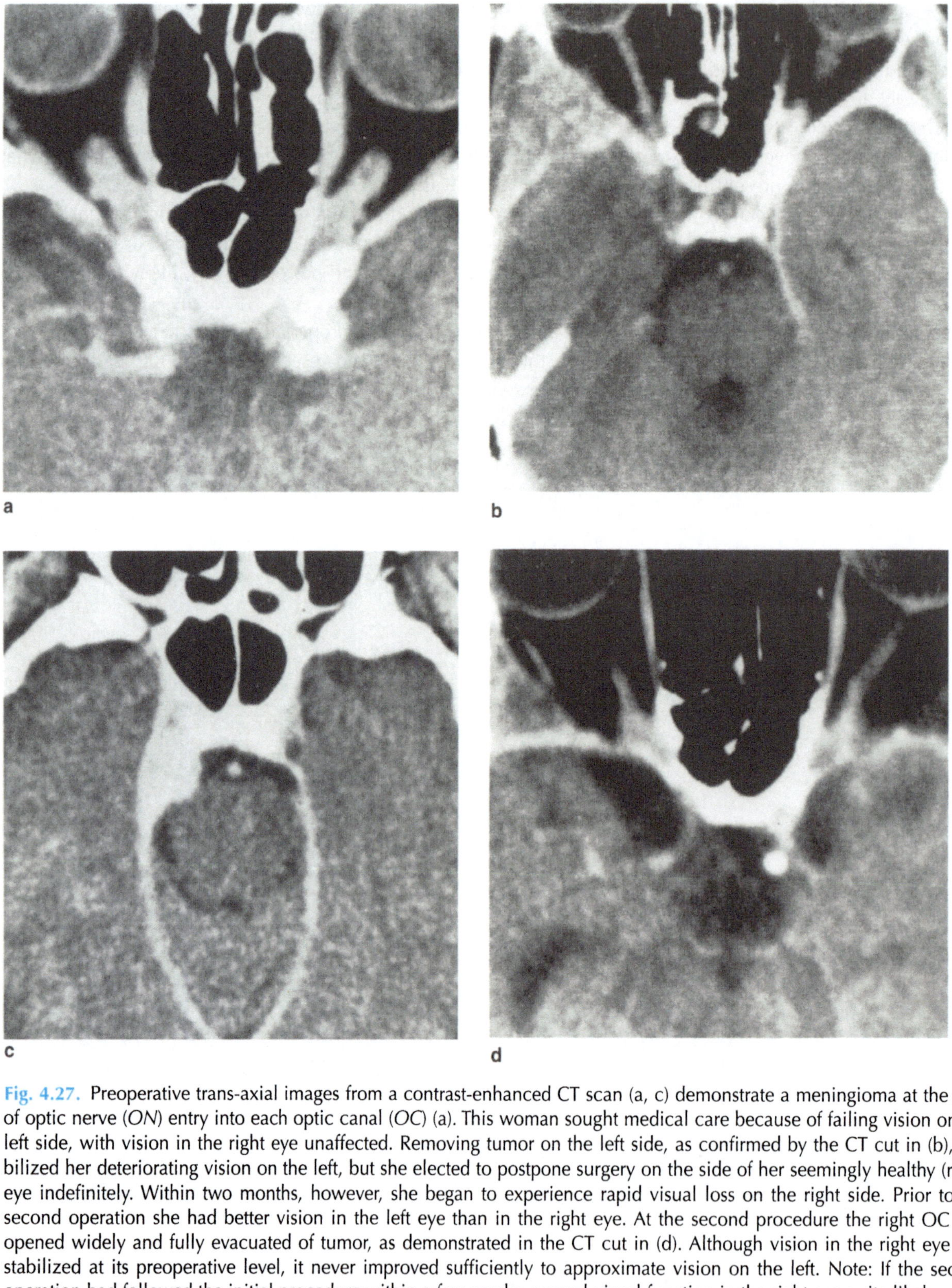

Fig. 4.27. Preoperative trans-axial images from a contrast-enhanced CT scan (a, c) demonstrate a meningioma at the sites of optic nerve (*ON*) entry into each optic canal (*OC*) (a). This woman sought medical care because of failing vision on the left side, with vision in the right eye unaffected. Removing tumor on the left side, as confirmed by the CT cut in (b), stabilized her deteriorating vision on the left, but she elected to postpone surgery on the side of her seemingly healthy (right) eye indefinitely. Within two months, however, she began to experience rapid visual loss on the right side. Prior to her second operation she had better vision in the left eye than in the right eye. At the second procedure the right OC was opened widely and fully evacuated of tumor, as demonstrated in the CT cut in (d). Although vision in the right eye was stabilized at its preoperative level, it never improved sufficiently to approximate vision on the left. Note: If the second operation had followed the initial procedure within a few weeks, normal visual function in the right eye quite likely could have been preserved. It is important to understand that the ON will be able to tolerate tumor within the OC only so long as space exists between the nerve and the bony walls of the canal. When ON compression results in failing visual acuity, blindness may follow in only a matter of weeks. Procrastination may lead to dire consequences

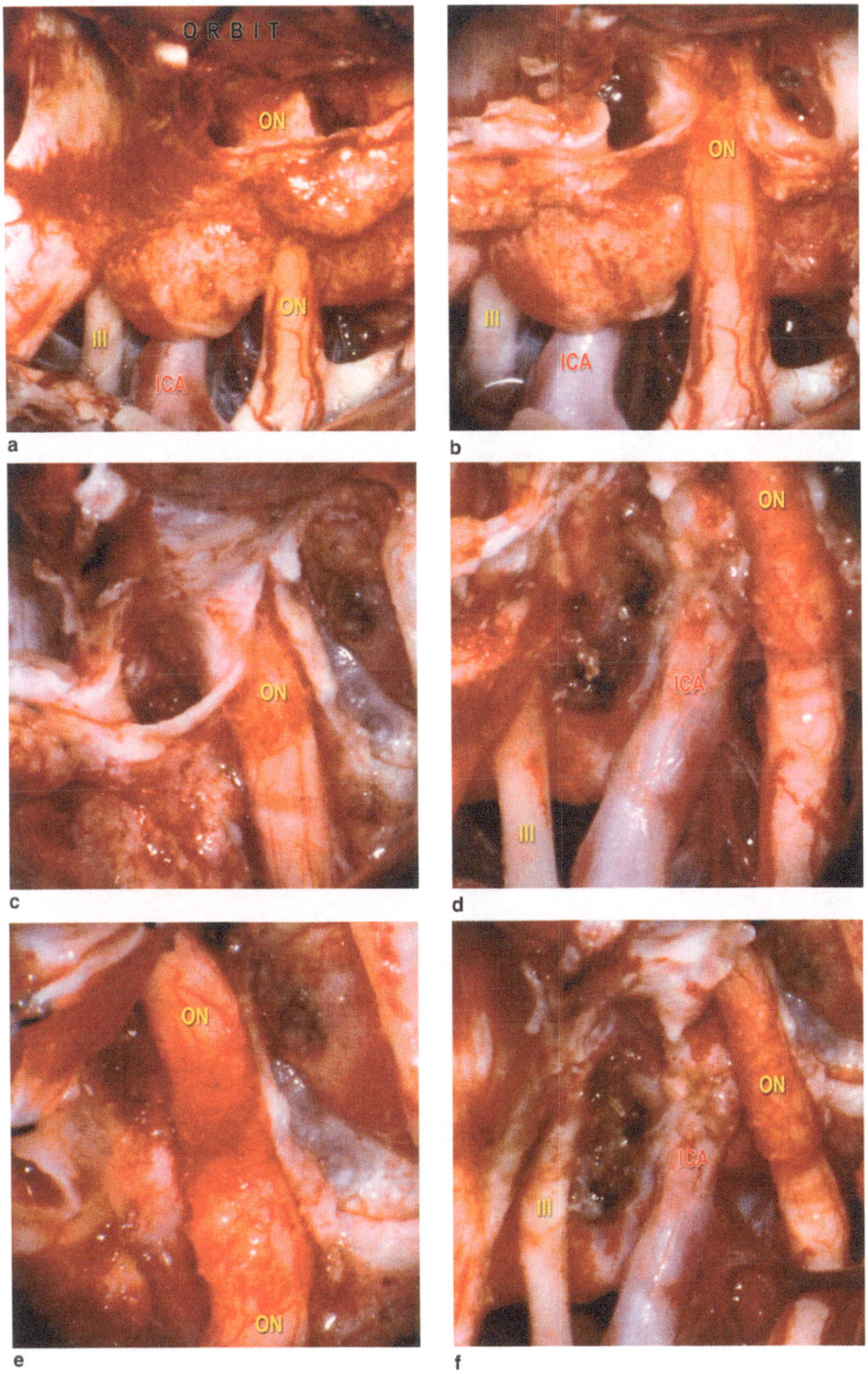

Fig. 4.28. An intraoperative photograph (a) of a meningioma invading the left optic canal (*OC*) in the patient presented in Fig. 4.27. The OC has been opened, and dura propria surrounding the optic nerve (*ON*) exposed, with tumor encircling the intradural ON at its point of entry into the OC. In (b), large piece of tumor is apparent between the intradural ON and the ICA laterally. In (c), dura propria has been incised parallel to the longitudinal axis of the ON, exposing tumor medial and lateral to the ON within the OC and circumferentially around the intradural ON. The photograph in (d) shows that tumor has been resected from the intradural ON and also from the OC, and from the anterior loop (*AL*) of the ICA in the antero-medial triangle. In (e) and (f) all intradural and OC tumor has been resected. Note: Photographs (e) and (f) indicate that the external surface of the intradural ON is of normal appearance, while at its site of the entry into the OC the ON is hyperemic and rough. These pathological surface changes are even more pronounced in the mid-portion of the ON dissected from the OC, while toward the peripheral end of the OC and at the apex of the orbit the ON surface (and its vasculature) once again becomes normal in appearance. Mucous membrane of the sphenoid sinus medial to the ON (e and f) is seen

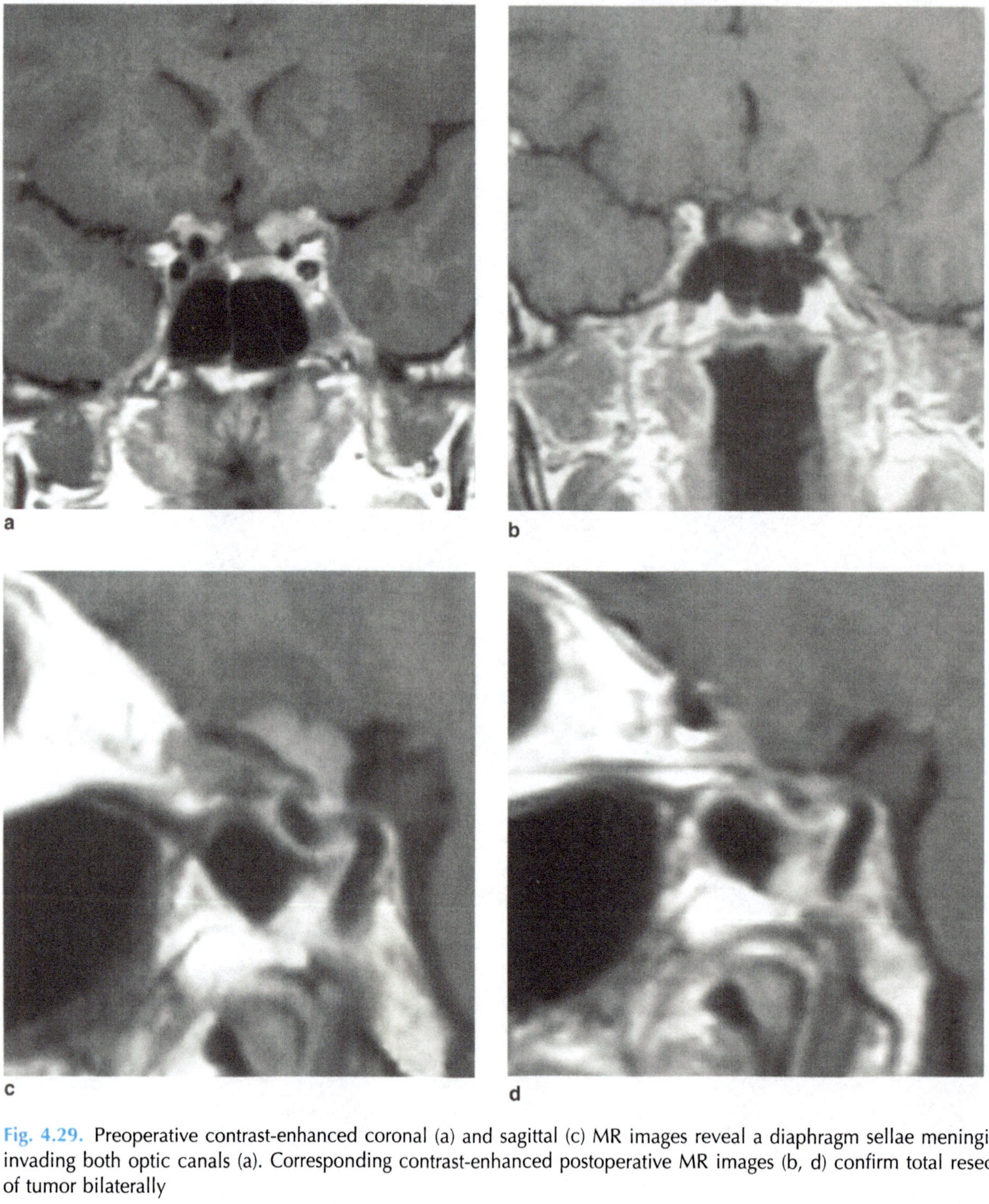

Fig. 4.29. Preoperative contrast-enhanced coronal (a) and sagittal (c) MR images reveal a diaphragm sellae meningioma invading both optic canals (a). Corresponding contrast-enhanced postoperative MR images (b, d) confirm total resection of tumor bilaterally

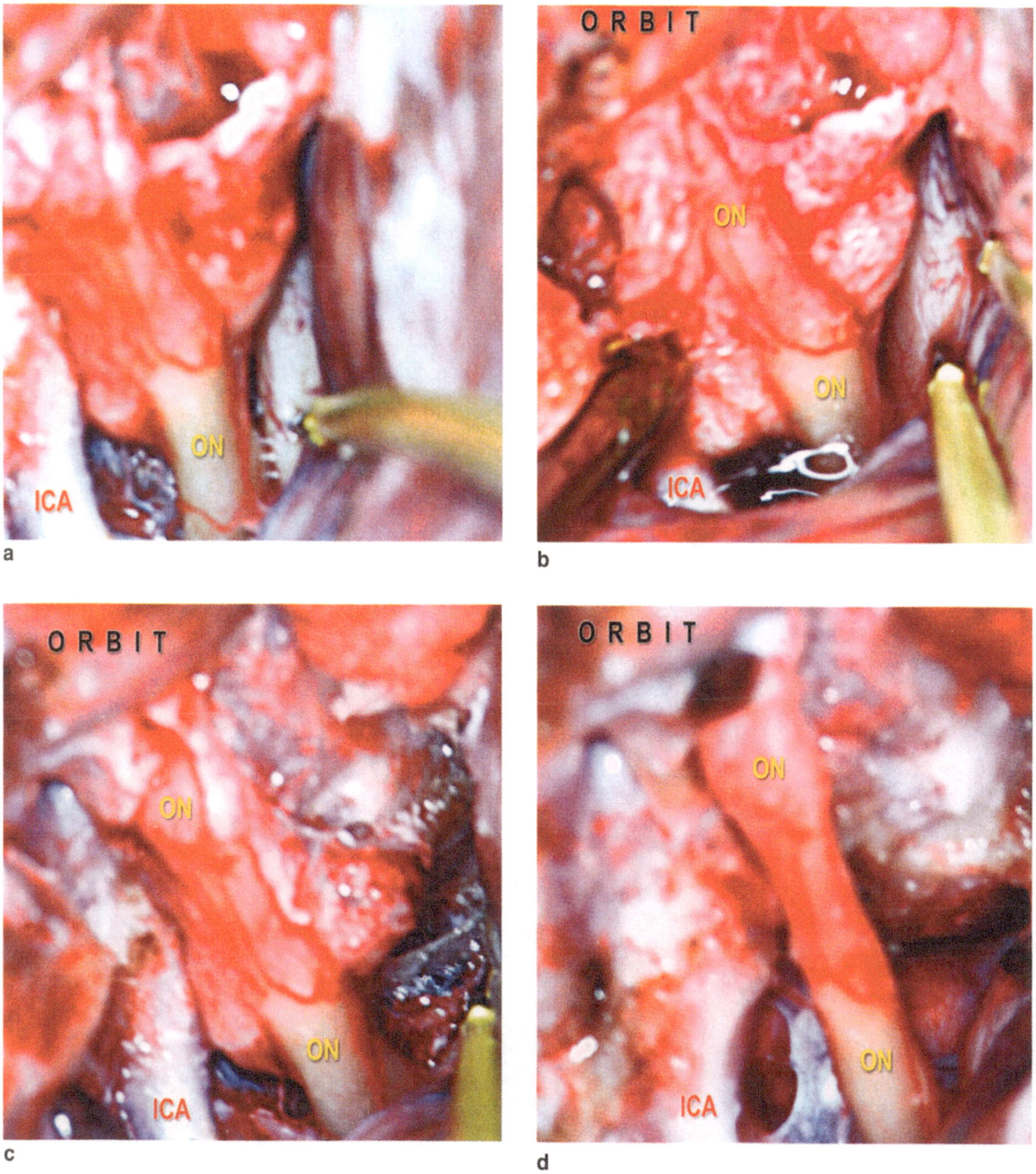

Fig. 4.30. An intraoperative photograph (a) of the diaphragm sellae meningioma invading the left optic canal (*OC*) in the patient presented in Fig. 4.29. The OC has been opened, and meningioma surrounding the optic nerve (*ON*) is apparent. The adjacent intradural ON is of normal appearance, and the extradural ON (within the OC) remains covered by dura propria. In (b), tumor surrounding the intradural ON and dura propria has been divided along the ON, exposing firm circumferential attachments to the ON throughout the OC. The photograph in (c) demonstrates that tumor has been partially resected from the lateral surface of the ON and that the ON peripheral to the tumor (within the OC) is of near normal appearance. In (d) all intradural meningioma and that surrounding the ON within the OC has been resected entirely. Note: The surface of the intradural ON appears normal, while that of the segment of the ON previously incased by tumor is clearly attenuated, regaining a more normal appearance at the peripheral end in the OC and at the apex of the orbit where the nerve was not affected by tumor

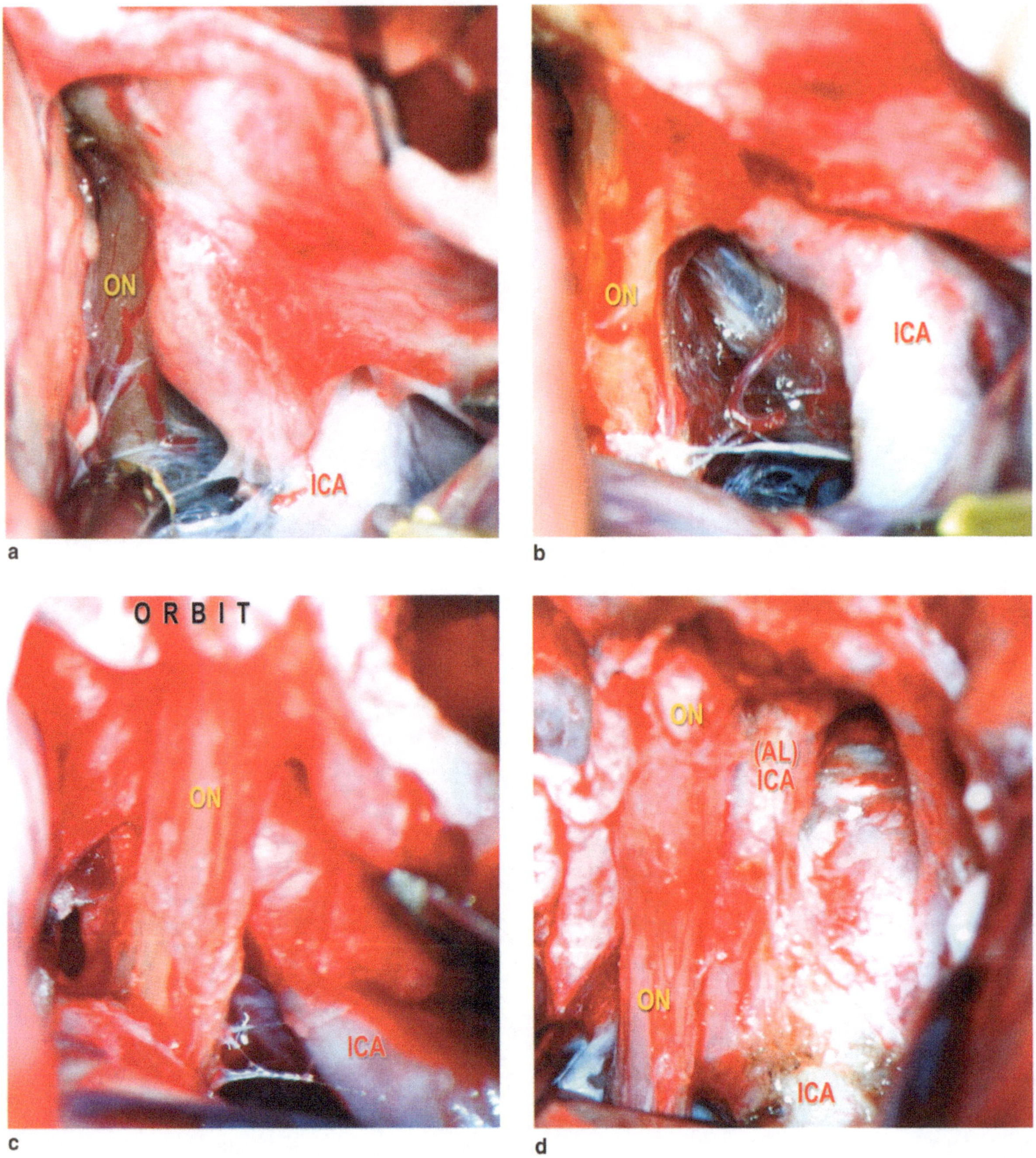

Fig. 4.31. An intraoperative photograph (a) of a meningioma invading the right optic canal (*OC*) in the patient presented in Fig. 4.29. The intradural ON is obviously degenerated. In (b), the intradural component of the tumor between the ON and the ICA laterally has been resected. In (c), the intradural ON and the ICA are now free of tumor, which had been resected from dura between these two structures, at the site the ON's entry into the OC, from the diaphragm sellae, and from beneath the ON. Dura propria was incised along the longitudinal axis of the ON, facilitating additional tumor resection. Notice that, in (d), despite blindness in the right eye, the ON continues to appear relatively healthy throughout the OC with nutritive arteries nicely aligned along its longitudinal axis. All intradural and OC tumor was resected. The ON surface throughout the OC segment has an abnormal appearance, where it had been encased by tumor. Invaded dura propria on the lateral surface of the ON throughout the OC was resected completely, and the distal leg of the AL of the ICA was exposed circumferentially. Note: Despite complete resection of meningioma surrounding the OC segment of the ON, and its relatively normal external appearance, vision did not improve within six months after surgery

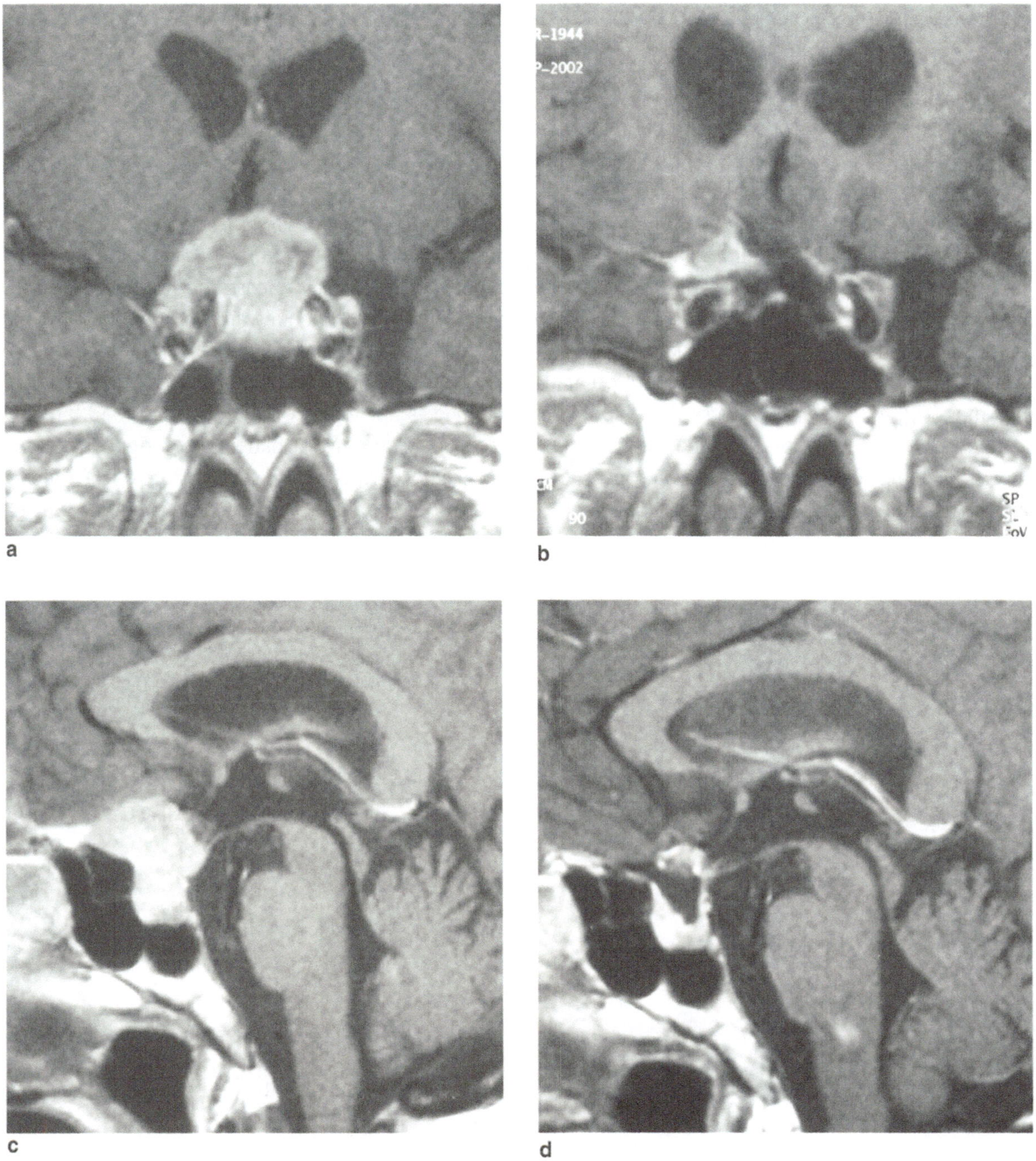

Fig. 4.32. Preoperative coronal (a) and sagittal (c) contrast-enhanced MR images reveal a meningioma of the diaphragm sellae within the sella, extending across the tuberculum sellae, invading both optic canals (*OCs*). Corresponding post-operative coronal (b) and sagittal (d) contrast-enhanced MR images demonstrate total resection of the tumor. Note: This patient remained blind for two months following surgery. In the third postoperative month a trace of vision returned. Within an additional two months the patient regained enough vision to make out fingers at 20 cm

Posterior clinoid process, dorsum sellae, and upper clival meningiomas

Pathological anatomy, clinical presentation, and diagnostic work-up

Relative to anterior clinoid process (ACP) meningiomas the number of lesions arising adjacent to the posterior clinoid process (PCP) is quite small, but their potential to spread in virtually any direction carries important implications for surrounding neural and vascular structures and neighboring parasellar space (PS) compartments. PCP meningiomas spreading laterally in the oculomotor trigone toward nerve III's site of entry into the PS may compress nerve III intradurally (against the sharp dural edge), along its course between the leaves of the lateral PS wall, or both. Those extending upward Figs. 4.33 and 4.35 may encase the ICA, the posterior communicating artery (PCom), and/or the anterior choroidal artery. Growth medialward and upward over the dorsum sellae may involve the pituitary stalk (PSt), while anterior extension is likely to displace the optic nerve (ON), even encase it and spread into the optic canal (OC). Lesions growing caudally become upper clival meningiomas (Fig. 4.34). Some traverse the diaphragm sellae (Fig. 4.35) into the sella and/or spread laterally to invade the PS through nerve III's site of entry or directly through dura. Most patients experience diplopia from nerve III compression, but other symptoms may occur, vague retro-orbital discomfort, even headache, and visual deterioration most common among them but also other clinical phenomena depending on the neural structures affected. Diagnosis is established through ophthalmological testing with CT and MR imaging.

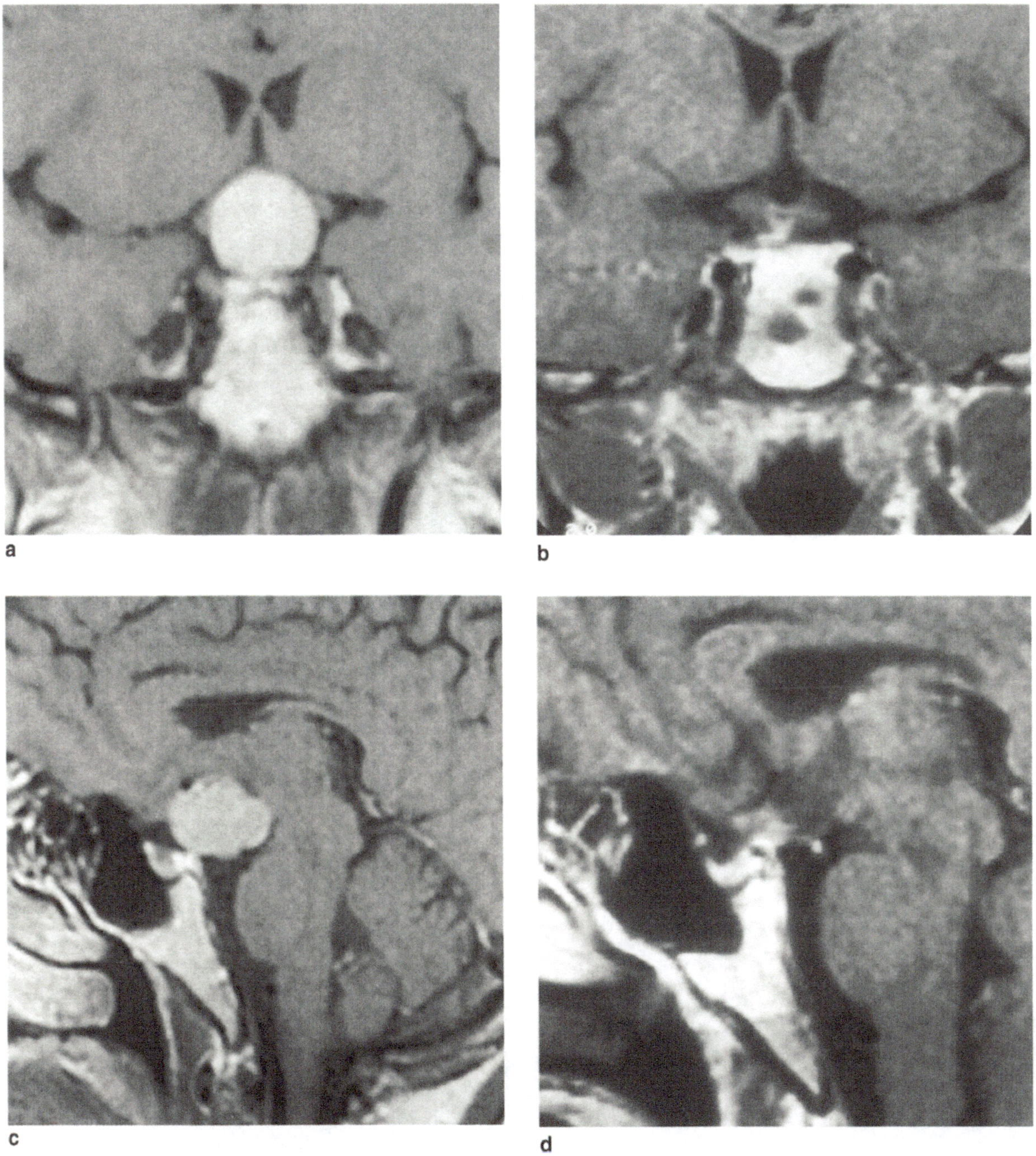

Fig. 4.33. Preoperative contrast-enhanced coronal (a) and sagittal (c) MR images demonstrate a typical dorsum sellae meningioma. Corresponding postoperative contrast-enhanced MR images, coronal (b) and sagittal (d), confirm its complete resection

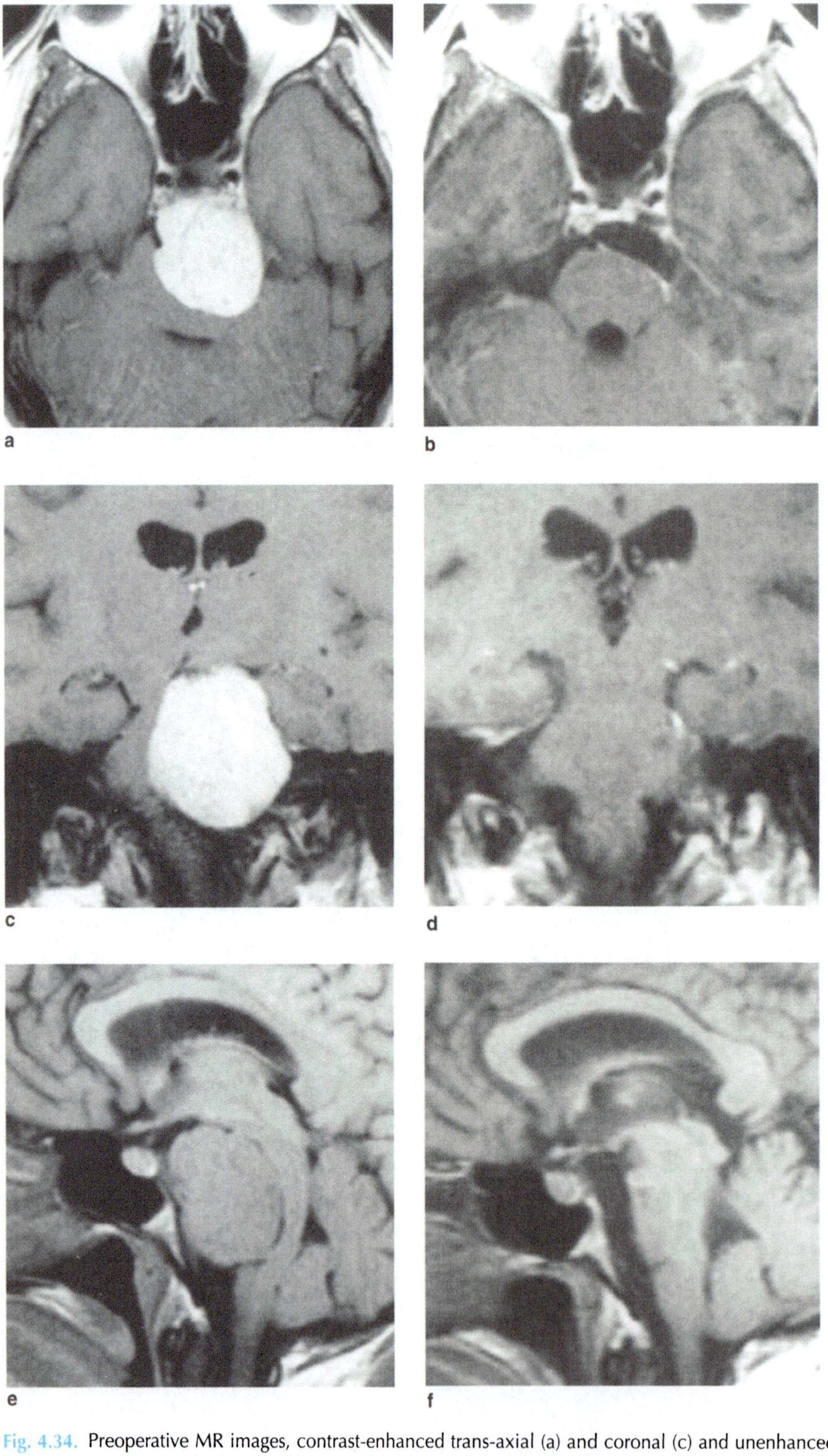

Fig. 4.34. Preoperative MR images, contrast-enhanced trans-axial (a) and coronal (c) and unenhanced sagittal (e), show a large meningioma attached to the dorsum sellae, the left posterior clinoid process (*PCP*), and upper clivus which severely compresses the brainstem (a, c). Corresponding postoperative contrast-enhanced images, trans-axial (b), coronal (d), and sagittal (f), demonstrate total resection of the lesion, which was achieved by the approach described in Chapter 2. Note, in (b), that the IV-th ventricle, compressed and displaced in (a), has regained its normal shape and size following surgery. The pons "re-expanded" after tumor resection and, importantly, there is no evidence of left temporal lobe injury

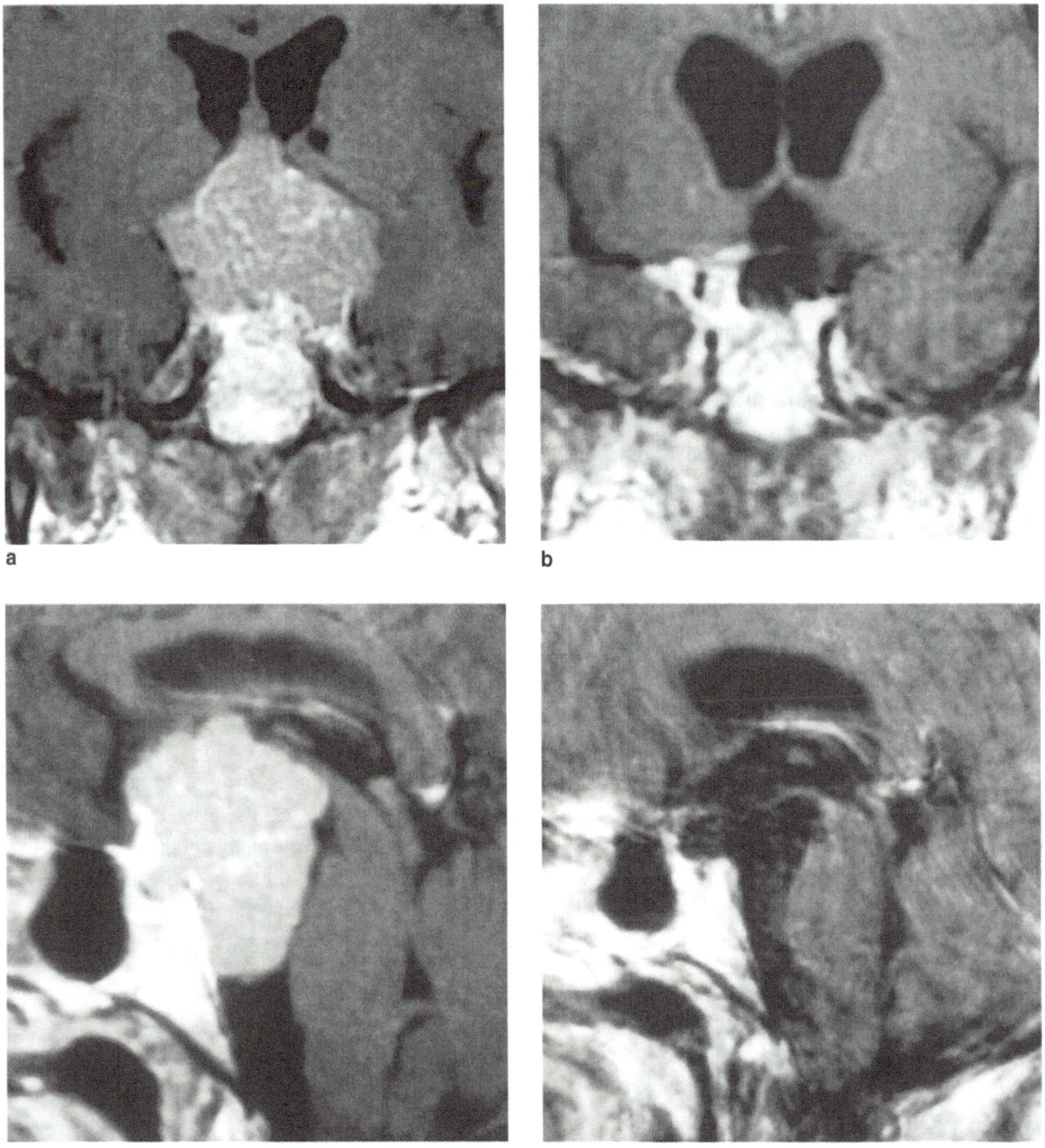

Fig. 4.35. Preoperative contrast-enhanced MR images, coronal (a) and sagittal (c), demonstrate a fairly large dorsum sellae meningioma extending caudally to the upper clival region and anteriorly over the diaphragm sellae into the sella, even farther to the tuberculum sellae (c). Corresponding postoperative images, coronal (b) and sagittal (d), demonstrate complete resection of the lesion. Note: Despite the size of meningioma – from left to right in (a) and from front to back in (c) – this tumor was resected totally via a unilateral approach

Surgical approach

Chapter 2 describes the techniques for patient positioning; osteoplastic craniotomy; unroofing the orbit; resecting the sphenoid wing, the ACP, and the walls of the OC (including the optic strut); also transecting dura at the superior orbital fissure (SOF) and splitting the leaves of the lateral wall of the PS. Only when all extradural bone work has been completed, the anterior half of the lateral PS wall divided into separate layers and nerves III, V1 and V2 exposed, may dura be incised along the longitudinal axis of the sylvian fissure centralward toward the distal (dural) ring (DR) of the ICA.

The sylvian fissure is divided throughout its entire length, allowing exposure of the ICA bifurcation and ICA branches (PCom and anterior choroidal artery) which are draped over the dorsomedial surface of the tumor. Nerve III is exposed laterally, where it enters the oculomotor canal, then dissected from tumor and mobilized slightly laterally, providing access to involved dura over the diaphragm sellae anterolateral to the PCP, which is coagulated. Dura dorsal to the PCP is incised, and the PCP is drilled away flush with the sellar floor, including the dorsum sellae to its contralateral surface or until encountering normal bone and dura. Bleeding typically occurs from venous sinus adjacent to the posterior sellar wall and the ipsilateral PS. These venous channels may occluded with Surgicel as necessary. Tumor-infiltrated dura is resected, then tumor adjacent to nerve III, the ICA and its branches, the diaphragm sellae, the ON, finally from the OC. After resecting all tumor and diseased dura, complete hemostasis is established and dura is closed in a watertight manner. Any dural defect in the vicinity of the DR of the ICA is covered with commercially available dural substitute, which is sutured to healthy dura and coated with fibrin glue.

Patients and results

PCP and dorsum sellae lesions account for fewer than 1% of meningiomas in the entire series. All but one presented as virgin tumors. Each was totally resected, along with adjacent infiltrated dura. Tumors were dissected from the ONs, the optic chiasm and the optic tracts, nerves III and IV, the ICA, the PCA, the PCom, Dawson's arteries, and the anterior choroidal artery, resulting in no new neurological deficits. Preoperative neurological symptoms and signs generally resolved shortly after surgery. The only deficits present after surgery were associated with two virgin lesions and the only tumor previously operated upon, all three partial nerve III paresis, which in every case was less severe than prior to surgery.

Clival meningiomas

Introduction

The clival region likely represents a destination of spread, rather than the true site of origin of most so-called clival meningiomas. Undoubtedly these tumors originate adjacent to the clivus, namely from the dorsum sellae, the posterior clinoid process (PCP), the parasellar space (PS), the petroclival region, and the lateral tentorial edge, but pure clival meningiomas are probably quite rare. Surgeon-authors' time-honored penchant for classifying them as of clival origin probably reflects a desire to pass along important pearls and bits of wisdom for the benefit of their colleagues. Since the clival region, like the sella, lies at the very center of the skull base, surgical approaches to it are demanding and success is clearly experience-dependent.

Relevant anatomy

The inferomedial triangle, the inferolateral (trigeminal) triangle, and, in its broad dimension, the cerebellopontine angle (CPA) are the most important paired landmarks of the paraclival region. Clearly they are the sites of origin of most clival meningiomas. The inferomedial triangle, circumscribed by nerve VI's point of entering dura (medially), the site of nerve IV's dural entry (laterally), and the PCP define a paraclival window through which the surgeon may best access the intradural clival region. The structures of primary importance (possible to expose both extradurally and intradurally) are the petroclinoid ligament and Dorello's space with nerve VI coursing through it.

The inferolateral (trigeminal) triangle, with its two sub-triangles (tentorial and osseous), may be exposed safely via the transpetrous approach rostral to the internal auditory canal (IAC). The inferolateral triangle provides access to extradural paraclival space by traversing its dural-tentorial subtriangle, while its osseous subtriangle defines the surgically safe segment of the petrous apex (i.e., anterior to the IAC). It is easier and safer to access the paraclival intradural space extradurally through the petrous apex (i.e. through the floor of Meckel's cave), than to attempt to approach this area by an intradural route. After resecting the petrous apex as described in Chapter 2, the clivus lies caudally and nerve VI's point of dural entry medially.

The larger CPA region, on the other hand, since it is immediately caudal and medial to the IAC and the jugular foramen, is more appropriately exposed by the suboccipital intradural approach.

Pathological anatomy, clinical presentation, and diagnostic work-up

Any attempt to divide the clivus into surgically relevant segments must respect the relationship between the sites of cranial nerves' traversing dura and extradural bone structures. Such an attempt should also define feasible approaches to the region's pathologies that minimize surgical destruction of the central skull base (CSB). Considering neural, dural, and bony structures, it seems appropriate to divide this region into the upper clivus, lower clivus, and foramen magnum subregions. Tumors adjacent to the upper clivus lie medial to the inferomedial triangle, while lesions adjacent to the lower clivus are located medial to the inferolateral triangle, including tumors of the CPA region above the level of nerve XII's dural entry.

The most common tumors of the upper clival subregion are meningiomas extending from the dorsum sellae, the PCP, and the PS. Lesions of the lower clival subregion are in most cases the so-called petroclival meningiomas. Generally they originate from meninges in the vicinity of Meckel's cave, spreading primarily toward the midline, even across it. Many such lesions extend caudally into the CPA subregion and into the IAC. Some paraclival meningiomas have wide dural attachments ("en plaque" lesions) while those with globular shapes have more limited dural involvement. A given tumor's size and shape dictates the vulnerability of adjacent neural and vascular structures, the likely source of symptoms and signs.

Brainstem compression may cause pyramidal tract signs in the absence of sensory loss while encasement of cranial nerves results in diplopia (nerves III, IV, VI), pain or sensory loss in the distribution of nerve V, also diminished hearing and/or facial palsy when nerves VII and VIII are compressed and/or stretched from the brainstem to the IAC or inside the IAC. Lower cranial nerve symptoms and signs result from compression or stretching of nerves IX through XII. Displacement of important structures, even the basilar artery (BA) and its branches, may be tolerated for fairly long periods of time. Early clinical manifestations of clival meningiomas typically appear where cranial nerves are tethered at their sites of traversing dura. They very rarely result from brainstem compression or BA displacement. Often symptoms and signs do not occur until a paraclival meningioma is quite large, as seen in Fig. 4.36.

The cornerstone of diagnosis remains CT and MRI. CT is particularly important to study the nature of skull base bone elements, especially the presence of pneumatization, which announces an increased risk of postoperative cerebrospinal fluid (CSF) leakage. Digital subtraction angiography (DSA) is useful to confirm the state and position of the BA and its branches. Any test of specific cranial nerve function is helpful not only preoperatively but for studying function following surgery.

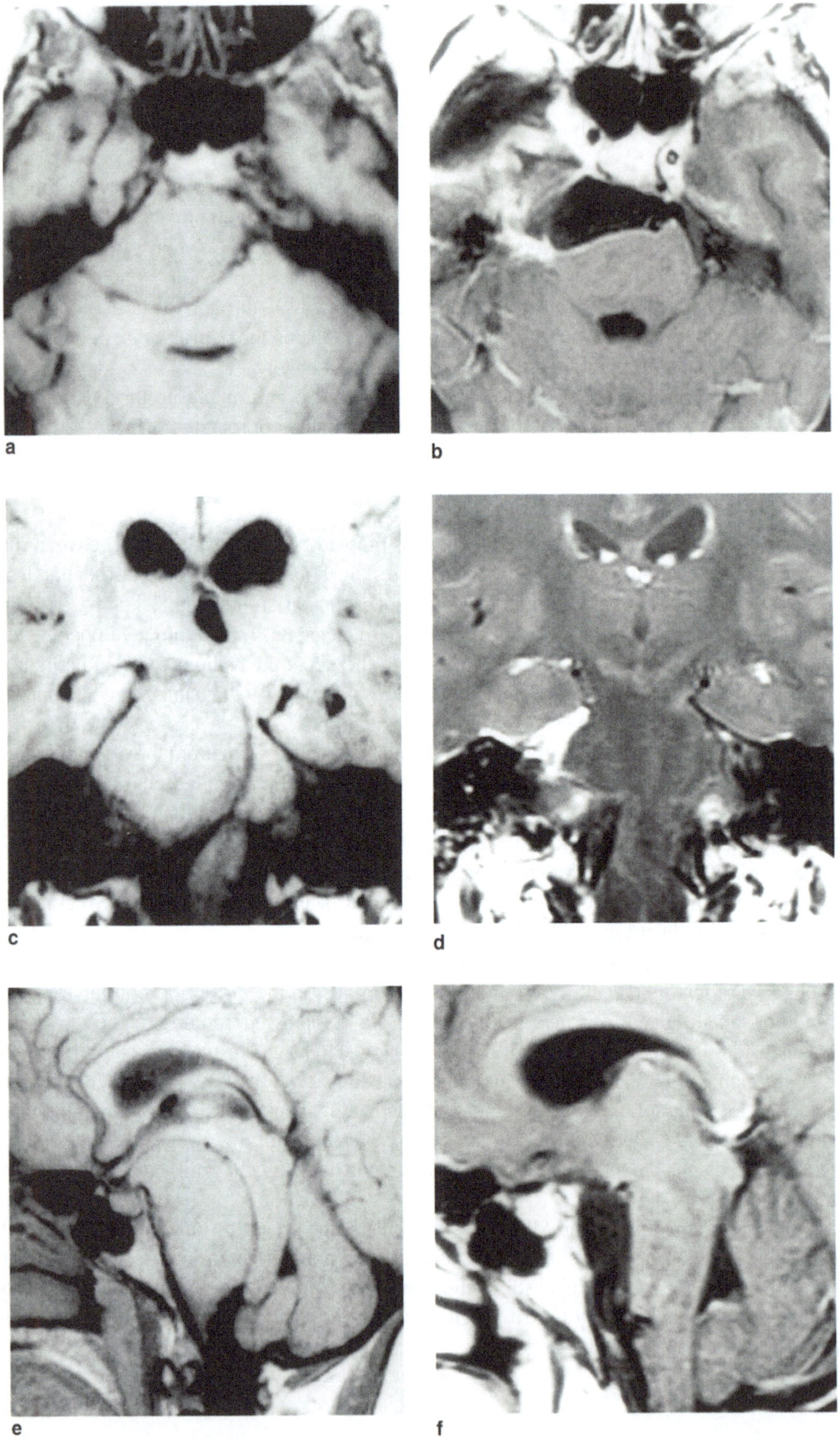

Fig. 4.36. Preoperative trans-axial (a), coronal (c), and sagittal (e) contrast-enhanced MR images reveal a huge meningioma arising from the right petroclival region. Note the brainstem compression and displacement in each cut, also the distorted fourth ventricle in (a). This tumor was removed in a single stage but required all elements of the general approach described in Chapter 2. Postoperative trans-axial (b), coronal (d), and sagittal (f) contrast-enhanced MR images confirm total resection of the lesion. Note that the brainstem returned to a more normal shape while the fourth ventricle remains slightly enlarged. The patient recovered nicely and was discharged from the hospital free of neurological deficit on the tenth postoperative day

Surgical approaches to clival meningiomas

Over the past two decades a number of surgical approaches to the clival region have resulted, owing largely to the explosion of interest in skull base surgery by various surgical subspecialties, including neurosurgery, ENT, and maxillofacial surgery. Ultimately, "surgical destruction" of the skull base must be justified by cost-benefit analysis, the value of tumor resection balanced against the cost of permanently altering adjacent neural and vascular structures. Our goal has always been to resect the lesion without creating permanent additional neurological deficits. Surgical approaches to the CSB must be tailored and executed in a manner which assures that all neural, arterial, and venous structures are preserved intact. This is particularly important and challenging for clival meningiomas.

Upper clival meningiomas

Patient positioning; scalp incision; designing and constructing an osteomuscular flap; unroofing the orbit; and resecting the sphenoid wing, the anterior clinoid process (ACP), and the superior, lateral, and inferior walls of the optic canal (OC) are all components of the general extradural approach to the CSB as described in Chapter 2. Additional steps with upper clival meningiomas depend on the size and type of tumor present.

Meningiomas of the diaphragm sellae, dorsum sellae, and PCP extending to the upper clivus require resection of the PCP and dorsum sellae to the floor of the sella. Tumors extending further caudally and those involving the medial PS wall require excision of infiltrated dura throughout the inferomedial triangle region, including the tentorial edge (from the points of entry of nerves III, IV, and VI). Tumor elements involving the PS and sella are resected as described previously.

Lower clival meningiomas

This approach is identical to that described previously for petroclival meningiomas. Pure petroclival lesions do not require removal of anterior bone structures – the lesser sphenoid wing, the ACP, and walls of the OC. If, however, tumor extends into the PS, only when the entire PS is opened and the petrous apex is resected may such lesions be extirpated totally. Nerve VI is the key neural structure lateral to the clivus. It must be exposed within the posterior fossa intradural space and dissected throughout Dorello's space into the PS. Once nerve VI is fully freed from the lesion, clival dura and bone must be resected until normal clival dura is encountered, sometimes well beyond the midline.

Cerebello-pontine angle and lower clival meningiomas

Lower clival meningiomas smaller than 3 cm in diameter (not extending into the middle cranial fossa) may be resected via the suboccipital intradural approach. The patient is positioned in the lateral park-bench position as for more common CPA tumors. A linear, retro-mastoid incision is made from the occipital region across the transverse sinus, exposing suboccipital muscles. Craniotomy extends from above the transverse sinus and below it sufficiently to expose the initial part of the sigmoid sinus. The dural incision extends from the transverse sinus obliquely inferiorly, toward the sigmoid sinus. At the midpoint of the dural opening, a second incision should be angled toward the junction of the transverse and sigmoid sinuses. The triangular dural flap based on the sigmoid sinus is reflected laterally, and the triangular flap based on the lower border of the transverse sinus is reflected superiorly. The tentorium is visualized from the transverse sinus and to the tentorial incisura at the ambiens cistern.

Tumor is detached from bone and from the under surface of the tentorium. All tumor-infiltrated tentorium must be removed with the mass. Nerves VII-XII typically are stretched over the dorso-lateral surface of these lesions. After identifying associated cranial nerves, the tumor may be gently detached from dura as it is debulked sequentially. Special attention should be paid to nerves IV and V, which are generally displaced and stretched upward (e.g., dorso-laterally). In most cases, dissecting these nerves from the tumor does not constitute a major problem. Isolating nerve VI from tumor, on the other hand, may be quite difficult. Dissecting nerve VI toward the brainstem is more easily accomplished when it is identified as it enters Dorello's space. After completely detaching the tumor from clival dura, it may be separated (dissected or resected) from the brainstem and the BA. It is important to remove a clival meningiomas piece by piece since brainstem perforators may be firmly adherent to its mass. Hasty mobilization of the tumor or aggressive coagulation may result in serious permanent deficits.

Staged resections

Particularly large clival meningiomas may be resected in two or more stages. For unilateral tumors extending the entire length of the clivus, the larger element, which generally displaces and stretches more neural structures, should be attacked in the initial stage, leaving the remainder to be addressed later. Since even a small remnant may grow to an unmanageable size, the second stage should not be unduly postponed, certainly not for years. This is also true for large bilateral clival tumors attached to the entire width of the clivus, whether predominately upper or lower. Any patient receiving radiosurgery for residual tumor must be followed closely with MRI.

Patients and results

Tumor resectability and outcome are summarized in Table 15. There were five patients with upper clival meningiomas, seven with lower lesions, and ten with tumors involving both the upper and lower clivus. Twenty-two patients were treated overall, 18 with tumors not previously operated on and four with residual lesions previously treated surgically elsewhere (Fig. 4.37). Total resection was achieved in nine of the virgin tumors. The remainder of the virgin lesions and all four of those residual from previous operations underwent gross-total resections.

All five upper clival meningiomas, six lower clival lesions, and five extending the entire length of the clivus were resected via the transpetrous trans-PS approach (A). One patient with a lower clival meningioma and three with tumors extending the length of the clivus required the suboccipital-retrosigmoid approach (B). Two patients with particularly large tumors, operated initially via the transpetrous approach, required suboccipital-retrosigmoid operations as a second stage (C).

All patients presented with symptoms and/or signs of cranial nerve compression. In addition, 13 had clinical manifestations of brainstem involvement. Seven patients experienced preoperative cranial nerve deficits which failed to clear following surgery, and an additional seven had other persisting symptoms and signs.

Table 15. Clival meningiomas: surgical approach, resectability, and outcome

Meningioma location	N° of cases		Approaches						Resectability				Deficits							
	Virgin												Pre-operative				Post-operative			
	Residual		A		B		C		Complete		Gross-total		Nerve(s)		Other		Nerve(s)		Other	
Upper clivus	4	18%	4	100%	–	–	–	–	3	75%	1	25%	4	100%	1	25%	–	–	–	–
	1	5%	1	100%	–	–	–	–	–	–	1	110%	1	100%	1	100%	1	100%	1	100%
Lower clivus	6	27%	5	83%	1	17%	–	–	3	50%	3	50%	6	100%	3	50%	1	17%	2	33%
	1	5%	1	100%	–	–	–	–	–	–	1	100%	1	100%	1	100%	1	100%	1	100%
Combined	8	36%	5	63%	2	25%	1	13%	3	38%	5	63%	8	100%	5	63%	2	25%	2	25%
	2	9%	–	–	1	50%	1	50%	–	–	2	100%	2	100%	2	100%	1	50%	1	50%
Collective	18	82%	14	78%	3	17%	1	6%	9	50%	9	50%	18	100%	9	50%	4	22%	4	22%
	4	18%	2	50%	1	25%	1	25%	–	–	4	100%	4	100%	4	100%	3	75%	3	75%
Total	22	100%	16	73%	4	18%	2	9%	9	41%	13	59%	22	100%	13	59%	7	32%	7	32%

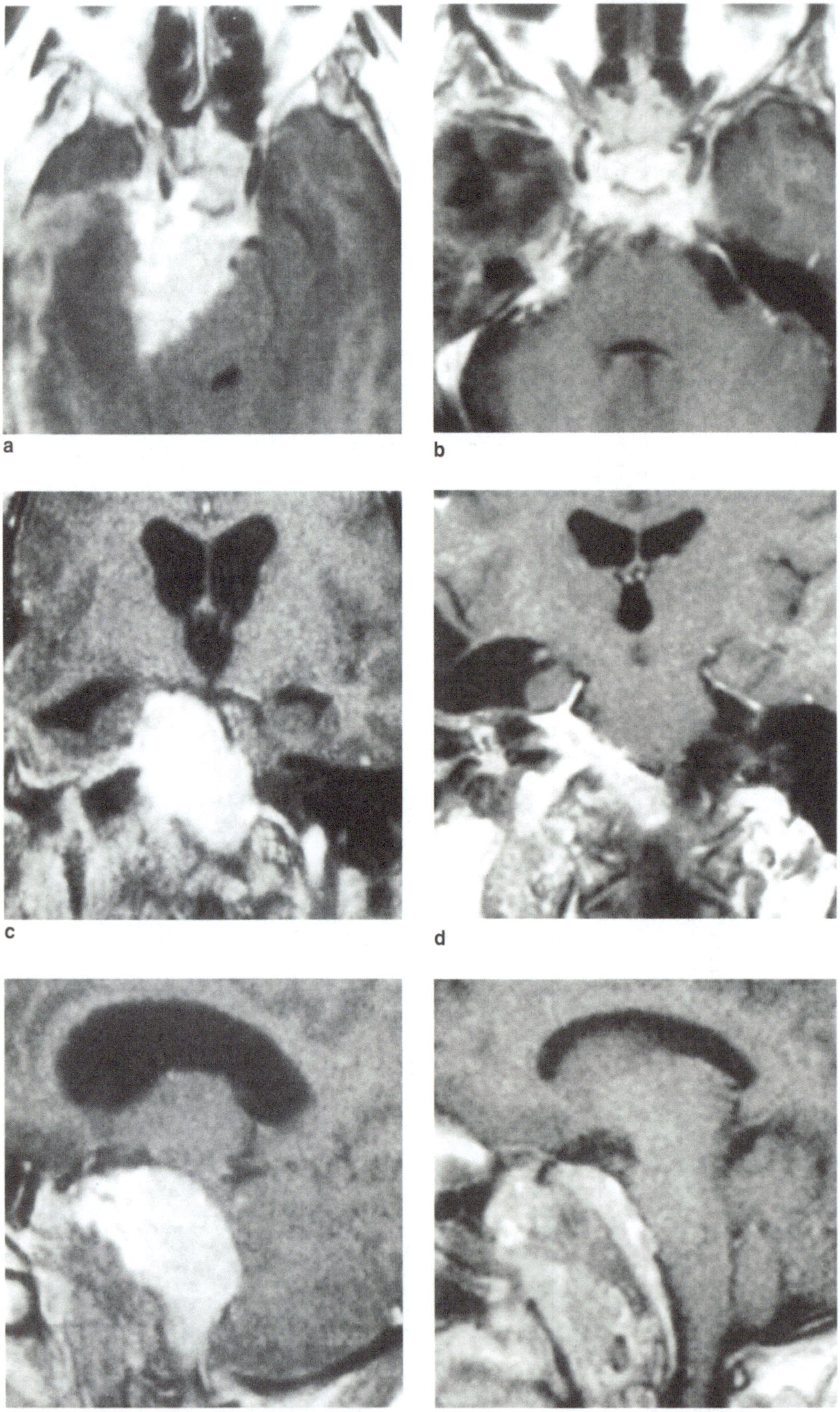

a b c d e f

Fig. 4.37. Preoperative trans-axial (a), coronal (c), and sagittal (e) contrast-enhanced MR images reveal a large right-sided petroclival meningioma residual from previous surgery. Note: The brainstem is displaced to the left, the basilar artery is encased, and tumor extends across the entire width of the clivus (a, c), also from the dorsum sellae to the cranio-cervical junction (e). The right temporal lobe and ipsilateral nerves II through VI were severely damaged at the first operation. Following a second operation, trans-axial (b), coronal (d), and sagittal (f) contrast-enhanced MR images demonstrate only subtotal resection of the remaining tumor. Note: Although (b), (d), and (f) confirm decompression of the brainstem and long tract deficits disappeared, nerves II through VI did not recover

Large central skull base meningiomas

Pathological anatomy, clinical presentation, and diagnostic work-up

Large central skull base (CSB) meningiomas are lesions involving the parasellar space (PS) unilaterally plus neighboring structures including the medial sphenoid wing, the anterior clinoid process (ACP), tuberculum sellae, upper clivus, and petroclival region, particularly when they extend across the midline. Typically these lesions arise on one side, then spread to the contralateral side Figs. 4.38–4.41. They do not originate at the clivus and extend bilaterally. In exceptional cases, an "en plaque" meningioma arising from the tuberculum sellae, diaphragm sellae, or upper clival region may spread bilaterally and invade each PS. Large CSB meningiomas present with symptoms and signs identical to the PS and adjacent areas meningiomas, but they generally traverse the midline. As with other PS tumors diagnosis is achieved by CT, MRI, angiographic studies plus ophthalmological and neurophysiological testing.

Surgical approach to the large CSB meningiomas

Generally, more complete resections are achieved with greater safety if meningiomas extending across the midline are treated in stages, surgery carried out initially on the side of greater tumor mass. The techniques and procedures described in Chapter 2 apply, specifically as altered for parasellar meningiomas. It is both reasonable and safe to interrupt the resection of bilateral tumors at the midline, especially with lesions involving the middle clival region. With large meningiomas of the upper clivus, it is possible to resect tumor, invaded bone, and infiltrated dura across the midline, but in many cases maximum resections are not possible in a single stage. The extent of contralateral extension indicated on preoperative studies usually allows the surgeon to determine at which point the first stage should be terminated. When significant tumor mass lies across the midline, no forced attempt should be made to remove it in a single stage Figs. 4.38–4.40. It is only prudent to operate again after a short interval. Under no circumstances should the second stage be delayed for years. Postponing it for too long may result in circumstances which have no reasonable solution Fig. 4.38. After surgery MR imaging should be repeated at regular intervals of no longer than six months.

Patients and results

The results of surgery are summarized in Table 16. Almost half the patients in this series harboring large meningiomas extending across upper midline (sella) and lower midline (clivus) had been operated previously elsewhere. Of 15 patients not previously operated on, only two resections were completed in a single stage. Of the remaining 13 with virgin lesions plus 12 with tumors residual from previous operations, additional surgery was required. Among the 13 with virgin tumors, two or three operations were performed. Of the 12 patients with residual lesions, two operations were carried out in six, and three for the remainder. In no patient was total tumor resection achieved. Only subtotal resections were possible in the majority, only four cases resulting in gross-total resections Figs. 4.39–4.41. Of 27 patients with large CSB meningiomas, only two had no neurological deficit prior to surgery. In three instances neurological deficits were worsened by surgery, 19 patients remaining the same postoperatively, and five experiencing complete relief of neurological impairment after surgery.

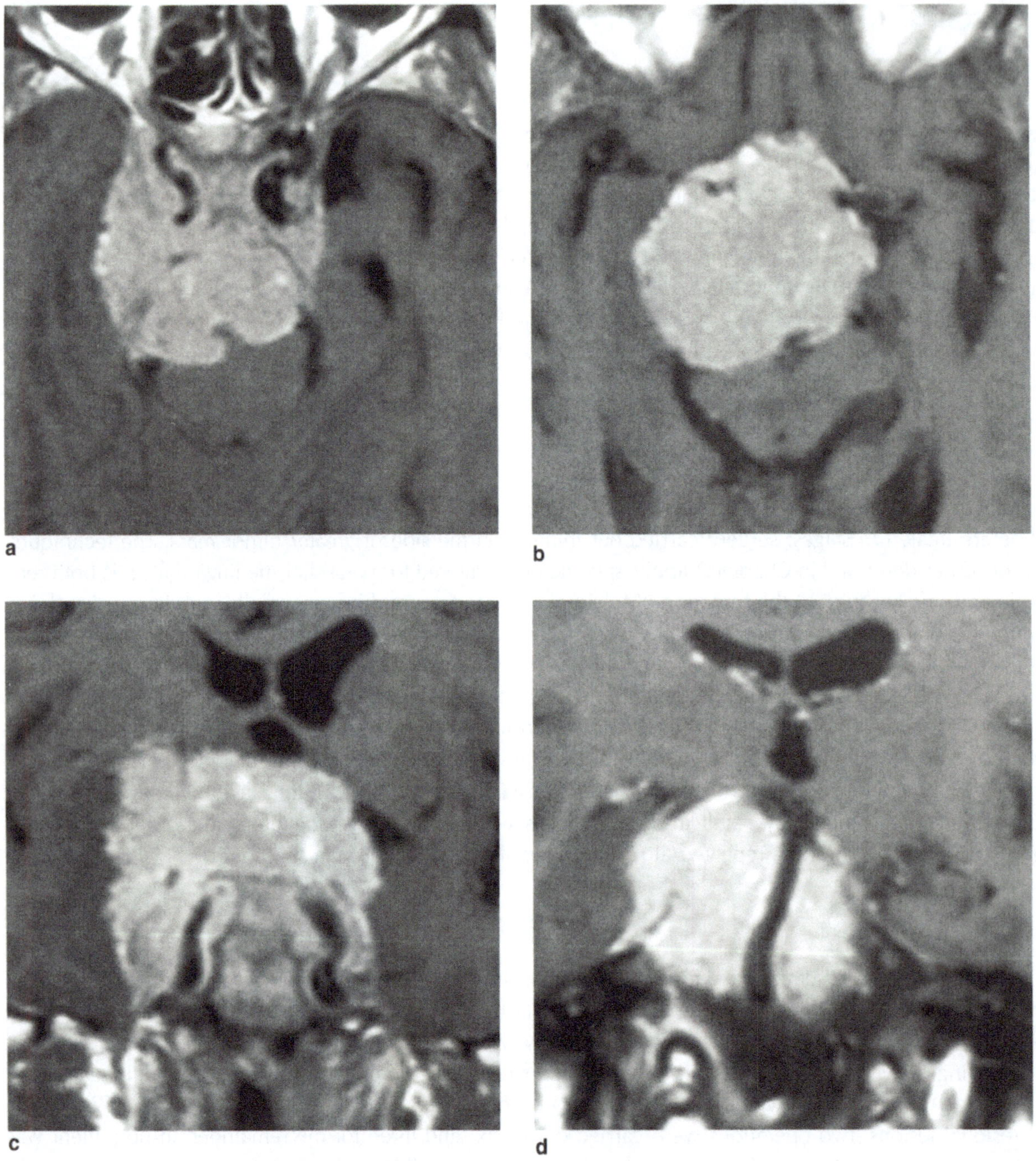

Fig. 4.38. Contrast-enhanced transaxial (a, b) and coronal (c, d) MR images demonstrate a large central skull base meningioma spreading from the right parasellar space (*PS*) and suprasellar and clival regions through the sella to the contralateral PS, invading its medial and dorsal compartments. Note, in (a), that the right PS-ICA is stenotic but not displaced. The narrowed right PS-ICA and tumor lining the tentorial edge indicate the lesion's origin and its contralateral extension. This patient recently presented for evaluation with surprisingly little visual loss and no deficits of nerves III through VI. He has not yet been operated because he is unwilling to accept the necessity of staged procedures and the transient postoperative cranial nerve deficits which seem inevitable

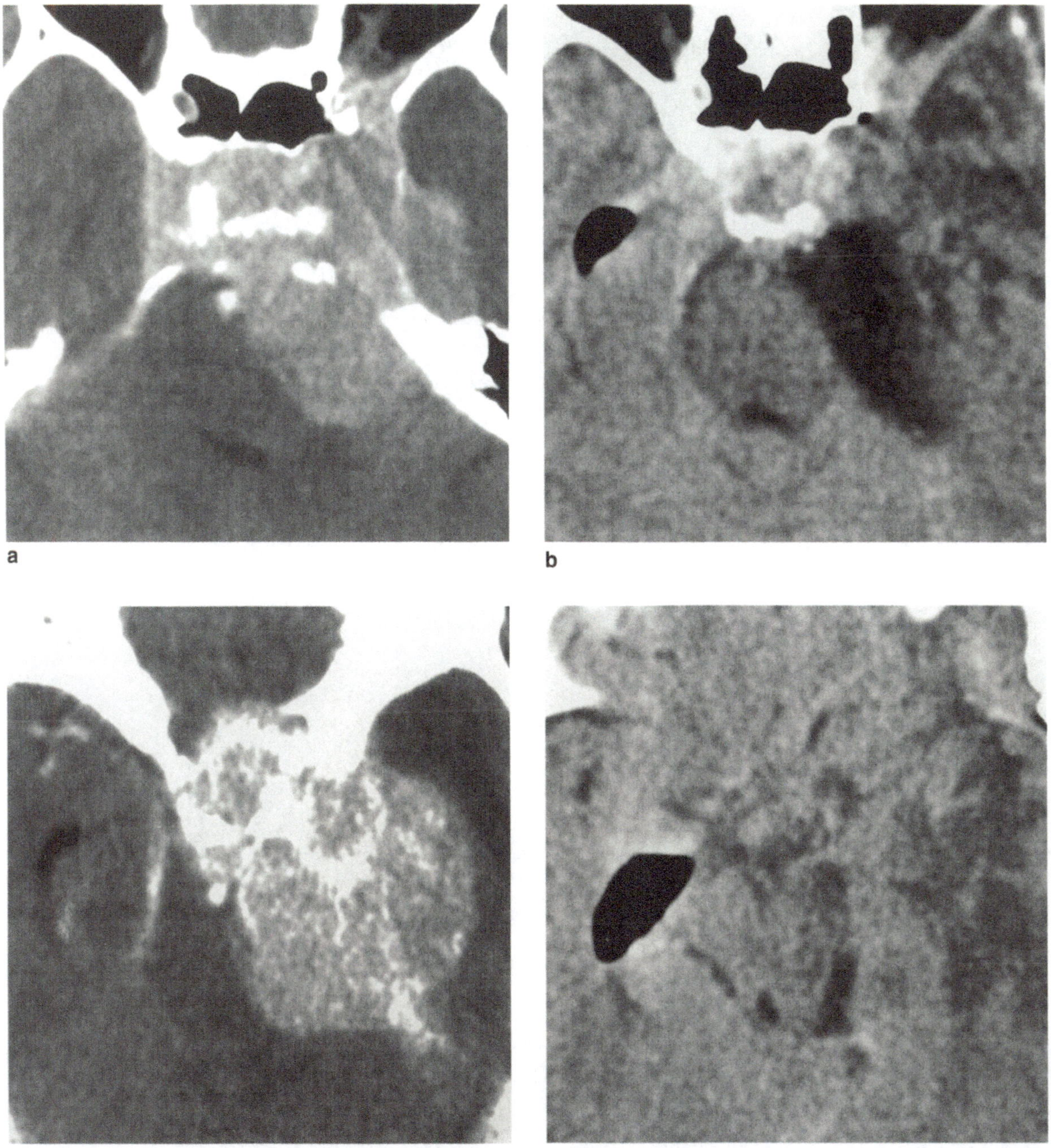

a b c d

Fig. 4.39. Preoperative trans-axial contrast-enhanced CT cuts (a, c) reveal a large left-sided petroclival meningioma involving the ipsilateral parasellar space (*PS*), the sella, and the upper clival region, extending into the contralateral PS. Corresponding postoperative CT images confirm resection of tumor from the left side (b, d). Note hyperostosis in (a) and (c) and the free particle of bone. Remaining tumor will be removed in a second stage, from the right side. Transaxial cuts demonstrate the extent of tumor invasion of the right PS at different levels

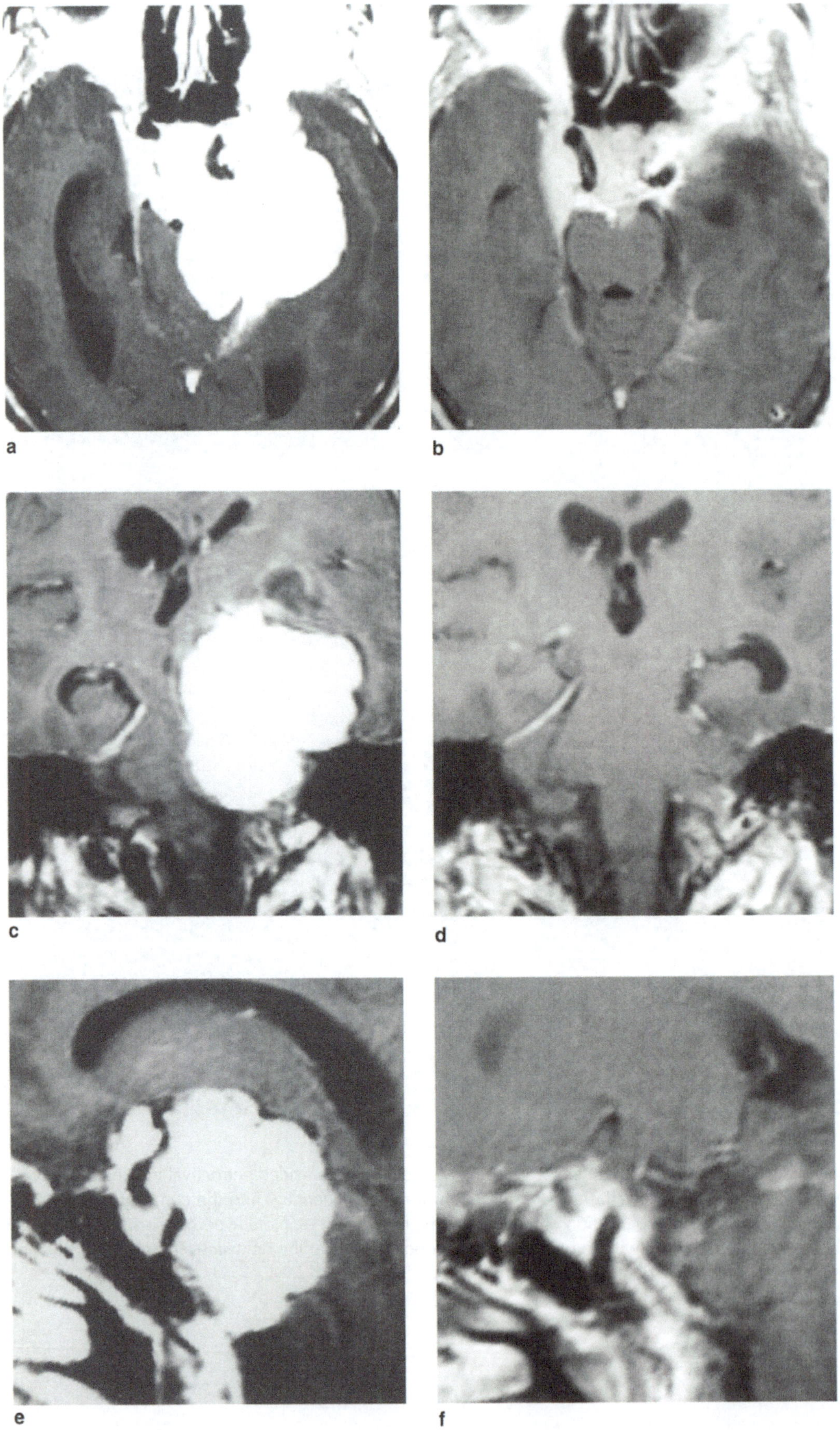

Fig. 4.40. Preoperative contrast-enhanced transaxial (a), coronal (c), and sagittal (e) MR images demonstrate a large central skull base (*CSB*) meningioma on the left which extends through the sella and suprasellar region along the upper clivus into the contralateral parasellar space (*PS*). Corresponding postoperative MR images (b, d, f) confirm that tumor has been removed from the left side. Incomplete paresis of nerves III and IV, the only postoperative deficits, cleared within three months. Tumor remaining in the upper clival region, inside the sella, and in the right PS (b) will be removed in a second operation at the patient's discretion, hopefully soon! Note: This is the same patient as presented with CT scans in Fig. 4.39

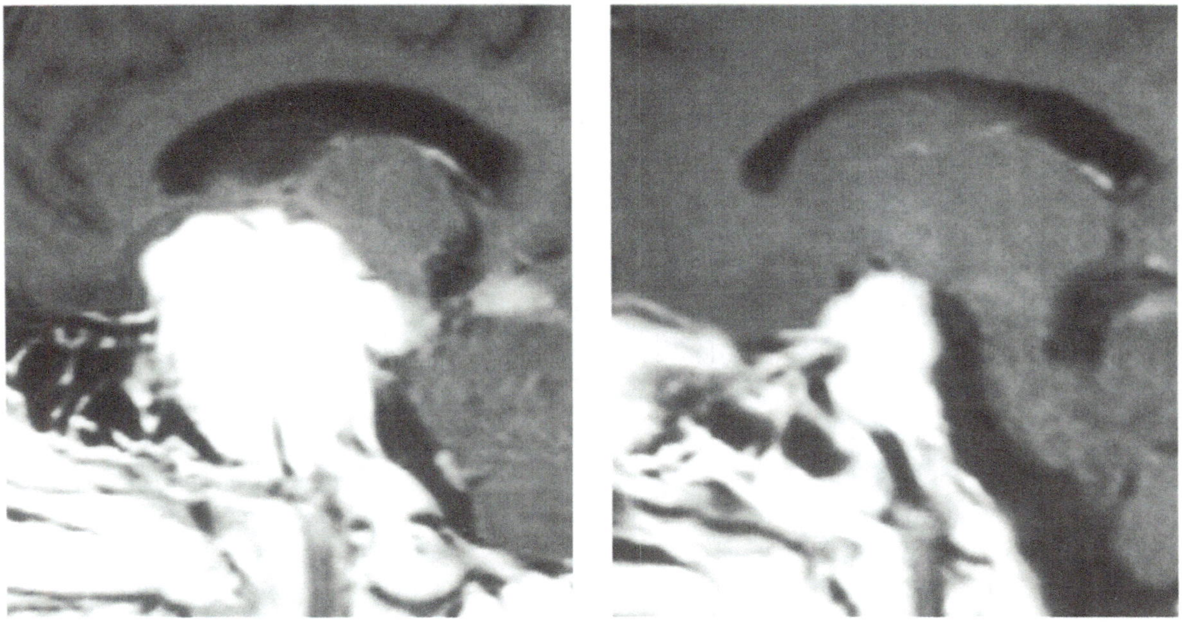

a b

Fig. 4.41. A preoperative sagittal contrast-enhanced MRI (a) reveals a large central skull base (*CSB*) meningioma in a 32-year-old lady who previously had been operated three times over two years at another institution. Following still another operation, a sagittal contrast-enhanced MRI (b) demonstrates only partial tumor resection. Note: Prior to the fourth surgical attempt this patient was totally blind in the right eye and was experiencing failing vision on the left, making the primary goal of surgery to preserve vision in the left (contralateral) eye. Owing to dense scarring and deranged anatomy from previous dissections, it was possible to remove only a limited amount of residual tumor, extensively cauterizing what remained. Ten days following surgery she developed a severe transnasal CSF leak, requiring a fifth operation. With that attempt total tumor excision was achieved, including the resection of nerves III through VI, which affected the right (already blind) eye, skeletonizing the parasellar space. The dural defect was closed with a large patch of fascia lata, and lumbar CSF drainage was maintained for ten days following surgery. No further CSF leakage occurred and visual function of left (contralateral) eye was preserved. This patient is an excellent example of the technical difficulties and diminishing expectations attendant to repeated operations

Table 16. Large central skull base meningiomas: number of operations, resectability and neurological outcome

Contralateral invasion of meningioma	N° of cases Virgin Residual		N° of surgeries 1		2		3		Tumor resection Gross-total		Sub-total		Preoperative deficits No		Yes		Postoperative deficits Same		Improvement		Worsening	
		%		%		%		%		%		%		%		%		%		%		%
Upper (sellar)	6	40	2	100	4	50	–	–	1	33.3	5	42	2	100	4	31	5	42	1	33.3	–	–
	5	42	–	–	5	83	–	–	1	100	4	36	–	–	5	42	3	42	1	50	1	33.3
Lower (clival)	4	27	–	–	2	25	2	40	1	33.3	3	25	–	–	4	31	3	25	1	33.3	–	–
	3	25	–	–	–	–	3	50	–	–	3	27	–	–	3	25	2	29	–	–	1	33.3
Combined	5	33	–	–	2	25	3	60	1	33.3	4	33	–	–	5	38	4	33	1	33.3	–	–
(sellar + clival)	4	33	–	–	1	17	3	50	–	–	4	37	–	–	4	33	2	29	1	50	1	33.3
Collective	15	100	2	100	8	100	5	100	3	20	12	80	2	100	13	100	12	100	3	100	–	–
	12	100	–	–	6	100	6	100	1	8	11	92	–	–	12	100	7	100	2	100	3	100
Total	27	100	2	100	14	100	11	100	4	100	23	100	2	100	25	100	19	100	5	100	3	100

Multiple: parasellar & distant meningiomas

This group consists of patients detected to have a central skull base (CSB) meningioma in combination with a meningioma at a distant intracranial site. Some harbored more than two lesions. In every case symptoms and signs resulting from the CSB lesion led to diagnosis, the other tumors being found incidentally. Sixteen patients underwent surgery, all but one with virgin tumors. In only four patients was total removal of every lesion possible. Resections were limited only with lesions of the parasellar space (PS), all associated distant lesions amenable to complete removal. Gross-total resection of the CSB meningioma was achieved in eight, only subtotal resections in four. Figs. 4.42–4.44 illustrate representative cases.

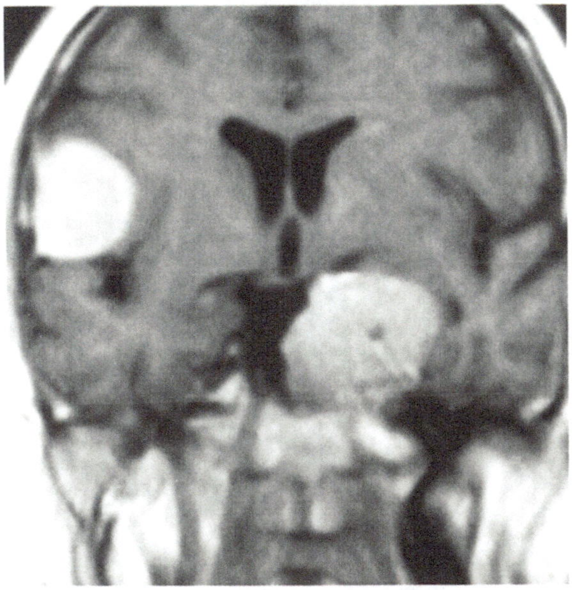

a

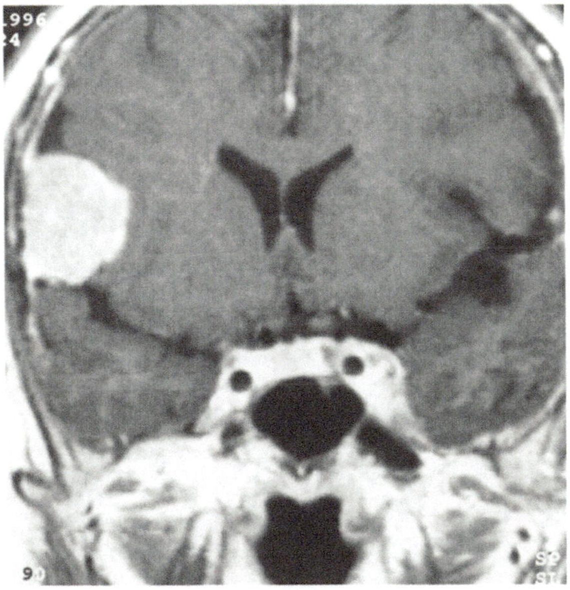

b

Fig. 4.42. A preoperative contrast-enhanced coronal (a) MR image shows the left PS lesion plus a second meningioma compressing contralateral temporal bone. Corresponding postoperative contrast-enhanced coronal (b) MR image confirm total resection of the PS lesion, with the contralateral convexity meningioma remaining. The contralateral tumor was removed in a second operation

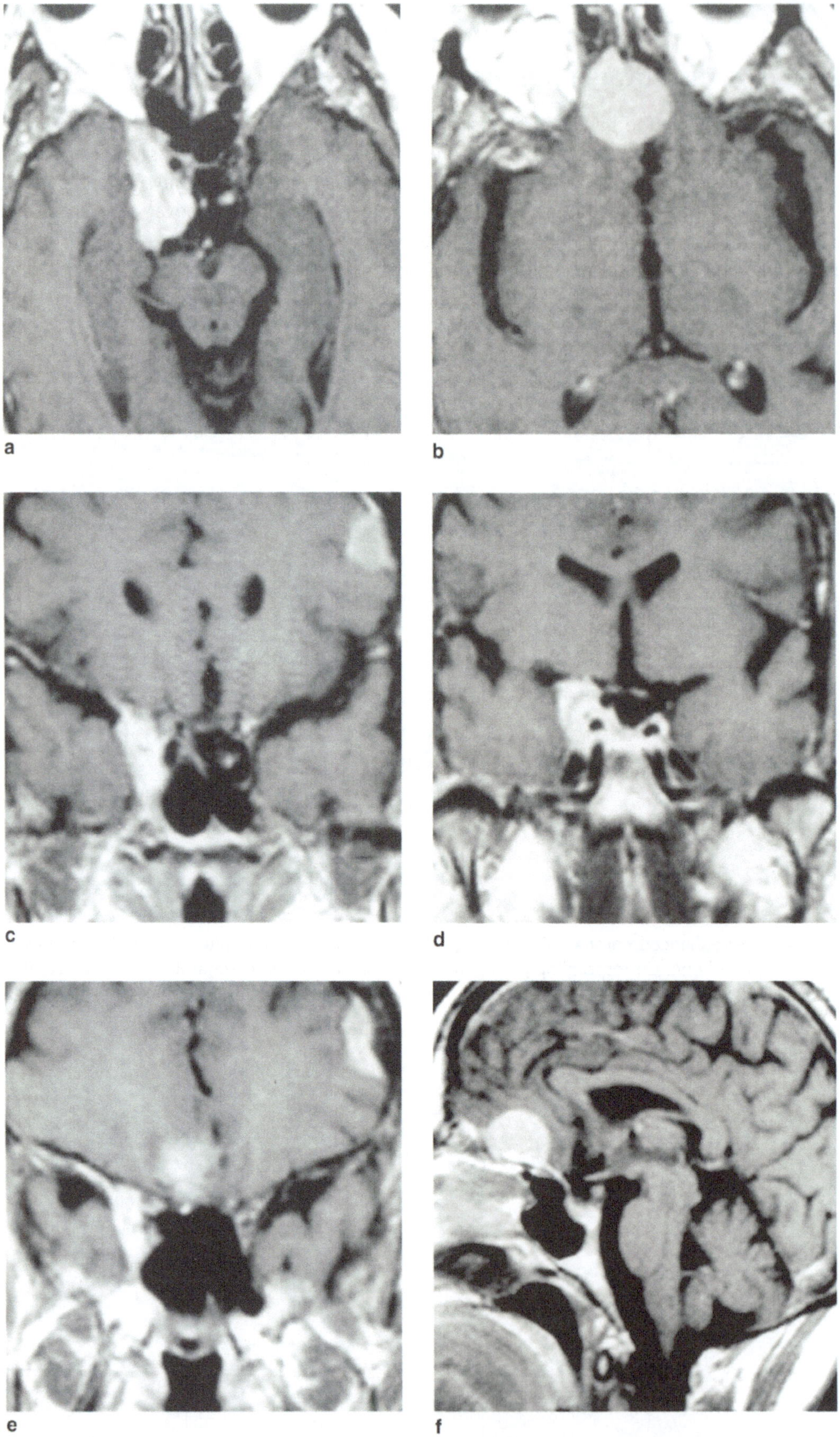

Fig. 4.43. A preoperative contrast-enhanced transaxial MR image (a) reveals a right parasellar space (*PS*) meningioma. A higher trans-axial cut (b) shows a second meningioma well anteriorly in the right anterior cranial fossa. Coronal (c, d, e) MR images in different planes demonstrate a third tumor, a contralateral convexity lesion. A right PS lesion plus a left temporoparietal lesion are seen in (c). The right PS lesion is apparent also in (d), and all three lesions more anteriorly in (e): the right PS and anterior cranial fossa lesions and the left-sided convexity meningioma. A contrast-enhanced sagittal MR image (f) shows the right anterior cranial fossa meningioma. The right-sided tumors were resected in a single stage, and the small contralateral lesion was removed later

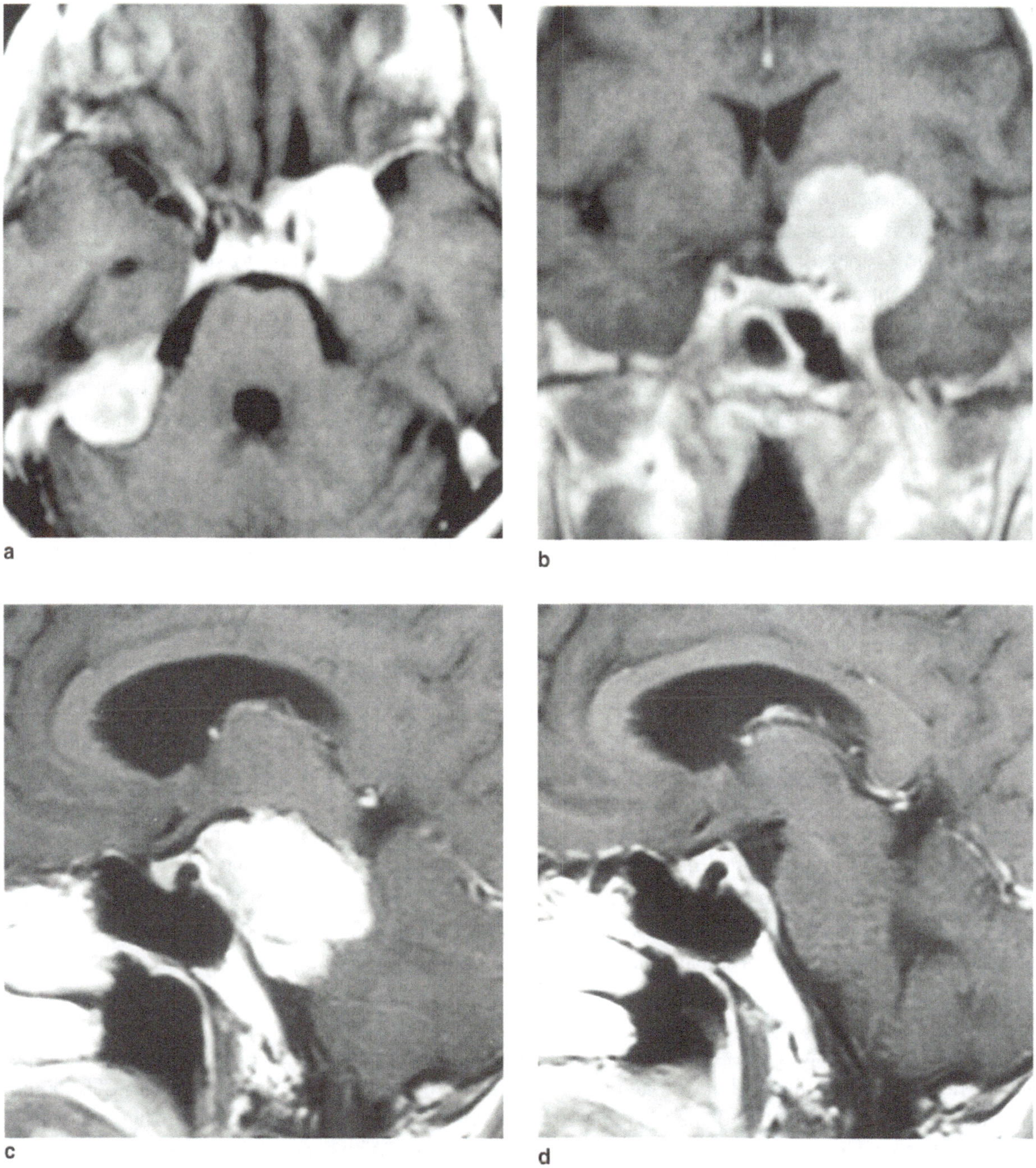

Fig. 4.44. A preoperative contrast-enhanced trans-axial MR image (a) reveals two lesions: one, a left anterior clinoid process (*ACP*) meningioma, the other, a right cerebello-pontine angle (*CPA*) meningioma. Contrast-enhanced coronal (b) and sagittal (c) images characterize the size and shape of the ACP lesion. A corresponding postoperative contrast-enhanced, sagittal MR image (d) confirms total resection of the ACP meningioma. Despite good result after resecting the ACP lesion, the patient did not come back for second surgery for the right CPA meningioma

Overall results in central skull base meningiomas surgery and conclusions

Central skull base (CSB) meningiomas are serious, often devastating neoplasms, occurring throughout the central skull base, always attached to dura. Their extent of dural attachment varies with growth type, "en plaque" lesions having widest spread, "global" meningiomas the least, with the majority involving dura to an intermediate extent. Since they respect no anatomical frontier, spreading easily from one compartment to neighboring regions, there are problems with discussing these lesions as tumors of specific anatomical sites. This notwithstanding, cases presented here (likely the largest meningioma series operated by a single neurosurgeon) were divided into lesions of 14 different sub-groups. Classifying them according to site of origin was not intended to point out clinical characteristics of lesions at various anatomical sites but to emphasize specific surgical requirements and problems – how to plan an ideal operative approach, one allowing the most complete resection possible while minimizing the risk of permanent neurological loss. Lesions at some sites reach considerable size prior to diagnosis, the incidence of preoperative neurological deficits varying remarkably (from 23% for sphenoid wing meningiomas to 100% for lesions of Dorello's space, the diaphragm sellae, and the optic canal).

Faced with patients in good neurological condition and clear MRI evidence of a CSB meningioma, neurosurgeons lacking requisite skill and experience may be tempted to attack such lesions, often to the peril of their patients. Unexpected complexities await the ill-prepared in the operating room. Less than adequate resections will result, often severe permanent neurological deficits. Furthermore, the destruction of important virgin landmarks and subsequent scarring are prone to convert potentially totally resectable meningiomas to lesions impossible to remove without serious neurological consequences.

To some extent neurological outcome is determined by neurological condition at the time of diagnosis. Of the 511 virgin cases coming to surgery in this series, 309 (60%) presented with signs of neurological impairment while 285 of 306 (93%) with residual tumors previously operated elsewhere had similar findings, an increased incidence of neurological findings of 33% among patients having surgery more than once. Determining the proportion of patients actually profiting from tumor resection is obtained by calculating the "neurological improvement index" (NII) (the number of patients who were neurologically improved after surgery divided by the total with preoperative deficits) for each subgroup of CSB meningiomas. These data are summarized in Table 17. Overall, the NII of virgin cases is 84% compared to 75% of those with residual or recurrent lesions. Clearly the first operation affords the best opportunity for safe, efficient tumor resection.

Extensive laboratory training is an absolute must for surgeons considering tackling CSB meningiomas. Without such investments of time and effort surgeons sooner or later will be confronted by insurmountable problems which almost certainly could have been avoided. A certain level of technical skill and knowledge of CSB anatomy is crucial for success. Numerous publications (intentionally not listed here) confirm that adequate basic neurosurgical skill plus knowledge of parasellar anatomy do not make one expert with CSB meningiomas. After operating on 100 CSB meningiomas, one becomes more realistic, less aggressive, and more cooperative with specialists proposing alternative treatment modalities. With each additional 100 meningiomas operated the surgeon becomes even more likely to recognize human limitations in dealing with these lesions. Experience teaches that meningiomas constitutes far from benign disease, both biologically and surgically. Our experience with totally resected meningiomas (histologically classified as totally benign) indicates that aggressive recurrence may occur even after a decade of no symptoms or MRI changes Fig. 4.45. One may never be certain that a given tumor will not recur. Currently, however, no treatment of meningiomas is superior to surgical resection, either total or gross-total/sub-total followed by adequate complementary measures (focused beam radiation, embolization, or conventional radiation). Experience with over eight hundred CSB meningioma operations confirms the reality that surgery is far from the final answer in the management of this nasty disease. One may only hope that the near future will bring biological and pharmacological solutions. In the meantime, committed, fully prepared surgeons must strive to excise CSB meningiomas as completely as possible while minimizing even temporary additional neurological deficits, providing the best possible conditions for the success of complementary treatment.

Table 17. Neurological outcome following surgery with respect to central skull base location

Meningioma location	Pre-op. deficits in virgin cases		Postoperative deficits: One week		Postoperative deficits: Six months		Pre-op. deficits in redue cases		Postoperative deficits: One week		Postoperative deficits: Six months		Improvement index: Virgin cases		Improvement index: Residual cases	
Sphenoid wing	11 47	23%	14 47	30%	1 47	2%	3 4	75%	3 4	75%	1 4	25%	10 11	91%	2 3	67%
Anterior clinoid process	24 54	44%	26 54	48%	2 54	4%	9 13	69%	12 13	92%	2 13	15%	22 24	92%	7 9	78%
Middle cranial fossa	– 12	–	1 12	8%	– 12	–	– –	–	– –	–	– –	–	– –	–	– –	–
Parasellar space	71 89	80%	80 89	90%	5 89	6%	34 34	100%	34 34	100%	27 34	79%	66 71	93%	7 34	20%
Parasellar space & adjacent regions	110 172	64%	165 172	96%	7 172	4%	201 216	93%	205 216	95%	27 216	12%	103 110	94%	174 201	86%
Superior orbital fissure	1 1	100%	1 1	100%	1 1	100%	6 6	100%	6 6	100%	1 1	100%	– –	–	– –	–
Dorello's space	4 4	100%	4 4	100%	2 4	50%	1 1	100%	1 1	100%	2 5	40%	2 4	50%	– –	–
Petroclival	21 36	58%	25 36	69%	3 36	8%	4 5	80%	5 5	100%	2 9	22%	18 21	86%	2 4	50%
Tuberculum sellae	21 35	60%	21 35	60%	4 35	11%	9 9	100%	9 9	100%	1 9	11%	17 21	81%	8 9	89%
Diaphragm sellae & optic canal(s)	7 7	100%	7 7	100%	2 7	28%	1 1	100%	1 1	100%	– –	–	5 7	71%	– –	–
PCP, dorsum sellae, and upper clival	2 6	33%	4 6	67%	1 6	17%	– –	–	– –	–	3 4	75%	1 2	50%	– –	–
Clival	7 18	39%	14 18	78%	2 18	11%	4 4	100%	4 4	100%	3 4	75%	5 7	71%	1 4	25%
Large CSB	15 15	100%	15 15	100%	9 15	60%	12 12	100%	12 12	100%	10 12	83%	6 15	40%	2 12	17%
Multiple: parasellar & distant	15 15	100%	15 15	100%	10 15	67%	1 1	100%	1 1	100%	1 1	100%	5 15	33%	– –	–
Total	309 511	60%	392 511	77%	49 511	9%	285 306	93%	293 306	96%	74 306	24%	260 309	84%	211 285	74%

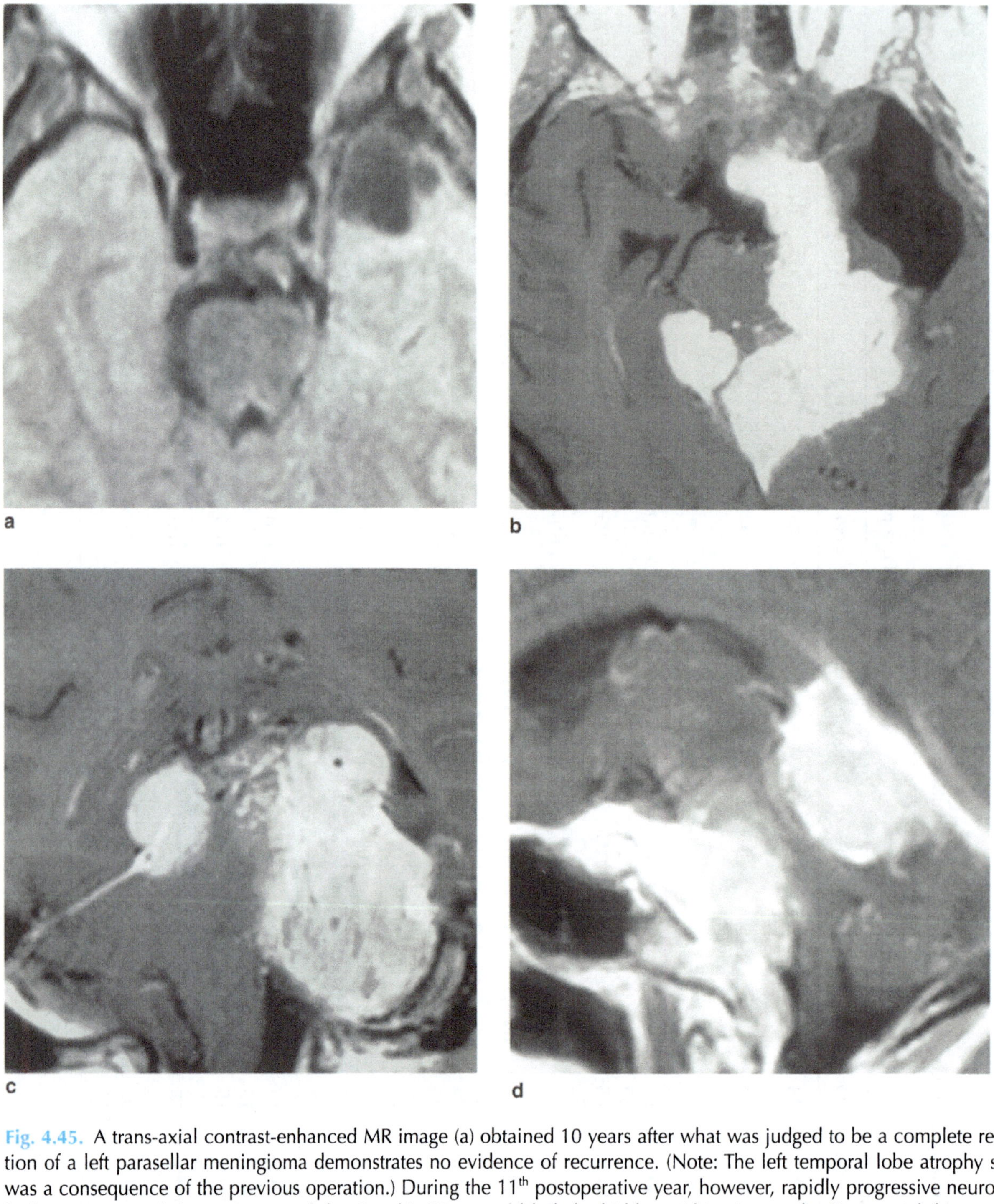

Fig. 4.45. A trans-axial contrast-enhanced MR image (a) obtained 10 years after what was judged to be a complete resection of a left parasellar meningioma demonstrates no evidence of recurrence. (Note: The left temporal lobe atrophy seen was a consequence of the previous operation.) During the 11th postoperative year, however, rapidly progressive neurological symptoms and signs reappeared, leaving this 50-year-old lady bedridden within six months. Trans-axial (b), coronal (c) and sagittal (d), contrast-enhanced MR images performed at that time revealed an inoperable recurrent lesion. Because of her poor general and neurological condition, additional surgery was not performed. She died within several weeks from pulmonary complications

References

1. Cushing H (1922) The meningiomas (dural endotheliomas): their source and favoured seats of origin. Brain 45: 282–316
2. Cushing H, Eisenhardt L (1938) The meningiomas. Charles C. Thomas, Springfield
3. Ghim TT, Seo JJ, O'Brien M, Meacham L, Crocker I, Krawiecki N (1993) Childhood intracranial meningiomas after high-dose irradiation. Cancer 71: 4091–4095
4. Granick MS, Martuza RL, Parker SW, Ojemann RG, Montgomery WW (1985) Cerebellopontine angle meningiomas: clinical manifestations and diagnosis. Ann Otol, Rhinol Laryngol 94: 34–38
5. Grunberg SM (1994) Role of antiprogestational therapy meningiomas. Human Reproduction [Suppl]9: 202–207
6. Grunberg SM, Weiss MH, Spitz IM, Ahmadi J, Sadun A, Russell CA, Lucci L, Stevenson LL (1991) Treatment of unresectable meningiomas with the antiprogesterone agent mifepristone. J Neurosurg 74: 861–866
7. Harrison MJ, Wolfe DE, Lau TS, Mitnick RJ, Sachdev VP (1991) Radiation-induced meningiomas: experience at the Mount Sinai Hospital and review of the literature. J Neurosurg 75: 564–574
8. Helseth A, Mork SJ, Johansen A, Tretli S (1989) Neoplasms of the central nervous system in Norway. IV. A population-based epidemiological study of meningiomas. APMIS 97: 646–654
9. Kepes JJ (1982) Meningiomas: biology, pathology and differential diagnosis. Masson, New York
10. Kepes JJ, Dunlap MD, O'Boynick P, Terreros D (1986) Meningioma with multiple rheumatoid nodules. Acta Neuropathologica 70: 314–319
11. Kleihues P, Burger PC, Scheitauer BW (1993) Histological typing of tumours of the central nervous system. Second edition. WHO international histological classification of tumors. Springer, Berlin Heidelberg
12. Longstreth WT Jr, Dennis LK, McGuire VM, Drangsholt MT, Koepsell TD (1993) Epidemiology of intracranial meningioma. Cancer 72: 639–648
13. Mack EE, Wilson CB (1993) Meningiomas induced by high-dose cranial irradiation. J Neurosurg 79: 28–31
14. Nakasu S, Hirano A, Shimura T, Llena JF (1987) Incidental meningiomas in autopsy study. Surg Neurol 27: 319–322
15. Preston-Martin S (1989) Descriptive epidemiology of primary tumors of the brain, cranial nerves and cranial meninges in Los Angeles County. Neuroepidemiology 8: 283–295
16. Rachlin JR, Rosenblum ML (1991) Etiology and biology of meningiomas. In: Al-Mefty O (ed) Meningiomas. Raven, New York, pp 22–37
17. Russell DS, Rubinstein LJ (1989) Pathology of tumorurs of the nervous system, 5th edn. Edward Arnold, London
18. Sedzimir CB, Frazer AK, Roberts JR (1973) Cranial and spinal meningiomas in a pair of incidental twin boys. J Neurol Neurosurg Psychiatry 36: 368–376
19. Van Buren JM, Poppen JL, Horrax G (1957) Unilateral exophthalmos: a consideration of symptom pathogenesis. Brain 80: 139–175
20. Wood MW, White RJ, Kernohan JW (1957) One hundred intracranial meningiomas found incidentally at necropsy. J Neuropathol Exp Neurol 16: 337–340

4.2

Pituitary tumors extending beyond the sella

Introduction

Continued growth of a pituitary tumor which has already reached the volume of the sella results in intrasellar pressure destined to erode the dorsum sellae, the posterior clinoid processes (PCPs), the anterior sellar wall (abutting the sphenoidal sinus), and the sellar floor, the clivus. Soft tissue sellar boundaries, the diaphragm sellae and incomplete lateral walls – which enable venous communication between the parasellar spaces (PSs) on either side – delay the tumor's escape from the sella, at least for some time. Eventually, however, these lateral wall "windows" allow benign tumors to enter one PS or the other, occasionally both (4).

Tumor invading the PS is easily demonstrated by MRI [1, 27], but such extension from the sella must be distinguished from the ballooning of sellar contents laterally into the PS by displacing, rather than traversing, the intervening membrane.

The presence of non-secreting adenomas, accounting for 1/3 of pituitary tumors, is primarily manifest by symptoms and signs of compressing normally functioning pituitary tissue (causing hypopituitarism) or extrasellar extension, including the compression of optic pathways by distending the diaphragm sellae or by compressing one or both oculomotor nerves [24, 26, 31]. The oculomotor nerve, particularly, has a surprising capacity to compensate its distension and stretching. Some tumors invade the PS and traverse the inner layer of its lateral wall, even extend intradurally, displacing and stretching nerve III upward and laterally against the sharp edge of the tentorium without producing occulomotor symptoms and signs. Intradural tumor nodules, covered by widened, paper-thin segments of nerve III, are encountered frequently.

Certain large intrasellar and parasellar tumors are not pituitary tumors at all, but metastatic lesions [14, 19, 22], tuberculomas [7], even abscesses [9]. Not only do pituitary tumors extend into the PS, the opposite may occur, a parasellar meningioma invading the sella, for instance [8]. In addition, pituitary tumors may occur concomitantly with craniopharyngiomas [32] or aneurysms [12, 15, 35, 37]. Recognizing prior to surgery when tumors and aneurysms coexist allows the surgeon to tailor his approach in order to deal with both lesions simultaneously [12]. When eliminating both lesions is not possible in a single operation, normally the aneurysm is treated first. When pituitary tumors and aneurysms are discovered incidentally during surgery, the aneurysm is normally clipped by the transcranial approach [37]. Aneurysms may result in disaster if they rupture during transsphenoidal surgery [35]. Rarely, carotid-ophthalmic aneurysms, typically large or giant ones, may simulate pituitary adenomas, even to the point causing hypopituitarism, visual deterioration, diplopia, or some combination of these symptoms and signs [18, 20]. Pituitary tumors, although occurring primarily in young adults and the middle-aged, are also seen in children and teenagers [11]. When large central skull base (CSB) tumors are discovered in children, one must be alert to the possibility of a rhabdomyosarcoma or a craniopharyngioma rather than a pituitary tumor. Pituitary tumors may also coexist with other tumors in the region [2].

Although the management of pituitary tumors confined to the sella will not be considered here, advances in the treatment of such lesions indirectly affect the management of those extending into the PS. Despite the success of Bromocriptine therapy for both prolactinomas (Fig. 4.46) and non-secreting pituitary tumors [34, 36], surgery still plays an important role in the treatment of intrasellar neoplasms [6, 21, 23, 31, 38]. The transsphenoidal route remains the preferred approach to adenomas causing visual loss [30, 37], but the incidence of residual and recurrent extrasellar pituitary tumors following transsphenoidal surgery is increasing [29, 32]. Treating such lesions with radiation therapy carries the risk of additional visual apparatus neuropathies [10, 13, 16, 17, 33], secondary brain tumors [3], and intrasellar pituitary fibrosarcomas [25, 28]. It is clear that pituitary tumors extending beyond the confines of the sella into the CSB should be totally resected so long as the surgical risk to adjacent structures and the threat of postoperative cerebrospinal fluid (CSF) fistula and ascending infection remains low. At present, the extradural approach discussed here amounts to the best opportunity for total tumor resection with minimal risk of permanent injury to neural and vascular structures while preserving CSB dural integrity.

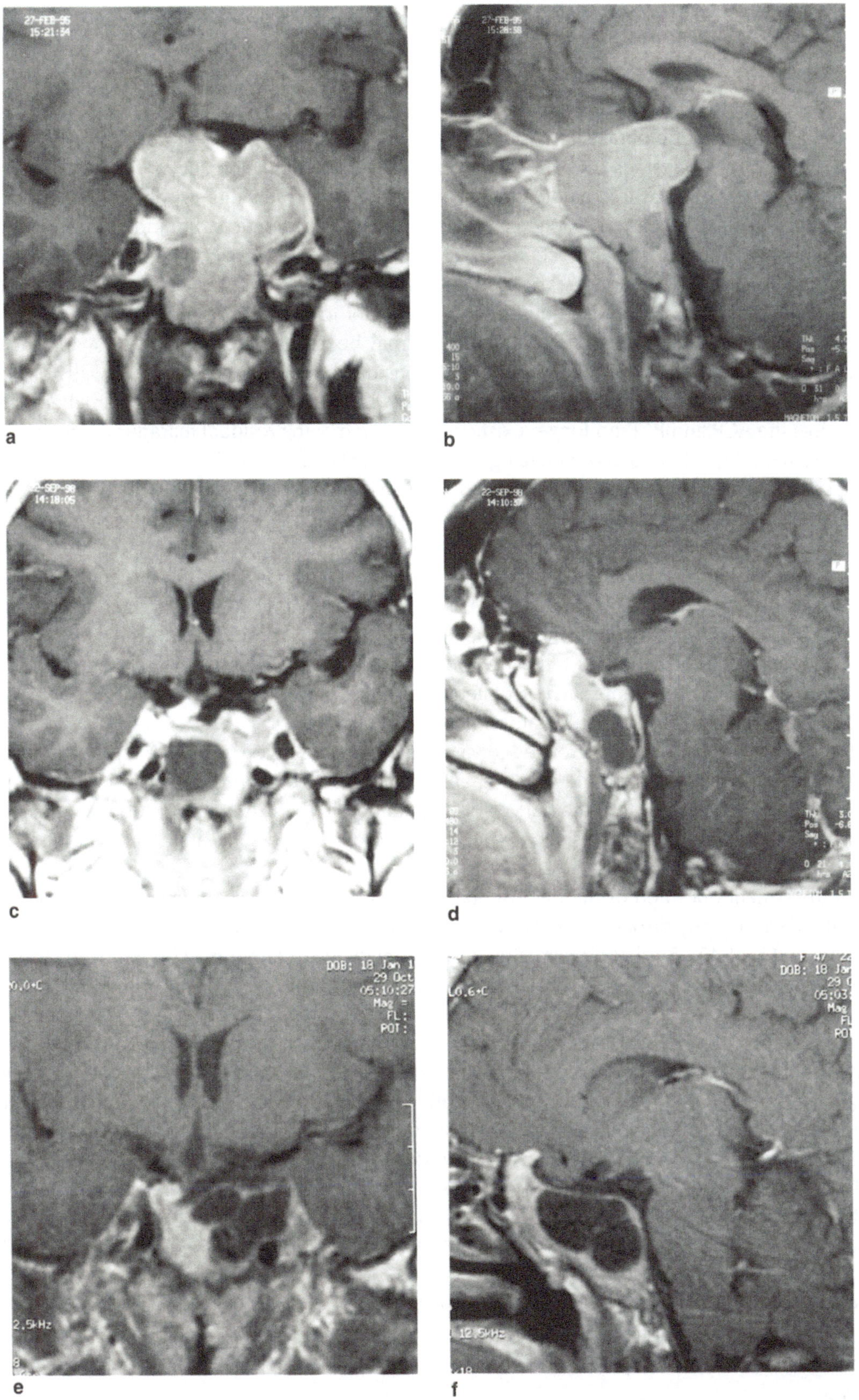

Fig. 4.46. Coronal (a) and sagittal (b) contrast-enhanced MR images demonstrate a large pituitary tumor extending beyond the confines of the sella. Pathological blood prolactin levels confirmed the nature of the lesion, and conservative treatment was instituted in 1995. The response was excellent, prolactin levels returning to near normal and the tumor mass shrinking significantly. Since visual impairment subsided, surgical treatment was not considered. Four years later, coronal (c) and sagittal (d) contrast-enhanced MR images revealed cystic tumor degeneration and a dramatically reduced mass. Contrast-enhanced MR images obtained seven years after diagnosis confirm continued control of the tumor (e, f). Note: Although initial MR images (a, b) demonstrate a surgically appealing lesion, subsequent studies (c, d, e, f) confirm that conservative treatment provides excellent results, emphasizing the importance of a team approach to pituitary tumors, especially prolactinomas

Surgical approach

The general extradural approach to the parasellar and adjacent CSB regions has been employed in our Institution for particularly large pituitary lesions over the past decade. The following elements are mandatory for every case: patient positioning, osteoplastic craniotomy, unroofing the orbit, resecting the sphenoid wing and the anterior clinoid process (ACP), opening the optic canal (OC), resecting the optic strut, and separating the outer layer of the lateral wall of the PS, exposing nerves III, IV and V (V1 and V2). Unroofing the intrapetrous ICA and resecting the petrous apex is not necessary. However, when such tumors invade paranasal sinuses (most commonly the sphenoidal and ethmoidal sinuses), it is necessary to detach dura from the tuberculum sellae in order that the tuberculum sellae and planum sphenoidale may be resected extradurally, allowing entry into the sphenoidal sinus. Large pituitary tumors often destroy considerable bone, including the walls of the sella and OC (including the optic strut). Also the ACP and PCP may be eroded, becoming much smaller than normal. In extreme cases the ACP and PCP no longer exist. Therefore, removing residual extradural bone must be accomplished carefully and gently to avoid damaging neural elements and the ICA.

Only when the general epidural approach is completed and the anteromedial, paramedial, Parkinson's, anterolateral, and lateral triangles are visualized entirely, may tumor resection be initiated. The tumor mass is entered by incising the membrane posterior to the anterior loop (AL) of the ICA in the anteromedial triangle (Fig. 4.47). If the posterior clinoid process (PCP) no longer exists, the window provided by the anteromedial triangle – lateral to the optic nerve (ON), posterior to the AL, and medial to nerve III – is sufficiently large to allow considerable debulking of the intrasellar tumor. Debulking the tumor is considered Phase I of its resection (Fig. 4.48). The tumor should be debulked from the *contralateral* half of the sella. With the head positioned for craniotomy and the extradural approach – rotated away from the side of the lesion – the corridor to the sella is diagonal, and therefore of little value after Phase I is completed.

To begin Phase II the operating table must be rotated toward the side of the lesion and the microscope redirected as appropriate. Entering the sella from the resulting angle allows the surgeon direct vision of the ipsilateral half of the sella, beneath and medial to the horizontal segment of the PS-ICA, without subjecting nerve III to the danger of lateral retraction. Phase II refers to debulking the tumor from the *ipsilateral* half of the sella, medial to the ipsilateral PS (Fig. 4.48).

Without changing the patient's position or the microscope, nerve III may be dissected from the inner layer of the lateral wall of the PS and gently displaced medially, exposing the horizontal segment of the PS-ICA deep in the paramedial triangle (Fig. 4.49). By slightly retracting nerve III medialward tumor may be removed farther laterally from the medial surface of the PS-ICA (Phase III), giving access to the floor of the ipsilateral PS. In order to continue resecting tumor more laterally, it is necessary to gently lift and displace the PS-ICA laterally. Before mobilizing the ICA, however, nerve IV must be dissected from the inner layer of the lateral wall of the PS and from V1, then mobilized medially, placing it adjacent to nerve III, thus enlarging Parkinson's triangle and exposing the medial loop (ML) of the ICA and tumor beneath it (Fig. 4.50). Freeing nerve IV from the inner wall of the PS avoids stretching it when the ICA is displaced laterally.

Phase IV is executed through the enlarged Parkinson's window. This affords exposure of the meningohypophyseal (MHT) and inferolateral (ILT) trunks, and, if necessary, their coagulation and transection. Tumor surrounding the ML of the PS-ICA is removed and nerve VI is exposed by gently lifting V1 (Fig. 4.50). Only when tumor is resected from the floor of the PS may the PS-ICA be gently displaced medially and Phase V begun.

In Phase V (Fig. 4.51), nerve VI is dissected from tumor throughout its PS course, from where it exits Dorello's space, proceeding to the superior orbital fissure (SOF). Once this is accomplished, the remainder of the tumor may be resected from the anterolateral compartment of the PS. Once the PS is emptied of tumor, precise hemostasis may be achieved.

In Phase VI, dura is opened along the sylvian fissure, which is then divided, and the diaphragm sellae is displaced gently anteriorly and downward while tumor is aspirated from the extradural side (Fig. 4.52). The pituitary stalk (PSt) and Dawson's arteries, which supply the optic chiasm and ipsilateral optic tract, must be identified and carefully dealt with in order not to cause vasospasm or rupture of some of them. The diaphragm sellae must be fully inspected, both its intradural and intrasellar surfaces. When all tumor is resected from beneath the diaphragm sellae, saline is introduced intradurally in order to test the integrity of CSB dura. Before closing the dura, any remaining tumor is resected from the opposite side of the sella, from the contralateral PS and the anterior sellar wall.

Where tumor has invaded the sphenoidal and/or ethmoidal sinuses, additional tuberculum sellae and planum sphenoidale bone must be removed and the paranasal sinuses entered. This is Phase VII (Fig. 4.53). Tumor is resected from the paranasal sinuses through the windows medial and lateral to the ON and anterior to the AL of the ICA. During Phase VII, tumor may easily be removed completely from the sphenoidal and ethmoidal sinuses. The presence of tumor in paranasal sinuses (and its subsequent removal) poses a serious danger of a postoperative CSF leak. Therefore, a watertight dural closure is extremely important (Fig. 4.54). To avoid CSF leakage through the diaphragm sellae or through the oculomotor slit (its point of entry into the PS), the entire extradural space at the CSB should be lined with fat, muscle, and fibrin glue, and lumbar drainage should be continued for five to seven days following surgery.

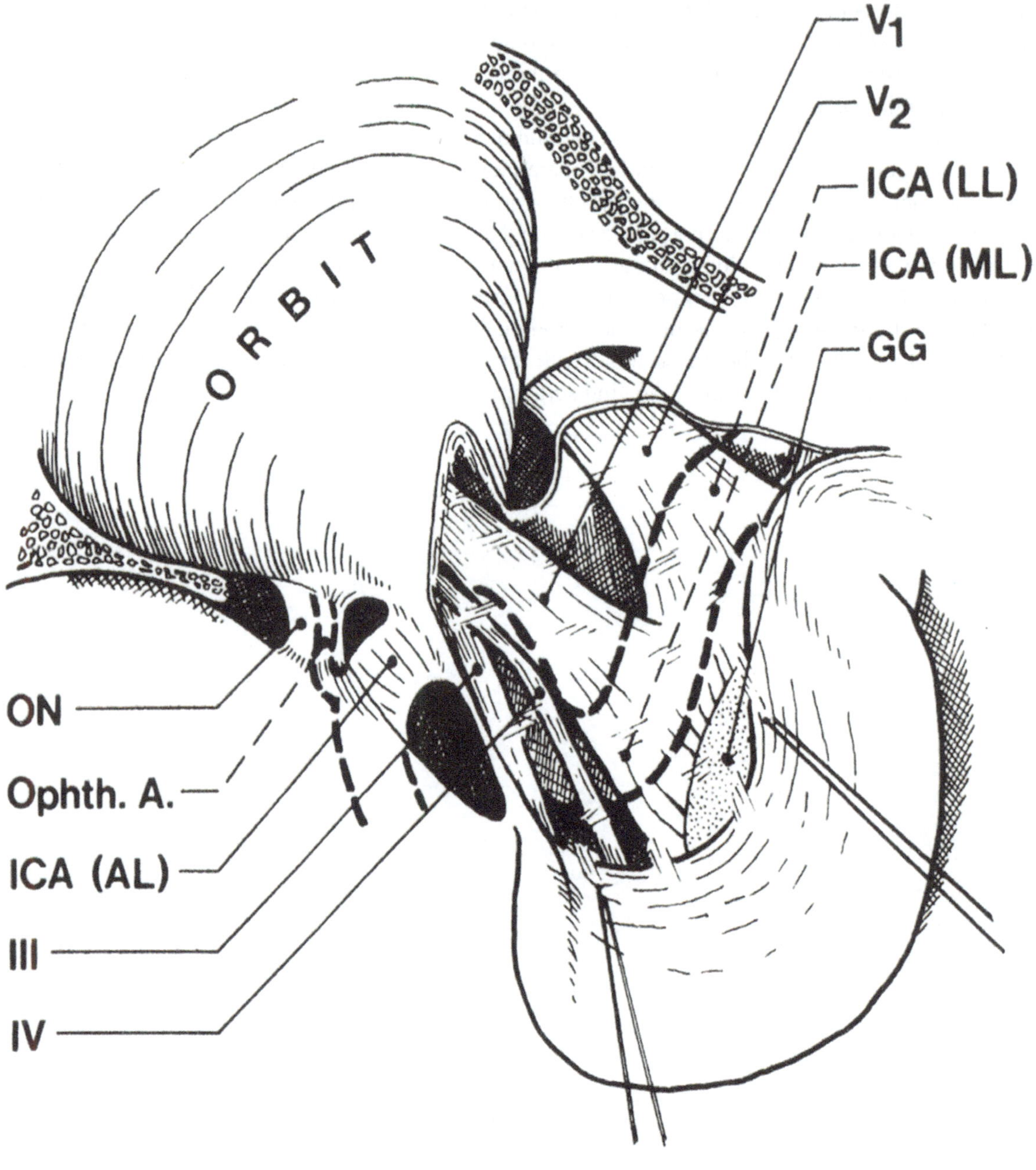

Fig. 4.47. A drawing of the right parasellar region after dissection of the outer layer of the lateral wall of the parasellar space (*PS*). Dura beneath the temporal lobe has been retracted with sutures. Removing the anterior clinoid process (*ACP*), the proximal and lateral walls of the optic canal (*OC*), and the optic strut between the OC and the anterior loop (*AL*) of the ICA exposes the dura propria covered optic nerve (*ON*), the ophthalmic artery arising from the AL of the ICA, nerves III, IV, V1, V2, and the anterior portion of the Gasserian ganglion (*GG*). The dotted lines indicate the course of the ICA within the PS. (With permission from V.V. Dolenc, Transcranial Epidural Approach to Pituitary Tumors Extending beyond the Sella, Neurosurgery, Vol. 41, No. 3, September 1997, p. 545)

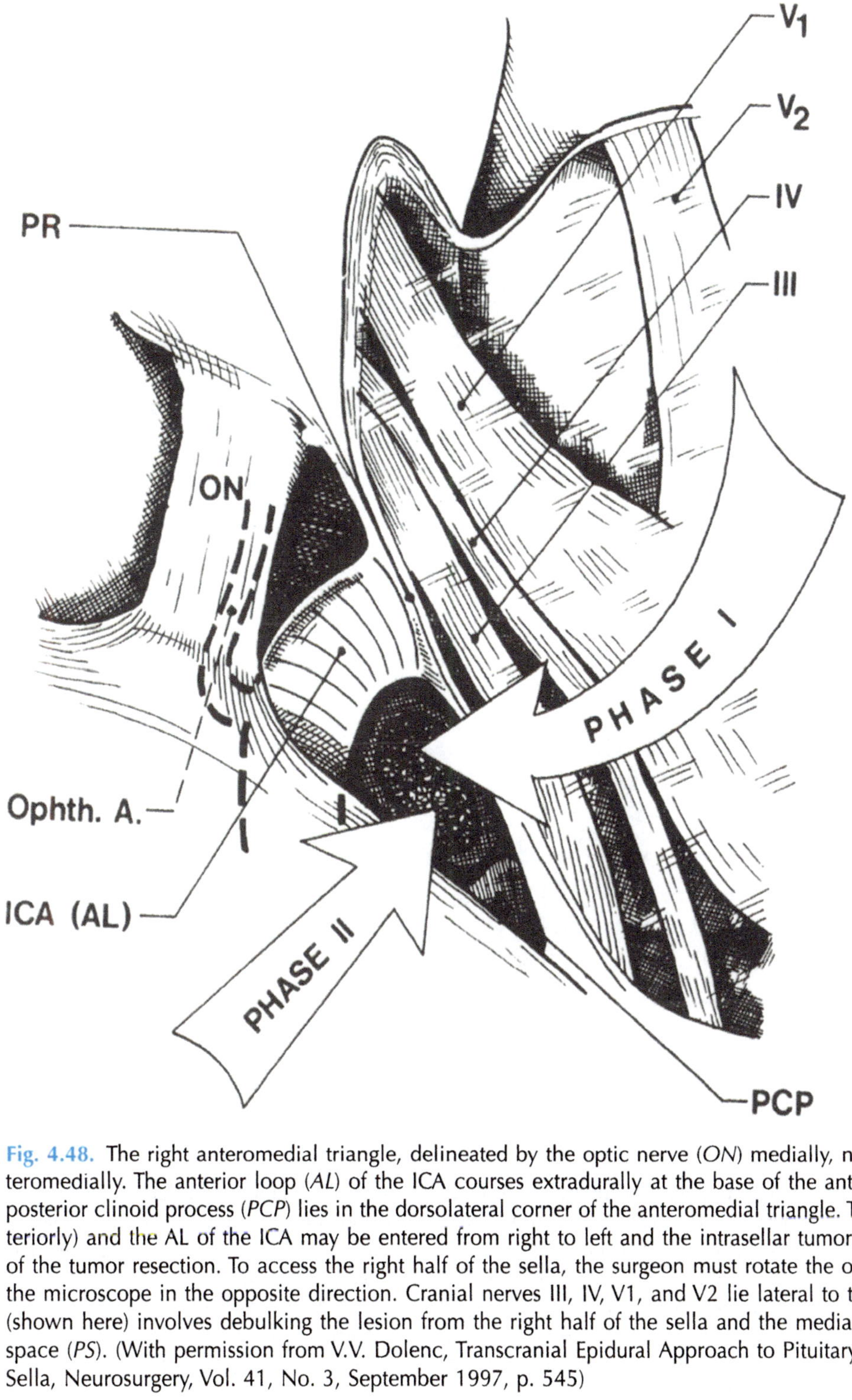

Fig. 4.48. The right anteromedial triangle, delineated by the optic nerve (*ON*) medially, nerve III laterally, and dura posteromedially. The anterior loop (*AL*) of the ICA courses extradurally at the base of the anteromedial triangle. The eroded posterior clinoid process (*PCP*) lies in the dorsolateral corner of the anteromedial triangle. The area between the PCP (posteriorly) and the AL of the ICA may be entered from right to left and the intrasellar tumor debulked, constituting Phase I of the tumor resection. To access the right half of the sella, the surgeon must rotate the operating table to the right, and the microscope in the opposite direction. Cranial nerves III, IV, V1, and V2 lie lateral to the proximal ring (*PR*). Phase II (shown here) involves debulking the lesion from the right half of the sella and the medial portion of the right parasellar space (*PS*). (With permission from V.V. Dolenc, Transcranial Epidural Approach to Pituitary Tumors Extending beyond the Sella, Neurosurgery, Vol. 41, No. 3, September 1997, p. 545)

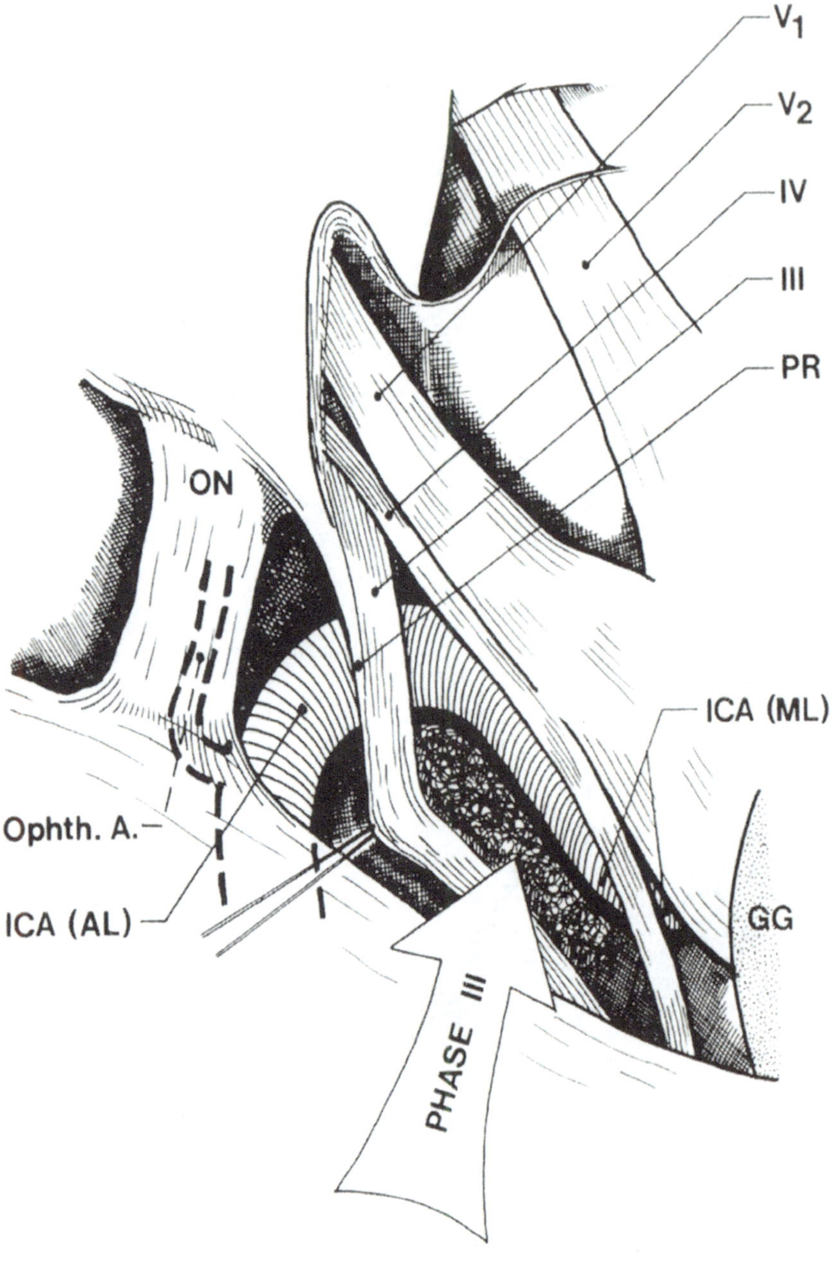

Fig. 4.49. The right paramedial triangle, circumscribed by nerve III medially, nerve IV laterally, and dura between their points of entry into the parasellar space (*PS*). Through this "window" the ICA is apparent, and additional tumor may be removed laterally and dissected from the proximal segment of the anterior loop (*AL*) of the PS-ICA, the horizontal segment of the ICA, and the distal segment of the medial loop (*ML*) of the ICA. The optic nerve (*ON*), the ophthalmic artery (*OA*), and the distal segment of the AL lie medial to nerve III, which is partially covered by the proximal (dural) ring (*PR*). Lateral to the paramedial triangle, nerves IV, V1, and V2 are exposed. The *arrow* indicates the direction in which the tumor is removed from medial to the PS-ICA in Phase III. (With permission from V.V. Dolenc, Transcranial Epidural Approach to Pituitary Tumors Extending beyond the Sella, Neurosurgery, Vol. 41, No. 3, September 1997, p. 546)

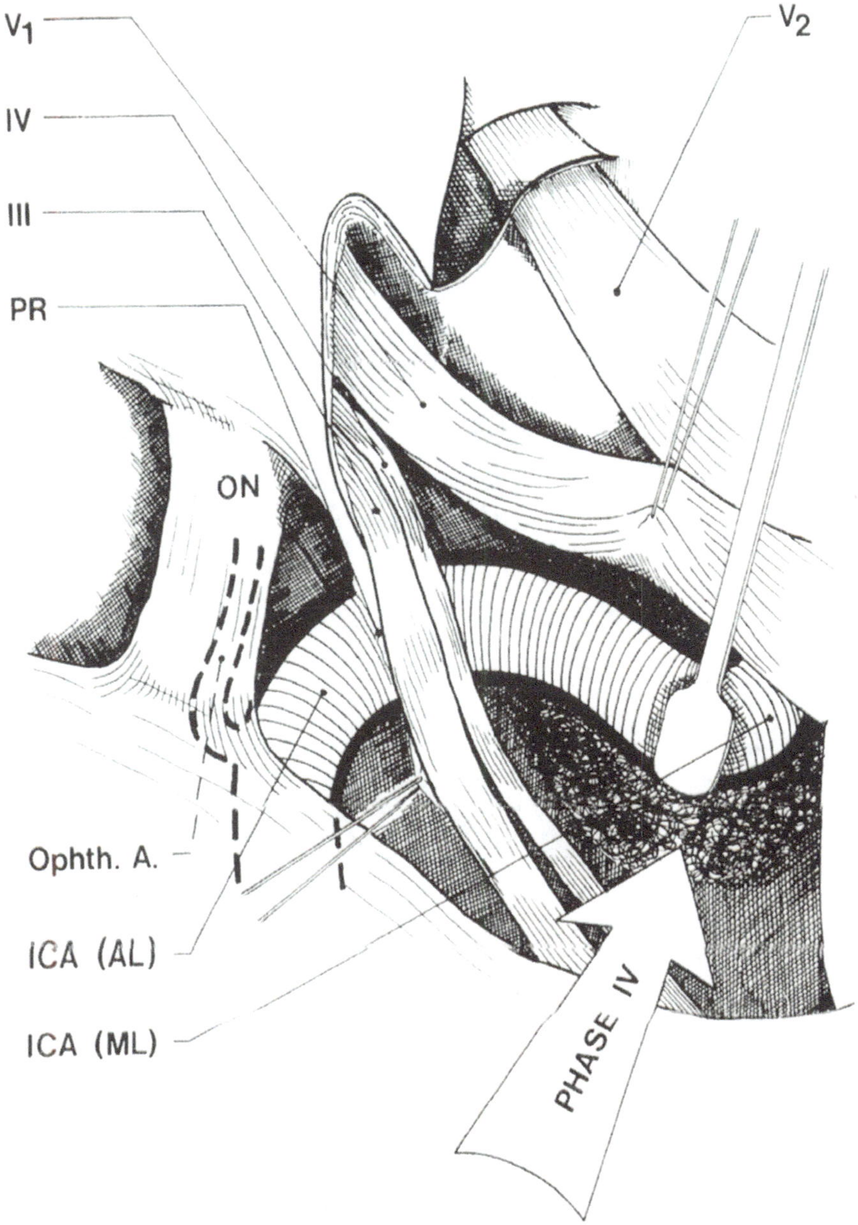

Fig. 4.50. A drawing of Parkinson's triangle on the right side, which has been enlarged by mobilizing nerve IV from V1. By displacing nerves III and IV medially, and V1 laterally, then slightly retracting the horizontal segment and medial loop (*ML*) of the ICA laterally, additional tumor may be removed from the ICA, the floor of the parasellar space (*PS*) medial to the ICA, and the petrous apex medial to Meckel's cave. The *arrow* indicates the route of Phase IV. (With permission from V.V. Dolenc, Transcranial Epidural Approach to Pituitary Tumors Extending beyond the Sella, Neurosurgery, Vol. 41, No. 3, September 1997, p. 546)

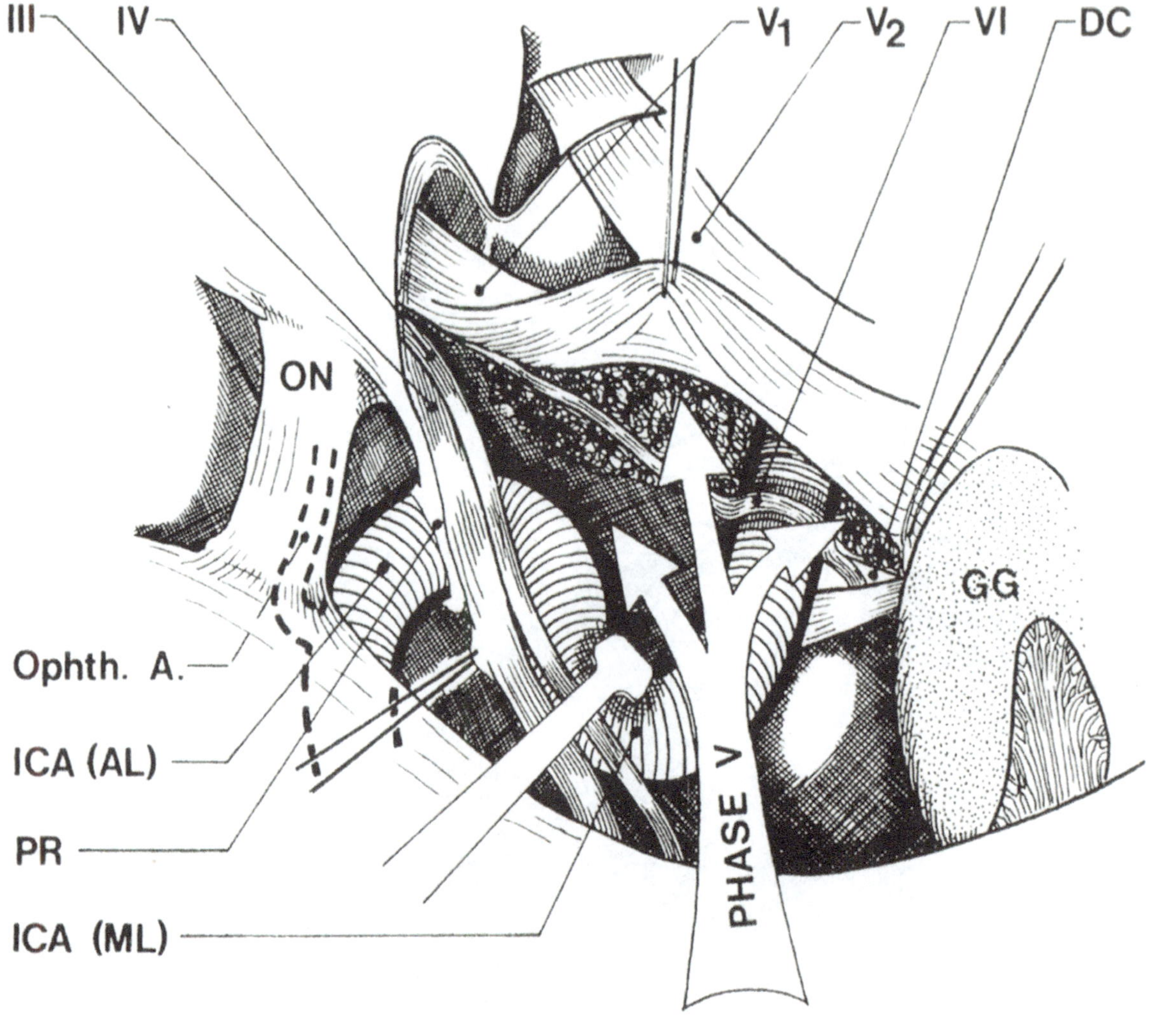

Fig. 4.51. The right Parkinson's triangle has been maximally enlarged by mobilizing cranial nerves III and IV medially, retracting the medial loop (*ML*) and the horizontal segment of the PS-ICA medially, lifting nerve V1 laterally, and displacing the Gasserian ganglion (*GG*) posterolaterally. Nerve VI is seen throughout the parasellar space (*PS*), from its exit from Dorello's space, crossing the ascending ICA, coursing beneath nerve V1 to the superior orbital fissure (*SOF*). During Phase V, remaining tumor is removed from the lateralmost PS compartments, from around nerve VI, medial to nerves V1 and V2, and from the posterior surface of the ascending ICA and from Dorello's space. (With permission from V.V. Dolenc, Transcranial Epidural Approach to Pituitary Tumors Extending beyond the Sella, Neurosurgery, Vol. 41, No. 3, September 1997, p. 546)

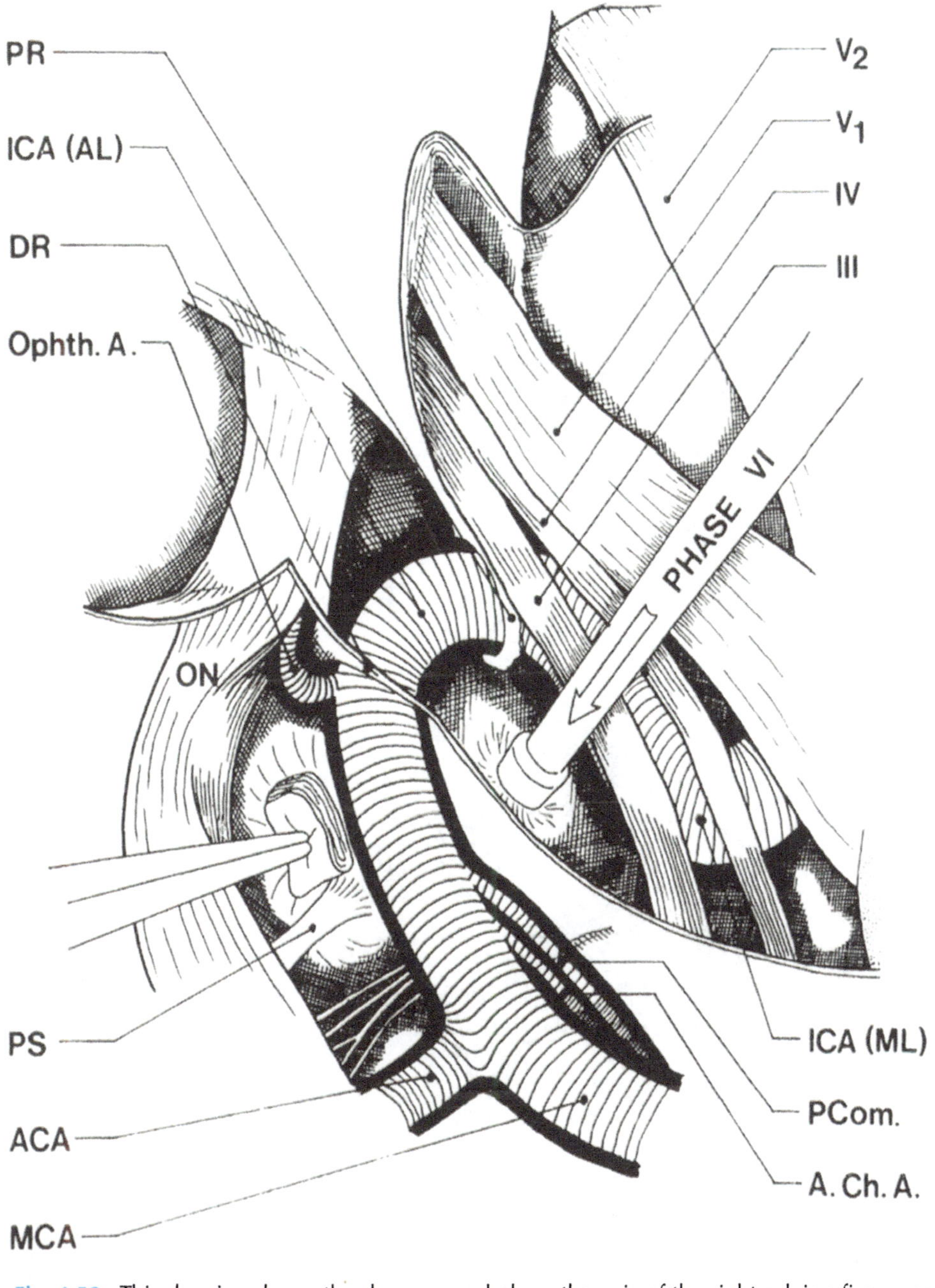

Fig. 4.52. This drawing shows the dura opened along the axis of the right sylvian fissure centrally to the distal (dural) ring (*DR*) and beyond, to dura propria enveloping the optic nerve (*ON*). The extradural sellar and parasellar regions are exposed with cranial nerves III, IV, and V1 and the ICA, the medial loop (*ML*), the horizontal segment, and the anterior loop (*AL*). Intradurally, the ICA bifurcation gives rise to the first segment of the anterior cerebral artery (*ACA1*) and the middle cerebral artery (*MCA*). The ON and ophthalmic artery (*OA*) are exposed. The diaphragm sellae and tumor beneath it are gently displaced downward (beneath the cottonoid) into the sella as tumor is being aspirated extradurally from below. Removing remaining tumor by displacing the diaphragm downward constitutes Phase VI of the tumor resection. The *arrow* on the suction tip indicates the direction of the extradural approach to the final part of the tumor, well beneath the diaphragm sellae. (With permission from V.V. Dolenc, Transcranial Epidural Approach to Pituitary Tumors Extending beyond the Sella, Neurosurgery, Vol. 41, No. 3, September 1997, p. 547)

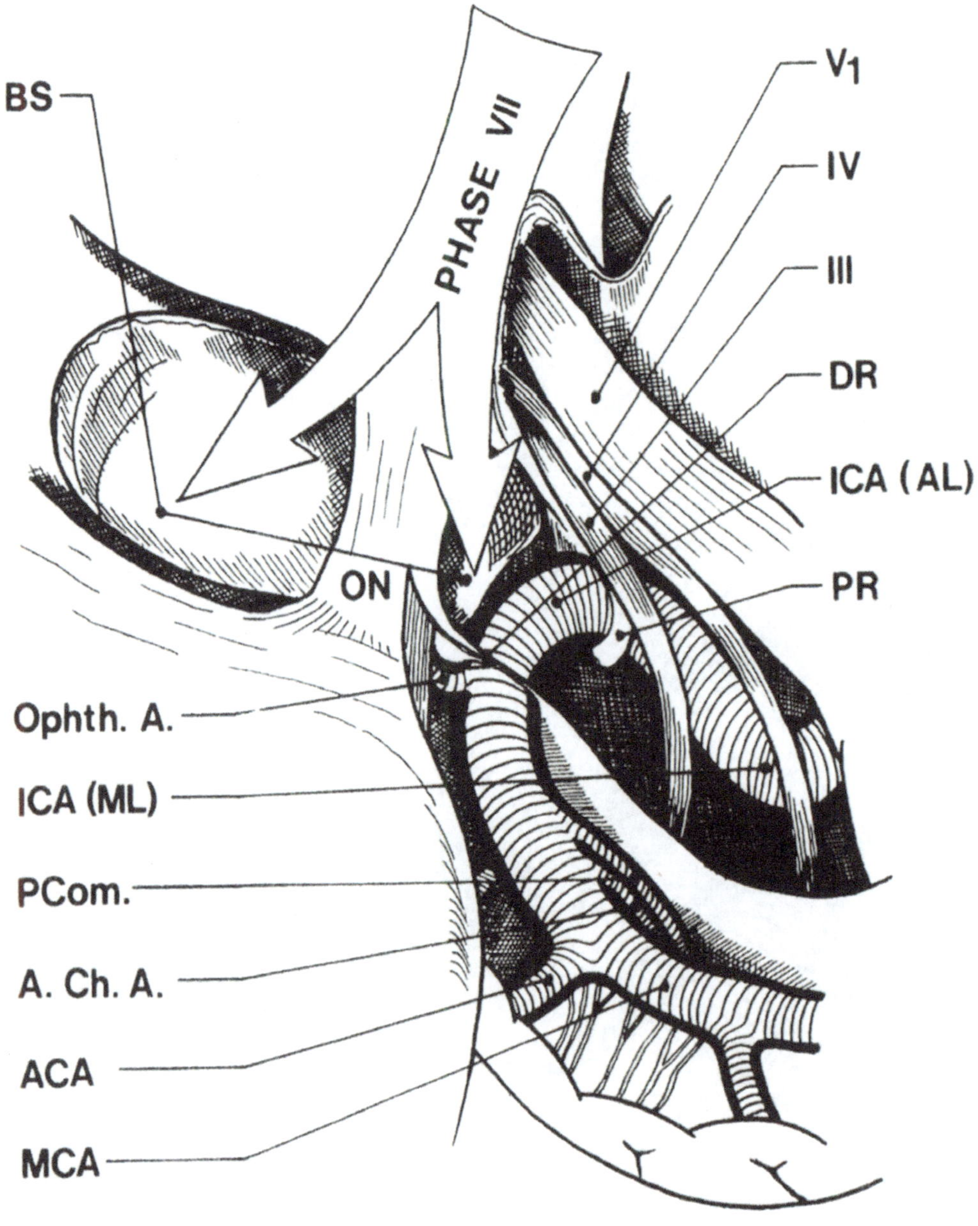

Fig. 4.53. The artist's rendering of Phase VII, including the right optic nerve (*ON*), the extradural and intradural ICA with its branches, and nerves III, IV, and V1. The tumor, extending into the sphenoidal and ethmoidal sinuses, is accessed through "windows" on either side of the ON. Phase VII applies only for those patients in whom the diaphragm sellae and dura at the floor of the central skull base (*CSB*) have been preserved intact. (With permission from V.V. Dolenc, Transcranial Epidural Approach to Pituitary Tumors Extending beyond the Sella, Neurosurgery, Vol. 41, No. 3, September 1997, p. 547)

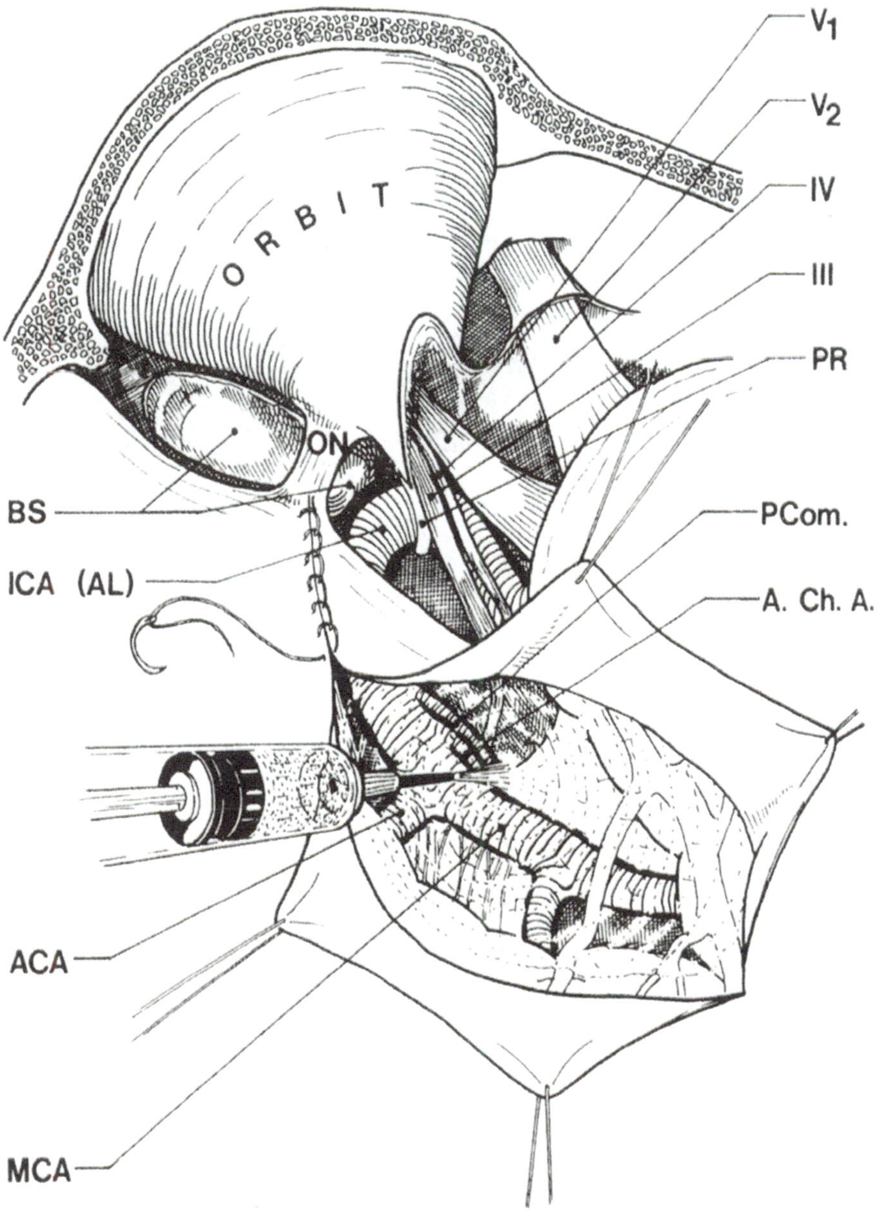

Fig. 4.54. The right parasellar and sellar regions are seen extradurally and the ICA (plus its branches) and sylvian region intradurally, including the sphenoidal sinus (*BS*), the optic nerve (*ON*) and its dura propria, the anterior loop (*AL*) of the ICA, and nerves III, IV, V1, and V2. The central part of the dural opening has been sutured. Filling the intradural space with saline confirms tshat dura covering the central skull base is intact. (With permission from V.V. Dolenc, Transcranial Epidural Approach to Pituitary Tumors Extending beyond the Sella, Neurosurgery, Vol. 41, No. 3, September 1997, p. 547)

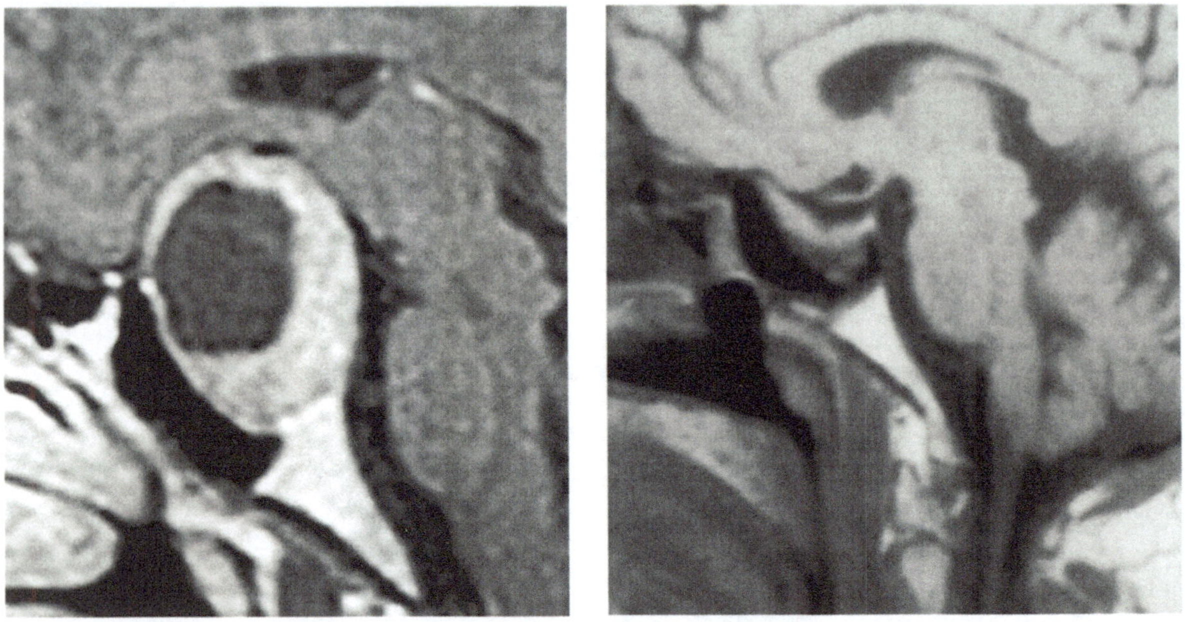

a b

Fig. 4.55. Preoperative, sagittal (a), contrast-enhanced MR image demonstrates a fairly large pituitary tumor extending into the suprasellar region. This patient had residual tumor after two transsphenoidal operations, resulting in a cavity within the tumor mass. Visual acuity temporarily improved following each operation but began to deteriorate after six months. Following the second transsphenoidal procedure, the patient agreed to have transcranial surgery which resulted in gross total tumor removal (b). Vision was restored to normal, and there has been no evidence of tumor recurrence after six years. (With permission from V.V. Dolenc, Transcranial Epidural Approach to Pituitary Tumors Extending beyond the Sella, Neurosurgery, Vol. 41, No. 3, September 1997, p. 548)

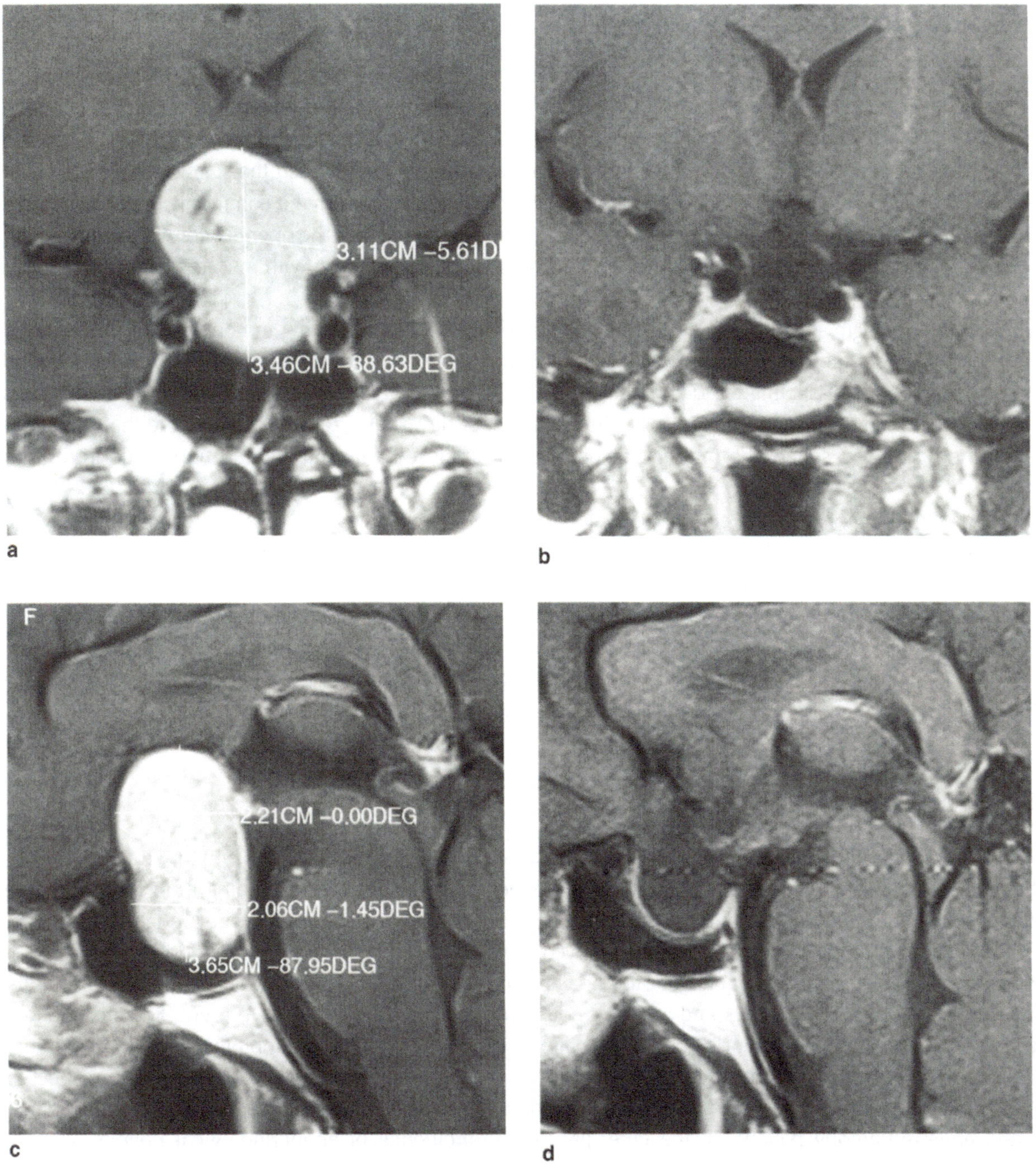

Fig. 4.56. Preoperative coronal (a) and sagittal (c) contrast-enhanced MR images demonstrate a pituitary tumor extending into the suprasellar space, displacing the visual apparatus upward. Corresponding postoperative coronal (b) and sagittal (d) contrast-enhanced MR images confirm total tumor resection. There was no diabetes insipidus following surgery

Special surgical circumstances

In cases where tumor protrudes intradurally through the oculomotor slit, it is extremely important to debulk it extradurally from beneath nerve III, introducing the suction tip through the anteromedial and paramedial windows while simultaneously controlling the tumor mass through the intradural exposure. Only when the mass has been reduced sufficiently that the central segment of nerve III (between the superior cerebellar and posterior cerebral arteries) is exposed, is it possible to dissect tumor safely from the commonly paper-thin nerve III throughout its intradural course to where it enters the lateral wall of the PS. The opening connecting intradural and extradural compartments at the site of nerve III's passage should be closed with muscle placed both inside and outside the dura.

Where the diaphragm sellae has been displaced extensively upward, resulting in compression of the optic chiasm, the ONs, and the hypothalamus, it is extremely important to avoid damaging Dawson's arteries and other perforators while dissecting the diaphragm sellae from the visual apparatus. Only if no bleeding occurs at this point may one be certain that these structures were not devascularized. In cases where the tumor has traversed the diaphragm sellae, the resulting defect should be closed with muscle and fibrin glue from both sides.

While removing tumor from the contralateral side of the sella, the surgeon must take special pains to avoid injuring the PS-ICA and nerves III and VI, also to avoid opening venous compartments of the PS.

If the diaphragm sellae and adjacent dura over the ipsilateral PS show signs of leakage after tumor removal, the entire extradural space should be lined with fat, muscle, and fibrin glue. By contrast, if dura over the CSB is demonstrated to be fully intact after tumor removal and hemostasis is perfect, these measures are unnecessary. If dura over the CSB requires closure with fat, muscle, and fibrin glue, patients should be treated with continuous lumbar drainage for 7 to 10 days after surgery.

Patients and results

Table 18 summarizes the surgical results of 411 consecutive patients harboring large pituitary tumors extending beyond the sella. During the first 10 years of our experience, 120 tumors extending from the sella were attacked by the transcranial intradural approach. These cases make up Group I. The updated approach, described previously [5] and now here, was employed in the remaining 291 patients (Group II).

Several conclusions seem apparent. First, the new approach resulted in a higher total tumor resection rate. This was true with both virgin cases and where residual and recurrent lesions were present. Second, the rate of postoperative visual improvement among those patients experiencing deterioration of vision prior to surgery was decidedly improved. Third, patients treated by the new method achieved better outcomes with respect to nerve III function. Fourth, there were fewer surgical complications (CSF leak, postoperative bleeding, infection, and epilepsy) among patients in the latter group. The incidence of endocrinological deficits was not particularly affected.

Preoperative diagnostic workups in all cases included cranial x-rays, CT scan, MRI, endocrinological evaluation by the Department of Endocrinology, and assessment of visual function (perimetry, visual acuity, and fundi) by ophthalmologists. Most virgin cases were referred either by ophthalmologists or endocrinologists. Patients with residual or recurrent tumors were transferred directly from other neurosurgical departments or from ophthalmologists and endocrinologists elsewhere. Digital subtraction angiography (DSA) was more exceptional than routine with virgin cases while it was virtually routine for most patients with residual lesions treated prior to the final three years of the series. From that point, all patients with residual and recurrent CSB pituitary tumors had MRI scans.

Table 18. Pituitary tumors treatment: resectability and results

Tumor resection, deficits, and surgical complications	Group I (120 patients) Virgin cases (49 patients)		Group I (120 patients) Residual and recurent tumors (71 patients)		Group II (291 patients) Virgin cases (141 patients)		Group II (291 patients) Residual and recurent tumors (150 patients)	
Tumor resection								
– Complete	31	(63%)	50	(70%)	136	(96%)	142	(95%)
– Incomplete	18	(37%)	21	(29%)	5	(3%)	8	(5%)
Visual field deficits and/or lower visual acuity								
– Preoperative	25	(51%)	55	(77%)	99	(70%)	125	(83%)
– Postoperative	20	(41%)	52	(73%)	7	(5%)	36	(24%)
– New	7	(14%)	14	(20%)	2	(1%)	4	(3%)
Cranial nerves III-VI deficits								
– Preoperative	27	(55%)	29	(41%)	70	(50%)	95	(63%)
– Postoperative	6	(12%)	7	(10%)	9	(6%)	10	(7%)
– New	1	(2%)	2	(3%)	3	(2%)	7	(5%)
Endocrinological deficits								
– Preoperative	45	(92%)	67	(94%)	120	(85%)	130	(87%)
– Postoperative	47	(96%)	70	(1%)	137	(97%)	143	(95%)
Surgical complications								
– CSF leak	3	6%	7	(10%)	–	–	4	(3%)
– Postoperative bleeding	1	2%	3	(4%)	–	–	2	(1%)
– Infection	–	–	1	(1%)	–	–	–	–
– Epilepsy	–	–	4	(6%)	–	–	–	–
– Other	–	–	–	–	–	–	–	–

Conclusions

The extradural-intradural approach to pituitary tumors extending beyond the sella allows total tumor excision with minimal risk of postoperative neurological worsening and other complications, obviating the need for postoperative radiation and other ancillary treatment. Nevertheless, the surgical challenges are sometimes great, particularly with very large tumors and those invading paranasal sinuses. Strict attention to detail is required.

References

1. Ahmadi J, North CM, Segall HD, Zee C-S, Weiss MH (1986) Cavernous sinus invasion by pituitary adenomas. AJR Am J Roentgenol 146: 257–262
2. Asari J, Yamanobe K, Sasaki T, Yamao N, Kodama N (1987) A case of prolactinoma associated with craniopharyngioma. No Shinkei Geka 15: 1313–1318
3. Brada M, Ford D, Ashley S, Bliss JM, Crowley S, Mason M, Rajan B, Traish D (1992) Risk of second brain tumour after conservative surgery and radiotherapy for pituitary adenoma. Br Med J 304: 1343–1346
4. Dolenc VV (1989) Anatomy and surgery of the cavernous sinus. Springer, Berlin Heidelberg New York Tokyo
5. Dolenc VV (1997) Transcranial epidural approach to pituitary tumors extending beyond the sella. Neurosurg 41(3): 542–552
6. Fusek I (1980) Remarks on the tactiPS and techniPS of transsphenoidal surgery for suprasellar pituitary adenomas. Zentralbl Neurochir 41: 209–214
7. Ghosh S, Chandy MJ (1992) Intrasellar tuberculoma. Clin Neurol Neurosurg 94: 251–252
8. Grisoli F, Vincentelli F, Raybaud C, Harter M, Guibout M, Baldini M (1983) Intrasellar meningioma. Surg Neurol 20: 36–41

9. Grosskopf D, Chamouard JM, Bosquet F, Billet R, Poisson M, Buge A (1987) Pituitary abscess: Study of a case and review of the literature. Neurochirurgie 33: 228–231
10. Guy J, Mancuso A, Beck R, Moster ML, Sedwick LA, Quisling RG, Rhoton Al Jr, Protzko EE, Schiffman J (1991) Radiation-induced optic neuropathy: A magnetic resonance imaging study. J Neurosurg 74: 426–432
11. Haddad SF, VanGilder JC, Menezes AH (1991) Pediatric pituitary tumors. Neurosurgery 29: 509–514
12. Hermier M, Turjman F, Tournut P, Laharotte JC, Sindou M, Froment JC, Doquesnel J (1994) Intracranial aneurysm associated with pituitary adenoma shown by MR angiography: Case report. Neuroradiology 36: 115–116
13. Jaffarin-Rea ML, Derome P, Bataini JP, Thomopoulos P, Bertagna X, Luton JP (1993) Influence of radiotherapy on long-term relapse in clinically non-secreting pituitary adenomas: a retrospective study (1970–1988). Eur J Med 2: 398–403
14. James RL Jr, Arsenis G, Stoler M, Neson C, Baran D (1984) Hypophyseal metastatic renal cell carcinoma and pituitary adenoma: Case report and review of the literature. Am J Med 76: 337–340
15. Jaunsolo MA, Aguirre M, Bellido D, Castro S, Ruiz-Vadlepenas MP, Hawkins FG (1986) Association of acromegaly and cerebral arterial aneurysm disclosed by a subarachnoid hemorrhage [in French]. Neurochirurgie 32: 266–268
16. Kline LB, Kim JY, Ceballos R (1985) Radiation optic neuropathy. Ophthalmology 92: 1118–1126
17. Kovacic JJ, Grigsby PW, Fineberg BB (1990) Recurrent pituitary adenomas after surgical resection: The role of radiation therapy. Radiology 177: 273–275
18. Krauss HR, Slamovits TL, Sibony PA, Verbalis JG, Nelson PB (1982) Carotid artery aneurysm simulating pituitary adenoma. J Clin Neuroophtalmol 2: 169–174
19. Nishio S, Tsukamoto H, Fukui M, Matsubara T (1992) Hypophyseal metastatic hypernephroma mimicking a pituitary adenoma: case report. Neurosurg Rev 15: 319–322
20. Ooi TC, Russell NA (1986) Hypopituitarism resulting from an intraselar carotid aneurysm. Can J Nurol Sci 13: 70–71
21. Pasztor E, Kemeny AA, Piffko P (1983) Transsphenoidal surgery for suprasellar pituitary adenomas. Acta Neurochir (Wien) 67: 11–17
22. Paulus P, Paridaens R, Mockel J, Coffernils M, Dhaens J, Baleriaux D, Rodesch G, Rutsaert J, Depierreux M, Flamment-Durand J (1990) Argyrophilic breast carcinoma, single metastasis to the pituitary gland. Bull Cancer (Paris) 77: 377–384
23. Ross DA, Wilson CB (1988) Results of transsphenoidal microsurgery for growth hormone-secreting pituitary adenoma in a series of 214 patients. J Neurosurg 68: 854–867
24. Sassolas G, Trouillas J, Treluyer C, Perrin G (1993) Management for nonfunctioning pituitary adenomas. Acta Endocrinol 1: 21–26
25. Sato K, Hayashi M, Komai T, Kubota T, Kawano H, Handa Y (1990) Clinical and histological study of pituitary fibrosarcoma following radiotherapy for pituitary adenoma: case report. Neurol Med Chir (Tokyo) 30: 888–892
26. Saul RF, Hilliker JK (1985) Third nerve palsy: The presenting sign of a pituitary adenoma in five patients and the only neurological sign in four patients. J Clin Neuroophtalmol 5: 185–193
27. Scotti G, Yu CY, Dillon WP, Norman D, Colombo N, Newton TH, De Groot J, Wilson CB (1988) MR imaging of cavernous sinus involvement by pituitary adenomas. AJR Am J Roentgenol 151: 799–806
28. Shi T, Farrell MA, Kaufmann JC (1984) Fibrosarcoma complicating irradiated pituitary adenoma. Surg Neurol 22: 277–284
29. Steiner E, Knosp E, Herold CJ, Kramer J, Stiglbauer R, Staniszewski K, Imhof H (1992) Pituitary adenomas: findings of postoperative MR imaging. Radiology 185: 521–527
30. Sullivan LJ, O'Day J, McNeill P (1991) Visual outcomes of pituitary adenoma surgery: St Vincent's Hospital 1968–1987. J Clin Neuroophtalmol 11: 262–267
31. Suzakawa K (1989) Evaluation of the transcranial approach to pituitary adenomas based on quantitative analysis of pre- and postoperative visual function. Neurol Med Chir (Tokyo) 29: 1012–1019
32. Symon L, Logue V, Mohanty S (1982) Recurrence of pituitary adenomas after transcranial operation. J Neurol Neurosurg Psychiatry 45: 780–785
33. Tachibana O, Yamaguchi N, Yamashima T, Yamashita J (1990) Radiation necrosis of the optic chiasm, optic tract, hypothalamus, and upper pons after radiotherapy for pituitary adenoma, detected by gadolinium-enhanced, T1-weighted magnetic resonance imaging: case report. Neurosurgery 27: 640–643
34. Takahashi T, Kuwayama A, Katoh T, Ichihara K, Kageyama N, Sasaki F (1984) Size reduction of a nonfunctioning pituitary adenoma after bromocritpine therapy: Case report. No Shinkei Geka 12: 71–80
35. Tsuchida T, Tanaka R, Yokoyama M, Sato H (1983) Rupture of anterior communicating artery aneurysm during transsphenoidal surgery for pituitary adenoma. Surg Neurol 20: 67–70
36. Van Schaardenburg D, Roelfsema F, van Seters AP, Vielvoye GJ (1989) Bromocriptine therapy for non-functioning pituitary adenoma. Clin Endocrinol 30: 475–484
37. Weir B (1992) Pituitary tumors and aneurysms: Case report and review of the literature. Neurosurgery 30: 585–591
38. Wilson CB (1990) Role of surgery in the management of pituitary tumors. Neurosurg Clin N Am 1: 139–159

4.3

Trigeminal neurinomas

Introduction

While neurinomas account for 2 to 3% of intracranial tumors, only 0.2% arise from the trigeminal nerve. They may occur at any age, but most are detected in the 4th decade. Men and women are affected equally [6, 27]. Bilateral trigeminal neurinomas are generally manifestations of the central form of Recklinghausen's disease [37].

Since these tumors may originate from different segments of nerve V, their precise location varies, those arising from the root of nerve V typically in the cerebello-pontine angle (CPA) and those arising from the Gasserian ganglion (GG) in Meckel's cave. Tumors originating from the nerve V root and the GG commonly remain intradural, and do not extend into the parasellar space (PS). The opposite is also true for lesions arising from one of the nerve V peripheral branches, such tumors generally remaining confined to the PS, failing to invade intradural space. Particularly large lesions may extend peripherally, eroding bone into one or more paranasal sinuses, the orbit, even extracranially beneath the skull base or within it laterally [2, 8, 15, 19, 24, 27, 38]. Trigeminal neurinomas invading the orbit may compress the optic nerve (ON) directly after eroding the anterior clinoid process (ACP) and bony wall(s) of the optic canal (OC), resulting in visual deterioration as the initial symptom [18, 31, 32]. Trigeminal neurinomas originating from the GG or the plexiform segment of the nerve V root may achieve significant size, extending in all directions while remaining confined to Meckel's cave, bowing its dura outward, resulting in diplopia from paresis of nerve VI [7, 9, 10, 21, 35, 39].

Symptoms and signs related to nerve V function usually occur subtly and progress steadily, but only occasionally are sensory deficits associated with trigeminal neuralgia [14, 15, 34]. In rare instances, the sudden onset of severe ipsilateral headache – due to intratumoral hemorrhage – announces the tumor's presence, and occasionally, lancinating pain in the distribution of the ipsilateral nerve V [3, 4, 20, 33].

Most of these lesions are entirely benign, but malignant lesions may occur in both adults and the young [18, 23]. Malignant schwannomas have been induced experimentally in rats with N-ethyl-N-nitrosourea [5, 17].

Currently, CT or MRI make the diagnosis of trigeminal neurinomas a relatively simple task, in stark contrast to an earlier era when atypical clinical presentations and slowly evolving clinical symptoms and signs made diagnosis extremely difficult [1, 13, 30, 32]. Sometimes plain x-rays suggest the presence of disease by erosions of the petrous apex and/or of the ACP, findings as important now as prior to the advent of CT and MRI. Today's computerized imaging techniques relegate angiographic studies, once so crucial for the diagnosis of these lesions [36], to largely historical significance.

Once diagnosis is established, the primary choice of treatment is surgical excision. Until recently a number of surgical approaches were advocated, primarily because of the protean locations of these lesions as dictated by their origin from different elements of nerve V [16, 22, 25, 26, 28, 29, 40]. The frontotemporal epidural approach we described previously [11, 12] offers the realistic possibility that such tumors, regardless of size or location, may be totally excised in a single stage with no additional neurological deficit. Since trigeminal neurinomas almost never require emergency surgery (even with intratumoral hemorrhage), it is far wiser to consider the full range of therapeutic options, even obtain a second neurosurgical opinion, than embark on an intradural approach which is destined to result in incomplete resection and the temptation to egregiously irradiate benign residual tumor. Our experience with a few patients who declined surgery suggests that the natural history of the disease is not yet clear, and will likely require cooperative studies for precise definition.

Surgical anatomy of nerve V

The root of nerve V and the GG are intradural, while the peripheral branches (V1, V2, V3, and the motor branch peripheral to the GG) are extradural, coursing in three different directions, V1 to the superior orbital fissure

(SOF), V2 to the foramen rotundum (FR), and V3 and the motor branch to the foramen ovale (FO). Originating from the pons, the motor branch of nerve V courses ventral to the sensory root through intradural space, bypassing the GG, then passing extradurally with V3 through FO to the muscles of mastication it enervates. V3 and the motor branch course between the inner and outer layers of the lateral wall of the PS, posteroinferior to the PS. V1 and V2 also pass from the GG between dural layers to their corresponding exit foramina. The segment of V3 from the GG to the FO is shorter than the V1 and V2 segments between the GG to the SOF and FR respectively.

The lateral 4–5 mm of the sensory root, its "plexiform segment," constitutes a curiosity of microsurgical anatomy that carries unique surgical significance. At this point fibers of the root abruptly lose their basically parallel orientation, becoming more braided in appearance, interwoven and diagonal as they exit from the GG. Dissecting lesions involving this part of the nerve typically result in additional neurological deficit since injury to microscopic fibers crossing perpendicular to the longitudinal axis of the nerve can hardly be avoided.

In order to fully understand CT and MR findings and assimilate them with the individual nerve V elements as is necessary to program an ideal surgical approach, it is important to know in detail the structure and location of each segment from the brainstem to their corresponding exit foramina at the skull base, including their precise extradural relationships to neighboring structures, particularly the bones of the skull base, dura, nerves III, IV, VI, the greater petrosal nerve (GPN), and the PS-ICA.

Since trigeminal neurinomas may arise from any segment of nerve V (from the pons to where branches traverse skull base foramina and beyond), it is also important for the surgeon to understand the pathological anatomy, which essentially amounts to correlating the CT and MR findings in individual cases to a precise knowledge of the normal anatomy of nerve V.

Pathological anatomy of nerve V

Trigeminal neurinoma, Type I: These tumors arise from one or more branches of nerve V or from the peripheral surface of the GG, then spread anterolaterally and peripherally along V1 and/or V2 and/or V3, thus involving the PS (Fig. 4.57). They may occupy the PS partially or entirely. Particularly large lesions typically exert continuous pressure on surrounding structures, resulting in characteristic clinical pictures, as well eroding the skull base, especially the ACP (Fig. 4.58a) and/or the apex of the petrous pyramid (Fig. 4.58b).

Trigeminal neurinoma, Type II: These lesions originate from the GG and/or the plexiform segment of the nerve V root and, depending on their size at the time of diagnosis, may remain completely or predominantly within Meckel's cave (Fig. 4.59). If further growth occurs, the petrous apex is eroded, followed by the spreading of tumor into the CPA as seen in Fig. 4.60a, b.

Trigeminal neurinoma, Type III: Tumors of this type originate from any segment of the nerve V root (Fig. 4.61) between the GG in Meckel's cave and the pons, then occupy the CPA as illustrated in Figs. 4.62a, b, 4.63b, and 4.64a, c, e. When such neurinomas become large, filling both the CPA and Meckel's cave, they may exert pressure on the brainstem, nerves IV, VII, VIII, even nerve VI.

Trigeminal neurinoma, Type IV: Tumors in this category may arise from different segments of nerve V, occupying the entire PS, Meckel's cave, and the CPA (Figs. 4.65 and 4.66). Such tumors arising from the GG or the plexiform segment of the root may become extremely large, occupying the PS as well as the CPA before diagnosis is established. They offer special surgical challenges.

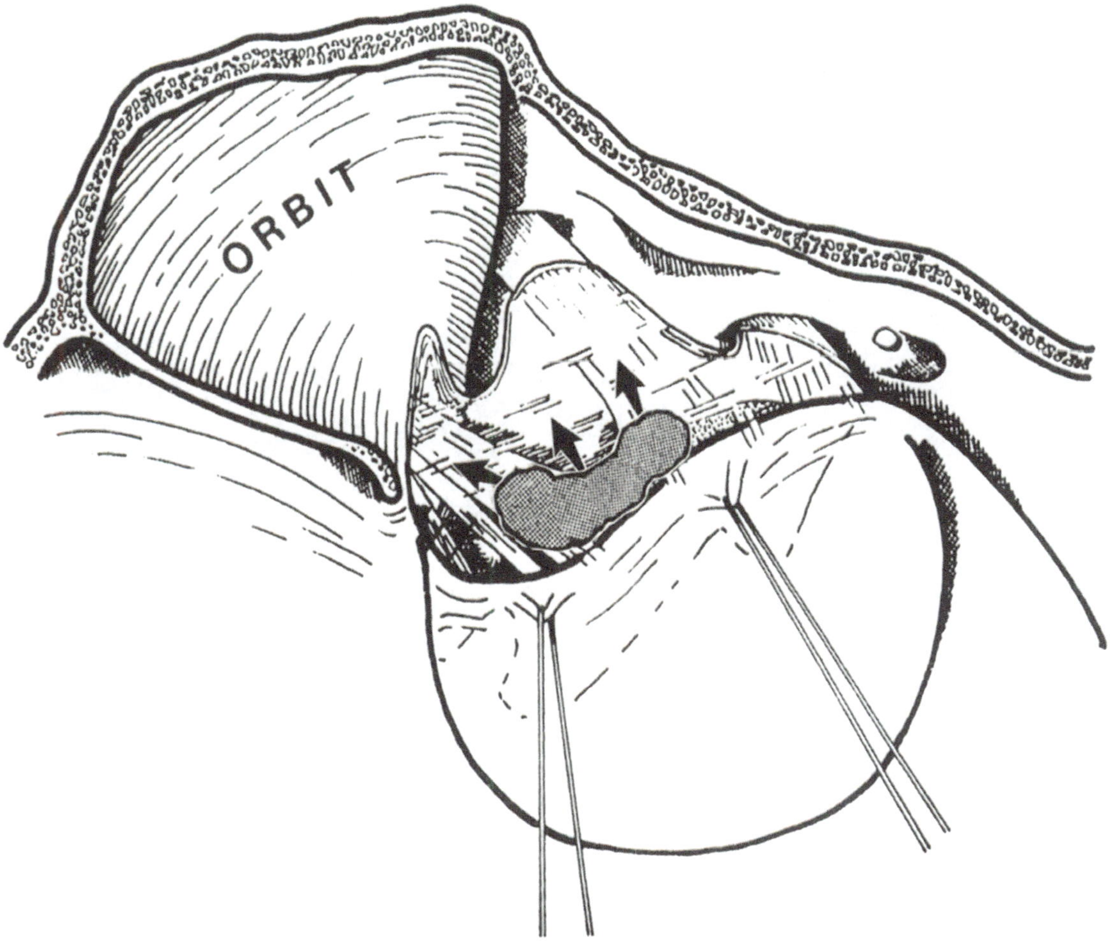

Fig. 4.57. A drawing of the extradural exposure of the parasellar region on the right side. The orbit has been unroofed, and nerves III, IV, V1, V2, V3, and the peripheral edge of the Gasserian ganglion (*GG*) are shown. Dura beneath the temporal lobe has been peeled from the lateral wall of the parasellar space (*PS*) and fixed with stay sutures. The middle meningeal artery (*MMA*) has been transected at the foramen spinosum (*FS*). This Type I trigeminal neurinomas arise from V1, V2, V3 and/or GG and are located between the outer (dural) and inner (areolar) layers of the lateral wall of the PS, invading the PS since the areolar layer is incomplete. They may grow along peripheral sensory branches (V1, V2, V3), even into the orbit and/or extracranially

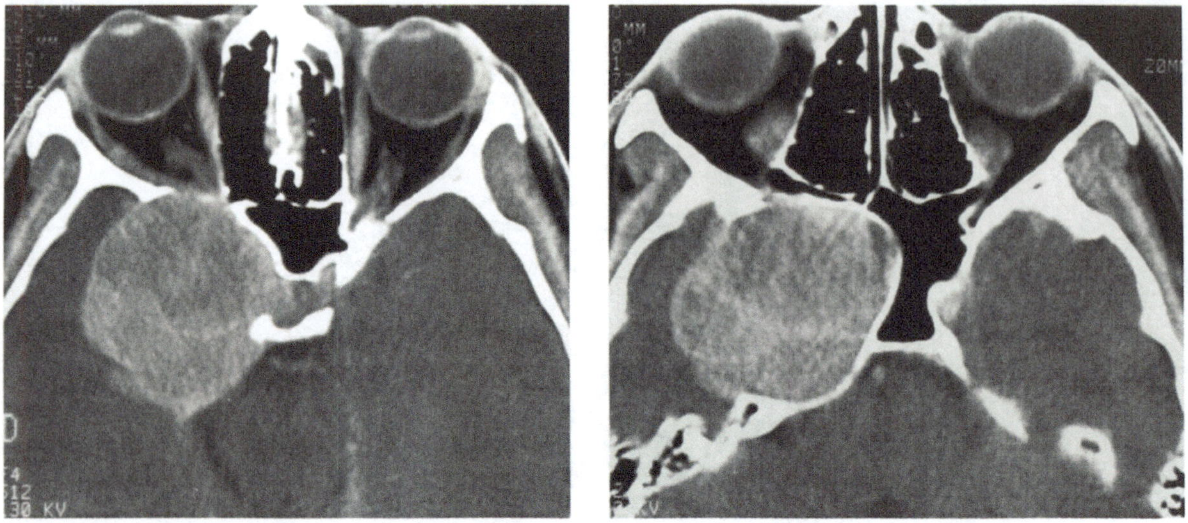

a b

Fig. 4.58. This CT scan reveals a large right-sided Type I parasellar neurinoma found incidentally in a 17-year-old student after routine screening testing of vision. Note, in (a), erosion of the right anterior clinoid process (*ACP*) (no longer visible while the left ACP remains intact), the lateral wall of the optic canal (*OC*), and the sphenoid wing. Vision in the right eye was impaired (although the patient was unaware of it) from compression of the optic nerve (*ON*). Erosion of the right petrous apex, dramatic inward bulging of the lateral wall of the sphenoidal sinus, and erosion of the sphenoid wing (b) reflect long-standing bone compression. Following complete resection of this lesion, the patient's visual function returned and has remained normal for 12 years without other evidence of tumor recurrence

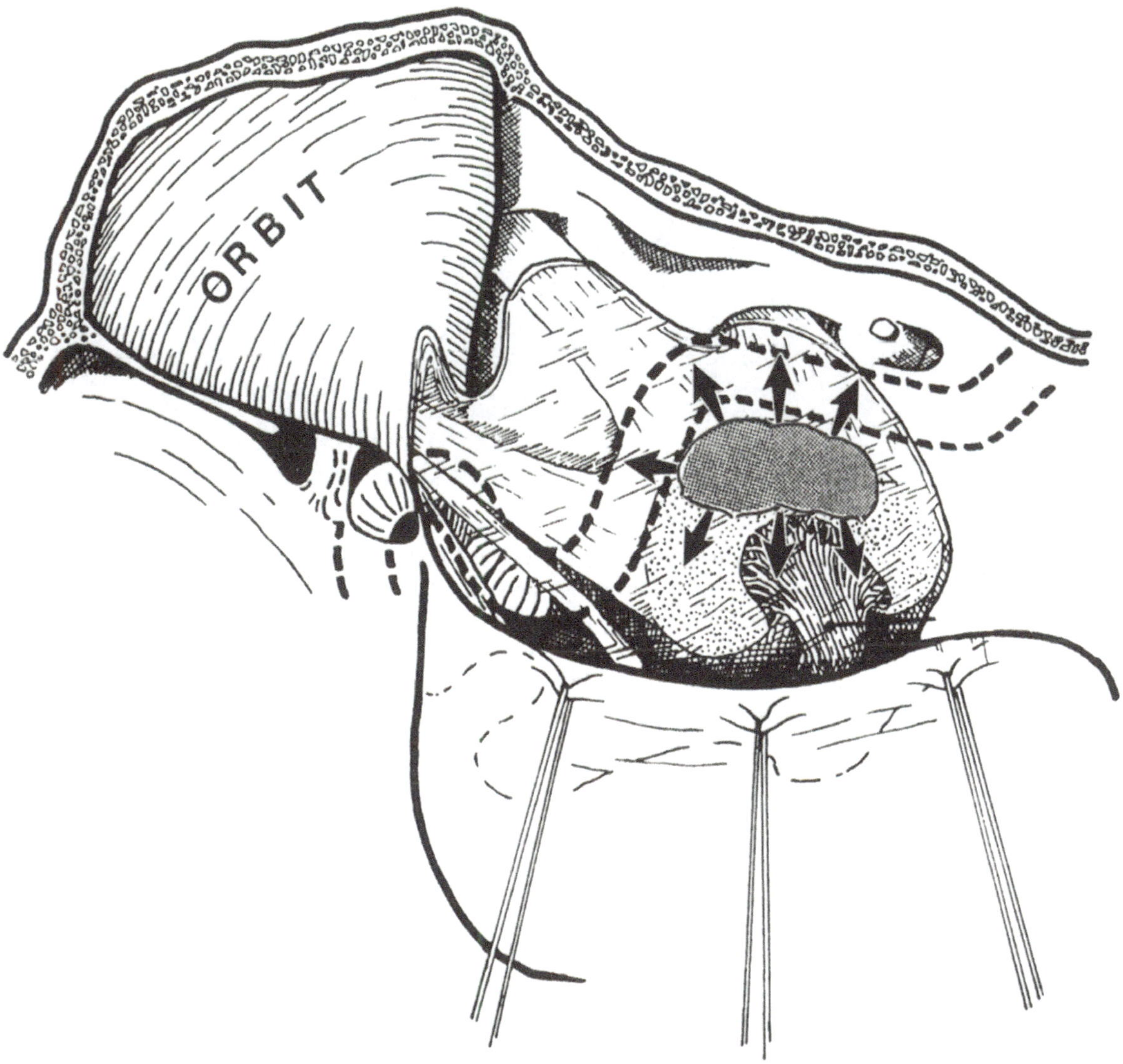

Fig. 4.59. A drawing of the surgical exposure of the right parasellar region, indicating nerves III, IV, V1, V2, V3, the Gasserian ganglion (*GG*), the plexiform segment of the root of nerve V, the four loops of the ICA in the skull base (dotted lines), and the central stump of the middle meningeal artery (*MMA*). The origin of the tumor from the GG is indicated. Type II trigeminal neurinomas grow within the GG, compress it, then fill Meckel's cave. As indicated here, such tumors expand anteriorly, laterally, medially, posteriorly, and may erode the petrous apex, extending into the cerebello-pontine angle (*CPA*). Anteromedial growth leads to compression of nerve VI against bone at its exit from Dorello's space where it enters the parasellar space (*PS*). Nerve VI paresis, resulting in diplopia, may be the first sign a trigeminal neurinoma is present

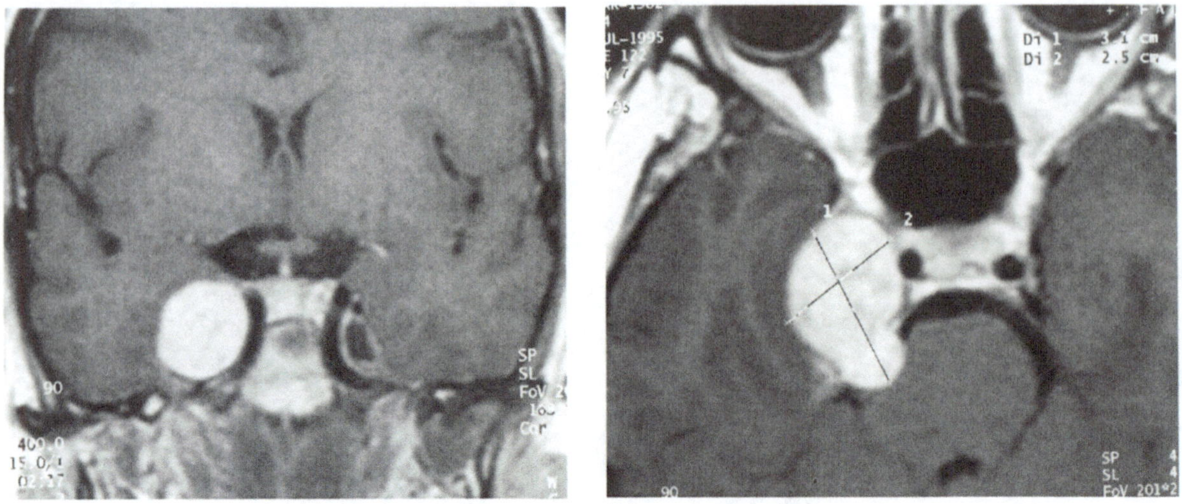

a b

Fig. 4.60. Coronal (a) and trans-axial (b) MR images demonstrate a typical Type II trigeminal neurinoma originating within the Gasserian ganglion (*GG*) and compressing (but not invading) the parasellar space (*PS*) anteriorly, extending into the posterior cranial fossa. The neurinoma is located lateral to the ascending segment of the ICA, which courses from its lateral loop (*LL*) toward its medial loop (*ML*), adjacent to the posterior clinoid process (*PCP*). In (a), comparing the adjacent ICA to the contralateral ICA indicates that its diameter is normal. Because the membrane between the PS and Meckel's cave has been displaced anterolaterally, nerve VI is compressed against bone where nerve VI exits Dorello's space, entering the PS. Although Meckel's cave is enlarged (the mass is 2.5 × 3.1 cm), tumor did not penetrate the PS. Note the lesion's rounded contour anteriorly and the small (compressed) PS anterior to it (b)

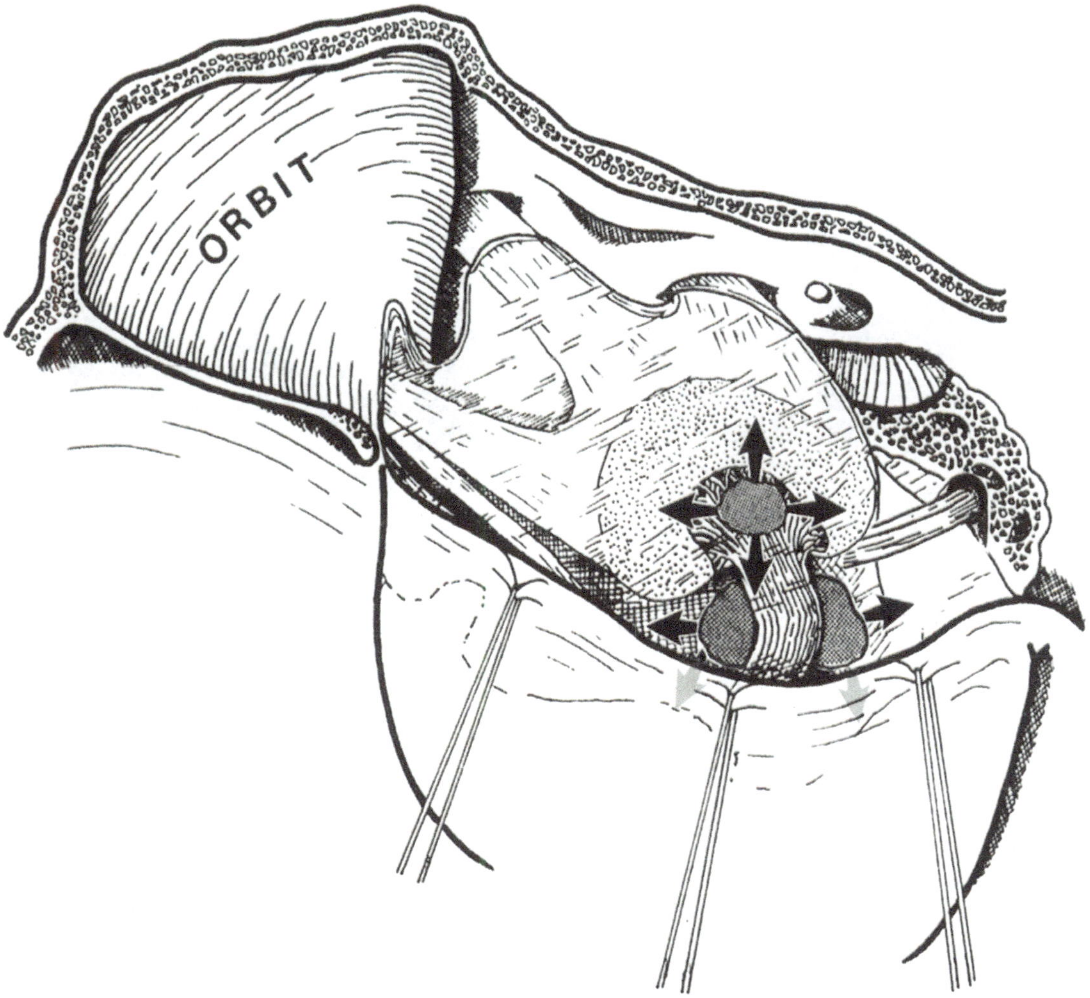

Fig. 4.61. A drawing of the right parasellar region after exposing nerves III, IV, V1, V2, V3, the Gasserian ganglion (*GG*), the root of nerve V, and nerves VII and VIII. Type III trigeminal neurinomas originate from the root of nerve V. They may remain limited to the cerebello-pontine angle (*CPA*) but may enter Meckel's cave by eroding the petrous apex

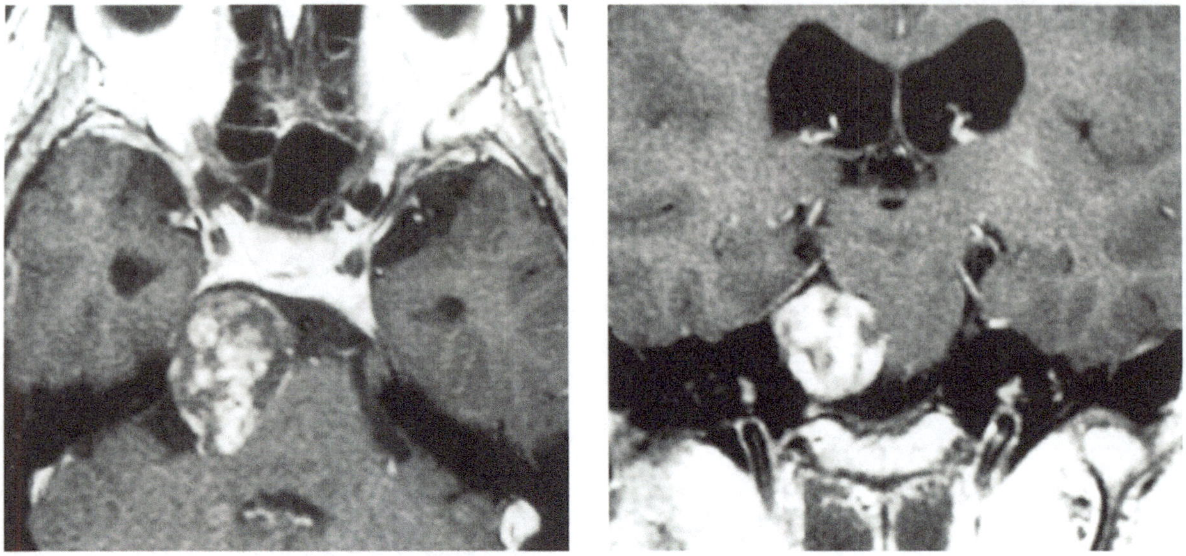

Fig. 4.62. Trans-axial (a) and coronal (b) MR images reveal a Type III trigeminal neurinoma within the cerebello-pontine angle (*CPA*). As in (a), Type III tumors originate from the root of nerve V and may compress the pons. Note the pontine indentation. This tumor does not extend across the petrous apex into Meckel's cave. In (b), the pontine "defect" from tumor growth is also apparent, the tumor remaining confined to the CPA

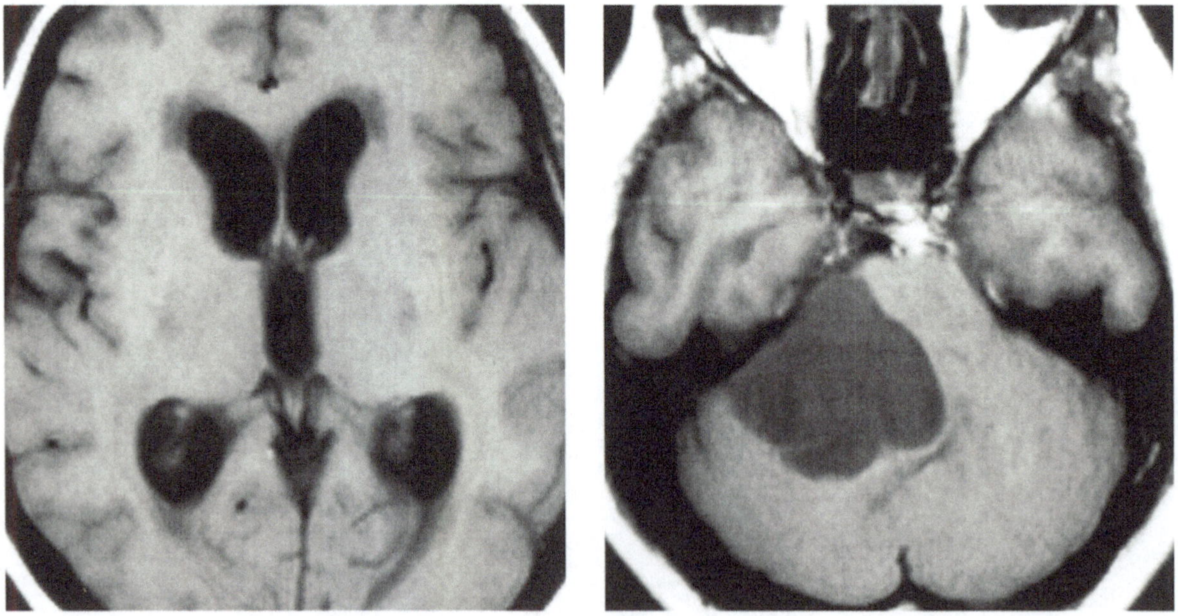

Fig. 4.63. Preoperative non-enhanced transaxial MR images at the level of the third ventricle (a) and at the level of CPA (b) show a large tumor in the right CPA which causes dilatation of the supratentorial ventricular system. In (a), periventricular lucences are apparent adjacent to the frontal horns of the lateral ventricles. This tumor is partially cystic. Although it compresses the brainstem and the right cerebellar hemisphere, no surrounding edema is present. In (b), the IV-th ventricle is almost entirely effaced, shifted away from the mass. Note: In such cases, draining the supratentorial ventricles while the tumor is present may be dangerous, posing a risk of upward herniation. Approaching lesions of this type directly, totally resecting them, simultaneously solves the problems of hydrocephalus and compression of the brainstem and cerebellum

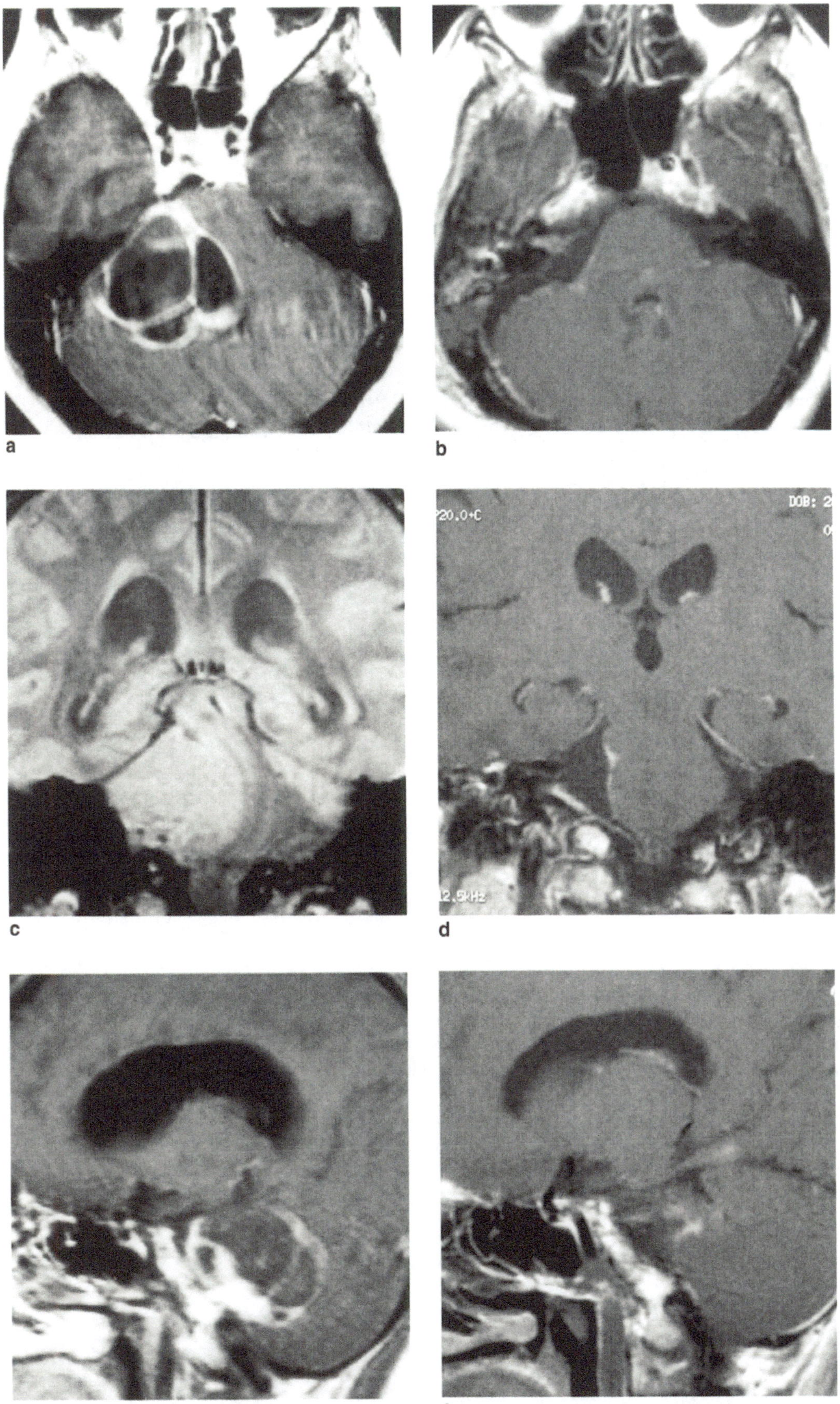

Fig. 4.64. Transaxial (a), coronal (c), and sagittal (e) contrast-enhanced MR images of the primarily cystic tumor shown in Fig. 4.63. Corresponding postoperative transaxial (b), coronal (d), and sagittal (f) contrast-enhanced MR images confirm total resection of the tumor. Hydrocephalus is no longer present, and the brainstem and cerebellum have resumed their normal shapes. The IV-th ventricle is fully patent. Note: This tumor originated from the root of nerve V, near the pons. Except for slight sensory loss in the distribution in all three branches of right trigeminal nerve, the patient made a complete recovery, leaving the hospital on the 7th postoperative day. There was no nerve V motor deficit

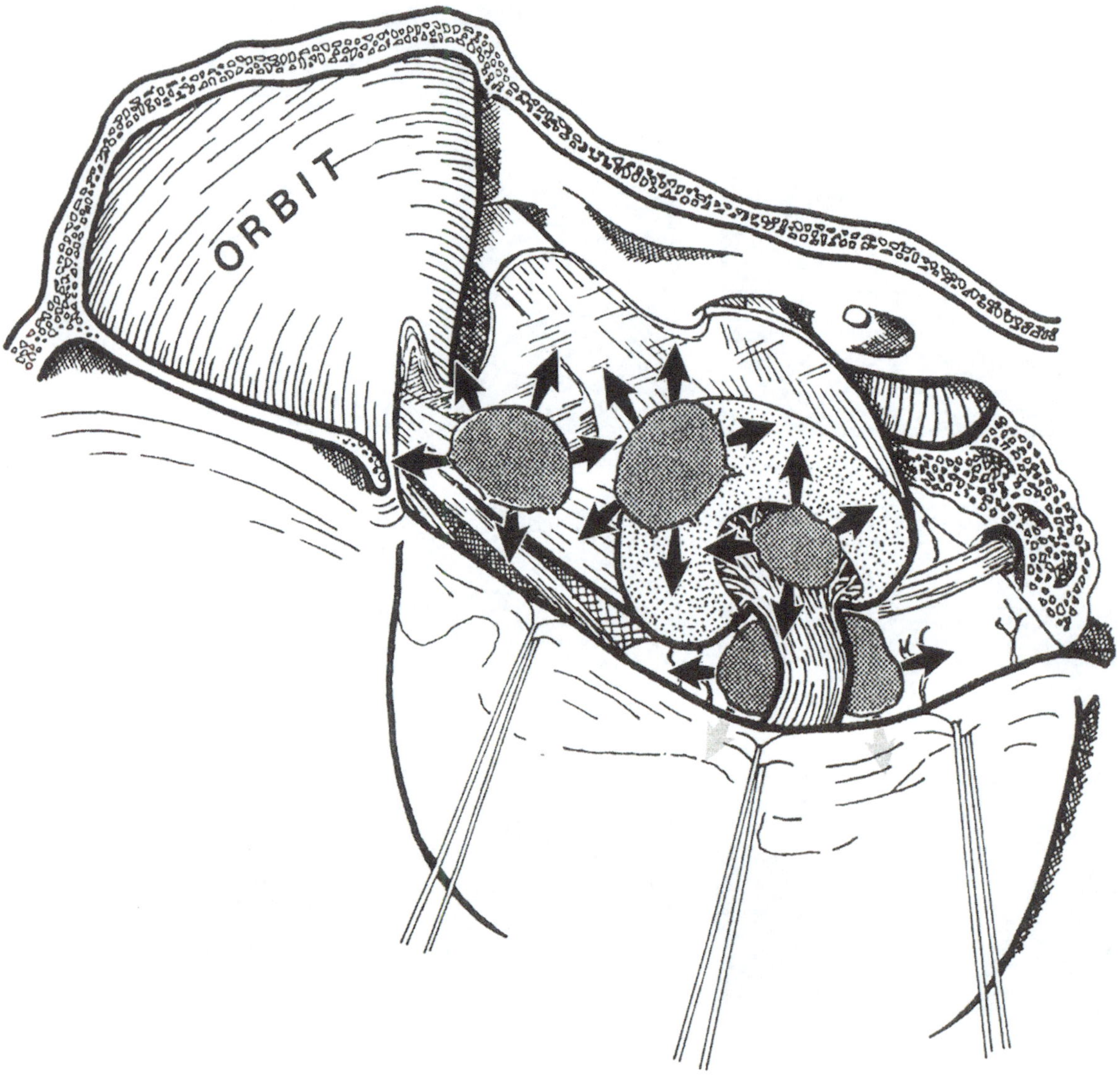

Fig. 4.65. A drawing of the right parasellar region after dura has been peeled from the lateral wall of the parasellar space (*PS*), the Gasserian ganglion (*GG*), and the plexiform segment of the nerve V root. The posterior loop (*PL*) of the intrapetrous carotid artery is indicated, along with nerves VII and VIII coursing into the internal auditory canal (*IAC*). The origin(s) of Type IV trigeminal neurinomas may be one or several, making them essentially a combination of Types I, II, and III

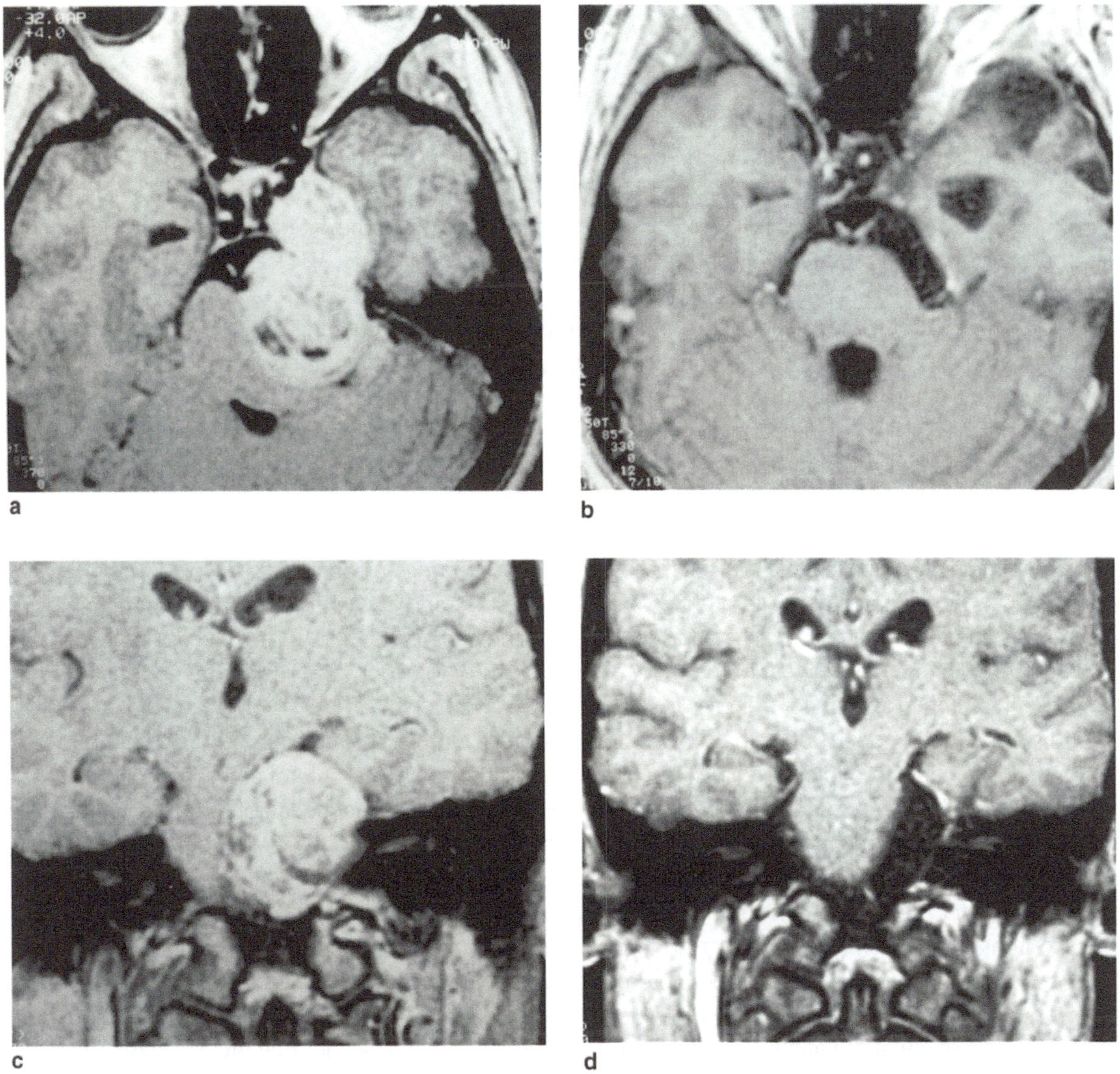

Fig. 4.66. Preoperative trans-axial (a) and coronal (c) and corresponding postoperative MR images (b,d) of a Type IV trigeminal neurinoma. The preoperative cuts indicate tumor in the cerebello-pontine angle (*CPA*) with a larger element in the parasellar region. The brainstem is severely compressed from the left side, and the proximal end of the ventricle IV has been shifted and distorted. This tumor was removed by the extradural approach in a single stage, its intradural (*CPA*) component resected by the transpetrosal route. The postoperative coronal cut (d) confirms total tumor removal with "re-expansion" of the brainstem

The surgical approach

Type I trigeminal neurinomas

Although the approach is generally that to the parasellar and adjacent central skull base regions as described in Chapter 2, specific alterations apply, depending on which type of tumor is to be operated. Patient positioning, craniotomy, unroofing the orbit, resecting the sphenoid wing to the lateral border of the SOF, dissecting the outer layer of the dural wall of the PS and exposing nerves III, IV, V1, V2 and V3 is the same for all four types.

With Type I neurinomas, those confined to the PS, resecting the ACP, opening the OC, and removing the optic strut should be accomplished as described in Chapter 2. Unroofing the petrous carotid canal is strongly recommended when these neurinomas displace or distort the PS ICA. When V1 function is absent preoperatively, indicating that the tumor probably originated from V1, this branch may be dissected vigorously, even destroyed, with impunity. When unroofing the ICA in the petrous bone, it is important that the greater petrosal nerve (GPN) be preserved at all cost. The GPN is crucial to protecting the ipsilateral eye. Large neurinomas originating from V1 are usually associated with significant erosion of the ACP (Fig. 4.58a). Partial destruction of the ACP facilitates its total removal but increases the risk of injuring surrounding structures, including the optic nerve (ON), the anterior loop (AL) of the ICA, and nerves III and IV when the ACP remnant is removed. With increasing surgical experience in the parasellar region, and because the ICA is only occasionally encased within or compressed by nerve V neurinomas, it is rarely necessary to expose the ICA in the petrous canal for the purpose of gaining proximal control.

When dissecting the lateral wall of the PS with most of these tumors, it is sufficient to expose the following anatomical triangles: the anteromedial, the paramedial, Parkinson's, the anterolateral, the lateral, Glasscock's, and Kawase-Shiobara's. After separating dura (the outer layer of the lateral wall of the PS) from the often incomplete fibrous membrane making up the inner layer, thus exposing nerves III, IV, V1, V2, V3 and GG, generally it becomes relatively easy to identify the most appropriate "window" between individual nerves through which one may enter the PS and remove the neurinoma. Some tumors may be resected primarily by suctioning, but the consistency of many requires resorting to the piece-meal technique. It is important to preserve what fibers of V1, V2 and/or V3 were spared by tumor growth. Great care must be taken also when dissecting tumors from the PS-ICA and nerve VI, which is typically displaced laterally, becoming located near bone at its exit from Dorello's space at the angle circumscribed by bone and the petroclinoid ligament. Nerve VI should be identified and dissected from tumor at this point, then followed along its course beneath V1 throughout the PS.

Resecting neurinomas from medial compartments of the PS may be achieved through Parkinson's triangle, which normally is the safest way to preserve cranial nerves III and IV function. Rarely these nerves are intricately involved with the tumor, requiring painstaking dissection. Once the tumor has been removed from the cranial nerves, it may be dissected from the ICA and its branches, the meningo-hypophyseal trunk (MHT) and the infero-lateral trunk (ILT). If the surgeon has the ability to preserve these two branches, in most cases he is sufficiently skilled to gently and completely skeletonize the ICA, making unroofing the ICA in the petrous bone unnecessary. One must be aware, however, that by not unroofing the ICA within the petrous bone the surgeon may face insurmountable difficulties should the ICA be injured within the PS. On the other hand, not disturbing the ICA in the petrous canal provides additional security for the GPN, hence minimizing the risk of eye injury following surgery.

After complete removal of the tumor, cranial nerves III, IV, V1, V2, V3, VI, and the ICA, including the MHT and ILT and sympathetic fibers attached to its wall, should be fully exposed throughout the PS. Venous bleeding from connecting sinuses between the left and the right PSs (inside the sella and dorsal to the dorsum selle and clivus) and the superior petrosal sinus (SPS) may be controlled with Surgicel. One is cautioned against placing excessive Surgicel into the PS in order to avoid displacing or compressing the ICA. Once complete hemostasis has been established, coating the PS cavity with fibrin glue is often a better alternative than lining it with Surgicel as a means of preventing postoperative bleeding.

Type II trigeminal neurinomas

Most neurinomas originating from the plexiform segment of the root of nerve V or from the GG are confined to Meckel's cave and are classified as Type II tumors (Fig. 4.59). With extension beyond Meckel's cave they invade the CPA and/or displace the dural membrane beneath the GG, which separates the PS from Meckel's cave anteriorly. Tumor within Meckel's cave compresses PS compartments and their structures anteriorly, thus compromising function of nerve VI by displacing it laterally, against bone. Such Type II neurinomas take on characteristics of Type IV lesions, and conceivably may be placed in either category.

In approaching small and medium-sized Type II neurinomas (even some of the larger lesions), the craniotomy, unroofing the orbit, and separating the layers of the lateral wall of the PS should be performed as described in Chapter 2. With these tumors it is unnecessary to resect the sphenoid wing, the ACP, or open the OC. By transecting dura at the SOF, the outer layer of the lateral wall of the PS may be identified and subsequently separated from the inner layer, thus exposing nerves III, IV, V1, V2 and V3. In most cases it is not necessary to dissect nerves III and IV from the inner layer of the lateral wall of the PS, but dissecting the outer (dural) layer from V1, V2, V3 and from GG is mandatory. In addition, dura must be peeled from the anterior surface of the petrous bone, where special attention is required in order to preserve the GPN. One must also be alert for the presence of the geniculate ganglion since it may not be covered by bone. In order to avoid damaging the geniculate ganglion while dissecting the GPN, dura must be peeled from the anterior surface of the petrous bone quite meticulously by sharp dissection. Dissecting dura from the anterior surface of the petrous bone should be continued to its superomedial margin, the site of the SPS. Finally, dura beneath the temporal lobe is peeled from the medial edge of the GG, exposing the medial surface of the tumor, the site of the plexiform segment of the nerve V root.

Since the GG in Meckel's cave is covered by dura and arachnoid, tumor is exposed by separating dura and incising the arachnoid. Debulking the tumor may be accomplished from its medial surface, after exposing the root of nerve V. Only then may one safely begin resecting tumor, slowly and carefully, insuring that the root of nerve V, the GG, V1, V2, V3, and the motor branch are preserved. Since the tumor is confined to Meckel's cave, it is not necessary to drill away the petrous apex. With most Type II trigeminal neurinomas, the tough dural membrane separating Meckel's cave from the PS dorsal compartments remains intact. It is not traversed by tumor. The PS may be simply displaced anteriorly, gently "squeezed." Opening it is not necessary. After complete removal of the tumor, the dural defect at Meckel's cave must be sealed with muscle and fibrin glue.

Type III trigeminal neurinomas

When the neurinoma extends to both sides of the petrous apex, into both the CPA and Meckel's cave, the initial approach is the same as with Type II tumors. In addition, after exposing V1, V2, V3, the GG, and tumor in Meckel's cave, the apex of the petrous bone is drilled away as described in Chapter 2. The extent of its removal depends on the amount of tumor in the CPA. The nerve V root must be exposed at the pons. From there it may be followed distally and dissected from tumor. If the lesion is large and nerve V is displaced extensively, either medially or posteriorly, the petrous apex should be drilled posterolaterally far enough to open the internal auditory canal (IAC). Whenever possible the SPS should be preserved intact to avoid interfering with the venous drainage of the PS, and, even more importantly, to avoid interfering with the petrous vein, which drains into the SPS. If the SPS is opened at surgery, it should be occluded both peripherally, through Parkinson's triangle, and centrally, at the medial border of the petrous bone, in such a manner as to avoid compromising flow from the petrosal vein into it. Following the complete removal of tumor and achieving precise hemostasis, the extensive dural defect along nerve V must be closed meticulously with muscle and fibrin glue.

Type IV trigeminal neurinomas

The approach in these cases is a combination of those employed for Type I and Type III trigeminal neurinomas. When the branches of nerve V are well exposed, together with tumor in the parasellar region, the tumor may be debulked through one or two of the triangular windows, facilitating further dissection of the outer layer of the lateral wall of the PS. Special attention must be paid to dissecting the GPN, particularly in cases where V1 is damaged or destroyed. Dissecting dura from the anterior surface of the petrous pyramid must be carried out as gently as possible. After identifying tumor in Meckel's cave, the overlying arachnoid is incised, then debulking begins, with the surgeon remaining ever mindful of the plexiform segment of the nerve V root and the GG, taking care to avoid injury to either.

After tumor in the PS and Meckel's cave is fully debulked, the petrous apex may be drilled away as described in Chapter 2, the amount removed depending on the size of the CPA component of the tumor. While debulking various elements of the tumor mass, and ultimately resecting it from nerve V structures, special care must be taken to avoid injuring nerves III, IV, VI, and the ICA (and its branches) in the PS. Neurinoma in the CPA is resected last, from nerve IV intradurally (in the ambiens cistern) and from the petrosal vein and nerves VII and VIII. Care must be taken not to occlude the petrosal vein. After skeletonizing the PS, it is particularly important to line all sites of potential venous bleeding with Surgicel as described for Type I lesions. With Type IV tumors postoperative bleeding may not only result in an epidural hematoma, but a potentially more devastating intradural clot, demanding that meticulous hemostasis be established.

If V1 and the GPN are completely lost, repairing V1 with a graft from the greater occipital nerve may be helpful. If this proves to be necessary, an adequate distal stump of V1 may be identified at the SOF. A watertight dural closure is achieved with muscle and fibrin glue. When the dural defect is large, the dorsal section of the temporalis muscle may be harvested with its fascia (or fascia lata) then sutured to the tentorium medially, and laterally to remnants of dura in the vicinity of V2 and V3, at the FR and FO respectively, finally covering the entire muscle graft and its suture lines with fibrin glue.

Patients, diagnostic procedures, and surgical results

Ninety-five patients were treated between 1981 and 2001. Eighty-five were operated as virgin cases, and five were re-operated for residual or recurrent tumor from operations performed elsewhere. One patient had a malignant tumor, and another, one with generalized Von Recklinghausen's disease, had bilateral tumors (a Type I lesion on the left and a Type IV tumor on the right). Three patients refused surgery.

Eighty-five patients with unilateral trigeminal neurinomas were operated as virgin cases and are presented in Table 19. Five had surgery for residual or recurrent tumors (Table 20), and the three refusing surgery are presented in Table 21. Patients with Von Recklinghausen's disease and malignant neurinomas are not included in Tables 19–21.

Table 19 indicates the preoperative nerve deficits among unilateral virgin cases with respect to tumor type and location. With parasellar tumors (Type I), diplopia was the first symptom in the large majority of cases, related to paresis of nerve VI and/or nerve III. Nerve IV or ON involvement was seldom the presenting symptom. Nerve V sensory deficits were detected in all patients, although some were not aware of impaired sensation prior to examination. Motor nerve V deficits were present in 16, associated with clinical atrophy of the masseter and temporalis muscles.

Only four patients presented with the sudden onset of facial pain. In all four hemorrhage was apparent on MR imaging, and evidence of bleeding was confirmed at surgery in three (Figs. 4.64 and 4.67). The fourth patient refused surgery, but her pain gradually subsided over two months (case I, Table 21). Exophthalmos was present in only four patients, and with the exception of one (Type I tumor, with orbital extension) it was mild. It is surprising that patients with even large parasellar trigeminal neurinomas were spared marked exophthalmos. One patient, operated for a recurrent neurinoma (case 3, Table 20), was initially bed-ridden with brainstem and cerebellar findings. After excising the tumor and decompressing his brainstem, this man became ambulatory and manifested a good outcome. One of the patients refusing surgery (Table 21, Case 2) presented with a CSF leak which led to meningitis.

With the exception of two patients experiencing transient postoperative corneal sensory deficits, of the 85 patients operated as virgin tumors were resected completely with no new neurological deficits apparent after

Table 19a. Preoperative neurological deficits among patients with previously unoperated trigeminal neurinomas

Tumor location			Preoperative cranial nerve(s) deficits													
									V							
Type	N° of cases		II		III		IV		Sensory		Motor		VI		Other	
Parasellar (I)	37	44%	3	8%	5	14%	2	5%	37	100%	3	8%	21	57%	1	3%
Meckel's cave (II)	22	26%	–	–	–	–	1	5%	22	100%	4	18%	12	55%	–	–
CPA (III)	15	18%	–	–	–	–	–	–	15	100%	3	20%	2	13%	4	27%
Combined (IV)	11	13%	1	9%	1	9%	1	9%	11	100%	3	27%	3	27%	4	36%
Total	85	100%	4	5%	6	7%	4	5%	85	100%	16	19%	38	45%	9	11%

Table 19b. Postoperative neurological outcome among patients having total resections of virgin neurinomas

Tumor location			Postoperative cranial nerve(s) deficits													
									V							
Type	N° of cases		II		III		IV		Sensory		Motor		VI		Other	
Parasellar (I)	37	44%	1	3%	2	5%	–	–	31	84%	1	3%	2	5%	–	–
Meckel's cave (II)	22	26%	–	–	–	–	–	–	19	86%	4	18%	3	14%	–	–
CPA (III)	15	18%	–	–	–	–	–	–	13	87%	4	27%	–	–	1	7%
Combined (IV)	11	13%	–	–	1	9%	–	–	9	82%	5	45%	2	18%	1	9%
Total	85	100%	1	1%	3	4%	–	–	72	85%	14	16%	8	9%	2	2%

Table 20. Patients operated for residual or recurrent trigeminal neurinomas

Case	Sex, age (yrs.)	Tumor location	N° of previous operations	Preoperative deficits: Nerve V Sensory	Nerve V Motor	Other cranial nerve(s)	Brain stem	Cerebellar	Postoperative deficits	Final outcome
1	M, 37	PS, MC, CPA	2	V_1, V_2, V_3	+	III, VII	–	+	same as preop.	good
2	M, 27	PS, MC, CPA	1	V_1+	–	–	–	–	same as preop. NO PAIN	excellent
3	M, 53	PS, MC, CPA	3	V_1+ V_2, V_3	+	VII, VII	+	+	same as preop. NO PAIN	good
4	F, 46	MC, CPA	2	V_1, V_2	–	–	+	–	same as preop.	excellent
5	F, 47	MC, CPA	1	V_1+, V_2	–	–	–	–	same as preop. NO PAIN	excellent

M Male, *F* female, *PS* parasellar space, *MC* Meckel's cave, *CPA* cerebello-pontine angle, V_1+ pain in distribution of V_1, *M*+ deficit.

Table 21. Patients with trigeminal neurinomas who refused surgical treatment

Case	Sex, age (yrs.)	Location	Clinical presentation	Diagnosis known for . . .
1	F, 53	parasellar	severe pain 3 months – now only stress related. slightly progressive exophthalmos	9 yrs.
2	F, 44	parasellar, CPA	first CSF leak 15 yrs. ago; meningitis 9 yrs. ago; after meningitis no CSF leak	15 yrs.
3	F, 53	CPA	continuous pain in V_1 and V_2 distribution; occasional severe pain (stress-related)	10 yrs.

F Female, *CPA* cerebello-pontine angle, *CSF* cerebro spinal fluid.

surgery. In general, preoperative neurological deficits cleared following surgery (Table 19) although those with atrophy of temporalis and masseter muscles (from involvement of the motor branch of nerve V) did not improve. Whether improving sensation following surgery indicates truly improved function or simply adaptation to partial sensory loss remains speculative. No patient experienced postoperative pain or unpleasant trigeminal dysesthesias. Since prompt appropriate measures were taken, neither of the patients with postoperative corneal sensory deficits developed visual loss. Tarsorrhaphy was rarely required.

The primary diagnostic tools employed were CT (Fig. 4.68) and MRI. Angiography was virtually never performed in this series. MRA was helpful at times, but DSA was not required.

In the patient with bilateral trigeminal neurinomas (Type IV on the right, Type I on the left) surgery was carried out initially on the left side, with a second operation planned for three months later. But the posterior fossa element of the untreated tumor resulted in critical brainstem compression within two months, leaving the patient bed ridden. She was hospitalized urgently and the residual tumor was explored intradurally (suboccipital approach) in order to decompress the brainstem. She did well following surgery, becoming ambulatory immediately. Understandably, she insisted that the parasellar component of her remaining tumor be

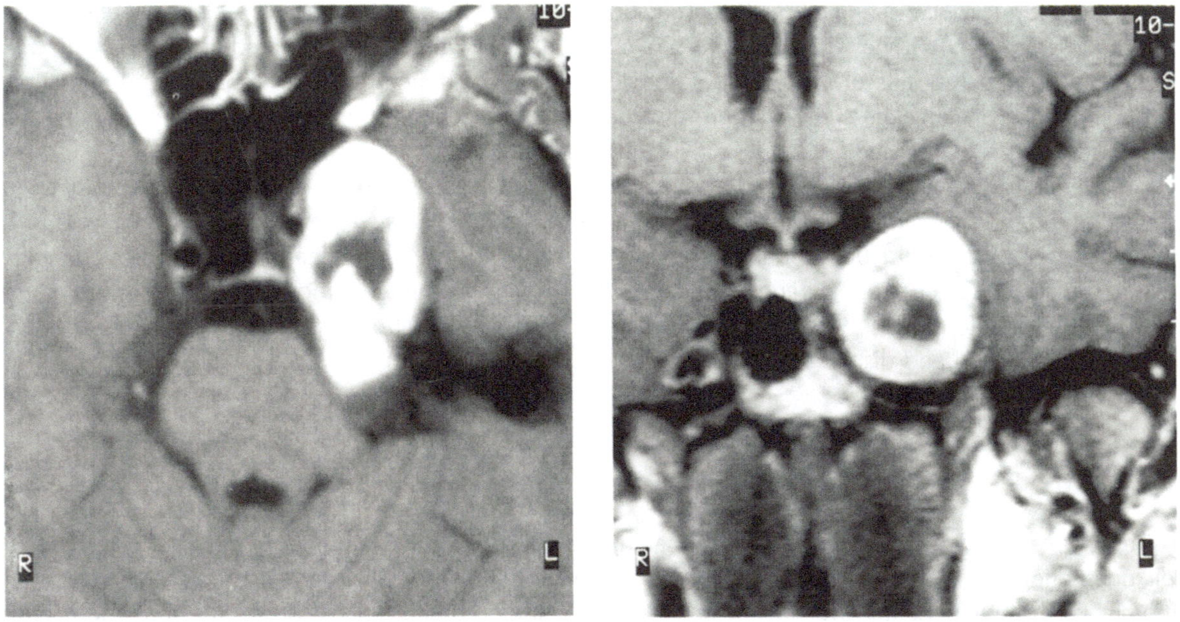

a b

Fig. 4.67. Trans-axial (a) and coronal (b) MR images of a Type IV trigeminal neurinoma in a patient who had been asymptomatic until three weeks prior to surgery, presenting with the sudden onset of left facial pain and mild diplopia from a mild nerve VI deficit. Neuroradiologists interpreted the cyst to represent "necrosis" within the tumor, but hemorrhage was confirmed at surgery. Nerve V, injured by sudden tumor expansion from the hemorrhage, likely had continued to compensate neurological loss until that time

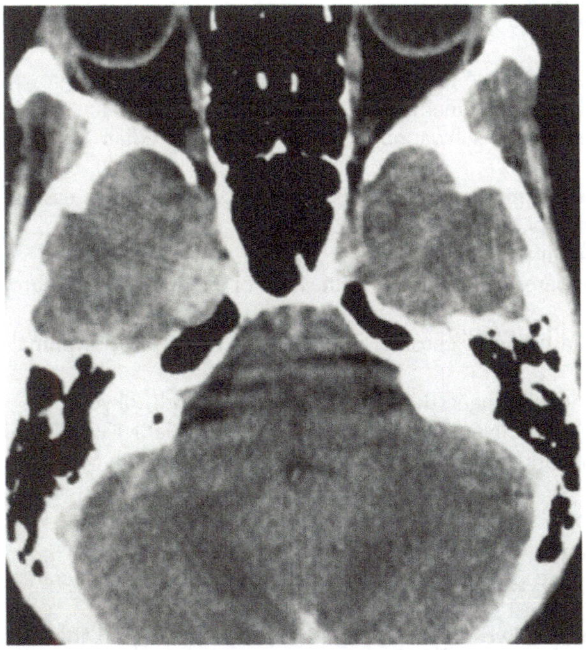

Fig. 4.68. This CT scan reveals a small right-sided trigeminal neurinoma. Note that both petrous pyramids are pneumatized, and the thin wall of petrous apex ipsilateral to the tumor has been eroded and destroyed. With tumors eroding pneumatized bone a cerebrospinal fluid (*CSF*) leak may be the first sign of disease. In cases like this one, where a CSF leak has not yet occurred, the surgeon must take special care to avoid CSF leakage following surgery. Generally, CT scans are more sensitive to bone erosion than MRI

excised without further delay, leading to a third operation after one week. Although no problems were encountered in the operating room, she was unable to move her jaw following surgery, related to gentle manipulation of the motor branch of nerve V. After a few sleepless nights, the deficit began to improve (and ultimately cleared completely), leaving the surgeon more aware than ever before of the importance of preserving the nerve V motor branch, particularly when it is involved bilaterally!

With the emergency exception described above, the frontotemporal extradural approach was employed in every case, and tumor resection was uniformly accomplished in a single stage, even with Type IV lesions. It is important to emphasize our belief that the key to removing Type IV trigeminal neurinomas is adequate removal of the apex of the petrous pyramid. It enables the surgeon to initiate tumor resection from the brainstem, making it possible to preserve all fibers of the nerve V root not destroyed previously by tumor growth, however much stretched and displaced.

Conclusions

These rare lesions, easily diagnosed by modern imaging studies, present in a number of different ways, owing to their originating from various elements of nerve V and the GG. Although surgical removal of trigeminal neurinomas can be challenging, even in the most skilled hands, complete excision by the trans-skull base extradural approach is generally possible with little risk of neurological morbidity and no mortality.

References

1. Amirchanjan SE, Lebedev AN (1979) The radiological diagnosis of trigeminus neurinomas of the gasserian ganglions. Radiologia Diagnostica 20: 78–85
2. Arena S, Hilal EY (1976) Neurilemomas of the infratemporal space: report of a case and review of the literature. Arch Otolaryngol 102: 108–184
3. Asaari S Tsuchida S, Fujiwara A, Yabuno N, Furuta T, Ohmoto T (1992) Trigeminal neurinoma presenting with intratumoral hemorrhage: report of two cases. Clin Neurol Neurosurg 94: 219–224
4. Asari S, Katayama S, Itoh T, Tsuchida S, Ohmoto T (1993) CT and MRI of haemorrhage into intracranial neuromas. Neuroradiology 35: 247–250
5. Camp RC, Koestner A, Vinores SA, Capen CC (1984) The effect of nerve growth factor and antibodies to nerve growth factor on ethylnitrosourea-induced neoplastic proliferation in rat trigeminal nerves. Vet Pathol 21: 67–73
6. Cantini R, Giorgetti W, Velleriani AM, Burchianti M (1987) Trigeminal schwannomas in adolescence. Pediatr Neurosci 13: 198–201
7. Chkili T, Medejel A, El Quarzazi A, Bellakhdar F (1979) Neurinomas of the trigeminal nerve. Apropos of a case of "hourglass" neurinoma occupying the posterior and middle fossa. Maroc Med 1: 223–228
8. Colin W (1990) Trigeminal intraoral schwannomas. Compendium 11: 672, 674–677
9. Currie J, Lubin JH, Lessell S (1983) Chronic isolated abducens paresis from tumors at the base of the brain. Arch Neurol 40: 226–229
10. Del Priore LV, Miller NR (1989) Trigeminal schwannoma as a cause of chronic, isolated sixth nerve palsy. Ophthalmologie 108: 726–729
11. Dolenc VV (1989) Anatomy and surgery of the cavernous sinus. Springer, Berlin Heidelberg New York Tokyo
12. Dolenc VV (1994) Frontotemporal Epidural Approach to Trigeminal Neurinomas. Acta Neurochir (Wien) 130: 55–65
13. Embry BM, Pritchard R (1989) Trigeminal neurilemoma: appearance on magnetic resonance imaging. South Med J 82: 653–656
14. Findler G, Feinsod M, Sahar A (1983) Trigeminal neurinoma with unusual presentation. Report of a case with trigeminal somatosensory-evoked response. Surg Neurol 19: 351–353
15. Fushimi S, Miura S, Suda Y, Fujii S, Kowada M (1987) A case of trigeminal neurinoma with marked extracranial extension. No Shinkei Geka 15: 1117–1122
16. Genka S, Terao H, Matsutani M, Hori T, Ikeda A (1981) A choice of two-stage operation and the microsurgical anatomy for large trigeminal neurinoma. No Shinkei Geka 9: 747–751
17. Hayashi K (1983) Tissue culture of trigeminal nerves from rats administered transplacentally with ethylnitrosourea. Acta Med Okayama 37: 301–320

18. Hedeman LS, Lewinsky BS, Lochridge Gk, Trevor R (1978) Primary malignant schwannoma of the Gasserian ganglion. Report of two cases. J Neurosurg 48: 279–283
19. Hitchon P, Graf CJ, McCabe BF, Namba K (1979) Trigeminal neurilemmoma presenting as a sphenoid sinus mass. Surg Neurol 11: 305–307
20. Ishiguro S, Kitabayashi M, Munemoto S, Ikeda M, Kimura A, Hayashi M (1986) Neurinoma of the trigeminal root with recurrent intratumoral hemorrhage. Case report. Neurol Med Chir 26: 734–737
21. Katoh M, Kadoya S, Emori T, Kwak R (1985) Trigeminal neurinoma presenting abducens nerve palsy as initial symptom: a report of two cases. No Shinkei Geka 13: 555–560
22. Lesoin F, Rousseaux M, Villette L, Autricque A, Dhellemmes P, Pellerin P, Vaneecloo JM, Leys D, Jomin M (1986) Neurinomas of the trigeminal nerve. Acta Neurochir (Wien) 82: 118–122
23. Levy WJ, Ansbacher L, Byer J, Nutkiewicz, Fratkin J (1983) Primary malignant nerve sheath tumor of the Gasserian ganglion: a report of two cases. Neurosurgery 13: 572–576
24. Lunardi P, Missori P, Gagliardi FM, Fraioli B (1989) Trigeminal schwannoma with infratemporal extension. Case report. J Neurol Sci 33: 293–295
25. McCormick PC, Bello JA, Post KD (1988) Trigeminal schwannoma. Surgical series of 14 cases with review of the literature. J Neurosurg 69: 850–860
26. Morniroli G (1970) Das Trigeminusneurinom. Inaugural Dissertation. University of Zurich, Neurosurgical Department
27. Nager GT (1984) Neurinomas of the trigeminal nerve. Am J Otolaryngol 5: 301–333
28. Pelissou I, Sindou M, Goutelle A, Pialat J, Duquesnel J (1988) Neurinoma of the trigeminal nerve. Excision by combined suboccipital and pteriono-temporal approach. Neurochirurgie 34: 8–16
29. Pollack IF, Sekhar LN, Jannetta PJ, Janecka IP (1989) Neurilemomas of the trigeminal nerve. J Neurosurg 70: 737–745
30. Rigamonti D, Spetzler RF, Shetter A, Drayer BP (1987) Magnetic resonance imaging and trigeminal schwannoma. Surg Neurol 28: 67–70
31. Ross DL, Tew JM Jr, Benton C, Eisentrout C (1984) Trigeminal schwannoma in a child. Neurosurgery 15: 108–110
32. Saito A, Nakazawa T, Matsuda M, Handa J (1989) Comparison of computed tomographic scanning and magnetic resonance imaging in the diagnosis of trigeminal schwannoma. Report of four cases. Neurol Med Chir 29: 1101–1106
33. Shimabukuro H, Masuzawa T, Miyagi K, Sato F (1983) Trigeminal neurinoma revealed by intratumoral hemorrhage following a minor head injury. Surg Neurol 19: 346–350
34. Tanaka A, Takaki T, Maruta Y (1987) Neurinoma of the trigeminal root presenting as atypical trigeminal neuralgia: diagnostic values of orbicularis oculi reflex and magnetic resonance imaging. A case report. Neurosurgery 21: 733–736
35. Vaquero J, Cabezudo JM, Lenuda G, Carrillo R, Bravo G (1981) Sumultaneous posterior and middle cranial fossa neurinomas. Acta Neurochir (Wien) 55: 321–327
36. Wullenweber R, Traupe H, Schwarz G (1976) Angiographic demonstration of a trigeminal neurinoma. Neuroradiology 10: 225–229
37. Yamada K, Ohta T, Miyamoto T (1992) Bilateral trigeminal schwannoma associated with von Recklinghausen disease. AJNR Am J Neuroradiol 13: 299–300
38. Yamada H, Sakai N, Ando T, Nishimura Y, Kishimoto Y (1980) Trigeminal neurinoma with massive extracranial extension: report of a case. No Shinkei Geka 8: 469–473
39. Yamashita J, Asato R, Handa H, Nakao S, Ogata M (1977) Abducens nerve palsy as initial symptom of trigeminal schwannoma. J Neurol Neurosurg Psychiatry 40: 1190–1197
40. Yasui T, Hakuba A, Kim SH, Nishimura S (1989) Trigeminal neurinomas: operative approach in eight cases. J Neurosurg 71: 506–511

4.4

Chordomas, chondrosarcomas, and chondromas

Introduction

Intracranial chordomas occur primarily at the central skull base (CSB), usually presenting as large, soft, grayish masses which displace CSB dura into brain. Some occupy the parasellar space (PS) or spread into the middle cranial fossa. They may grow into naso-pharyngeal recesses. Rarely, however, do chordomas traverse skull base dura and invade the intradural space, even less often do they invade brain tissue. Typically benign histologically, chordomas are invasive, portending a poor prognosis. Yet some are truly malignant and have the capacity to metastasize [2, 3, 6, 8, 10, 13, 15, 16, 18, 19].

The appearance of the term "chondroid chordoma" [9], suggesting a connection between chondromas and chordomas, has resulted in a certain amount of confusion. "Chondroid chordomas" and chondrosarcomas are sometimes difficult to distinguish [4, 10, 17]. The literature is replete with discussions of chondrosarcomas [1, 3, 5, 7, 12, 14], and chondromas have also been reported [11, 20]. Although a detailed discussion of the histological characteristics of these tumors is beyond the scope of this chapter, it is important to point out common characteristics they possess. First, all arise from bones possessing a cartilaginous matrix, those which are not "hematopoietic". Typically, they originate from synchondroses, where bones join, specifically at the spheno-ethmoidal, spheno-occipital, and temporo-occipital junctions. Spreading from the central and lateral clivus, also from the petro-clival and petro-occipital junctions, all occur at the CSB. Secondly, they often become quite large before affecting neural and/or vascular structures. In most cases, particularly where chordomas are present, initial operations are not technically difficult since the neoplasm, seeking a path of least resistance, spreads in an exophytic fashion rather than burrowing into the depths of the CSB. Since little anatomical knowledge is required to remove the bulk of such tumors, the surgeon is deceived into thinking that a satisfactory resection has been achieved. The caveat is that important elements uniformly remain hidden as tiny phalanxes within the CSB. Any overlooked or inaccessible tumor continues to grow, ultimately rendering resection virtually impossible. Finally, each of these tumors requires additional treatment after surgery, however certain the surgeon may be that total resection has been achieved. Our own experience with nine cases confirms the conclusion of reports that Proton Radiation Treatment (PRT) is the most effective complementary treatment currently available.

Pathological anatomy, clinical presentation, and diagnostic procedures

Each of these tumors arises from the CSB, destroying normal bone but generally remaining extradural, however large the mass it finally attains. Eventually, cranial nerves coursing through the skull base are affected, nerve VI generally first. If a timely diagnosis is not made, other cranial nerves become involved, even bilaterally. Headaches, more commonly in the form of a vague discomfort, generally precede cranial neuropathies. Only occasionally, when such tumors penetrate dura or compress clival dura against the brainstem, do long tract symptoms and signs become part of the clinical picture (Figs. 4.69–4.72). These neoplasms may expand not only caudally but anteriorly, entering paranasal sinuses and resulting finally in nasal obstruction; even laterally, causing eustachian tube dysfunction; or invading the sella, resulting in hormonal dysfunction. In some cases they are the basis for early childhood squints which commonly are ascribed to other causes. Following seemingly successful eye-muscle(s) rehabilitation or even operation(s), identical symptoms and signs re-appear in adolescence and beyond, only then resulting in the correct diagnosis (Fig. 4.73). Growth rates of chondrosarcomas are not necessarily low, but these lesions must achieve considerable size before sentinel symptoms and signs appear.

The single best diagnostic procedure is MRI, but precise characterization of bone destruction and paranasal sinus involvement requires CT scanning. Chordomas may displace the skull base ICA considerably, but vessel size is rarely affected. Although chordomas seldom invade the ICA wall, chondrosarcomas may, sometimes resulting in stenosis in addition to displacement. Preoperative angiography should be considered with both chordomas and chondrosarcomas, but it is indispensable in the case of the latter.

Surgical approach to chordomas, chondrosarcomas and chondromas

Head positioning, scalp incision, and osteomuscular craniotomy are carried out as described in Chapter 2. Since these tumors are predominantly extradural, accessing the parasellar space (PS), and through the PS the clivus, is accomplished as described previously. The approach is tailored depending on which parts of the clivus (upper, middle, or lower) and paraclival region(s) are involved. It is never advisable to enter the intradural space only to have to incise dura again to access the paraclival region extradurally. The extradural component should be removed first, then, but only if the lesion invades dura and bulges into the brainstem, should dura be opened. In such cases clival dura must be carefully dissected from the brainstem and other vital structures, continually displacing it toward the clivus (Figs. 4.69 and 4.72). Leaving clival dura intact when possible minimizes the risk of postoperative cerebrospinal (CSF) leak.

Since total resection of these tumors is not possible, extensive resection of CSB bone serves no useful purpose. The surgeon's goal should be to remove the bulk of the tumor and decompress the brainstem and other neural structures, especially affected cranial nerves and the visual apparatus. Decompressing critical neural structures and providing sufficient free space between tumor and the visual apparatus and brainstem is important in order to avoid increasing neurological deficits by treating them with PRT (Fig. 4.72).

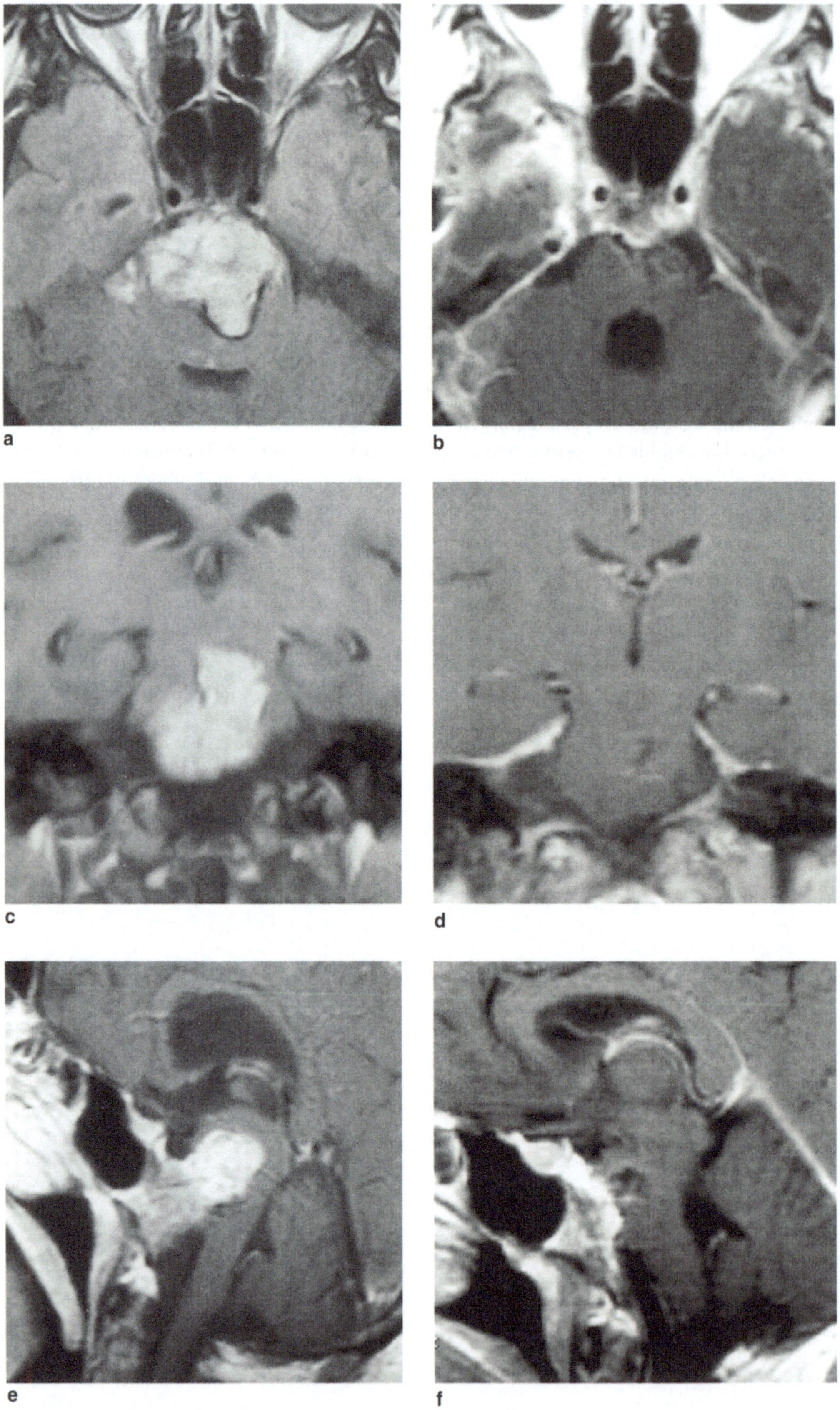

Fig. 4.69. Preoperative contrast-enhanced, transaxial (a), coronal (c) and sagittal (e) MR images demonstrate a large, centrally placed chordoma with significant brainstem encroachment. Brainstem symptoms and signs dominated the clinical picture. Following sub-total resection the patient's neurological status returned essentially to normal within six months. After one year, routine MRI demonstrated a small tumor re-growth but no hydrocephalus. She was doing well clinically and refused further surgery. After an additional three years, however, brainstem symptoms recurred and MRI confirmed advancing tumor growth. MRI six months after repeat surgery (a much more radical resection, but still sub-total) demonstrated no evidence of tumor recurrence (b, d, f). Chordoma persists in the clivus (f) has remained quiescent for two years after the second operation. Although this patient refused conventional radiation therapy and PRT, she remains asymptomatic

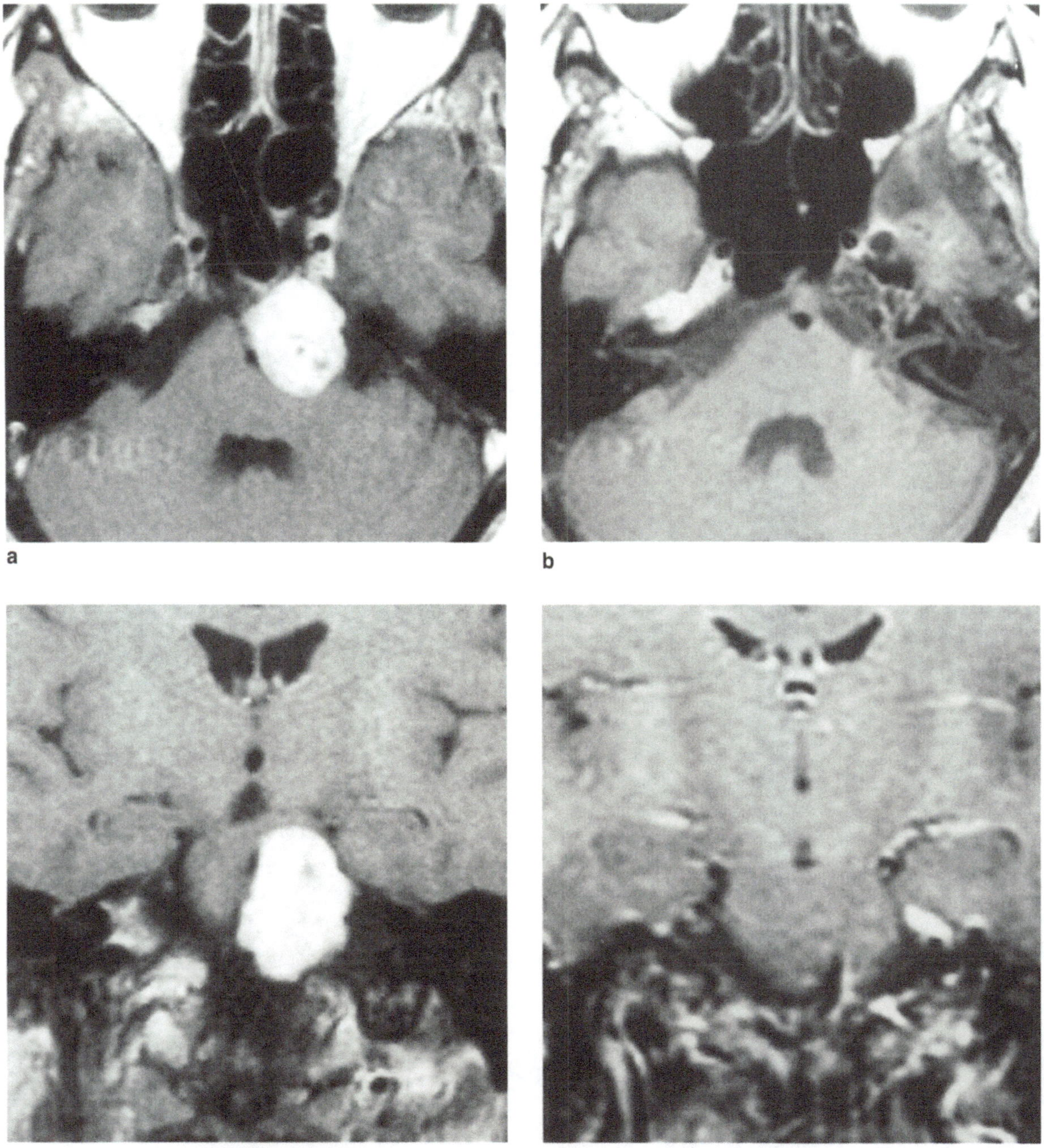

a b c d

Fig. 4.70. Preoperative, transverse (a) and coronal (c), gadolinium-enhanced MR images demonstrate a left-sided chondrosarcoma which originated from the clivus and petrous apex, extending upward and posteriorly, compressing the pons. Gross-total resection was confirmed by enhanced transaxial (b) and coronal (d) MR images. This patient refused PRT but changed her mind 18 months later when a repeat MRI revealed a small recurrent tumor. Two years after PRT she remains symptom-free and without evidence of tumor recurrence

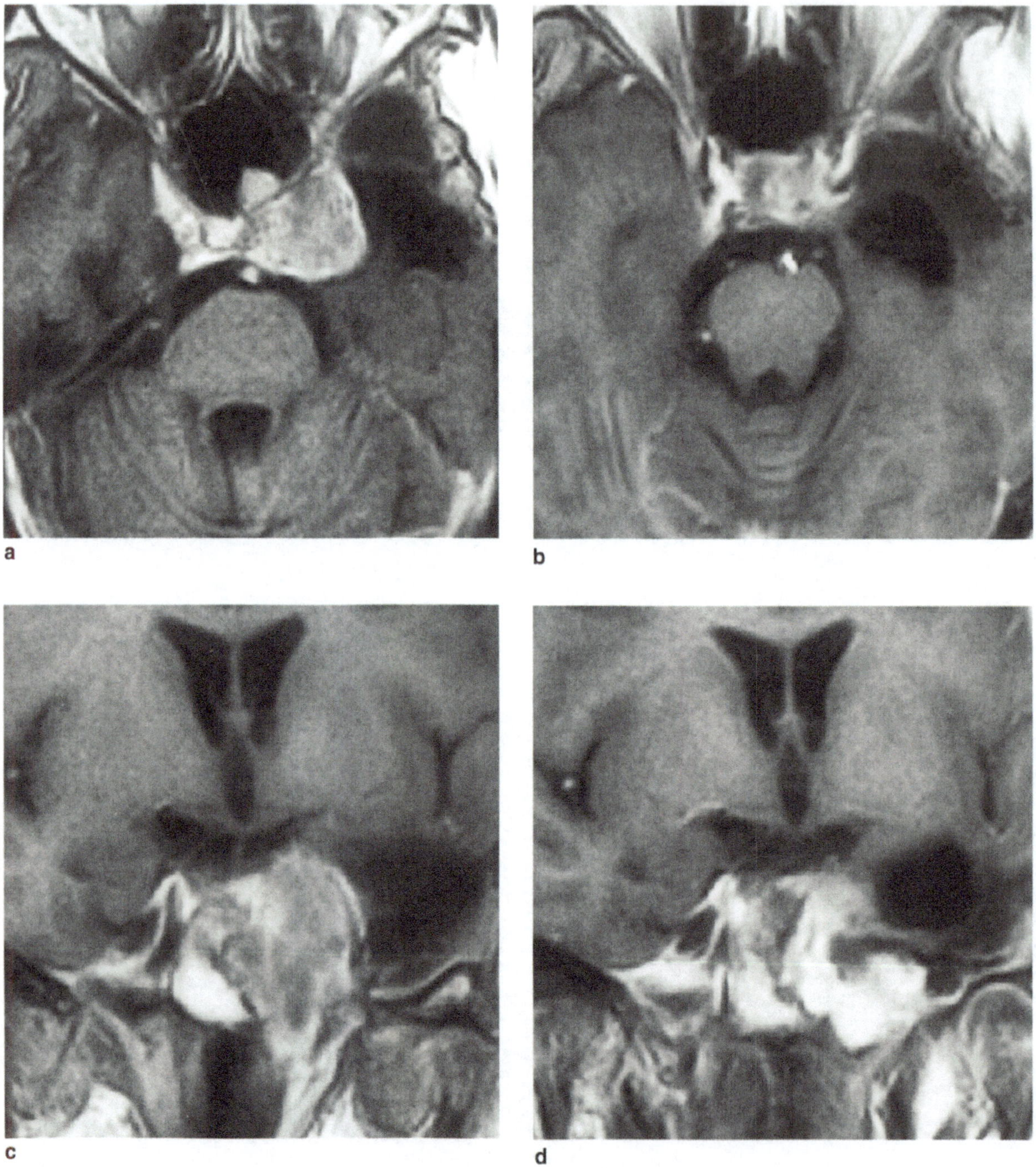

Fig. 4.71. This 55-year old woman was operated for a chondroid chordoma in the left parasellar region. The tumor was not totally resected, and she was irradiated postoperatively. For a period of 10 years she enjoyed normal life without any neurological deficits. Ten years after surgery she noticed double vision, and left nerve VI palsy was diagnosed. Control contrast-enhanced, transaxial (a) and coronal (c), MR images revealed a recurrent tumor located in the left parasellar region. Corresponding postoperative contrast-enhanced, transaxial (b) and coronal (d), MR images demonstrate subtotal resection of the lesion. Additional complementary treatment with linear accelerator has been conducted. The impaired function of nerve VI did not improve during the first six months postoperatively. Except diplopia, she has no other neurological deficits. Note: The atrophy of the anterior part of the left temporal lobe (a and c) due to intradural approach at first surgery

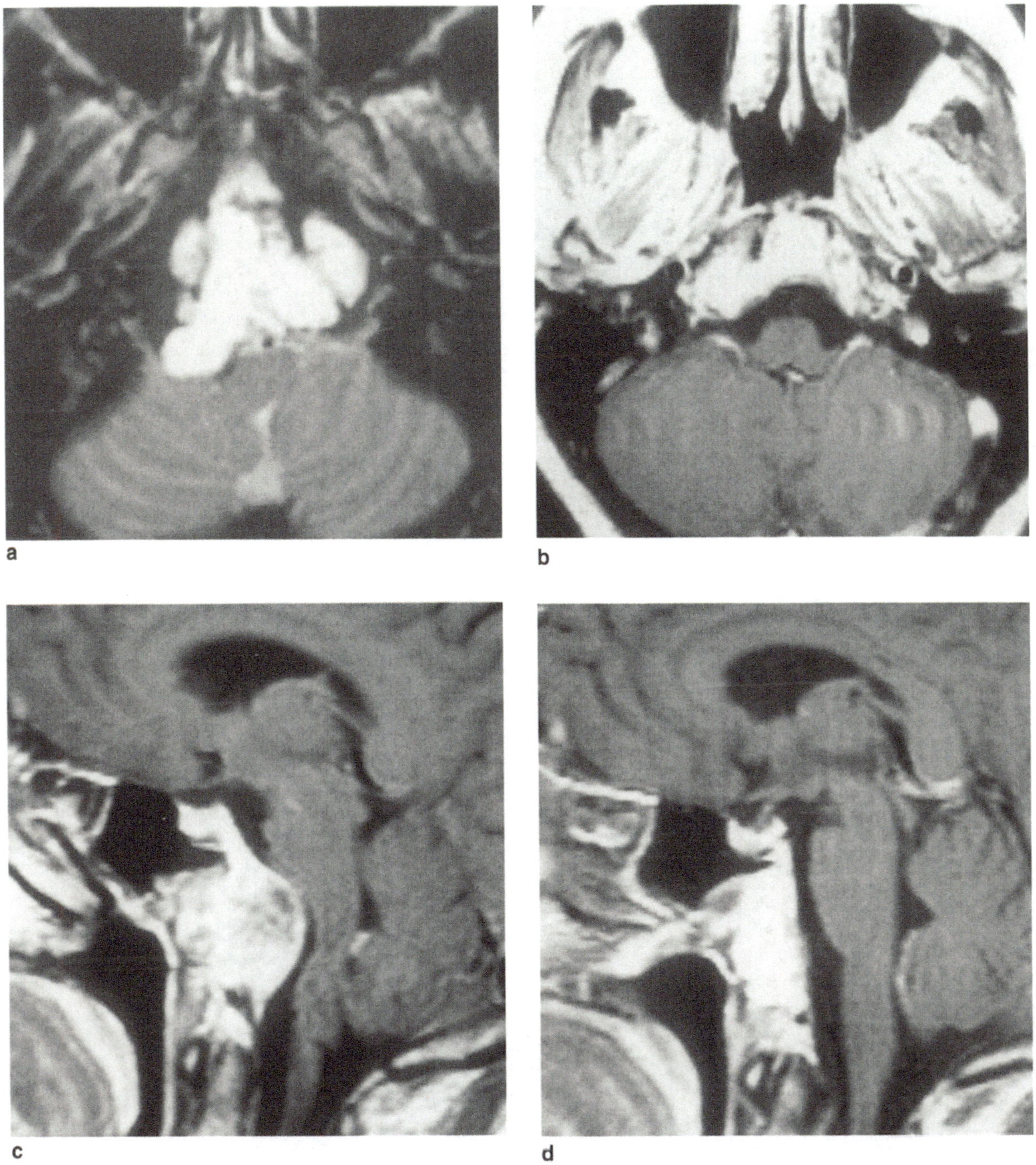

Fig. 4.72. Preoperative contrast-enhanced, transaxial (a) and sagittal (c), MR images reveal a large central skull base (CSB) chondrosarcoma which was partially resected as demonstrated by corresponding postoperative contrast-enhanced MR images (b and d). This patient's left nerve VI palsy improved immediately after surgery. Following Proton Radiation Treatment (*PRT*) she remains symptom free with no evidence of tumor recurrence after one year

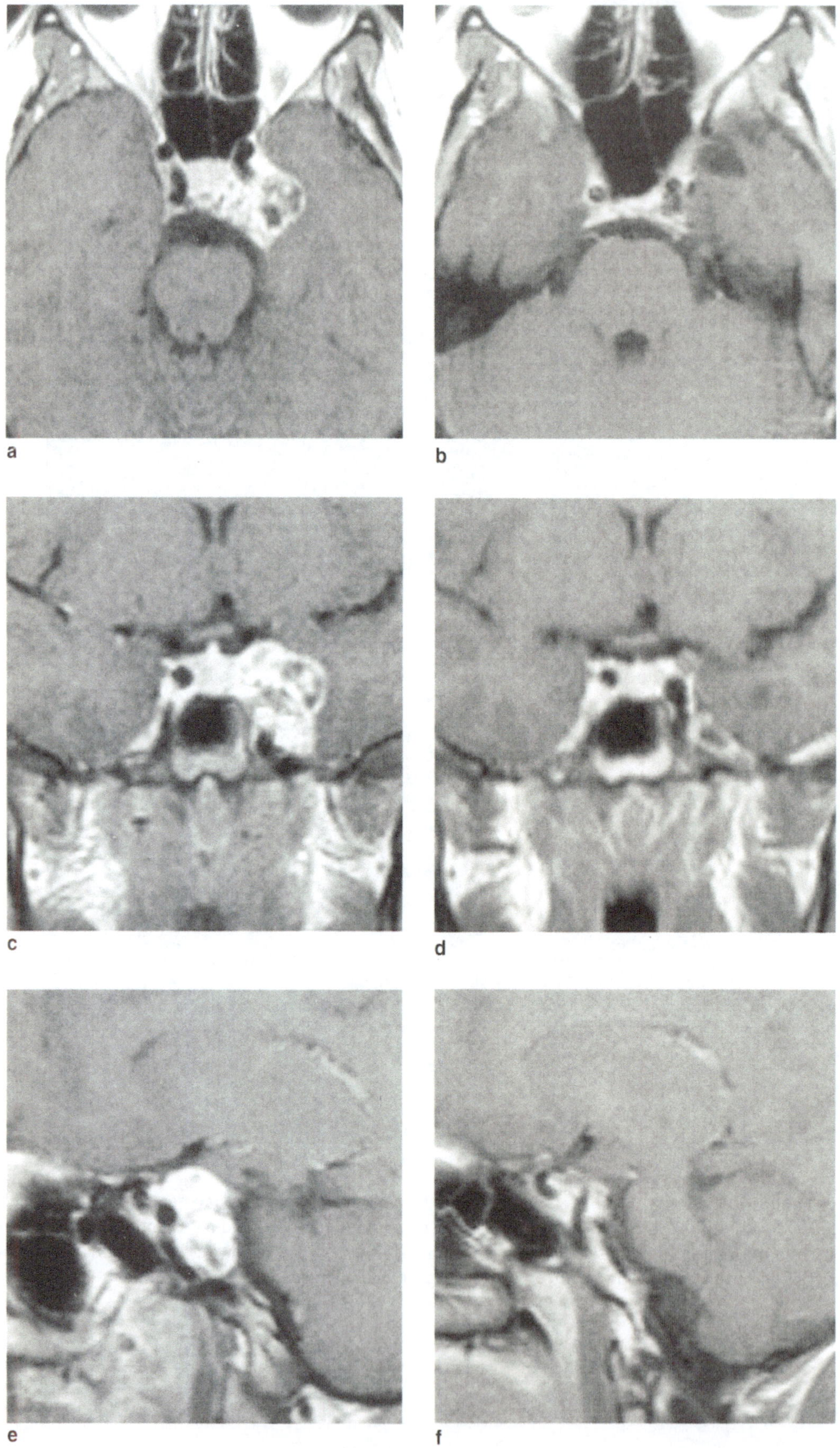

Fig. 4.73. This 17-year-old girl had left nerve VI paresis when she was three years old. Due to "successful rehabilitation", paresis of nerve VI disappeared after one year. When she was 17 years old, the same VI nerve paresis re-appear and MR images revealed left-parasellar chondrosarcoma (a, c, e). Following surgical gross-total resection (b, d, f), nerve VI regained normal function again. Three months after surgery, the PRT was conducted. At six months after surgery the patient is free of neurological symptoms

Patients and results

Surgical results in 50 patients are presented in Table 22. In none was total resection achieved. Gross-total resections were possible in 31 patients and only sub-total in 19. All but one patient had cranial nerve deficits prior to surgery which persisted for at least one week postoperatively. Three patients with preoperative long-tract signs remained unchanged following surgery, and three additional patients developed pyramidal tract findings after surgery. Six months after surgery only 20 patients manifested persisting cranial nerve deficits, and three continued to have long tract deficits.

Two patients died, one with a chordoma, another with a chondrosarcoma. The patient with the chordoma required four operations. Tumor continued to grow despite initial gross-total resection followed by conventional radiation. The histological findings remained unchanged with each subsequent operation, the patient dying after the fourth attempt, at which time massive tumor was encountered throughout the entire CSB. The patient dying from a chondrosarcoma was a young woman who was bed-ridden prior to her first operation early in the series. During surgery, her encased ICA ruptured and was occluded, since repair was not feasible at that time.

A 22-year-old male with a chondrosarcoma was operated four times, the interval between operations becoming shorter and shorter despite repeated gross-total tumor resections. The interval between the third and fourth operations was only six months. Following his final operation, he and eight subsequent patients with chondrosarcomas over the past decade received PRT. To date, no evidence of recurrence has been detected in any of them.

Chordomas were treated with conventional radiation after gross-total or sub-total tumor resections. In four cases the tumor recurred, but in every case the histological examination remained unchanged, seeming to imply that the histological picture has no relationship with the biological nature of chordomas.

In addition to the 50 patients treated surgically, four with radiological characteristics typical of chondrosarcomas were not treated (Fig. 4.73). Follow-up with periodic MRI has revealed no change in tumor volume over the past three years. In two of them, unfortunately, the associated mild diplopia (from abducens nerve palsy) has not changed either.

Conclusions

Although usually chordomas and chondrosarcomas are histologically benign, occasionally they behave as highly malignant lesions. All of these tumors are fairly easy to treat surgically, but they are probably incurable. Although additional follow-up is needed, it appears that the combination of gross-total tumor resection and postoperative PRT amounts to highly effective treatment for chordomas and chondrosarcomas.

Table 22. Tumor type, resectability, complementary treatment employed, and outcome

N° and type of tumor		Operation		Preoperative deficits		Extent of resectin				Postoperative deficits after six months		
		1st	2nd	Cranial nerve(s)	Other	Gross-total	Sub-total	PRT	GK or fracionated irradiation	Cranial nerve(s)	Other	Deaths
Chordoma	29	20	9	29	2	27	2	1	15	11	2	1
Chondrosarcoma	19	17	2	19	1	4	15	8	–	9	1	1
Chondroma	2	2	–	1	–	–	2	–	–	–	–	–
Total	**50**	39	11	49	3	31	19	9	15	20	3	2

References

1. Adegbite ABO, McQueen JD, Paine KWE, Rozdilsky B (1985) Primary intracranial chondrosarcoma: a report of two cases. Neurosurgery 17: 490–494
2. Bottles K, Beckstead JH (1984) Enzyme histochemical characterization of chordomas. Am J Surg Pathol 8: 443–447
3. Bourgouin PM, Tampieri D, Robitaille Y, Robert F, Bergeron D, del Carpio R, Melancon D, Either R (1992) Low-grade myxoid chondrosarcoma of the base of the skull: CT, MR, and histopathology. J Comp Ass Tomograph 16: 268–273
4. Brooks JJ, LiVolsi VA, Trojanowski JQ (1987) Does chondroid chordoma exist? Acta Neuropathologica 72: 229–235
5. Brown E, Hug EB, Weber AL (1994) Chondrosarcoma of the skull base. Neuroimaging Cin North Am 4: 529–541
6. Coffin CM, Swanson PE, Wick MR, Dehner LP (1993) Chordoma in childhood and adolescence. A clinicopathologic analysis of 12 cases. Arch Pathol Laboratory Med 117: 927–933
7. Coltrera MD, Googe PB, Harrist TJ, Hyams VJ, Schiller AL, Goodman ML (1989) Chondrosarcoma of the temporal bone. Diagnosis and treatment of 13 cases and review of the literature. Cancer 58: 2689–2696
8. Fox JE, Batsakis JG, Owano LR (1968) Unusual manifestations of chordoma: report of two cases. J Bone Joint Surg 50A: 1618–1628
9. Heffelfinger MJ, Dahlin DC, MacCarty CS, Beabout JW (1973) Chordomas and cartilaginous tumors at the skull base. Cancer 32: 410–420
10. Ishida T, Dorfman H (1994) Chondroid chordoma versus low-grade chondrosarcoma of the base of the skull: can immunohistochemistry resolve the controversy? J Neuro-Oncol 18: 199–206
11. Ishii R, Sato S, Ueki K, Oyake Y (1974) Myxoosteochondroma in the pons. Journal of Neurosurgery 41: 240–243
12. Leedham PW, Swash M (1972) Chondrosarcoma with subarachnoid dissemination. J Pathol 107: 59–61
13. Meis JM, Raymond AK, Evans HL, Charles RE, Giraldo AA (1987 "Dedifferentiated" chordoma. A clinicopathologic and immunohistochemical study of three cases. Am J Surg Pathol 11: 516–525
14. Morimoto T, Sasaki T,Tokakura K, Ishida T (1992) Chondrosarcoma of the skull base. Report of six cases. Skull Base Surg 2: 177–185
15. Morris AA, Rabinovitch R (1947) Malignant chordoma of the lumbar region: report of a case with autopsy: comment on unusual metastasis of the brain, lungs, pancreas, sacrum and axillary and iliac lymph nodes. Arch Neurol Psychiatry 57: 547–564
16. O'Connell JX, Renard LG, Liebsch NJ, Erfird JT, Munzenrider JE, Rosenberg AE (1994) Base of skull chordoma. A correlative study of histologic and clinical features of 62 cases. Cancer 74: 2261–2267
17. Walker WP, Landas SK, Bromely CM, Sturm MTM (1991) Immunohistochemical distinction of classic and chondroid condromas. Modern Pathol 4: 661–666
18. Wang CC, Everette J. Jr (1968) Chordoma: brief review of the literature and report of a case with widespread metastasis. Cancer 22: 162–167
19. Watkins L, khudados ES, Kaleoglu M, Revesz T, Sacares P, Crockard HA (1993) Skull base chordomas: a review of 38 patients, 1958–1988. Brit J Neurosurg 7: 241–248
20. Wu WO, Lapi A (1970) Primary non-skeletal intracranial cartilaginous neoplasms: report of a chondroma and mesenchymal chondrosarcoma. J Neurol Neurosurg Psychiatry 33: 469–470

4.5

Parasellar space cavernomas

Introduction

Intra-axial cavernomas, occurring both supratentorially and infratentorially (in the brain stem), constitute a not uncommon, well described vascular entity. By contrast, extra-axial cavernomas are rare [1, 2, 5–17]. Extra-axial cavernomas arising in the parasellar space (PS) are extremely rare. Our entire series of over 1700 central skull base (CSB) tumors includes only 19 PS cavernomas, representing slightly greater than 1% of all CSB lesions. Microscopic characteristics of PS cavernomas are similar to intra-axial lesions, but their surgical requirements are quite different, justifying their distinction as "cavernous hemangiomas". When large, such lesions pose real problems for the surgeon, sometimes defying total resection.

Pathological anatomy

Cavernous angiomas are soft, highly vascular lesions, growing around the ICA (and its branches) and nerve VI, typically displacing the walls of the PS outward, resulting in stretching nerves III, IV, and V. Commonly, nerve VI is encased by the lesion, as though locked in putty. The ICA may be displaced but not compressed (Fig. 4.74).

Cavernous angiomas commonly present with headaches and cranial nerve paresis that come on with increased physical activity. Greater cardiac exertion increases blood flow to the lesion, which becomes turgid, reminiscent of a sponge taking up fluid. Increased lesion mass typically results in diminished nerve VI function. The resulting diplopia often is a very disturbing initial symptom. Nerve VI palsy is often detectable on examination, but the deficit normally subsides as the patient relaxes. Following their onset, however, both diplopia and detectable nerve VI palsy rarely disappear entirely, often assuming an undulating pattern, varying with physical activity. Diagnosis is easily made by MR imaging, although it may not be apparent by non-enhanced CT [4]. Once the diagnosis has been established most patients elect to be treated as soon as possible, but this is not entirely necessary since the symptoms and signs of PS cavernous angiomas do not progress so rapidly that urgent surgery is needed.

Larger cavernous angiomas may be associated with adjacent bone erosion resulting from long-standing pulsations. This phenomenon plus their characteristically intense contrast-enhancement distinguish them from meningiomas, the most common PS tumor. Neither do cavernous angiomas possess "tails" extending along the tentorial edge as is so typical of meningiomas.

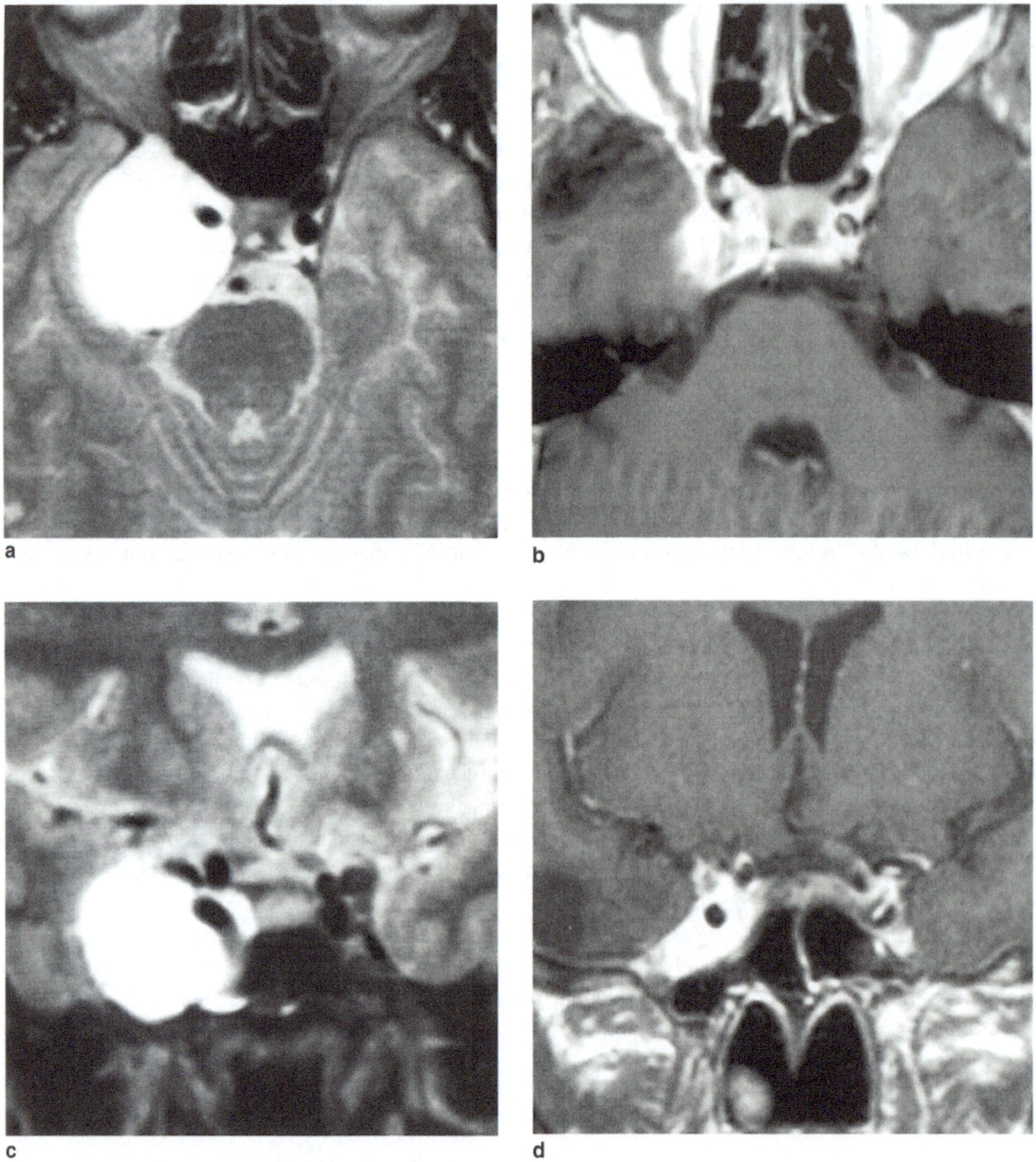

Fig. 4.74. Preoperative trans-axial (a) and coronal (c) contrast-enhanced MR images show a cavernoma of the right parasellar space. Corresponding postoperative contrast-enhanced MR images (b, d) demonstrate gross-total resection of the tumor. Note: No attempt at total resection was made because of the lesion's particular adherence to nerve VI. Diplopia, from abducens paresis, disappeared within three months of surgery

Surgical treatment

Patient positioning, craniotomy, and the initial steps are the same as described in Chapter 2, including the dissection of the outer (dural) layer of the lateral wall of the PS. Unroofing the orbit, removing the anterior clinoid process (ACP), and unroofing the optic canal (OC) should also be carried out as described previously. Once the inner layer of the lateral PS is exposed, nerves III, IV, V1, and V2 come into view, then, depending on the size and location of the cavernous angioma, tactics for its removal may be determined.

With all large cavernous angiomas the ICA should be exposed through Glascock's triangle, allowing it to be temporarily occluded if necessary. Similarly, the anterior loop (AL) of the ICA is exposed in anteromedial triangle and prepared for temporary clipping. Since the primary blood flow of PS cavernous angiomas is provided by the PS-ICA, it is prudent to sacrifice principal branches feeding the lesion at the outset. After coagulating and transecting the infero-lateral trunk (ILT) and the meningo-hypophyseal trunk (MHT), the lesion usually shrinks. If it continues to pulsate forcefully, it is advisable to place temporary clips on the ICA proximally and distally. Since even this does not completely devascularize cavernous angiomas, they should not be resected piece by piece. After dissecting adherent cranial nerves, the lesion usually shrinks further and may be occluded with the bipolar coagulator from its periphery.

Although dissecting nerves III, IV, V1, and V2 from a cavernous angioma is rarely difficult, dissecting nerve VI can be more demanding. In order to avoid damaging it, one must not coagulate blindly within the depths of Parkinson's triangle. Once nerve VI is fully dissected from the lesion, occluding its venous component generally has been achieved, leaving only the simple matter of removing the lesion. Dissecting it from the ICA is never difficult. After resecting the lesion, temporary ICA clips are released, and hemostasis confirmed. The empty PS should then be filled with Surgicel and fibrin glue to prevent postoperative bleeding.

Patients and results

Presenting symptoms and signs and surgical outcome of 19 patients are presented in Table 23. Four had been operated elsewhere. Fifteen lesions were totally excised, removals being gross-total in three and sub-total in one. Headaches uniformly subsided within a few days after surgery, but preoperative nerve deficits persisted and new deficits sometimes occurred. Two patients became blind in the ipsilateral eye [3], and no satisfactory explanation was forthcoming since little coagulation was necessary in the vicinity of the optic nerve (ON) and the walls of the OC had been resected. Both exhibited preserved vision during the afternoon and evening following surgery, losing ipsilateral vision totally the following day.

Table 23. Surgical treatment of cavernous angiomas of the PS

Presenting symptom(s) and sign(s)		Preoperative	Postoperative		
			First week	After 6 months	New
Headache		19	–	–	
Diplopia		17	19	4	2
Nerve(s) deficits	II	1	3	2	2
	III	7	11	1	4
	IV	4	8	1	4
	V	4	9	–	5
	VI	12	16	2	4

Conclusion

Cavernous hemangiomas of the PS result in slow but steadily progressive symptoms and signs. Their surgical resection is never urgent, allowing sufficient time for appropriate diagnostic study and second opinions when desired. In contrast to other PS lesions, piece-meal techniques should not be employed unless the lesion has been fully devascularized, coagulated, and separated from adjacent nerves. To resect them otherwise is destined to result in huge blood loss and dysfunction of nerves III through VI after the surgeon has struggled blindly and mightily in a sea of blood. Employing the techniques described here, however, assures the bloodless resection of these lesions with preservation of neurological function. Managing PS cavernous angiomas by endovascular methods or by conventional or focused beam radiotherapy remains controversial. In skilled hands surgical resection remains superior to all other treatment modalities.

References

1. Finkemeyer VH, Kautzky R (1968) Das Kavernom des sinus cavernosus. Zentabl Neurochir 29: 23–30
2. Harada T, Aoyama N, Terada H, Tsujimoto S, Hosogi Y, Iwasaki S (1982) A case of extracerebral hemangioma of the middle fossa. Prog Comput Tomogr (Tokyo) 4: 460–465
3. Harper DG, Buck DR, Early DO (1982) Visual loss from cavernous hemangiomas of the middle cranial fossa. Arch Neurol 39: 252–254
4. Ichikizaki K, Shiobara R, Shizana H, Izaka Y, Toya S, Ohuchi T, Suzuki S, Shiga H, Ohtani M, Mikama K, Izumi G (1980) The usefulness of computed tomography for the diagnosis of intracranial cavernous hemangiomas. Prog Comput Tomogr (Tokyo) 2: 319–327
5. Ishijima Y, Matsumura H, Kageyma N (1966) Intracranial cavernous hemangioma: Report of two cases. Arch Jpn Chir 35: 748–754
6. Ito J, Takahashi M, Saito A, Honda H, Yajima K, Ueki K (1977) A case of cavernous hemangioma of the middle fossa with visualization of the feeding artery and faint tumor stain in cerebral angiography. Jpn J Clin Radiol 22: 339–344
7. Kawaguchi S, Ohsawa T (1965) A case of cavernous hemangioma at the base of the skull. Clin Neurol (Tokyo) 5: 705–708
8. Kawai K, Fukui M, Tanaka A, Kuramoto S, Kitamura K (1978) Extracerebral cavernous hemangioma of the middle fossa. Surg Neurol 8: 19–25
9. Kudo T, Ueki S, Kobayashi H, Torige H, Tadokoro M (1989) Experience with the ultrasonic aspirator in a cavernous hemangioma of the cavernous sinus. Neurosurgery 23: 628–631
10. Meyer FB, Lombardi D, Scheithauer B, Nichols DA (1990) Extra-axial cavernous hemangiomas involving the dural sinuses. J Neurosurg 73: 187–192
11. Namba S (1983) Extracerebral cavernous hemangioma of the middle cranial fossa. Surg Neurol 19: 379–388
12. Odake G, Tanaka K (1986) Cavernous hemangiomas of the middle fossa: case report and review of the literature. Neurol Med Chir (Tokyo) 26: 58–67
13. Ogasawara S, Nakazasa K, Katsura S, Oizumis (1973) Cavernoma on the intracranial base: report of a case. Brain Nerve (Tokyo) 6: 735–737
14. Rigamonti D, Pappas CTE, Spetzler RF, Johnson PC (1990) Extracerebral cavernous angiomas of the middle fossa. Neurosurgery 27: 306–310
15. Rosenblum B, Rothman AS, Ianzieri C, Song S (1986) A cavernous sinus hemangioma. J Neurosurg 65: 716–718
16. Sepehrnia A, Tatagiba M, Brandis A, Samii M, Prawitz RH (1990) Cavernous angioma of the cavernous sinus: case report. Neurosurgery 27: 151–155
17. Tsuiki K, Abe S, Chiba M, Murakami H, Saiki I, Kanaya H (1987) A case of cavernous angioma of the middle cranial fossa originating in the cavernous sinus with hyperprolactinemia. Neurol Surg (Tokyo) 15: 301–305

4.6

Rare benign parasellar space lesions

Cholesteatoma

Among the 12 rare benign parasellar lesions in this series, four were cholesteatomas [1]. Intracranial cholestheatomas are generally easily excised, but total removal is sometimes difficult because of their adherence to neural and vascular structures, especially those occurring in the parasellar space (PS). Simply opening the tumor gives access to the PS and often the primary mass may be removed without injuring neural structures or the PS-ICA and its branches. But it is extremely difficult to resect such lesions completely without a precise understanding of normal PS anatomy and how it has been altered by the tumor, particularly with respect to PS-ICA position and the displacement of nerves III through VI. Tumors will recur if not totally excised. Owing to their particularly slow growth rates, cholesteatomas typically recur only after several years, in our series between three (Case 1) and 15 years (Case 2). Case 1 recurred a second time 5 years after re-operation, and Case 2 has not yet recurred after10 years following re-operation. Neither of the remaining cholesteatomas (two) has recurred after 5 and 7 years.

Thrombosis of the parasellar space

Thrombosis of the PS occurred in four cases, two of which were operated, the remaining four followed without surgery. Acute PS thrombosis often resembles a neoplasm clinically, resulting in compression of neural and vascular structures. Even MRI characteristics are quite similar. Depending on the extent of compression of adjacent structures, acute PS thrombosis may present with sudden retro-orbital pain extending to the ipsilateral face and orbit, associated with deficits of nerves III through VI. With partial thrombosis, only mild pain or discomfort may be present in the absence of neurological loss.

Venous thrombosis of the PS does not require surgery. But one must be alert to determine its presence. Serial MR studies are often necessary to confirm the diagnosis. In the absence of continuing pain, one cannot rely on a single study. We learned this lesson the hard way!

With our first two cases, we mistakenly failed to repeat MRI studies performed at another institution soon after the onset of thrombosis. One was a 9 year-old child who had experienced sudden retro-orbital pain associated with paresis of nerves III, IV, and VI. His MRI, performed three months earlier, revealed a PS "tumor". Although the MR images were of good quality, they seemed somewhat strange, the PS being only slightly enlarged, with its lateral wall bulging outward. Another curious finding, which could have led to the correct diagnosis, was the fact that contrast-enhanced MR images were little different from unenhanced images. Although the child was said to be in pain when the MRI was performed, at the time of our assessment he was pain-free but deficits of nerves III, IV, and VI remained. Having no previous experience with venous thrombosis within the PS at that time, we took the child to the operating room only to find that the PS was empty! Obviously his thrombus had resolved since the MRI was performed. Fortunately, there were no surgical complications, and all neurological deficits cleared within an additional three months.

Two years later, having forgotten this experience, we admitted another patient, an adult, who had developed acute (virtually complete) paresis of nerves III through VI in the absence of pain. At surgery the PS was also found to be empty, resulting in another "successful" but entirely unnecessary PS exploration. Within tree months all neurological function returned.

Finally, we began to catch on! Subsequently, two patients were admitted whose outside MR films demonstrated no change after contrast injection. Both had extensive cranial nerve deficits but no continuing pain. Repeat MRI demonstrated total disappearance of both of these "pseudotumors," one within three months, the other within four months. Cranial nerve deficits recovered spontaneously and completely.

Under no circumstances should patients exhibiting persistent external opththalmoplegias after the disappearance of pain be subjected to surgery without follow-up diagnostic study.

Tolosa-Hunt syndrome

The sudden onset of retro-orbital pain associated with paresis of one or more cranial nerves which does not subside spontaneously with or without analgesics likely indicates inflammation within the PS. Post-contrast MR imaging demonstrating a vague shadow in the PS but no true mass effect suggests the presence of Tolosa-Hunt syndrome [2, 3]. The resolution of pain with a trial of corticosteroid therapy confirms the diagnosis (Fig. 4.75). Although pain subsides dramatically, neurological function recovers more slowly, if at all. Unfortunately, if ipsilateral visual function is affected, our experience indicates that it will not improve. Pain control persists after tapering and discontinuing corticosteroids over two to three weeks but the neurological deficit remains unchanged. Pain may recur after days to weeks, sometimes even months. Surgical exploration in such circumstances results in no improvement. Occasionally, surgical exploration or simple aspiration of the PS is helpful to confirm the presence of inflammation and rule out malignant cells infiltrating the PS. Two of our four Tolosa-Hunt cases were explored surgically. In both instances granulomatous tissue was recovered from the PS which manifested typical changes of inflammation on histological examination. Such cases commonly profit from corticosteroid treatment.

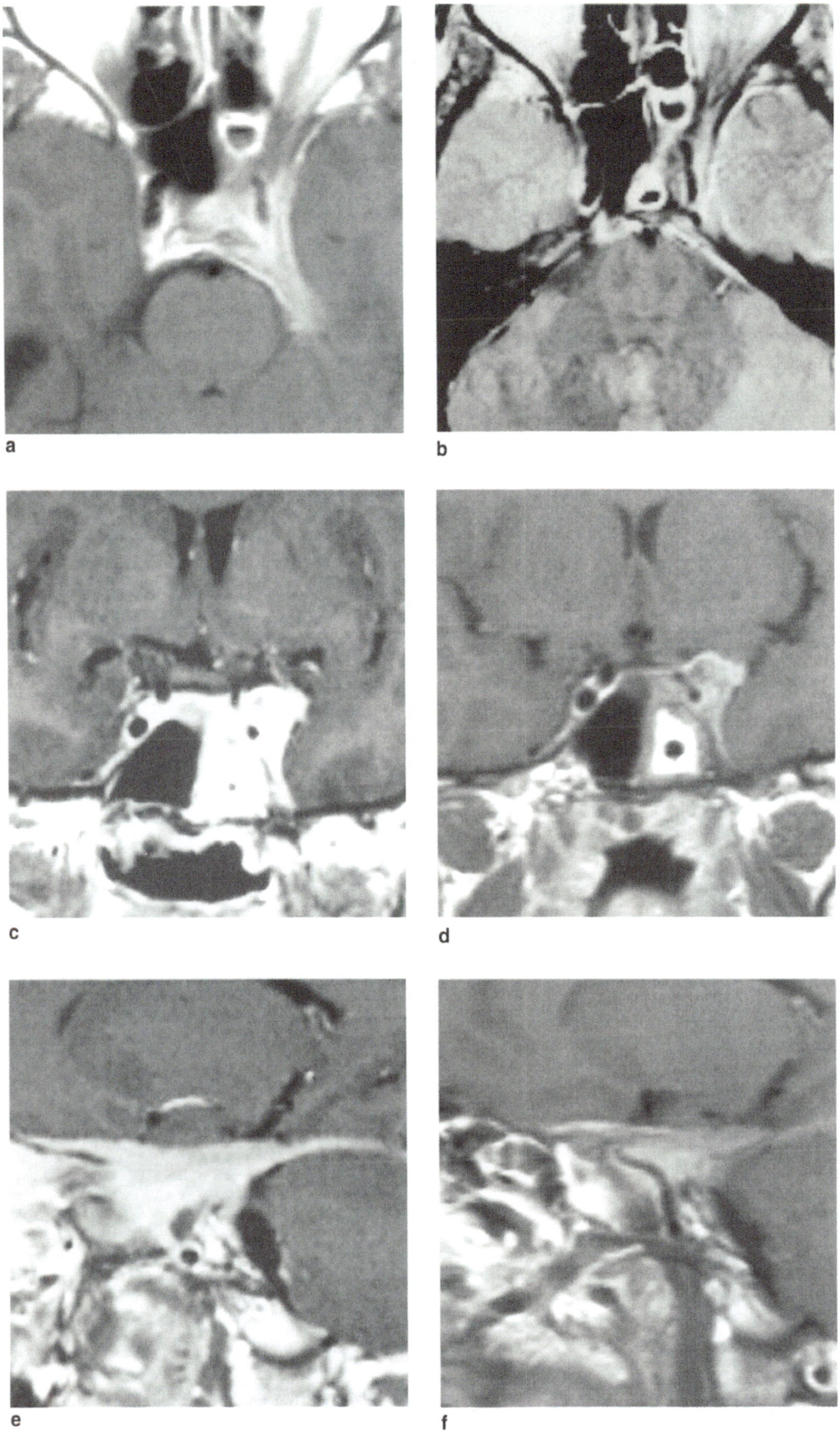

Fig. 4.75. Transaxial (a), coronal (c), and sagittal (e) contrast-enhanced MR images reveal a vague lesion in the left parasellar space which is heterogeneous and extending irregularly along the dura. Corresponding transaxial (b), coronal (d), and sagittal (f) contrast-enhanced MR images, performed one month after the initiation of corticosteroid therapy, confirm significant reduction in the lesion's size and degree of contrast-enhancement. Pain subsided considerably within three days of treatment, disappearing after one week. Function of nerves III through VI began improving during the second week after corticosteroid treatment was initiated. Six months later the patient remained pain-free, but blind in the ipsilateral eye with only partial recovery of other cranial nerve function

Conclusions

Each patient with a CSB cholesteatoma required repeated operations, only gross-total resections possible in all four. But significant clinical improvement resulted in each one. Surgery is not required with PS venous thrombosis. Two delayed surgical explorations, of the four patients with acute thromboses, revealed no clot or other abnormality, the same confirmed in two others by subsequent MRI. Of four instances of Tolosa-Hunt syndrome, an inflammatory process of the CSB, surgical exploration in two confirmed the presence of granulomatous tissue. Since totally removing inflammatory tissue is neither desirable nor possible, surgical exploration is not indicated unless a malignancy is suspected. Of these 12 patients, four surgical explorations (two for venous thrombosis and two for Tolosa-Hunt syndrome) were not only unnecessary but no therapeutic resection was achieved.

References

1. Baumann CHH, Bucy PC (1956) Paratrigeminal epidermoid tumors. J Neurosurg 13: 455–468
2. Hunt WE, Meagher JN, LeFever HE, Yeman W (1961) Painful ophthalmoplegia: its relation to indolent inflammation of the cavernous sinus. Neurology 11: 56–62
3. Tolosa E (1954) Periarteric lesions of the carotid siphon with the clinical features of a carotid infraclinoidal aneurysm. J Neurol Neurosurg Psychiatry 17: 300–302

4.7

Craniopharyngiomas

Introduction

Only complete surgical resection of the craniopharyngioma offers consistent promise in both children and adults [3, 7, 10, 14, 15, 17, 18, 20–22, 24, 35, 37–39]. The first attempt must be as radical as possible, but preserving adjacent vascular and neural structures is equally important. Postoperative radiation should be reserved for tumors remaining or recurring after secondary operations, never following the initial craniotomy [2, 11, 16, 19, 28, 32, 36]. Radioactive yttrium implantation, employed at one time with enthusiasm for patients with cystic lesions, has been abandoned, not only because evidence of tumor growth control was not forthcoming but because of the devastating amount of scar it induced, precluding serious surgical attempts in the future. Post-yttrium operations have been largely doomed to failure, the only consistent outcome being additional neurological loss.

Pathological anatomy

Craniopharyngiomas occur in the midline between the sella and ventricle III, some occupying ventricle III (Figs. 4.76–4.80). Depending on size and location, they primarily affect neuroendocrinological structures and visual pathways. Tumors confined to the sella create endocrine disturbances by irritating and/or destroying the pituitary gland or its means of communication with the hypothalamus (Figs. 4.79 and 4.80). Suprasellar midline lesions abut the pituitary stalk and the visual apparatus, most commonly generating visual symptoms and signs (Figs. 4.77–4.80). It is surprising, however, how much neuroendocrinological structures, even visual pathways, resist compression and distension from tumor expansion. Additional upward growth further affects the optic chiasm, pituitary stalk, and hypothalamus. Continued growth is likely to displace the hypothalamus laterally in either or both directions and shove the floor of ventricle III further upward. Such tumor expansion often results in hydrocephalus, requiring urgent surgical intervention (Figs. 4.76, 4.77, 4.79).

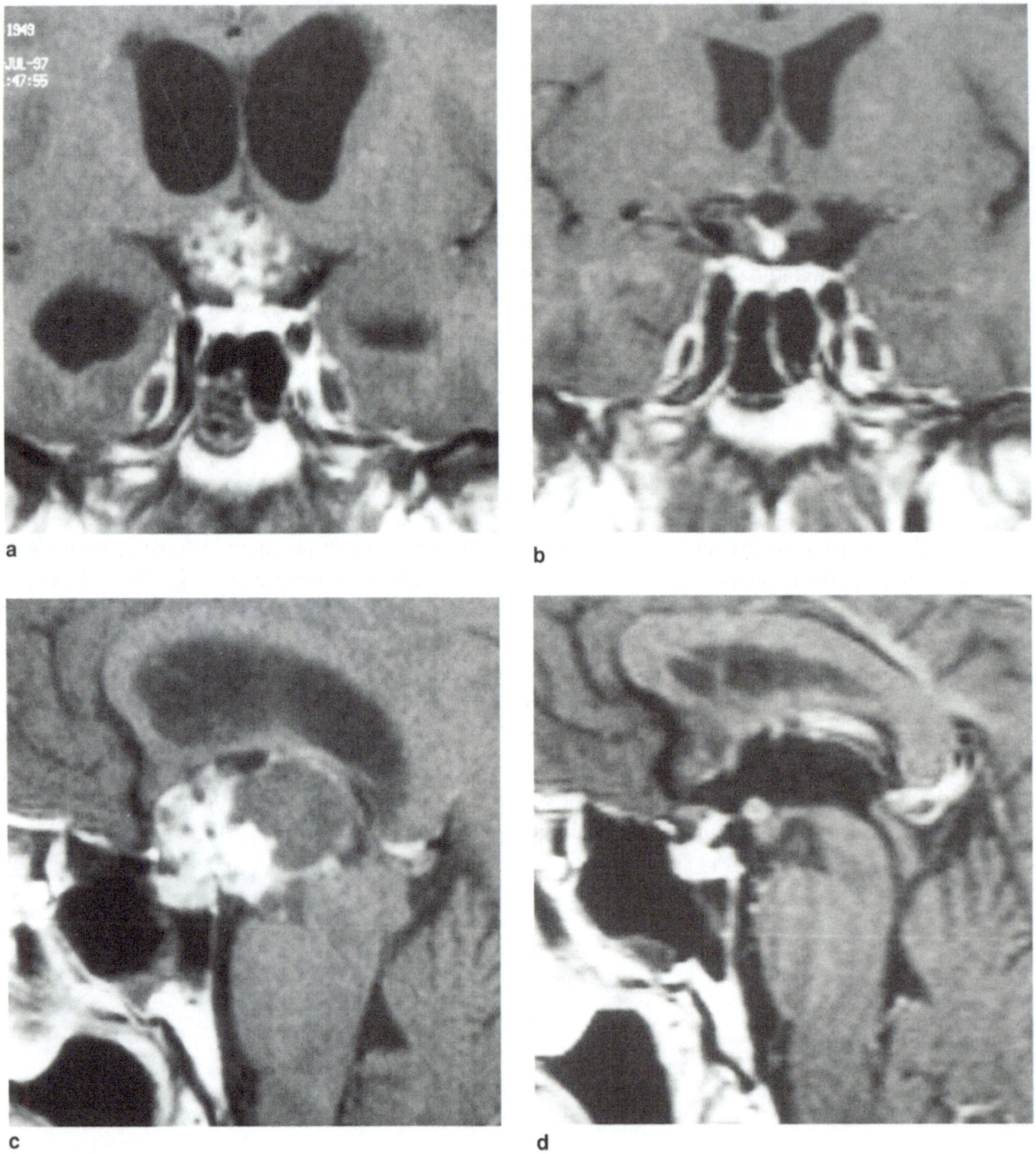

Fig. 4.76. Preoperative coronal (a) and sagittal (c) contrast-enhanced MR images demonstrate a suprasellar multicystic craniopharyngioma originating from the sella, its large cystic component bulging into the III-rd ventricle. Corresponding contrast-enhanced MR images obtained two years after surgery (b, d,) confirm complete resection and provide no evidence of tumor recurrence

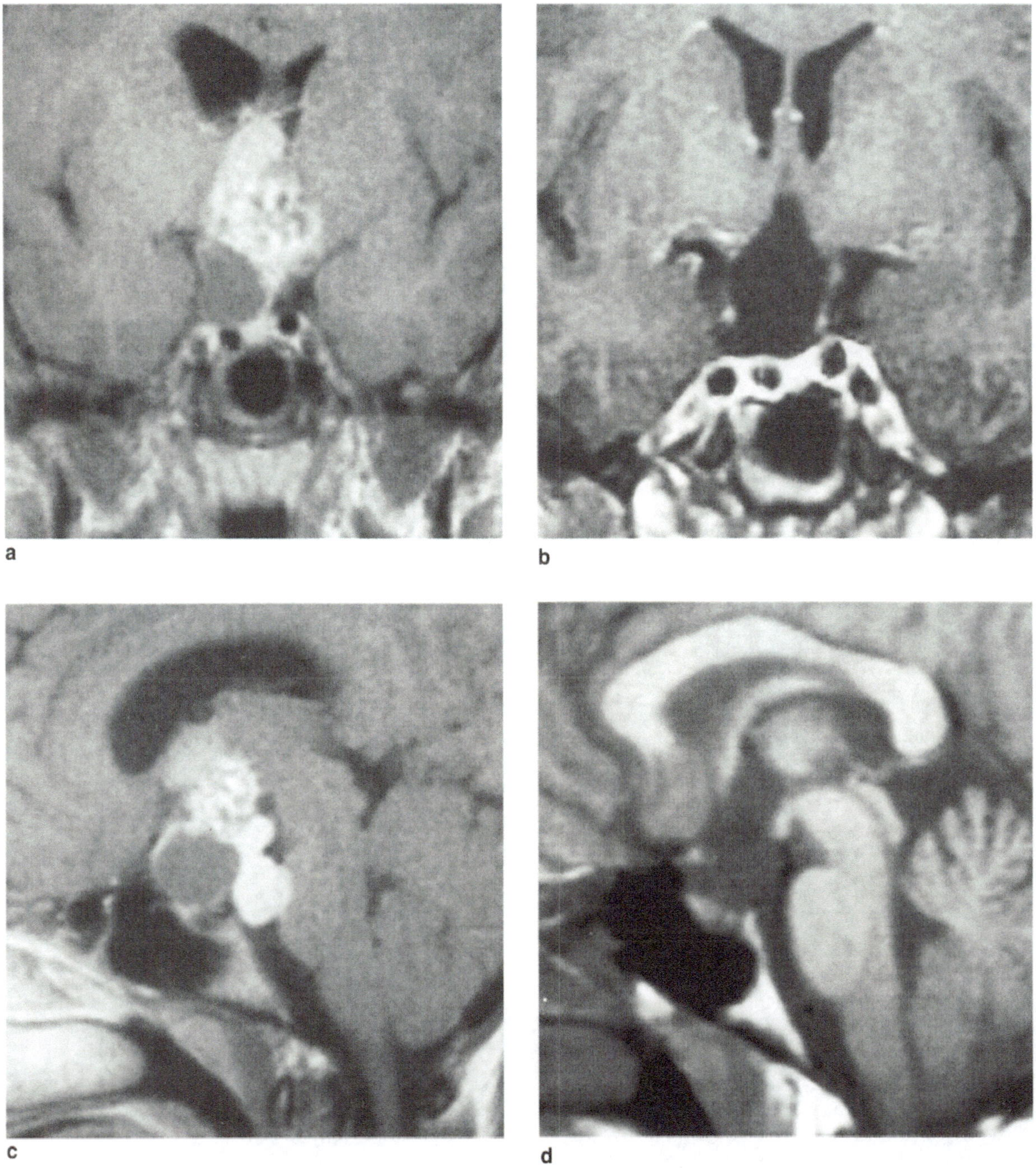

Fig. 4.77. Preoperative coronal (a) and sagittal (c) contrast-enhanced MR images reveal a suprasellar and retrosellar multicystic craniopharyngioma bulging into the III-rd ventricle and displacing each half of the hypothalamus laterally. Corresponding contrast-enhanced MR images obtained five years postoperatively (b, d) show no residual or recurrent tumor

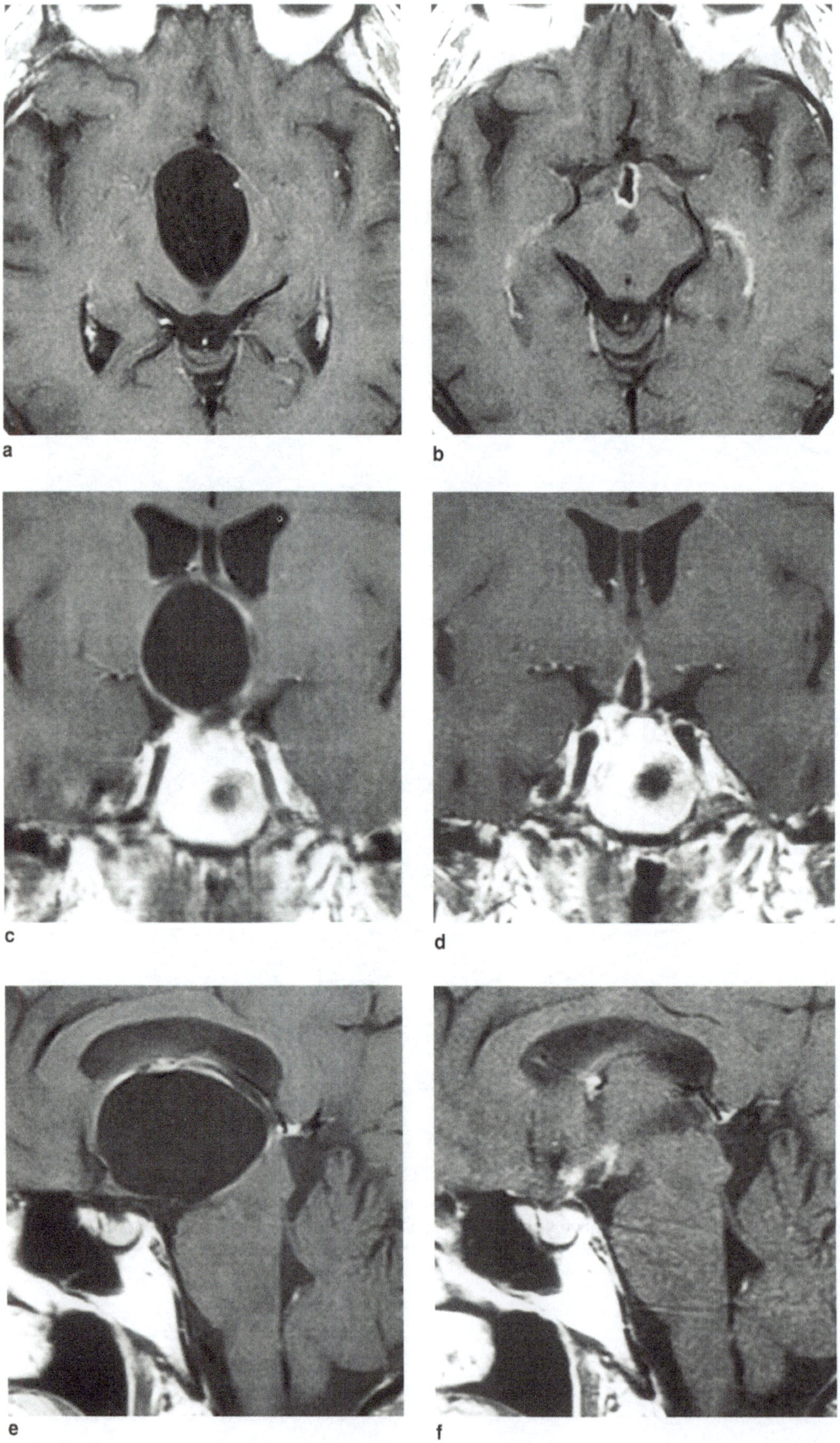

Fig. 4.78. Preoperative trans-axial (a), coronal (c), and sagittal (e) contrast-enhanced MR cuts demonstrate a large suprasellar cystic lesion with a slightly enhancing capsule but little mass-producing solid component. Six months after surgery, corresponding contrast-enhanced MR images (b, d, f) reveal a cystic lesion with mildly enhanced borders, only several mm in diameter

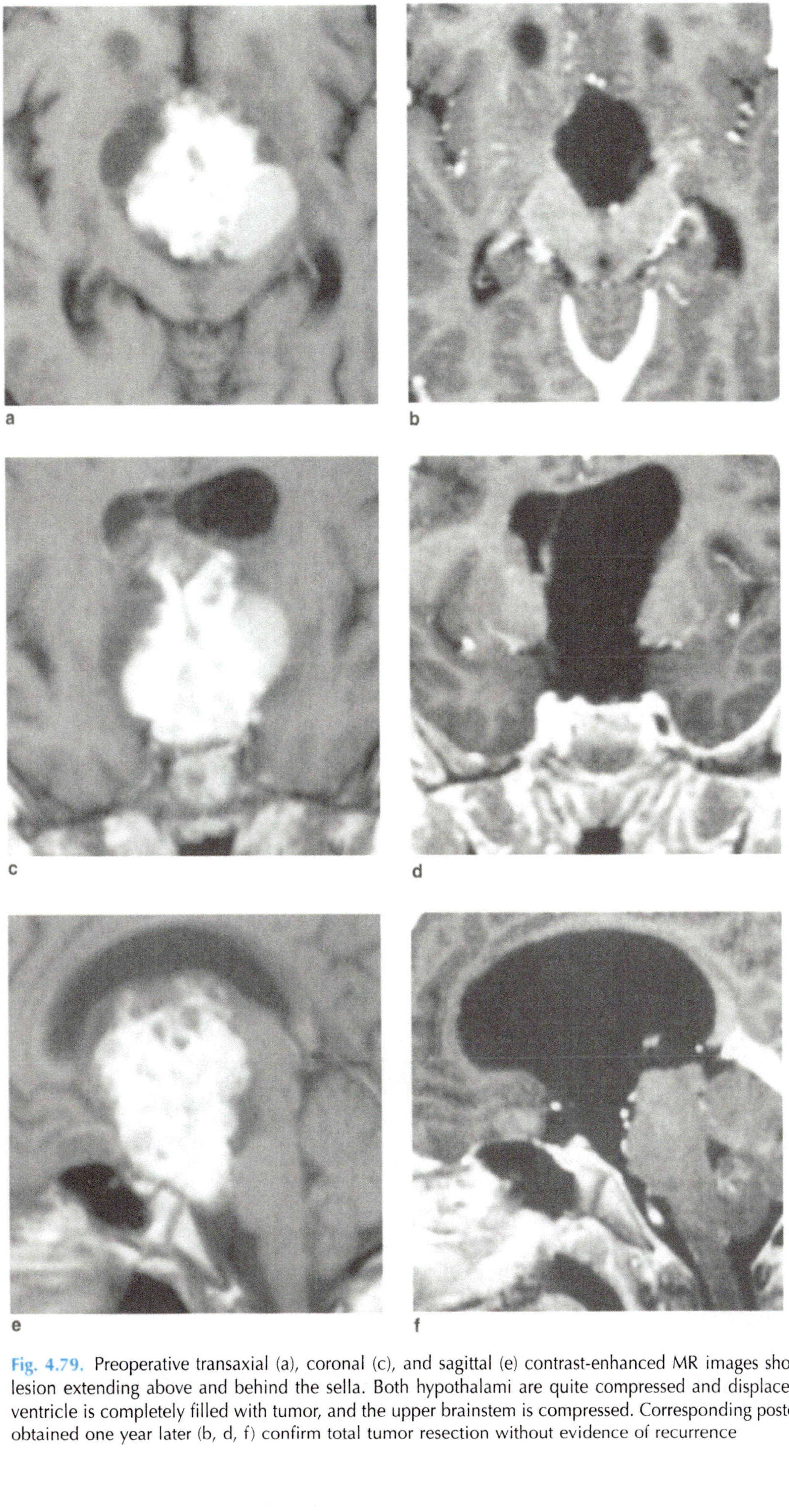

Fig. 4.79. Preoperative transaxial (a), coronal (c), and sagittal (e) contrast-enhanced MR images show a huge polycystic lesion extending above and behind the sella. Both hypothalami are quite compressed and displaced laterally. The III-rd ventricle is completely filled with tumor, and the upper brainstem is compressed. Corresponding postoperative MR images obtained one year later (b, d, f) confirm total tumor resection without evidence of recurrence

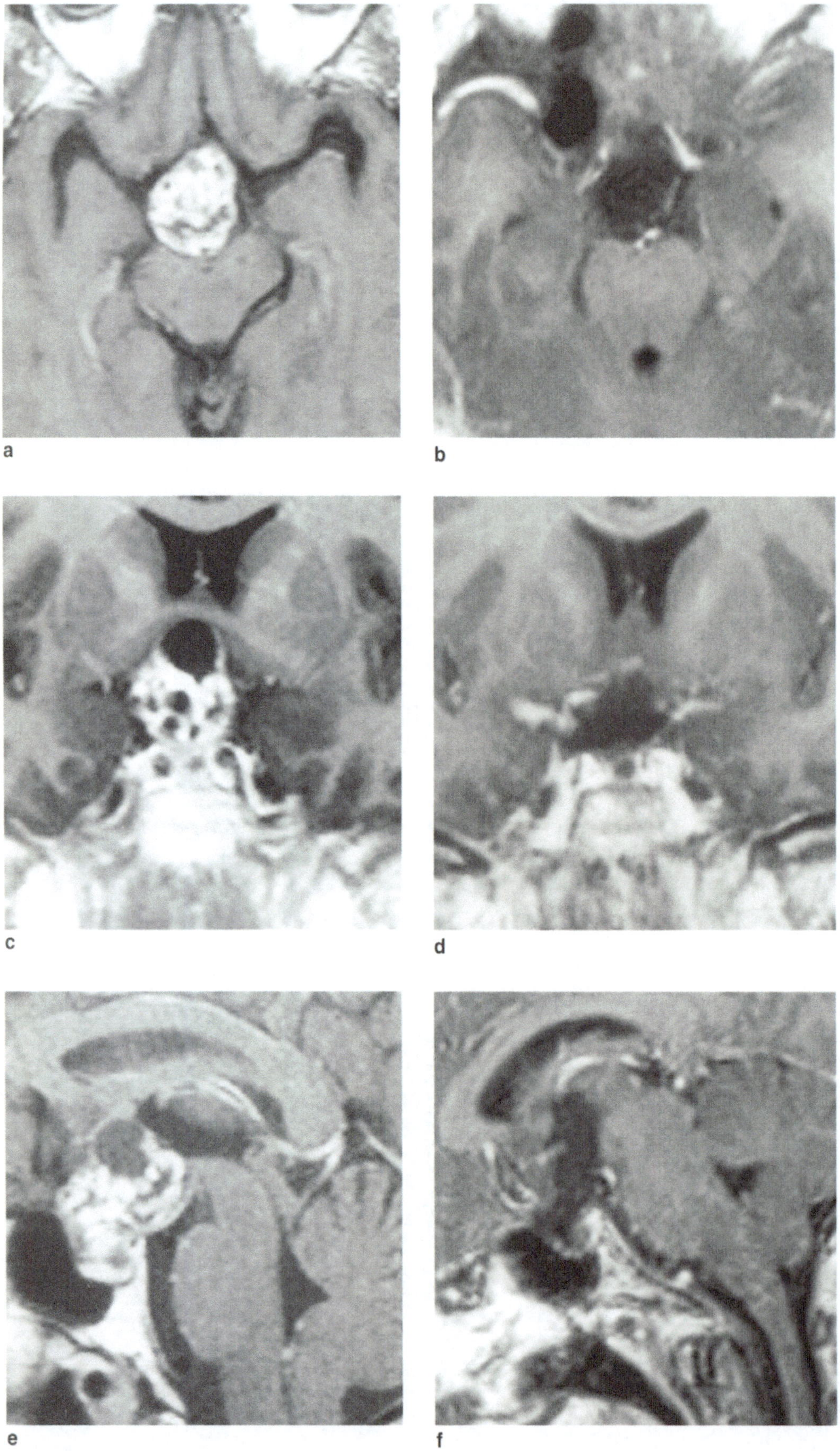

Fig. 4.80. Preoperative trans-axial (a), coronal (c), and sagittal (e) contrast-enhanced MR images demonstrate a medium-sized polycystic craniopharyngioma occupying the entire sella and extending above it. Corresponding postoperative MR images after three months (b, d, f) confirm total tumor resection

Surgical technique

In general, the CSB approach described in Chapter 2 applies to craniopharyngiomas with good effect. Intrasellar lesions require completely resecting the anterior clinoid process (ACP) and opening the optic canal (OC). Tumors occupying primarily the suprasellar space allow the roof of the orbit, the sphenoid wing, and the ACP to be left intact unless the chiasm is displaced significantly upward, resulting in hypothalamic compression. Where tumor compresses the floor of ventricle III, extending upward to the level of foramen Monroe bilaterally, the ventricular component should be removed through the lamina terminalis [4, 25–27, 30]. Even larger tumors, those filling the majority of ventricle III and occluding both foramina of Monroe, typically require combining the method described in Chapter 2 with interhemispheric, transcallosal, and transventricular approaches.

Under no circumstances should the optic tract be retracted. Tumor remnants sequestered within the floor or at the base of ventricle III may not be accessed laterally with safety, but from beneath the optic tract or through the lamina terminalis via the interhemispheric transcallosal approach. This is true also for some other tumors invading the hypothalamus and ventricle III, those affecting the superior end of the pituitary stalk, and those whose inferior elements surround either or both optic tracts. The final remnant of tumor beneath the optic tract(s) should never be forcefully delivered from around a bank of vital hypothalamic or visual fibers. It is far better to combine approaches from different directions. Removing tumor from a remote location through a corridor walled by vital brain introduces the risk of venous bleeding at a site where direct control is not possible. To stuff large amounts of Surgicel blindly into such cavities may only complicate the problem, however gently it is placed. Forcefully extracting tumor from its bed may damage important perforators arising from the circle of Willis. Mechanical irritation may send such tiny vessels into vasospasm, creating an area of hypothalamic and brainstem ischemia (including the pituitary stalk and visual apparatus) which is wider than the area damaged by long-standing tumor presence and direct surgical trauma. To avoid such ischemia, one must select precisely the correct approach or combination of approaches; strictly adhering to traction-free, piecemeal resection techniques, employing endoscopy as method of seeing around corners; and remaining constantly aware that typically it is the final tumor element to be removed that results in the greatest amount of neurological injury.

Solid components of craniopharyngiomas vary from quite soft to rock-hard, requiring drilling to be removed. Totally solid lesions are uncommon, most being partially cystic (Figs. 4.76, 4.77 and 4.79, 4.80). Cystic elements may vary from multiple small cysts to a large single cyst. Emptying cystic cavities enhances working conditions by increasing dissecting space more quickly and more safely than by removing solid tumor piece by piece. Occasionally imaging studies suggest that the lesion is totally cystic (Fig. 4.78). In such cases simply opening the cyst is not adequate. Neither should the cyst wall be removed by displacing it blindly from remote corners. Such techniques risk damaging vital veins and arteries, as well as delicate neural structures. The walls of such cysts are ideally resected, but only under direct vision in order to avoid stretching perforating arteries, tearing veins, and distorting adherent neural tissue. Cyst walls typically contain tumor cells which almost certainly assure reformation of the cyst if left behind.

When neurological symptoms and signs are present prior to surgery, it is virtually impossible to achieve total tumor resection without affecting the neurological picture adversely, at least temporarily [4]. Such is the nature of the vascular and neural structures involved. The slightest injury to the pituitary stalk results in postoperative diabetes insipidus, which may become serious if overlooked. Yet, when intensivists are aware that diabetes insipidus may be eminent six to eight hours following surgery, its treatment is simple and generally successful.

Surgery for residual and/or recurrent craniopharyngiomas is always more difficult than removing those not previously operated upon [8]. One faces the difficult task of dissecting neural and vascular structures from adhesions. In addition to directly damaging important neural structures by dissection, depriving them of vascular supply will affect their subsequent function. In order to preserve preoperative neurological and endocrine function, it is much preferred to leave scar and tumor capsule intact, even tumor parenchyma likely to be densely adherent to important structures. It is equally important to preserve small perforators arising from the circle of Willis.

Patients and results

Two hundred seventy-five consecutive patients were operated for craniopharyngiomas over two decades. Preoperative clinical status and outcome are summarized in Table 24. Preoperative visual disturbances were detected in 201 cases, and 249 had endocrinological deficits. Forty patients manifested preoperative hydrocephalus from bilateral occlusion of the foramina of Monroe. In general, preoperative visual deficits improved after surgery, complete resolution occurring in 76. Endocrinological deficits were much less likely to subside, only 17 of the 249 clearing completely after surgery.

Of the entire series, 183 were virgin cases at the time of surgery, 92 patients having been operated previously at another institution. Over all, total resection was achieved in 161 cases, gross-total in 91, and subtotal in 23. Re-operation was required in 53, 20 patients requiring two operations and six having three or more secondary operations. These data are summarized in Table 25.

There was one death, a nine-year-old girl with a large suprasellar tumor extending into the III-rd ventricle, compressing the hypothalamus bilaterally. She awakened after surgery and CT performed within hours revealed no tumor, hydrocephalus, or evidence of bleeding. After speaking with her parents during the evening of the day of surgery, she died the following morning after an uncontrollable generalized epileptic seizure. Her sodium had increased from 140 mEq/L to 180 mEq/L during the early morning hours.

Table 24. Craniopharyngioma surgery: preoperative neurosurgical and endocrinological status and outcome

		Preoperative deficits			Postoperative deficits	
Tumor location	N°	Visual	Endocrinological	Hydrocephalus	Visual	Endocrinological
Intrasellar	7	–	5	–	–	5
Suprasellar	138	75	129	–	36	107
Suprasellar & III-rd ventricle	117	117	103	32	82	109
III-rd ventricle & Hydrocephalus	11	9	11	8	7	10
Other	2	–	1	–	–	1
Total	275	201	249	40	125	232

Table 25. Craniopharyngioma surgery: resectability and secondary operations

				Resection			Reoperation		
Tumor location	N°	Virgin cases	Residual and/or recurrent	Total	Gross-total	Sub-total	1	2	3+
Intrasellar	7	6	1	5	2	–	1	–	–
Suprasellar	138	89	49	87	48	3	28	11	5
Suprasellar and III-rd ventricle	117	79	38	61	38	18	22	8	1
III–rd ventricle and Hypothalamus	11	9	2	8	2	1	1	1	–
Other	2	–	2	–	1	1	1	–	–
Total	275	183	92	161	91	23	53	20	6

Hydrocephalus was considered an indication for urgent surgical intervention, both to relieve hydrocephalus and resect tumor. In no case was a shunt placed. Radiation therapy was reserved for those patients requiring a second re-operation (the third operation in patients with virgin tumors, the fourth for those having their initial surgery at another institution). We believe that secondary operations result in less trauma than irradiation. Gamma Knife therapy was employed after the second operation in two patients with intrasellar craniopharyngiomas, conventional radiation in the remaining 24 cases.

Conclusions

Craniopharyngiomas in children and in adults, being primarily solid or cystic, rarely pose a difficult task for the skilled neurosurgeon unless they are too large to be resected via a single route, either the lateral approach or the interhemispheric-transcallosal-transventricular approach. While physically resecting such tumors may be "easily" accomplished, the post-operative neurological outcome may be devastating. The unwary surgeon may unwittingly sacrifice his patient's well being by resecting the tumor capsule, cyst wall, and/or the solid component from its bed, peeling it from the visual apparatus, the hypothalamus, and out of the III-rd ventricle. More than direct trauma to surrounding neural tissue, coagulating bleeding veins and stretching small arteries perforating the hypothalamus and visual apparatus lead to severe neurological [1, 6, 29, 31] and endocrine deficits [9, 12, 13, 23, 33, 34]. Infarction of tissue surrounding the tumor bed is primary responsible for poor surgical results. Removing the tumor mass is one thing. Doing it atraumatically, preserving all adjacent vessels is quite another! Perhaps as no other operation in this region, craniopharyngioma resections are extremely dangerous. Unfortunately, the price for neurosurgeons' "heroic" actions must be paid by his patients.

References

1. Abrams LS, Repka MX (1997) Visual outcome of craniopharyngioma in children. J Pediatr Ophthalmol Strabismus 34: 223–228
2. Amacher AL (1980) Craniopharyngioma: the controversy regarding radiotherapy. Childs Brain 6: 57–64
3. Baskin DS, Wilson CB (1986) Surgical management of craniopharyngiomas. A review of 74 cases. J Neurosurg 65: 22–27
4. Bhagwati SN, Deopujari CE, Parulekar GD (1990) Lamina terminalis approach for retrochiasmal craniopharyngiomas. Childs Nerv Syst 6: 425–429
5. Bucci MN, Chin LS, Hoff JT (1987) Preoperative morbidity associated with operative resection of craniopharyngioma: a review of ten years experience. Neurochirurgia 30: 135–138
6. Cabezudo Artero JM, Vaquero Crespo J, Bravo Zabalgoitia G (1987) Status of vision following surgical treatment of craniopharyngiomas. Acta Neurochir (Wien) 73: 165–177
7. Cabezudo JM, Vaquero J, Areitio E, et al (1981) Craniopharyngiomas: a critical approach to treatment. J Neurosurg 55: 371–375
8. Caldarelli M, di Rocco C, Papacci F, et al (1998) Management of recurrent craniopharyngioma. Acta Neurochir (Wien) 140: 447–454
9. Carmel PW (1985) Vegetative dysfunctions of the hypothalamus. Acta Neurochir (Wien) 75: 113–121
10. Carmel PW, Antunes JL, Chang CH (1982) Craniopharyngiomas in children. Neurosurgery 11: 382–389
11. Chung WY, Pan DH, Shiau CY, et al (2000) Gamma knife radiosurgery for craniopharyngiomas. J Neurosurg 93 (Suppl 3): 47–56
12. Curtis J, Daneman D, Hoffman HJ, et al (1994) The endocrine outcome after surgical removal of craniopharyngiomas. Pediatr Neurosurg 21 [Suppl]1: 24–27
13. DeVile CJ, Grant DB, Hayward RD, et al (1996) Growth and endocrine sequelae of craniopharyngioma. Arch Dis Child 75: 108–114
14. DeVile CJ, Grant DB, Kendall BE, et al (1996) Management of childhood craniopharyngioma: can the morbidity of radical surgery be predicted? J Neurosurg 85: 73–81
15. Fahlbusch R, Hongger J, Paulus W, et al (1999) Surgical treatment of craniopharyngiomas: experience with 168 patients. J Neurosurg 90: 237–250

16. Fisher EG, Welch K, Shillito J Jr, et al (1990) Craniopharyngiomas in children. Long-term effects of conservative surgical procedures combined with radiation therapy. J Neurosurg 73: 534–540
17. Garcia-Uria J (1978) Surgical experience with craniopharyngioma in adults. Surg Neurol 9: 11–14
18. Graham PH, Gattamaneni HR, Birch JM (1992) Pediatric craniopharyngiomas: a regional review. Br J Neurosurg 6: 187–193
19. Habrand JL, Ganry O, Couanet D, et al (1999) The role of radiation therapy in the management of craniopharyngioma: a 25-year experience and review of the literature. Int J Radiat Oncol Biol Phys 44: 255–263
20. Hayward R (1999) The present and future management of childhood craniopharyngioma. Childs Nerv Syst 15: 764–769
21. Hoff JT, Patterson RH Jr (1972) Craniopharyngiomas in children and adults. J Neurosurg 36: 299–302
22. Hoffman HJ, De Silva M, Humphreys RP, et al (1992) Aggressive surgical management of craniopharyngiomas in children. J Neurosurg 76: 47–52
23. Honegger J, Buchfelder M, Fahlbusch R (1999) Surgical treatment of craniopharyngiomas: endocrinological results. J Neurosurg 90: 251–257
24. Katz EL (1975) Late results of radical excision of craniopharyngiomas in children. J Neurosurg 42: 86–93
25. King TT (1979) Removal of intraventricular craniopharyngiomas through the lamina terminalis. Acta Neurochir 45: 277–286
26. Klein HJ, Rath SA (1989) Removal of tumors in the III ventricle using the lamina terminalis approach. Three cases of isolated growth of craniopharyngiomas in the III ventricle. Childs Nerv Syst 5: 144–147
27. Maira G, Anile C, Colosimo C, et al (2000) Craniopharyngiomas of the third ventricle: trans-lamina terminalis approach. Neurosurgery 47: 857–865
28. Mokry M (1999) Craniopharyngiomas: a six year experience with Gamma Knife radiosurgery. Stereotact Funct Neurosurg 72 [Suppl]1: 140–149
29. Mutlukan E, Cullen JF (1990) Visual outcome after craniopharyngioma. Ophthalmology 97: 539–540
30. Patterson RH Jr, Danylevich A (1980) Surgical removal of craniopharyngiomas by a transcranial approach through the lamina terminalis and sphenoid sinus. Neurosurgery 7: 111–117
31. Repka MX, Miller NR, Miller M (1989) Visual outcome after surgical removal of craniopharyngiomas. Ophthalmology 96: 195–199
32. Richmond IL, Wara WM, Wilson CB (1980) Role of radiation therapy in the management of craniopharyngiomas in children. Neurosurgery 6: 513–517
33. Sklar CA (1994): Craniopharyngioma: encodrine sequellae of treatment. Pediatr Neurosurg 21 [Suppl]1: 120–123
34. Sorva R, Heiskanen O, Perheentupa J (1988) Craniopharyngioma surgery in children: endocrine and visual outcome. Childs Nerv Syst 4: 97–99
35. Thomsett MJ, Conte FA, Kaplan SL, et al (1980) Endocrine and neurologic outcome in childhood craniopharyngioma: review of effect of treatment in 42 patients. J Pediatr 97: 728–735
36. Weiss M, Sutton L, Marcial V, et al (1989) The role of radiation therapy in the management of childhood craniopharyngioma. Int J Radiant Oncol Biol Phys 17: 1313–1321
37. Wen DY, Seljeskog EL, Haines SJ (1992) Microsurgical management of craniopharyngiomas. Br J Neurosurg 6: 467–474
38. Yaşargil MG, Curcic M, Kis M, et al (1990) Total removal of craniopharyngiomas. Approaches and long-term results in 144 patients. J Neurosurg 73: 3–11
39. Zuccaro G, Jamovich R, Mantese B, et al (1996) Complications in paediatric craniopharyngioma treatment. Child Nerv Syst 12: 385–391

4.8

Malignant central skull base tumors

Introduction

Tumors of the central skull base (CSB) with radiographic and microscopic characteristics promising "malignant" biological behavior may be of local origin or the result of metastasis of distant disease. Furthermore, facial (splanchnocranial) and cervical malignancies invade the CSB by direct spread. As with malignancies presenting elsewhere, CSB lesions are rarely curable by surgery, but "complete" resections, combined with other treatment modalities, sometimes result in good outcomes that may be measured in years.

Our surgical concept of treating malignant CSB lesions is much different from dealing with benign lesions, in some ways antagonistic, even opposite. With *benign* lesions, since life itself is only rarely at stake, surgical excision and preservation of associated neural and vascular structures are of virtually equal importance, the primary goal always being to preserve the patient's quality of life. The surgeon is forced to give up the goal of total resection only when the risk of impairment of vital neural and vascular function increases sufficiently to jeopardize quality of life. Total excision of *malignant* CSB lesions, on the other hand, typically cannot be achieved without resecting tumor-invaded vessels and nerves. Palliative surgical methods exist, but salvaging life to the maximum requires a very aggressive approach, commonly making considerable postoperative neurological loss inevitable, always more extensive than that present prior to surgery.

To enhance the preservation of life in the presence of malignant CSB lesions, quality of life must often be sacrificed. Protecting adjacent neural structures simply affords too much sanctuary for the malignant tumor. Consequently, the choice between aggressive and conservative treatment becomes a matter of individual preference, requiring extensive discussion with the patient and his or her family. He natural history of untreated or variously treated disease becomes of great importance. Countless questions arise, many of which may be answered differently by various clinical authorities. The opinions of different specialties (oncologists, ENT and maxillofacial surgeons, plus confirming consultations as desired by the patient) becomes a must in order that fully informed decisions are possible. For many, more conservative management is a better fit with personal philosophies and religious values. When the decision is made to radically excise a malignant CSB tumor, the preservation of quality of life necessarily becomes of secondary importance. This is never a decision to be made by the operating surgeon. The length of the surgeon's life is not at risk. Neither is the surgeon faced with living with the consequences of aggressive treatment.

Pathological anatomy

Malignant CSB lesions may be divided into three groups. A) Lesions originating within [4, 7, 9] and remaining confined to the CSB at the time of diagnosis (Fig. 4.81). B) Lesions arising from within the CSB which infiltrate adjacent and facial and/or neck region(s) – among them are hemangiopericytomas (Fig. 4.82), malignant trigeminal neurinomas, pituitary carcinomas, and parasellar space (PS) sarcomas, and lymphomas [2] – or vice versa carcinomas of the splanchnocranium and oropharyngeal structures (Fig. 4.83) and other neck compartments which infiltrate the CSB [8, 10]. C) Metastatic diseases to the CSB are not frequent but they do exist [1, 3, 5, 6]. Among them the most frequent metastatic carcinoma to the CSB is the breast carcinoma.

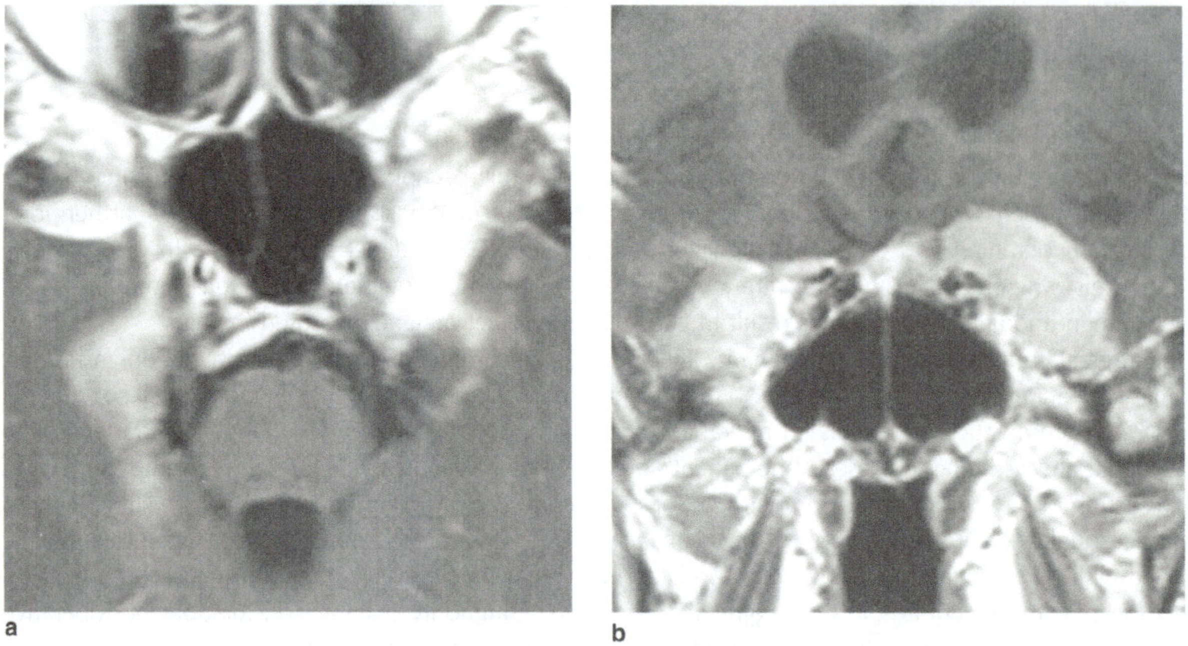

Fig. 4.81. This 60-year-old patient presented with dull retro-orbital aching and diminished vision in the left eye. Transaxial (a) and coronal (b) contrast-enhanced MR images revealed a bilateral parasellar tumor of unusual shape and density. Surgery was carried out from the left side, and non-Hodgkin's lymphoma was removed gross-totally from the parasellar space (*PS*) and the optic canal (*OC*). Tumor on the right side was subsequently treated by only non-surgical measures. To our surprise, vision in the left eye improved, and the neurological impairment present immediately after surgery (nerves III through VI on the left) subsided within two months

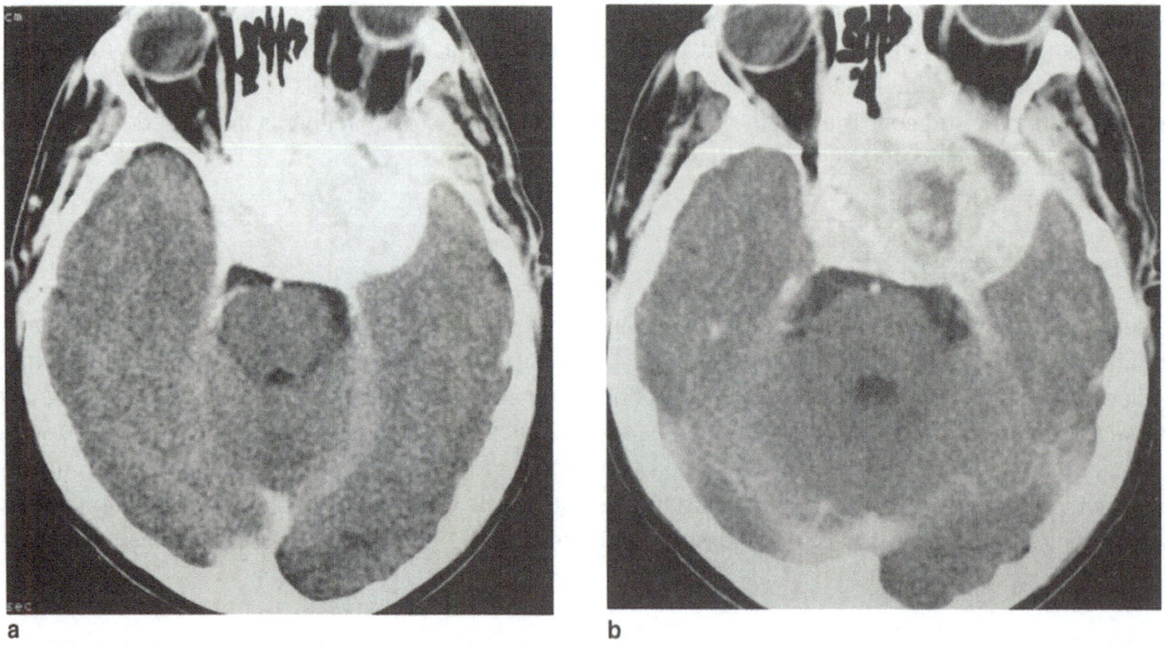

Fig. 4.82. This 22-year-old man had four previous operations over a three-year period for a large central skull base (*CSB*) hemangiopericytoma. A preoperative CT scan (a) reveals a large residual tumor occupying the sella, the retrosellar space, the left parasellar space, the sphenoidal and ethmoidal sinuses, and the left orbit. The corresponding postoperative CT scan (b) demonstrates only partial reduction of the tumor mass. Its vascularity and the presence of an abscess (related to previous surgery) precluded a more radical resection at that time. We had planned to make another surgical attempt after treating the residual infection, but the patient and his family subsequently chose to have only radiation therapy. He died one year later

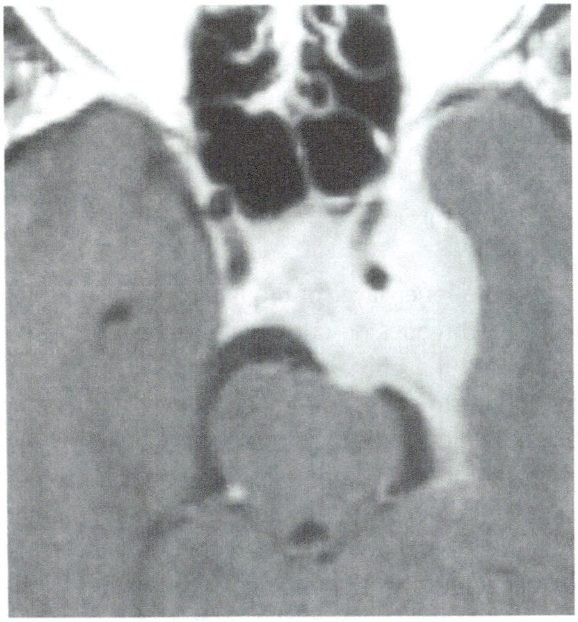

a

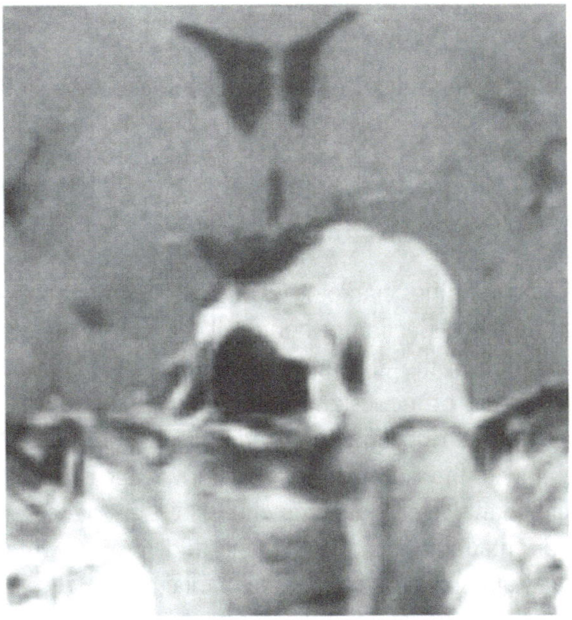

b

Fig. 4.83. This 58-year-old man had been "successfully" operated and irradiated for a left-sided pharyngeal carcinoma two years prior to the onset of symptoms and signs which predicted involvement of the left parasellar space. Preoperative transaxial (a) and coronal (b) contrast-enhanced MR images revealed tumor within the sella and the left parasellar space (*PS*), extending into the posterior cranial fossa, along the tentorial edge (a), and into the left side of the pharynx (b). Since he was unwilling to accept the risk of complete ophthalmoplegia if total excision was not possible, prior to surgery he specifically requested that no attempt be made to aggressively excise tumor adherent to cranial nerves II through VI if its nature was identical to his previous pharyngeal cancer. At surgery tumor was encountered which was densely adherent to all neural and vascular structures in the region, allowing only a very limited resection after biopsy

Surgery

The surgical approach to malignant CSB lesions is dictated by their location and sites of origin.

Techniques and strategies for dealing with malignant lesions limited to the CSB (tipically invading the PS)

Patient positioning, craniotomy, and the extradural approach to the inner layer of the lateral wall of the PS, including the dissection of nerves III through V, are identical to techniques described in Chapter 2. The size and location of the CSB malignancy dictates the point of entry into the CSB and the lesion itself. The histological nature of the lesion determines the extent that neighboring structures (dura, periosteum, and cranial nerves) require resection. In general, the PS should not be entered unless radical extirpation is anticipated. We disagree with the view that detailed knowledge of normal and pathological anatomy is not necessary for resecting malignant lesions involving the PS. Quite to the contrary: considerable experience is required to maintain one's orientation within the CSB and to determine which structures are normal and which are diseased. Determining if and when one or several cranial nerves necessitate resection requires considerable experience. While malignant tumors tend to draw the surgeon into the PS without demanding contemplation of ideal avenues of approach, critical questions remain. How far may one proceed with safety? The ICA is always a consideration. Should adjacent structures be excised in addition to the malignant trigeminal neurinoma, the sarcoma, the metastatic lesion, or other malignancy? If so, which neural elements may be spared and how?

Strategies for treating malignant CSB lesions extending into the splanchnocranium and/or the neck

We prefer to operate on these tumors in two stages, dealing first with the CSB component, later the element involving facial bones and/or the neck via a different approach. "Complete" resection of extradural and intradural components (including infiltrated dura) must be followed by a watertight dural closure with fascia lata in order to minimize not only the risk of a CSF leak but the possibility of an intradural infection (which is enhanced by contamination from exposing adjacent paranasal sinuses). Establishing a tissue diagnosis during the first stage may lead to the initiation of complementary non-surgical treatment and/or facilitate the planning of the second stage. The second stage, if necessary, should be carried out as soon as possible after uncomplicated healing of dura from the first operation, ideally within two or three weeks. The institution of complementary treatment need not be delayed following the initial procedure.

Strategies for managing malignant tumors involving facial bones and/or neck which extend to the CSB

For malignancies invading the CSB from sites of origin in the splanchnocranium and neck we favor resecting the bulk of peripheral elements first, to the borders of the CSB. Often orbital exenteration or exenteration of the ethmoidal, sphenoidal, and/or maxillary sinuses or resection of the entire area deep to the middle cranial fossa is performed concomitantly. In some cases this constitutes sufficient surgical treatment.

Depending on the tissue diagnosis, minimal invasion of the PS may not require surgery, and the patient is scheduled immediately for complementary treatment. In cases where radical tumor removal seems desirable, total resection of tumor from the PS may be accomplished in a second stage, ideally within two weeks of the initial operation. The entire lesion should be resected, along with all adherent cranial nerves. It is extremely important that dura not directly involved with tumor be preserved. Since a watertight dural closure is not 100% reliable unless the dura is not diseased or damaged, only when dura is clearly infiltrated by tumor should it be resected and reconstructed. Postoperative CSF leaks in such patients may prove to be fatal.

Combined neurosurgical and ENT/maxillofacial surgical strategies for treating malignant skull base tumors

If the team approach to splanchnocranial, neck, and intracranial malignant tumors is anticipated as a single stage, either the neurosurgeons or the head-and-neck surgeons should begin the operation, guided by the protocol outlined in II. and III. above for surgical staging. Such collaboration carries three primary disadvantages. First, the risk of postoperative CSF leak and intracranial infection is increased. Second, once the histological diagnosis is obtained, the services of the second surgical team may not be required.

Third, postoperative responsibilities of different surgeons are sometimes obscured, resulting in distracting political problems. Obviously, all treating surgeons have a stake in the outcome of surgery and must be directly involved in clinical management until the patient has recovered sufficiently for transfer to the Oncology Department. So long as the patient's recovery remains uncomplicated, problems are not likely to arise. Troubles begin when a surgically related complication occurs (CSF leak, hematoma, infection, etc.). Fortunately, such circumstances arise only occasionally. Responsibilities equally shared in the operating room instantly become more complex. One surgical group or the other is expected to accept responsibility for the complication. Finger-pointing results. Rather than dispassionately searching for the source of the complication and concentrating on its management, precious time and energy are expended in an effort to prove that their individual surgical actions were optimal in every way. Serious misunderstandings may result. Patients never benefit from such controversy. Anyone's grandmother could have predicted this kind of conflict. "Too many cooks spoil the broth". Let the surgeon organizing a multi-specialty team beware. Know the competency and character of every surgeon selected, and do not fail to recognize the importance of Murphy's Law. If something *can* go wrong it *will* go wrong! Any surgeon selected must have the capacity and willingness to engage in a personal relationship with the patient. Patients have a tendency to fear the doctor, however well qualified, trying to steer them from dire circumstances, since he or she holds that surgeon responsible for the problems in the first place.

Results

The results of surgery for malignant CSB tumors are summarized in Table 26. Thirty-one patients were treated for malignant CSB lesions, the malignancy confined to the CSB in 16. Nine were treated for large splanchnocranial and neck carcinomas invading the CSB, and six for malignant lesions extending from the CSB to the facial skeleton.

"Total" removal was achieved in 16, resections being gross-total in eight and only partial in seven. The most complete resections were achieved among patients with hemangiopericytomas confined to the CSB, even some extending beyond the CSB. Despite such extensive resections, however, hemangiopericytomas recurred at the same site (some extending to adjacent regions), postoperative complementary oncological treatment notwithstanding. Such patients required repeated surgical intervention and chemotherapy. One patient harboring a hemangiopericytoma has survived 17 years and continues to do well. Another, after having his life-expectancy estimated to be no more than 6 months after diagnosis, underwent surgery plus Gamma Knife treatment and remains alive after nine years. A third patient with a hemangiopericytoma, however, was treated after four previous operations prior to transfer, dying after our partial resection plus additional conventional radiation therapy (Fig. 4.82).

One patient required three secondary operations for distant metastases of a pituitary carcinoma, both intracranial and intraspinal. Following surgery and conventional radiation for his intrasellar and parasellar tumor, he developed a cerebellar metastasis after five years which required re-operation and another course of radiation therapy. After an additional two years, he developed an intramedullary cervical metastasis. Following "total" resection of that lesion, he came to still a fourth operation two years later for a metastatic lesion in the lumbar spinal canal. He continues to live independently, now nine years after the initial diagnosis, manifesting no evidence of recurrence of his pituitary malignancy.

Table 26. Resectability of malignant central skull base tumors

Tumor	N° of cases	Virgin cases	Rest/recurrent tumors	Tumor resection			Re-operations		
				"Complete"	Gross-total	Partial	1	2	3
Hemangiopericytomas	12	9	3	10	1	1	5	3	3
Pituitary carcinoma	2	2	–	2	–	–	–	–	1
PS-sarcoma	3	3	–	–	2	1	–	–	–
CSB metastatic lesion	5	5	–	4	1	–	–	–	–
Splanchno-cranial carcinomas	9	3	6	–	4	5	–	–	–
Total	31	22	9	16	8	7	5	3	4

Despite initial enthusiasm on the part of our ENT and maxillofacial colleagues, total resection of a malignant CSB lesion extending from a splanchnocranial carcinoma was never achieved. In five of nine cases, only partial resection of the CSB component was possible despite excellent responses to previous extracranial operations plus conventional radiation.

Complementary non-surgical treatment was required with all malignant CSB tumors. With one exception, hemangiopericytomas required additional surgery despite postoperative conventional radiation. All secondary operations for hemangiopericytomas were for recurrences within the CSB.

References

1. Greenberg HS, Deck MD, Vikram B, Chu FCH, Posner JB (1981) Metastasis to the base of the skull: clinical findings in 43 patients. Neurology 31: 530–537
2. Jääskeläinen J, Servo A, Haltia M, Wahlström T, Valtonen S (1985a) Intracranial hemangiopericytoma: radiology, surgery, radiotherapy, and outcome in 21 patients. Surg Neurol 23: 227–236
3. Lam S, Margo CE, Beck R, Pusateri TJ, Pasucci S (1987) Cavernous sinus syndrome as the initial manifestation of multiple myeloma. J Clin Neuroophthalmol 7: 135–138
4. Nakatomi H, Sasaki T, Kawamoto S, Fujimaki T, Furuya K, Kirino T (1996) Primary cavernous sinus malignant lymphoma treated by gamma knife radiosurgery: case report and review of the literature. Surg Neurol 46: 272–279
5. Nogami K, Nishijima M, Endoh S, Takaku A (1992) Malignant melanoma metastatic to cavernous sinus and skull with an unknown primary origin: report of a case. Neurol Surg 20: 1017–1020
6. Post MJD, Mendez DR, Kline LB, Acker JD, Glaser JS (1985) Metastatic disease to the cavernous sinus: clinical syndrome and CT diagnosis. J Comput Assist Tomogr 9: 115–120
7. Reynard M, Brinkley JR (1983) Cavernous sinus syndrome caused by rhabdomyosarcoma. Ann Ophthalmol 15: 94–97
8. Sham JST, Cheung YK, Choy D, Chan FL, Leong L (1991) Nasopharyngeal carcinoma: CT evaluation of patterns of tumor spread. AJNR Am J Neuroradiol 12: 265–270
9. Thomas HG, Dolman CL, Berry K (1981) Malignant meningioma: clinical and pathological features. J Neurosurg 55: 929–934
10. Traserra J, Comas J, Conde C, Cuchi A, Cardesa A (1990) Metastatic involvement of the cavernous sinus from primary pharyngolaryngeal tumors. Head Neck 12: 426–429

Subject index

GPSR Compliance

The European Union's (EU) General Product Safety Regulation (GPSR) is a set of rules that requires consumer products to be safe and our obligations to ensure this.

If you have any concerns about our products, you can contact us on ProductSafety@springernature.com

In case Publisher is established outside the EU, the EU authorized representative is:

Springer Nature Customer Service Center GmbH
Europaplatz 3
69115 Heidelberg, Germany

Batch number: 10731594

Printed by Printforce, the Netherlands